POTASSIUM CHANNELS
Structure, Classification, Function and Therapeutic Potential

ELLIS HORWOOD SERIES IN PHARMACEUTICAL TECHNOLOGY
(incorporating Pharmacological Sciences)

Editor: Professor M. H. RUBINSTEIN, School of Health Sciences, Liverpool Polytechnic

EXPERIMENTAL DRUG DESIGN: Pharmaceutical Applications
N. A. Armstrong & K. C. James
MICROBIAL QUALITY ASSURANCE IN PHARMACEUTICALS, COSMETICS AND TOILETRIES
Edited by S. Bloomfield *et al.*
DRUG DISCOVERY TECHNOLOGIES
C. Clark & W. H. Moos
PHARMACEUTICAL PRODUCTION FACILITIES: Design and Applications
G. Cole
PHARMACEUTICAL TABLET AND PELLET COATING
G. Cole
THE PHARMACY AND PHARMACOTHERAPY OF ASTHMA
Edited by P. F. D'Arcy & J. C. McElnay
PHARMACEUTICAL THERMAL ANALYSIS: Techniques and Applications
J. L. Ford and P. Timmins
PHYSICO-CHEMICAL PROPERTIES OF DRUGS: A Handbook for Pharmaceutical Scientists
P. Gould
DRUG DELIVERY TO THE GASTROINTESTINAL TRACT
Edited by J. G. Hardy, S. S. Davis and C. G. Wilson
HANDBOOK OF PHARMACOKINETICS: Toxicity Assessment of Chemicals
J. P. Labaune
TABLET MACHINE INSTRUMENTATION IN PHARMACEUTICS: Principles and Practice
P. Ridgway Watt
PHARMACEUTICAL CHEMISTRY, Volume 1 Drug Synthesis
H. J. Roth *et al.*
PHARMACEUTICAL CHEMISTRY, Volume 2 Drug Analysis*
H. J. Roth *et al.*
PHARMACEUTICAL TECHNOLOGY: Controlled Drug Release, Volume 1
Edited by M. H. Rubinstein
PHARMACEUTICAL TECHNOLOGY: Controlled Drug Release, Volume 2*
Edited by M. H. Rubinstein
PHARMACEUTICAL TECHNOLOGY: Tableting Technology, Volume 1
Edited by M. H. Rubinstein
PHARMACEUTICAL TECHNOLOGY: Tableting Technology, Volume 2*
Edited by M. H. Rubinstein
PHARMACEUTICAL TECHNOLOGY: Drug Stability
Edited by M. H. Rubinstein
PHARMACEUTICAL TECHNOLOGY: Drug Targeting*
Edited by M. H. Rubinstein
UNDERSTANDING ANTIBACTERIAL ACTION AND RESISTANCE
D. A. Russell and I. Chopra
RADIOPHARMACEUTICALS USING RADIOACTIVE COMPOUNDS IN PHARMACEUTICS AND MEDICINE
Edited by A. Theobald
PHARMACEUTICAL PREFORMULATION: The Physicochemical Properties of Drug Substances
J. I. Wells
PHYSIOLOGICAL PHARMACEUTICS: Biological Barriers to Drug Absorption
C. G. Wilson & N. Washington

* *In preparation*

POTASSIUM CHANNELS

Structure, Classification, Function and Therapeutic Potential

Editor:
NIGEL S. COOK
Cardiovascular Department, Preclinical Resarch
Sandoz Limited, Basel, Switzerland

ELLIS HORWOOD LIMITED
Publishers · Chichester

Halsted Press: a division of
JOHN WILEY & SONS
New York · Chichester · Brisbane · Toronto

First published in 1990 by
ELLIS HORWOOD LIMITED
Market Cross House, Cooper Street,
Chichester, West Sussex, PO19 1EB, England
The publisher's colophon is reproduced from James Gillison's drawing of the ancient Market Cross, Chichester.

Distributors:
Australia and New Zealand:
JACARANDA WILEY LIMITED
GPO Box 859, Brisbane, Queensland 4001, Australia
Canada:
JOHN WILEY & SONS CANADA LIMITED
22 Worcester Road, Rexdale, Ontario, Canada
Europe and Africa:
JOHN WILEY & SONS LIMITED
Baffins Lane, Chichester, West Sussex, England
North and South America and the rest of the world:
Halsted Press: a division of
JOHN WILEY & SONS
605 Third Avenue, New York, NY 10158, USA
South-East Asia
JOHN WILEY & SONS (SEA) PTE LIMITED
37 Jalan Pemimpin # 05–04
Block B, Union Industrial Building, Singapore 2057
Indian Subcontinent
WILEY EASTERN LIMITED
4835/24 Ansari Road
Daryaganj, New Delhi 110002, India

British Library Cataloguing in Publication Data
Potassium channels.
1. Medicine. Drug therapy. Potassium
I. Cook, Nigel S.
615'.2383

Library of Congress Card No. 89–24435

ISBN 0–7458–0624–4 (Ellis Horwood Limited)
ISBN 0–470–21605–0 (Halsted Press)

Typeset in Times by Ellis Horwood Limited
Printed in Great Britain by The Camelot Press, Southampton

Table of Contents

Contributors

D. J. Adams, Department of Pharmacology, University of Miami School of Medicine, Miami, Florida 33101, USA

M. L. J. Ashford, Department of Pharmacology, University of Cambridge, Tennis Court Road, Cambridge CB2 1QJ, UK

C. R. Bader, Département de Physiologie, Centre Médical Universitaire, CH-1211 Genève 4, Switzerland

L. Bernheim, Département de Physiologie, Centre Médical Universitaire, CH-1211 Genève 4, Switzerland

D. Bertrand, Département de Physiologie, Centre Médical Universitaire, CH-1211 Genève 4, Switzerland

R. E. Buckingham, Beecham Pharmaceutical Research Division, Medicinal Research Centre, Coldharbour Road, The Pinnacles, Harlow, Essex CM19 5AD, UK

N. S. Cook, Cardiovascular Department, Preclinical Research, Sandoz AG, CH-4002 Basel, Switzerland

P. Gardner, Department of Medicine, Stanford University, Stanford, California 94305, USA

C. Haimann, Center Peripheral Neuropathies 'B Ceccarelli', Department of Pharmacology, University of Milano, Milan, Italy

J. V. Halliwell, Department of Physiology, Royal Free Hospital, School of Medicine, University of London, Rowland Hill Street, London NW3 2PF, UK

D. G. Haylett, MRC Receptor Mechanisms Group, Department of Pharmacology, University College London, Gower Street, London WC1E 6BT, UK

D. H. Jenkinson, MRC Receptor Mechanisms Group, Department of Pharmacology, University College London, Gower Street, London WC1E 6BT, UK

J. Longmore, Department of Physiological Sciences, Stopford Building, University of Manchester, Manchester M13 9PT, UK

J. D. McCann, Laboratory of Membrane Transport, Department of Internal Medicine, University of Iowa College of Medicine, Iowa City, Iowa 52242, USA

W. Nonner, Department of Physiology and Biophysics, University of Miami School of Medicine, Miami, Florida 33101, USA

W. Osterrieder, Pharmaceutical Research Department, F. Hoffmann-LaRoche & Co. Ltd, CH-4002 Basel, Switzerland

L. D. Partridge, University of New Mexico, School of Medicine, Albuquerque, New Mexico 87131, USA

P. J. Pfaffinger, Center for Neurobiology and Behavior, Department of Pharmacology, Howard Hughes Medical Institute, Columbia University, New York, NY 10032, USA

U. Quast, Cardiovascular Department, Preclinical Research, Sandoz AG, CH-4002 Basel, Switzerland

P. Rorsman, Gothenburg University, Department of Medical Physics, POB33031, S-400 33 Göteborg, Sweden

S. A. Siegelbaum, Center for Neurobiology and Behavior, Department of Pharmacology, Howard Hughes Medical Institute, Columbia University, New York, NY 10032, USA

P. N. Strong, Jerry Lewis Muscle Research Centre, Department of Paediatrics and Neonatal Medicine, Royal Postgraduate Medical School, London W12 0NN, UK

D. Swandulla, Max-Planck-Institut für Psychiatrie, Am Klopferspitz 18A, D-8033 Martinsried, West Germany

G. Trube, Pharmaceutical Research Department, F. Hoffmann-LaRoche & Co. Ltd, CH-4002 Basel, Switzerland

J. F. Waterfall, Department of Cardiovascular Biology, Roche Products Limited, Welwyn Garden City, Hertfordshire AL7 3AY, UK

M. J. Welsh, Laboratory of Membrane Transport, Department of Internal Medicine, University of Iowa College of Medicine, Iowa City, Iowa 52242, USA

A. H. Weston, Department of Physiological Sciences, Stopford Building, University of Manchester, Manchester M13 9PT, UK

Preface

It is now approaching 40 years since Hodgkin and Huxley first described the current carried by potassium ions in the squid giant axon and the contribution of this current to the excitability of the nerve. For many years thereafter, K^+ channels remained the territory of the electrophysiologist. Whilst scientists working with Na^+ and Ca^{2+} channels had available potent and selective ligands like tetrodotoxin and the organic Ca^{2+} channel blockers, those wishing to study K^+ channels and their function were limited to less potent and selective agents such as tetraethylammonium and the aminopyridines. The lack of a high affinity ligand for K^+ channels certainly contributed to the relatively slow progress in our biochemical understanding of these channels.

An important development came towards the end of the 1970s, with the identification of apamin as the first toxin which potently and selectively blocked K^+ channels. Apamin was soon joined by an impressive array of other toxins, and together with the application of elegant electrophysiological techniques such as the single channel and whole cell voltage clamp, we now have a much clearer understanding of the involvement of K^+ channels in specific physiological processes. However, as seems often to be the case, with an appreciation of the function of different K^+ channels has come a realization of both their complexity and diversity. Around ten different classes of K^+ channels have now been described; moreover inhomogeneity of the properties of a specific channel type between different tissues suggests the need for further sub-classification. This highlights a general problem of classifying K^+ channels, namely that one is dealing with a group or family of channels which show certain unifying properties but at the same time exhibit distinct pharmacological or electrophysiological differences. The molecular biology of K^+ channels is still in its infancy, only one channel (the transient outward or A-channel) having so far been cloned. We are therefore not yet at a stage where the different channels can be classified according to molecular criteria, although it is widely anticipated that advances on this front will very soon contribute to a more rational basis for their classification.

This is the first book devoted solely to the subject of K^+ channels. Any initial reservations one might have about the 'specialized' nature of this subject, should be allayed when one considers that K^+ channels probably exist in every cell in the body (presumably contributing to the physiological function of the cell). Hence, whatever field of biological research the reader might come from, almost without exception a case can be made for consideration of the important modulatory role of K^+ channels within that discipline. This book is divided into two sections, the first dealing with the structure and classification of K^+ channels and the second with their physiological function, their possible alteration in pathological situations, and their potential as therapeutic targets.

It is fair to say that much of the current interest in K^+ channels has been provoked by the realization that a number of therapeutic agents already used clinically or at present undergoing development, appear to act (directly of otherwise) by opening or closing K^+ channels. This awakening of the pharmaceutical industry to the potential of K^+ channels as drug targets would appear to herald a new era in the history of these channels. With the introduction of medicinal chemistry, K^+ channels have clearly moved on from the time when they were considered only the domain of the electrophysiologist. It is hoped that this book will appeal to a wide audience, biochemists, physiologists and pharmacologists, both in academia and in the pharmaceutical industry. The comprehensive treatment of the subject should be of value both to those already working in the field and to those for whom K^+ channels are a new area of interest.

Finally, I would like to express my gratitude to the authors of the various chapters, who have helped make this book a reality; my role as editor has certainly been made easier by the quality of their contributions. At a time when many books are branded as outdated by the time they reach the bookshelves, we have strived to bring everything together within as short a time span as possible. Thus, a number of new developments which at the time of writing were still 'in the press' have been included, to provide the reader with a truly up-to-date account of the world of K^+ channels. Once again I thank all those who have contributed and who have put up with my incessant pestering along the way.

Nigel Cook

Basel
August 1989

For Sharron, Stephen and James

Abbreviations

Potassium channels are abbreviated K_{ATP}, K_{ACh}, etc., with the currents through the respective channels shown in parentheses ($I_{K(ATP)}$, $I_{K(ACh)}$). The subscript 'i' or 'o' refers to intracellular or extracellular effects (e.g. Ca_i^{2+} = intracellular (cytosolic) Ca^{2+}). Abbreviations not listed below are defined as they appear in the text.

$0\,Ca^{2+}$	essentially Ca^{2+}-free (EGTA-containing) medium
5-HT	5-hydroxytryptamine
5-HETE	5-hydroxyeicosatetraenoic acid
8-HEpETE	8-hydroxy-11,12,epoxy-5,9,14-eicosatrienoic acid
8Br-cGMP	8-bromo cyclic guanosine 3′,5′-monophosphate (cell-permeable form of cGMP)
12-HETE	12-hydroxyeicosatetraenoic acid
12-HPETE	12-hydroperoxyeicosatetraenoic acid
12-KETE	12-keto-5,8,10,14-eicosatetraenoic acid
γ	single-channel conductance
A23187	calcium ionophore
APD	action potential duration
AHP	after-hyperpolarization
ADP	adenosine 5′-diphosphate
AMP	adenosine 5′-monophosphate
AMP–PCP	adenylyl(β,γ-methylene)-diphosphonate (a non-hydrolysable ATP analogue)
AMP–PNP	5″-adenylyl-imidodiphosphonate (a non-hydrolysable ATP analogue)
4-AP	4-aminopyridine
ATP	adenosine 5′-triphosphate
ATPγS	adenosine-5″-*O*-(3-thiotriphosphate)
BK_{Ca}	large-conductance ('big') Ca^{2+}-activated K^+ channels (current: $I_{BK(Ca)}$)

BP	blood pressure
BRL 34915	cromakalim (K^+ channel opener, Beecham)
β-BTX	β-bungarotoxin
C9	nonyltriethylammonium ion
cAMP	adenosine 3′,5′ cyclic monophosphate
CAN	Ca^{2+}-activated non-specific cation channels (also abbreviated K/Na_{Ca})
cAPK	cAMP-dependent protein kinase
cGMP	guanosine 3′,5′ cyclic monophosphate
CNS	central nervous system
CTX	charybdotoxin
3,4DAP	3,4-diaminopyridine
cDNA	complementary deoxyribonucleic acid
cNAN	cyclic nucleotide-activated non-specific cation channels
CRK	cromakalim (K^+ channel opener, Beecham)
DAG	diacylglycerol
DC_{10}	1,2-didecanoylglycerol (protein kinase C activator)
DHP	dihydropyridine (voltage-sensitive Ca^{2+} channel blocker)
DNA	deoxyribonucleic acid
dTC	*d*-tubocurarine chloride
DTX	dendrotoxin
EC_{50}	50% effective concentration
ECG	electrocardiogram
EDRF	endothelial-derived relaxing factor(s)
EDHF	endothelial-derived hyperpolarizing factor(s)
EDTA	ethylenediaminetetraacetic acid
EGTA	ethyleneglycol bis(aminoethylether)tetraacetic acid
E_K	potassium equilibrium potential
EPSP	excitatory post-synaptic potential
FMRFamide	molluscan tetrapeptide (Phe-Met-Arg-Phe-NH_2) which causes activation of the K^+ current $I_{K(S)}$ in certain neurones
GABA	γ-aminobutyric acid
G_0	whole cell conductance
G protein	guanine nucleotide regulatory binding protein
GPP(NH)P	5′guanylylimidodiphosphate (non-hydrolysable GTP analogue)
GTP	guanosine 5′-triphosphate
GTPγS	guanosine 5′-*O*-(3-thio)triphosphate (non-hydrolysable GTP analogue)
HeLa cells	human cervix epithelial cell line
HEPES	*N*-(2-hydroxyethyl)piperazine-*N*′-(2-ethanesulphonic acid) (buffer, $pK_a = 7.5$ at 25°C)
HR	heart rate
I_A	transient outward K^+ current (channel: K_A)
I_{AHP}	slow Ca^{2+}-activated K^+ current, sensitive to apamin (also referred to as $I_{K(Ca)S}$ and $I_{SK(Ca)}$
IC_{50}	concentration causing 50% inhibition
i.c.v.	intra-cerebroventricular injection

i.d.	intraduodenal injection
I_D	slowly inactivating outward K^+ current, responsible for the extended delay (hence I_D) of the first action potential
I_K	terminology often used for the delayed rectifier K^+ current (channel: K_V)
$I_{K(ATP)}$	ATP-sensitive K^+ current (channel: K_{ATP})
I_{K1}	commonly used abbreviation for the inward (or 'anomalous') rectifier K^+ current (also known as I_R, I_{AR}, I_Q, I_H in different tissues)
$I_{K(Ca)}$	Ca^{2+}-activated K^+ current (channel: K_{ca})
IK_{Ca}	Intermediate-conductance Ca^{2+}-activated K^+ channels (current: $I_{IK(Ca)}$)
I_{Kf}	the 'fast' part of the depolarization-activated outward K^+ current in certain neurones, sometimes sub-divided further into I_{Kf1} and I_{Kf2} (see also I_{Ks})
I_{Kf1}	see I_{Kf}
I_{Kf2}	see I_{Kf}
$I_{K(s)}$	serotonin-inactivated K^+ current (channel: K_s)
I_{Ks}	the 'slow' part of the depolarization-activated outward K^+ current in certain neurones (see also I_{Kf})
$I_{K(SR)}$	current through sarcoplasmic reticulum 'maxi-K' (high-conductance K^+) channel
$I_{K(V)}$	delayed rectifier outward K^+ current (channel: K_V)
I_M	M (muscarinic-inhibited) K^+ current
i.p.	intraperitoneal injection
IP_3	myo-inositol 1,4,5-trisphosphate
IPSP	inhibitory postsynaptic potential
I_R	inward ('anomalous') rectifier K^+ current (also known as I_{AR}, I_{K1}, I_Q, I_H or (in heart) I_X
ITP	inosine 5′-triphosphate
i.v.	intravenous injection
I–V	current–voltage (relationship)
k_{+1}	forward rate constant
k_{-1}	reverse rate constant
K_A	K^+ channel responsible for the transient outward or A current (I_A)
K_{ACh}	Atrial muscarinic-activated (and adenosine-activated) K^+ channel (current: $I_{K(ACh)}$)
K_{ATP}	ATP-sensitive K^+ channel (current: $I_{K(ATP)}$)
K_{Ca}	Ca^{2+}-activated K^+ channel (current: $I_{K(Ca)}$), subdivided into BK_{Ca}, IK_{Ca} and SK_{Ca}
K_D	equilibrium dissociation constant
kDa	kilodaltons
K_I	equilibrium dissociation constant for inhibitor
K_i^+	intracellular K^+
KIC	α-ketoisocaproate
K_{Na}	Na^+-activated K^+ channel (current: $I_{K(Na)}$)
K_0^+:	extracellular K^+ concentration
KCO	K^+ channel opener
K_S	serotonin-inactivated K^+ channel (current: $I_{K(S)}$)

K_V	delayed (outward) rectifier K^+ channel (current: $I_{K(V)}$, also often referred to simply as I_K)
LHRH	luteinizing hormone releasing hormone
LQV	*Leiurus quinquestriatus hebraeus* venom
LTX	leiurotoxin I
LV d*P*/d*t*	first derivative of left ventricular pressure with time
MCD peptide	mast cell degranulating peptide (peptide 401)
M current	muscarinic ACh-inactivated K^+ current (see I_M)
M_r	relative molecular mass
mRNA	messenger ribonucleic acid
N	number of functional channels (e.g. in a membrane patch)
NAD	α-nicotinamide adenine dinucleotide
NADP	nicotinamide adenine dinucleotide phosphate
NDGA	nordihydroguaiaretic acid (lipoxygenase blocker)
NTX	noxiustoxin
PCP	phencyclidine
pA_2	negative logarithm of the molar concentration of competitive antagonist causing a two-fold rightward shift of the agonist dose–response curve
PC12 cells	phaeochromocytoma cell line
pD'_2	Negative logarithm of the molar concentration of a non-competitive antagonist causing a 50% reduction in the maximum of the agonist dose–response curve
PHA	phytohaemagglutinin (mitogen)
PI	phosphatidyl inositol
pIC_{50}	negative logarithm of the molar concentration causing a 50% inhibition
P_K	K^+ permeability
PKA	cAMP-dependent protein kinase
PKC	protein kinase C
PKI	small heat-stable protein kinase inhibitory protein
PMA	12-phorbolmyristate 13-acetate (protein kinase C activator)
pO_2	oxygen tension
P_o	open probability of channel
P_{Rb}	Rb^+ permeability
PTX	pertussis toxin (islet-activating protein)
QA	quarternary ammonium ion
RBF	renal blood flow
RINm5F cells	insulin-secreting pancreatic β-cell line
ROC	receptor-operated channel
RVD	regulatory volume decrease (of cells which have swollen in a hypotonic media)
S current	serotonin-inactivated K^+ current (see K_S)
SD	standard deviation
s.e.m.	standard error of the mean
SHR	spontaneously hypertensive rat
SK_{Ca}	small conductance Ca^{2+}-activated K^+ channels (current: $I_{SK(Ca)}$)
SPSHR	stroke-prone spontaneously hypertensive rat

TEA	tetraethylammonium ion
TPA	12-*O*-tetradecanoyl-phorbol-13-acetate (protein kinase C activator)
TRIS	tris(hydroxymethyl)aminomethane (buffer)
TTX	tetrodotoxin (Na^+ channel blocker)
UTP	uridine 5′-triphosphate
VF	ventricular fibrillation
VOC	voltage-sensitive ('operated') channel

Part I
Structure and classification

1

The molecular structure of potassium channels

P. N. Strong

1.1 INTRODUCTION

Our knowledge of the molecular structure of ion channels has expanded rapidly, as cloning techniques have enabled the complete amino acid sequences of large membrane proteins to be determined. The existence of two superfamilies of ion channels has been established: those ion channels activated by membrane voltage and those activated by external ligands. Examples of the first superfamily include sodium channels (Noda *et al.*, 1984) and the dihydropyridine-binding protein (putative voltage-dependent calcium channel; Tanabe *et al.*, 1987). Examples of the second superfamily include the nicotinic acetylcholine receptor (Numa *et al.*, 1983), the $GABA_A$ receptor (Schofield *et al.*, 1987) and the glycine receptor (Grenningloh *et al.*, 1987).

All of these structures have been elucidated after the ion channel protein in question was purified and characterized. The determination of partial N-terminal amino acid sequences enabled synthetic oligonucleotide probes to be constructed which in turn were able to identify clones from cDNA libraries. The isolation and sequencing of full length cDNA clones subsequently allowed prediction of the corresponding complete amino acid sequence.

Biochemical purification techniques require (1) an abundant source of the protein to be isolated and (2) the availability of specific, high-affinity ligands or toxins with which to monitor purification and to provide affinity purification supports. Until recently, our knowledge of the molecular structure of potassium channels was meagre in comparison with information about other ion channel proteins. There is a great diversity of potassium channel subtypes; probably more so than with any other ion channel. In the broad category of currents carried by voltage-gated potassium channels are the delayed rectifier, anomalous rectifier and transient, A-type currents. In the other category of ligand-gated potassium channels are those gated by calcium, ATP, neurotransmitters, intracellular kinases and GTP-binding proteins (for a review, see Rudy, 1988). Tissues rich in potassium channels have not been discovered and toxins with a high affinity for potassium channels have only recently been identified (for reviews, see Moczydlowski *et al.*, 1988; Castle *et al.*, 1989; Strong, 1990). It has only been in the last year that the purification of the first putative potassium ion channel protein has been reported (a dendrotoxin-binding, A-type channel protein; Rehm and Lazdunski, 1988; Parcej and Dolly, 1989). The only other types of potassium channel that have been solubilized and partially purified are calcium-activated potassium channel proteins.

Another recent approach that has dramatically altered our knowledge of potassium channel structure has been the application of the techniques of molecular genetics (Baumann *et al.*, 1987; Kamb *et al.*, 1987; Papazian *et al.*, 1987). Indeed the power of this approach has opened a new chapter in determining the structure of ion channel proteins.

This review will initially discuss the two approaches to determining the molecular structure of A-type potassium channels: (1) the biochemical toxin-binding protein approach, using dendrotoxin; and (2) the genetic, cloning approach, isolating the defective gene (an A-type potassium channel) from the *Drosophila* mutant, *Shaker*. The review will then turn to discuss our present knowledge about the molecular structure of calcium-activated potassium channels. Finally, possibilities for future research in areas already mentioned, as well as prospects for elucidating the structures of other types of potassium channels, will be discussed.

1.2 A-TYPE POTASSIUM CHANNELS

1.2.1 Dendrotoxin-binding proteins

Dendrotoxins are a group of snake toxins isolated from African mambas that block, both selectively and with high affinity, A-type potassium channels in the vertebrate nervous system (for a review, see Harvey and Anderson, 1985). Dendrotoxin and its homologue protein I are therefore promising tools for use in the biochemical identification and isolation of A-type channel proteins. Toxin-binding sites can be solubilized in the presence of most detergents with retention of toxin binding and no loss of binding affinity (K_D=0.08–0.4 nM; Mehraban *et al.*, 1984; Rehm *et al.*, 1988; Black *et al.*, 1989; Rehm and Lazdunski, 1988). Binding sites are remarkably stable (half-life 7 days at 4°C; Parcej and Dolly, 1989). Omission of potassium ions from the solubilization medium leads to a complete loss in specific binding. An estimate for the molecular weight of the solubilized binding protein of 405 000–465 000 has been obtained from gel filtration and sedimentation analysis (Black *et al.*, 1989) and this compares with estimates of 240 000–265 000 obtained from radiation inactivation experiments on the toxin-bound membranes (Dolly *et al.*, 1984). The purified channel complex appears to be a multimeric protein and subunits of 76 kDa, 38 kDa and 35 kDa have been identified (Rehm and Lazdunski, 1988; Parcej and Dolly, 1989). This compares with cross-linking studies, which have identified a 65-kDa ^{125}I-dendrotoxin-binding polypeptide in rat brain and two binding polypeptides (68 kDa and 75 kDa) in chick brain. The most highly purified channel-binding protein, obtained using sequential toxin I and wheatgerm lectin affinity columns and achieving a several thousand-fold enrichment, has realized a purity of 0.4–1.6 nmol toxin-binding sites/mg protein (cf. calculated value for 'pure' binding protein of 2.2 nmol/mg) (Rehm and Lazdunski, 1988). It will be seen later (sections 1.2.3 and 1.2.5) that components of A-type channels cloned from *Shaker* mutants are proteins of 60–70 kDa and 30–40 kDa, producing a satisfying measure of agreement between the two approaches.

1.2.2 *Shaker* mutants

Shaker mutants of the fruitfly *Drosophila melanogaster* were originally selected from flies that continued to shake their legs vigorously under ether anaesthetic. Nerve fibres of *Shaker* mutants show broadened action potentials and multiple firings, while mutant neuromuscular junctions show prolonged transmitter release (Jan *et al.*, 1977). These observations have led several investigators to suggest that the *Shaker* gene encodes a potassium channel (see Salkoff and Tanouye, 1986, for a review). Subsequent voltage clamp experiments have shown that *Shaker* mutations selectively affect fast transient potassium currents (A currents) in longitudinal flight muscles, without affecting other ionic currents. Some *Shaker* alleles have a complete

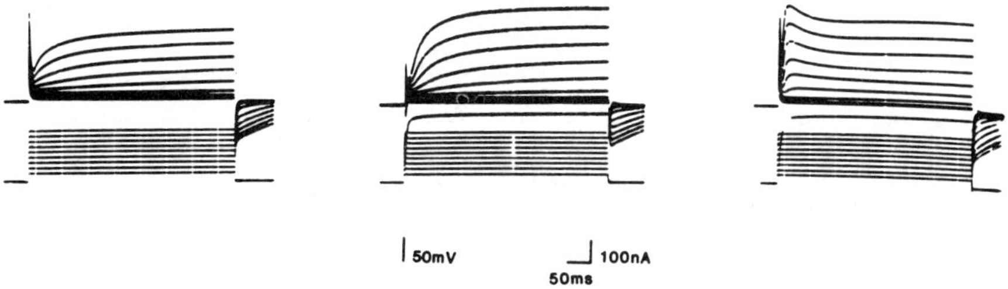

Fig. 1.1 — Potassium currents in a *Shaker* mutant. Currents from dorsal flight muscle of *Drosophila* pupae, measured with a two-electrode voltage clamp. The voltage jump protocol is displayed beneath each current trace. The transient, outward, A current is present in the wild-type (right) but completely missing in the $Shaker^{KS133}$ mutant (centre) and inhibited by 4-aminopyridine in the wild type (left). The delayed rectifying potassium current is not affected by the mutation, or by 4-aminopyridine. Adapted from Jan *et al.* (1983), with permission.

absence of A currents (Fig. 1.1), while others have reduced levels of A currents or exhibit subtle alterations in voltage sensitivity or kinetic properties (Salkoff, 1983; Jan *et al.*, 1983; Wu and Haugland, 1985). Gene dosage studies indicate that some mutations encode defective gene products, while other mutations cause channel underproduction (Tanouye *et al.*, 1981; Salkoff, 1983; Timpe and Jan, 1987). Thus, taken altogether, there is much evidence to suggest that the *Shaker* locus encodes a structural component of the A channel.

1.2.3 Identification of cDNA clones and characterization of the transcription unit

Having identified (on cytological grounds) chromosomal translocations in several *Shaker* mutations, the *Shaker* gene can be precisely localized by mapping the molecular positions of these mutations with cloned genomic DNA isolated from this region (chromosomal 'walking'). Coding sequences can then be identified by using the genomic DNA to isolate cDNA clones. These cDNA clones hybridize to different patterns of genomic sites, suggesting that the *Shaker* locus is a complex transcription unit, subject to alternative exon splicing. In fact the transcription unit is more than 120 kilobases and multiple and alternative splicing mechanisms give rise to a family of proteins as different combinations of exons are selected as the primary transcript is processed into the mature message.

Many different cDNA clones have been independently isolated by three different groups of workers. A European collaboration has resulted in the identification of four different clones (Baumann *et al.*, 1987; Pongs *et al.*, 1988) and the alignment of the cDNA sequences with genomic DNA indicates 21 regions of co-linearity, belonging to 21 exons. With the exception of exons 4/5 and 12/13, all exons contain intervening sequences which are spliced out to generate transcripts with alternative exons. The differential use of a splice site generates either exon 4 or 4/5 and either 12 or 12/13, respectively.

Two classes of K channel proteins have been deduced from sequence analysis of these cDNA clones (Fig. 1.2); class I transcripts encode long proteins (class I

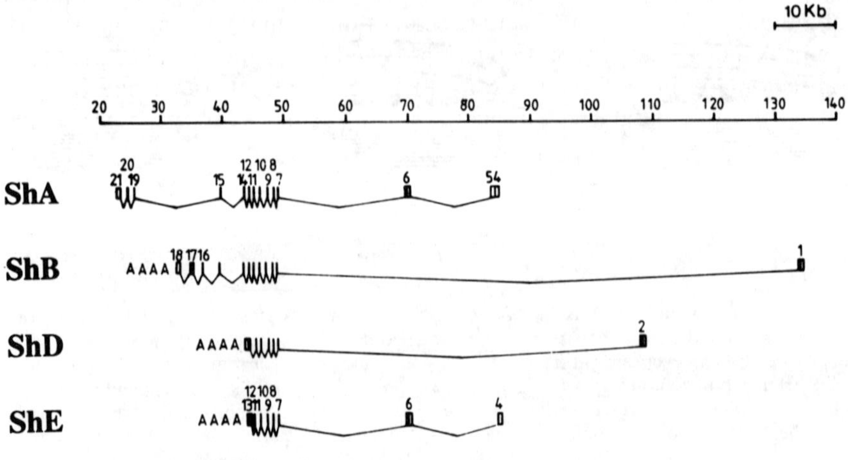

Fig. 1.2 — Structures of *Shaker* transcripts. Transcripts aligned to genomic DNA by sequence comparison of exons. The 21 exons are indicated as boxes or vertical bars (number from right to left in the direction of transcription), while the introns are diagrammed as intervening lines. Open boxes are untranslated regions; solid boxes encompass the derived open reading frame. AAA indicates 3′ termini of class II transcripts; transcripts are named after the protein which they encode. Adapted from Pongs *et al.* (1988), with permission.

proteins, mol. wt 60–70 kDa), while class II transcripts encode a truncated version (class II proteins, 30–40 kDa). Both class I and II transcripts have a central core (exons 7–12, clustered within 10 kilobases and occupying less than 10% of the genome); the differential use of the donor splice site at exon 12 determines the class of transcript. If the donor splice site at exon 12 is not used, then the exon becomes extended (12/13) and an alternative polyadenylation site leads to a truncated protein. The 5′ ends of the four different cDNA clones are generated by the addition of either exons 1, 2, 4 and 6 or 4/5 and 6 to the central core, respectively. Proteins derived from class I transcripts (ShA, M_r=70 300; ShB, M_r=72 500) have different N-termini; ShA is much more hydrophilic than ShB. Although the two transcripts are encoded in different exons at the end of the central core region (ShA, 19; ShB, 16) they show considerable homology at the amino acid level, differing only in a single amino acid substitution (Val for Ala); at the C-terminus, ShA is less hydrophilic than ShB. Proteins derived from class II transcripts (ShD, M_r=35 000; ShE, M_r=40 500) have a common C-terminal and a single splice site at the 5′ end of the common core which leads to either a 16 or a 61 amino acid N-terminus (ShD and ShE, respectively). Except for the last amino acid, the sequence of ShE is identical to the first 348 residues of ShA. Apart from ShD, all *Shaker* proteins contain polyglutamine tracts (*opa* elements; Wharton *et al.*, 1985).

The group of Jan and Jan (Papazian *et al.*, 1987; Tempel *et al.*, 1987; Schwarz *et al.*, 1988) have also identified a number of clones, all of which appear to be class I transcripts, using the previous nomenclature. They have a common central eight exons, (7–12, 14) and can be divided into four subgroups, A–D. To avoid confusion, *Shaker* proteins originating from the Jan group will be suffixed '(J)'. There are two sets of mutually exclusive exons at the 3′ end of the common region: one set shared by groups A and C, the other set by groups B and D. At the 5′ end of the common central region are three different exons: that of group C, that of group D and that shared by groups A and B (Fig. 1.3). Of the proteins encoded by these cDNA clones,

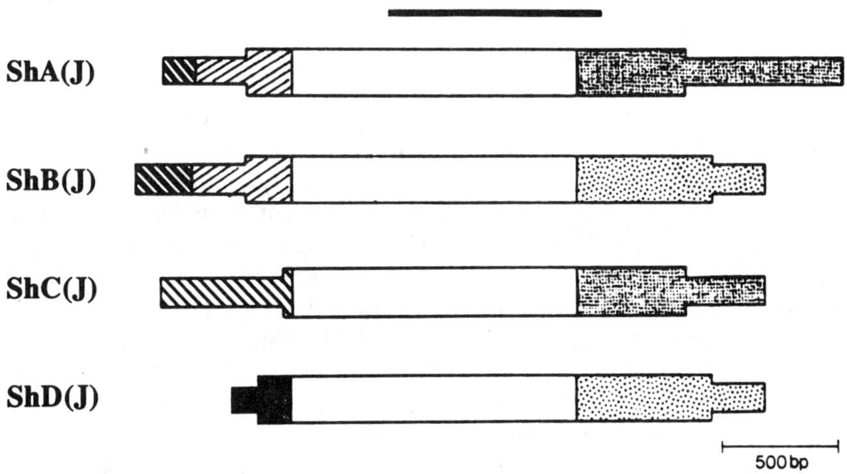

Fig. 1.3—Structure of *Shaker*(J) transcripts. The four classes of transcripts isolated by Jan, Jan and colleagues are shown and named according to the proteins encoded by the transcripts. Regions of identity are similarly shaded and regions encoding for protein are drawn wider. The location of the central hydrophobic, membrane-spanning domain is indicated at the top of the figure with a bar. Adapted from Schwarz *et al.* (1988), with permission.

their amino terminals diverge at a single splice site; ShA(J) (M_r=70 200) and ShB(J) (M_r=74 000) share a 60 amino acid sequence, whereas ShC(J) (M_r=64 000) has a distinct eight-residue amino terminal sequence. The N-terminus of ShD(J) is at present uncompleted. ShA(J)/ShC(J) and ShB(J)/ShD(J) diverge into alternative exons two-thirds of the way through the central core regions, although the amino acid sequences of the two groups within these distinct exons differ by only three amino acids. Beyond the central core, the two groups still show substantial similarities in the C-terminal region; most amino acid differences are conservative substitutions. At the C-terminal ends of the proteins, 13 out of the 14 last residues are identical.

Comparing the proteins identified by these two groups, the cDNAs encoding ShA and ShA(J) have only two A/G exchanges with the longest open reading frame, leading to one conservative amino acid substitution (Val/Ile). Although very similar (both class I), proteins coding B transcripts are different, since the N-terminus of ShB(J) is identical to that of ShA/ShA(J), while ShB is not. However, in distinction, ShD(J) belongs to class I, whereas ShD belongs to class II.

A third group have isolated another set of four clones (Kamb *et al.*, 1987; Iverson *et al.*, 1988), coding for both class I and class II proteins. Of the two class I clones, one (clone H4) appears to be identical to that coding for ShB(J), while another (clone H37) is very similar but not identical to ShA(J), differing slightly in the N-terminal region. The two clones coding for class II proteins (H2 and E1) both differ from the class II transcripts isolated by the European group.

Considering the diversity of cDNA clones already isolated from the *Shaker* locus by three independent groups, it is probable that more will subsequently be identified in the near future.

1.2.4 Structure of voltage-dependent potassium channel proteins

One would like to be able to use the amino acid sequence data, deduced from cDNA sequences, to construct models for these ion channel proteins, particularly to identify those segments of the protein that span the bilayer and might be involved in the formation of the ion-selective pore. Models of transmembrane pores all involve peptides with hydrophobic side chains forming membrane-spanning α-helices, packed in such a way that they form a hydrophilic pore which is lined by the polar faces of several of these helices. The minimum number of amino acids that can span a bilayer as an α-helix is 22; thus it is necessary to search protein sequences for hydrophobic segments, at least 22 residues in length. These segments can be conveniently identified by scanning sequences using an algorithm (Kyte and Doolittle, 1982; Guy, 1985); the resultant plot, known as a hydropathy profile, can be used to predict membrane-spanning regions.

Using hydropathy analysis on the amino acid sequences derived from cDNAs, all the putative potassium channel proteins are, as would be predicted, membrane proteins with a central hydrophobic core. The central hydrophobic core is surrounded by hydrophilic domains at the carboxyl and amino termini of the proteins. Proteins derived from class I transcripts have six clearly defined hydrophobic transmembrane regions (S1–S6) (Fig. 1.4), while proteins derived from class II transcripts only possess transmembrane regions S1–S3. The three transmembrane regions common to all of these proteins are encoded by the region encompassing

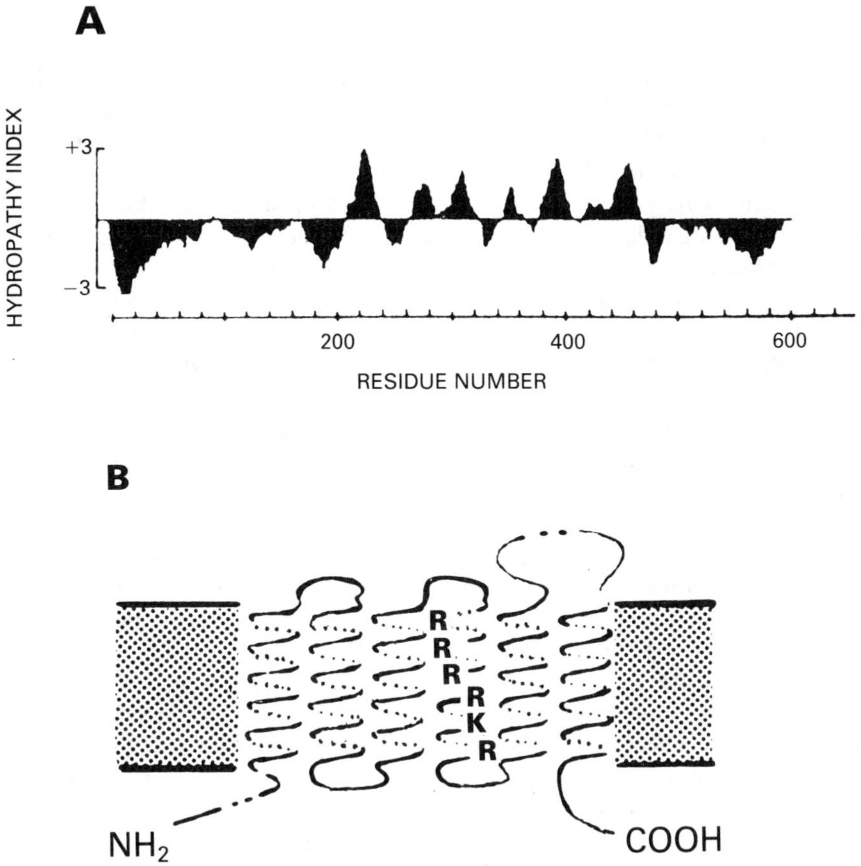

Fig. 1.4 — Structural model for the *Shaker* potassium channel. Hydropathy analysis of a class I Shaker potassium channel protein (ShA) and the predicted structure of a single unit of this channel, based on the same analysis. (A) Hydropathy indices are plotted versus amino acid residues, using the Kyte and Doolittle (1982) algorithm with a 19 amino acid window size. Positive index values represent hydrophobic groups and negative index values indicate hydrophilic groups. (B) Predicted channel structure, showing six transmembrane segments and the sequence of positively charged Lys (K) and arginine (R) residues in S4, comprising the hypothetical voltage sensor. The voltage-activated sodium channel protein and the dihydropyridine-binding protein are made up of four repeating units of this basic structure. Adapted from Pongs *et al.* (1988), with permission.

exons 10–12. The similarities between class I proteins and the repeat structure of the voltage-dependent sodium channel and the dihydropyridine-binding protein (putative voltage-dependent calcium channel) are striking. All possess the same six transmembrane α-helical segments. However, in contrast to the structures of the sodium and calcium channel proteins, which have four internal repeats, each repeat possessing the familiar six transmembrane-spanning regions, *Shaker* proteins do not have a repeat structure. Best-fit sequence homologies are found between *Shaker* segments S1–S3 and repeat I, between S4 and repeat II and between S5–S6 and repeat IV.

Based on this analogy with other voltage-dependent vertebrate ion channel proteins, similar models have been proposed for the structure of class I A-type potassium channels. The European group have provided a model (Pongs *et al.*, 1988) which orients the transmembrane regions S1–S6 in a pseudosymmetric manner, with both amino and carboxyl termini on the same (cytoplasmic) side of the membrane (Fig. 1.4). This cytoplasmic orientation derives from the observation that these proteins probably do not have a hydrophobic signal sequence.

The model is also supported by three predicted *N*-glycosylation sites (Hubbard and Ivatt, 1981), being located on the extracellular faces of the proteins, and two possible cAMP-dependent phosphorylation sites (Krebs and Beavo, 1979), being located on the cytoplasmic face of the proteins. Each membrane-spanning region is encoded in a separate exon: S1(exon10), S2(11), S3(12), S4(14), S5(14/15) and S6(16 or 19). In S6, although the cDNA clones are encoded by different exons, they show considerable homology at the amino acid level, differing only in a single amino acid substitution (Val for Ala). The transmembrane segment S4 also contains a characteristic sequence of positive charges (six Lys or Arg residues every third residue), which is also found in both sodium and calcium voltage-dependent ion channel proteins. This alignment of positive charges in S4 has been proposed to be the voltage sensor of voltage-dependent ion channel proteins and to carry the gating currents (Catterall, 1985; Noda *et al.*, 1986).

This model also speculates that segments S1–S3 form the pore of the channel while segments S4–S6 regulate the gating of the channel by a voltage-sensing mechanism. The model would predict that *Shaker* class II proteins would not have a voltage sensor. From an evolutionary point of view, the homologies referred to earlier between various segments of *Shaker* proteins and other vertebrate ion channel proteins would suggest that the proposed segments which line the inner wall of the channels evolved from a common ancestor via a different route from that taken by segments suggested to be part of the voltage sensor (Pongs *et al.*, 1988).

A slightly different model for class I proteins has been proposed (Tempel *et al.*, 1987), in which the voltage sensor is given its own membrane-spanning region, thereby bringing the total number of such regions to seven. The other main structural implication of this model is that an odd number of transmembrane regions places amino and carboxyl termini on opposite sides of the membrane.

It should be pointed out that predictive models are just that and the tentative structures made from such predictions should be treated with caution, as indeed the Jans and their colleagues have subsequently pointed out (Tempel *et al.*, 1988). The shortcomings of this approach as a predictive tool have recently been admirably discussed by Lodish (1988); there are many examples of transmembrane proteins (e.g. porins) which lack pronounced hydrophobic domains and consist predominantly of β-sheeted structures (Paul and Rosenbusch, 1985). Membrane-spanning segments made up of β-sheeted structures are 'invisible' to conventional hydropathy analysis.

Direct evidence that the *Shaker* locus encodes a potassium channel has been obtained by demonstrating that mRNA transcribed *in vitro* from individual cDNA clones is able to direct the synthesis of a functional potassium channel in the

Xenopus oocyte translation system. (Timpe *et al.*, 1988b; Iverson *et al.*, 1988). Oocytes injected with mRNA coding for class I proteins (ShA1(J), ShB1(J)=H4, H37), displayed transient outward currents under voltage clamp conditions (Fig. 1.5). These currents had the properties of A currents as judged by their voltage

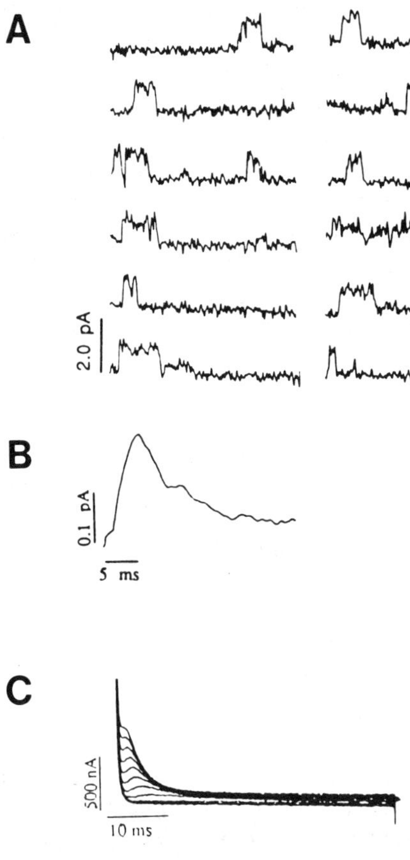

Fig. 1.5 — Potassium A channels expressed in *Xenopus* oocytes. Single A channels recorded in an outside-out patch from an oocyte injected with H4 RNA. (A) Representative sweeps containing channel openings during the test pulse (holding potential −80 mV, test potential +20 mV, pulse duration, 30 ms, interval between pulses, 5 s; data filtered at 2 kHz). Leak and uncompensated capacitance digitally subtracted. (B) Ensemble average of 64 consecutive sweeps, digitally filtered at 800 Hz (about 15% of the sweeps contained no channel openings). (C) Macroscopic currents from oocytes injected with the same message. Using a two-microelectrode voltage clamp, the membrane potential was held at −90 mV followed by a 1 s hyperpolarizing prepulse to −120 mV, then stepped to test potentials ranging from −50 mV to +50 mV in 10-mV increments. Test pulses applied at a frequency of 1 every 5 s. Only currents during the test depolarization are shown. Adapted from Iverson *et al.* (1988).

sensitivity, inactivation kinetics, ion selectivity and sensitivity to 4-aminopyridine. However, the mRNAs from the different clones induced different kinetics of activation and inactivation in oocytes, indicating that both the N-terminal and

C-terminal domains of the A channel protein can affect channel kinetics. These experiments also demonstrate that a single *Shaker* protein can act as a functional potassium channel. Since the protein resembles one of four similar domains in the other voltage-activated ion channel proteins, this leads to the speculation that the functional channel expressed in *Xenopus* oocytes is a homo-oligomeric assembly of the single *Shaker* protein (Timpe *et al.*, 1988b). No evidence for the formation of hetero-oligomeric channels was found when oocytes were injected with ShA(J) mRNA, mixed with ShB(J) or ShC(J) mRNAs (Timpe *et al.*, 1988a).

The mRNA from a truncated form of ShB1(J) (83 amino acids deleted from the carboxyl terminal, thereby retaining the transmembrane core of the protein intact) was completely unable to generate A currents. Some class II transcripts (corresponding to H2 and E1) were also unable to generate A currents (Iverson *et al.*, 1988) although the message encoding another class II protein (ShD) has been shown to form functional channels in oocytes that behave like outward rectifiers and only inactivate in the time range of seconds (Pongs, 1989). The question of subunit structure (monomeric versus multimeric, homo-oligomeric versus hetero-oligomeric and, in particular, the role played by class II proteins) in the assembly of the native channel must now await biochemical characterization.

1.2.5 Cloning potassium channel genes in mammalian brain

Using low-stringency, cross-hybridization strategies, rat and mouse brain cDNA libraries have been screened with cDNA probes encoding the transmembrane sequences S1–S6 of *Shaker* proteins. A mouse brain protein (MBK1; Tempel *et al.*, 1988) has been identified (M_r=56 400) by screening with ShA(J). Many regions of MBK1 are conserved with the *Shaker* proteins. In general the hydrophobic transmembrane segments are more highly conserved than the intervening hydrophilic regions. In particular, there is 95% homology between the S4 segments and between the positively charged tracts presumed to represent the voltage sensor. Another highly conserved stretch (>95%) is found in a region of over 50 amino acids immediately preceding the first transmembrane segment, S1. There is little sequence similarity at either end of the proteins and the *opa* elements are completely missing in the vertebrate sequence. In contrast to homology at the protein level, nucleotide sequences are much more divergent; in the region of the voltage sensor, there is only 33% homology at the nucleotide level. Thus the evolutionary pressures to conserve the potassium channel (considering flies and mice diverged over 600 million years ago) are quite remarkable (Hille, 1984). Other ion channel proteins, whether voltage-gated or ligand-gated, show nothing like this degree of conserved structure between arthropod and vertebrate (Hermans-Borgmeyer *et al.*, 1986; Schofield *et al.*, 1987; Bossy *et al.*, 1988) and further indicate that the highly conserved domains in these two potassium channels serve important, channel-specific functions.

A very similar protein to MBK1 has been identified in rat cortex (RCK1, Baumann *et al.*, 1988), obtained by screening a cDNA library with ShB. MBK1 and RCK1 only differ in 3 out of 495 amino acids. Although MBK1 mRNA has not been shown to direct the synthesis of functional potassium channels, RCK1 mRNA was successful in producing potassium currents in *Xenopus* oocytes, with properties similar to those of the non-inactivating delayed rectifier channel found in mammalian neurons (Stuhmer *et al.*, 1988). The currents were also sensitive to potassium channel-blocking agents such as tetraethylammonium, 4-aminopyridine, MCD peptide and dendrotoxin (but not to apamin). The main differences in A channels expressed from *Shaker* mutants and those from RCK1 mRNA is in their activation behaviour. The former are inactivated over the millisecond time scale, while those channels of mammalian origin are inactivated over tens of seconds.

The deduced amino acid sequences of these cloned A-type channel proteins from mammalian brain are of comparable size (56 kDa) to those of the dendrotoxin-binding proteins purified from mammalian brain (65–75 kDa) discussed earlier (section 1.2.1), if one considers that the molecular weights of the latter are probably increased by glycosylation.

Using a rat hippocampal cDNA library and screening with a synthetic oligonucleotide probe based on conserved sequences between *Shaker* proteins and MBK1, similar successful expression of a functional A-type channel has recently been reported (Christie *et al.*, 1989).

Antibodies against the *Shaker* locus have been raised by immunizing rabbits with a synthetic peptide (ROJO peptide) derived from the highly conserved cytoplasmic N-terminal region immediately preceding the first transmembrane segment S1 (Barbas *et al.*, 1989). This antibody recognizes a 71-kDa band on Western blots of *Drosophila* membrane proteins as well as single bands on blots of *Helix* neural membrane proteins and yeast membrane proteins. In blots of mammalian brain membranes, a number of bands are observed between 35 and 65 kDa, with a major component of 38 kDa. The sizes of the proteins recognized on blots are again in agreement with the sizes found for dendrotoxin-binding proteins in brain.

1.3 CALCIUM-ACTIVATED POTASSIUM CHANNELS

1.3.1 Apamin-binding proteins

The bee venom toxin apamin is a centrally acting neurotoxin that blocks specifically, and with high affinity, certain calcium-activated potassium channels (Habermann, 1984; Moczydlowski *et al.*, 1988; Strong, 1990). The toxin appears to selectively target a subgroup of low-conductance channels, none of which have single channel conductance values in excess of 20 pS. Although the toxin binds with very high affinity (K_Ds typically in the range 10–400 pM), the density of binding sites on all tissues examined is very low indeed (B_{max}<30 fmol/mg protein) — two orders of magnitude less than that found for tetrodotoxin-sensitive sodium channels. This has made the purification of apamin-sensitive calcium-activated potassium ion channel proteins a formidable biochemical task.

A variety of apamin-binding proteins have been identified using either chemical cross-linking or photoaffinity labelling. A peptide of 33 kDa has been found on brain membranes by different laboratories (Hugues *et al.*, 1982; Schmid-Antomarchi *et al.*, 1984; Wu *et al.*, 1985) using straightforward chemical cross-linking. With photoactivatable derivatives of apamin, proteins of 22–33 kDa, 57–59 kDa and 85–87 kDa have been identified on a variety of tissues, including brain membranes, cultured neurones, heart, smooth muscle and liver membranes (Seagar *et al.*, 1985, 1986; Marqueze *et al.*, 1987). Radiation inactivation analysis has produced divergent molecular weight claims for the intact ion channel protein, dependent on whether lyophilized and rehydrated rat brain (M_r=250 000; Schmid-Antomarchi *et al.*, 1984) or frozen rat brain (M_r=84 000–115 000; Seagar *et al.*, 1986) was used. How many of these identified proteins are *bona fide* components of a putative ion channel complex remains to be seen and the possible oligomeric nature of the channel is still to be resolved.

Apamin-binding proteins can be effectively solubilized from rat brain using either sodium cholate or a zwitterionic detergent, CHAPS (Schmid-Antomarchi *et al.*, 1984; Seagar *et al.*, 1987). Binding constants (K_D=25–40 pM) are similar to those found for the membrane-bound protein; the solubilized protein retains other pharmacological properties such as a requirement for potassium ions and the ability of neuromuscular blockers to compete for the binding site. The best preparations reported have a binding capacity of 17 fmol/mg protein and a half-life of 50 h at 1°C (Seagar *et al.*, 1987). From density gradient sedimentation analysis, the detergent/protein complex has an $S_{20,w}$ of 20, which corresponds to a molecular weight of 700 000, although as much as half of this could be contributed by detergent. Assuming a molecular weight of 350 000 the homogeneous protein should have a specific activity of 3 000 000 fmol/mg protein and therefore a purification factor of a further five orders of magnitude is needed!

1.3.2 Other studies

The calcium-activated potassium channel (or to be more precise, a calmodulin-binding protein) from the outer renal medulla of pig kidney has been solubilized, partially purified and reconstituted into phospholipid vesicles (Klaerke *et al.*, 1987). Kidney membranes, solubilized in the detergent CHAPS, were purified on a calmodulin afffinity column. Two proteins, 36 kDa and 51 kDa (representing 0.7% of soluble protein), were eluted from the column with EGTA. This affinity-purified preparation, reconstituted into phospholipid vesicles, was able to take up $^{86}Rb^+$ tracer in a calmodulin-dependent manner, although not as efficiently as native membrane vesicles.

1.4 CONCLUSIONS

The direct cloning of A-type potassium channels from the *Shaker* mutant has revolutionized our experimental approach to the structure of potassium channel proteins. Cross-hybridization experiments with *Shaker* probes has led directly to the characterization of A-type potassium channels in brain, and using similar techniques it should be possible to clone other types of potassium channels that share homologies at the cDNA level. However, that is not to consign the classical biochemical approach to the waste basket. The genetic approach does not allow identification of essential parts of the structure that are encoded for in other parts of the genome. These other components are often key elements in the biological assembly, regulation and modulation of the structure. Many questions that have arisen about A channel structure, as a result of the cloning approach, now require the actual isolation of the functional protein, in order to be answered. Foremost among these questions is the subunit structure of the A channel: Is it a single polypeptide unit or is it an oligomeric structure, made up of unique/different subunits? To help in this matter, biochemists can use the antibodies raised against the synthetic ROJO peptide, as an affinity support; the antibody appears to recognize proteins across yeast, invertebrates, arthropods and mammals.

The use of the *Xenopus* oocyte translation system, coupled with site-directed mutagenesis studies, are important techniques for studying the function of particular domains of A-channel proteins in the future. Similarly, antibodies directed against different regions of the molecule will help to confirm structural predictions made on the basis of hydropathy analysis concerning extracellular and cytoplasmic domains of the channel.

Unless a more suitable source of channel protein appears (and the PC12 cell might prove the answer here; Schmid-Antomarchi *et al.*, 1986), the purification of low conductance, apamin-sensitive calcium-activated potassium channels will prove to be a Herculean task. Biochemical studies of high-conductance, calcium-activated potassium channels are in their infancy and, again, the problem of choice of tissue for isolation is a major one. One of the greatest difficulties in any purification scheme — that of solubilization with retention of biological activity — seems to have been overcome with the kidney medulla. The use of this promising preparation, with an additional affinity purification step employing charybdotoxin (Smith *et al.*, 1986; Gimenez-Gallego *et al.*, 1988) — a potent blocker of these channels — might be an attractive approach. The *Shaker* probes, at the time of writing, have not been successful in pulling out cross-hybridizing cDNA clones which encode calcium-activated potassium ion channels; however, this situation may change at any moment. However, a *Paramecium* mutant with defective calcium-activated potassium channels has been described (Schaeffer *et al.*, 1987) and a possible next step would be to define and clone the locus of this mutation.

A high-affinity probe for the delayed rectifier channel, noxiustoxin, has been characterized (Possani *et al.*, 1982; Carbone *et al.*, 1987) but no studies have yet been reported on noxiustoxin-binding proteins. For the molecular characterization of other potassium channel subtypes, we are still in uncharted waters and must await the discovery of new toxins or the isolation of new mutants. Future progress in elucidating potassium channel structure will not depend on one experimental approach in isolation but the application of diverse techniques, both new and old. If the rate of progress of the last two years is maintained, then our molecular knowledge of potassium channels will soon rival that of sodium ion channels and the acetylcholine receptor ion channel.

REFERENCES

Barbas, J. A., Rubio, N., Pedroso, E., Pongs, O., and Ferrus, A. (1989) Antibodies against *Drosophila* potassium channels identify membrane proteins across species. *Mol. Brain Res.* **5** 171–176.

Baumann, A., Krah-Jentgens, I., Muller, R., Muller-Holtkamp, F., Seidel, R., Kecskemethy, N., Casal, J., Ferrus, A., and Pongs, O. (1987) Molecular organization of the maternal effect region of the *Shaker* complex of *Drosophila*: characterization of an I_A channel transcript with homology to vertebrate Na^+ channel. *EMBO J.* **6** 3419–3429.

Baumann, A., Grupe, A., Ackermann, A., and Pongs, O. (1988) Structure of the voltage dependent potassium channel is highly conserved from *Drosophila* to vertebrate central nervous systems. *EMBO J.* **7** 2457–2463.

Black, A. R., Donegan, C. M., Denny, B. J., and Dolly, J. O. (1989) Solubilization and physical characterization of acceptors for dendrotoxin and β-bungarotoxin from synaptic membranes of rat brain. *Biochemistry* **27** 6814–6820.

Bossy, B., Ballivet, M., and Spierer, P. (1988) Conservation of neural nicotinic acetylcholine receptors from *Drosophila* to vertebrate central nervous systems. *EMBO J.* **7** 611–618.

Carbone, E., Prestipino, G., Spadavecchia, L., Francolini, F., Possani, L. D. (1987) Blocking of the squid giant axon K+ channel by noxiustoxin: a toxin from the venom of the scorpion *Centruroides noxius. Pflugers Archiv.* **408** 423–431.

Castle, N. A., Haylett, D. G., and Jenkinson, D. H. (1989) Toxins in the characterization of potassium channels. *Trends Neurosci.* **12** 59–65.

Catterall, W. A. (1985) The electroplax sodium channel revealed. *Trends Neurosci.* **8** 39–41.

Christie, M. J., Adelman, J. P., Douglass, J., and North, R. A. (1989) Expression of a cloned rat brain potassium channel in *Xenopus* oocytes. *Science* **244** 221–224.

Dolly, J. O., Halliwell, J. V., Black, J. D., Williams, R. S., Pelchen-Mathews, A., Breeze, A. L., Mehraban, F., Othman, I. B., and Black, A. R. (1984) Botulinum toxin and dendrotoxin as probes for studies on transmitter release. *J. Physiol. Paris* **79** 280–303.

Gimenez-Gallego, G., Navia, M. A., Reuben, J. P., Katz, G. M., Kaczorowski, G. J., and Garcia, M. L. (1988) Purification, sequence and model structure of charybdotoxin, a potent selective inhibitor of calcium-activated potassium channels. *Proc. Natl. Acad. Sci. USA* **85** 3329–3333.

Grenningloh, G., Rienitz, A., Schmidt, B., Methfessel, C., Zensen, M., Beyreuther, K., Gundelfinger, E. D., and Betz, H. (1987) The strychnine-binding subunit of the glycine receptor shows homology with nicotinic acetylcholine receptors. *Nature* **328** 215–220.

Guy, H. R. (1985) Amino acid side-chain partition energies and distribution of residues in soluble proteins. *Biophys. J.* **47** 61–70.

Habermann, E., (1984) Apamin. *Pharmacol. Ther.* **25** 255–270.

Harvey, A. L., and Anderson, A. J. (1985) Dendrotoxins: snake toxins that block potassium channels and facilitate neurotransmitter release. *Pharmacol. Ther.* **31** 33–55.

Hermans-Borgmeyer, I., Zopf, D., Rysek, R. P., Hovemann, B., Betz, H., and Gundelfinger, E. D. (1986) Primary structure of a developmentally regulated nicotinic acetylcholine receptor protein from *Drosophila. EMBO J.* **5** 1503–1508.

Hille, B. (1984) *Ionic channels of excitable membranes.* Sinauer, Sunderland, Massachusetts.

Hubbard, S. C., and Ivatt, R. J. A. (1981) Synthesis and processing of asparagine-linked oligosaccharides. *Ann. Rev. Biochem.* **50** 555–583.

Hugues, M., Schmid, H., and Lazdunski, M., (1982) Identification of a protein component of the Ca^{2+}-dependent K^+ channel by affinity labellling with apamin. *Biochem. Biophys. Res. Comm.* **107** 1577–1582.

Iverson, L. E., Tanouye, M. A., Lester, H. A., Davidson, N., and Rudy, B. (1988)

A-type potassium channels expressed from *Shaker* locus cDNA. *Proc. Natl. Acad. Sci. USA* **85** 5723–5727.

Jan, Y. N., Jan, L. Y., and Dennis, M. (1977) Two mutations of synaptic transmission in *Drosophila. Proc. Roy. Soc. Lond. B* **198** 87–108.

Jan, L. Y., Barbel, S., Timpe, L., Laffe, C., Salkoff, L., O'Farrell, P., and Jan, Y. N. (1983) Mutating a gene for a potassium channel by hybrid dysgenesis: an approach to the cloning of the *Shaker* locus in *Drosophila. Cold Spring Harbor Symp. Quant. Biol.* **48** 233–245.

Kamb, A., Iverson, L. E., and Tanouye, M. A. (1987) Molecular characterization of *Shaker*, a *Drosophila* gene that encodes a potassium channel. *Cell* **50** 405–413.

Klaerke, D. A., Petersen, J., and Jørgensen, P. L. (1987) Purification of Ca^{2+}-activated K^+ channel protein on calmodulin affinity columns after detergent solubilization of luminal membranes from outer renal medulla. *FEBS Lett.* **216** 211–216.

Krebs, E. G., and Beavo, J. A. (1979) Phosphorylation–dephosphorylation of enzymes. *Ann. Rev. Biochem.* **48** 923–959.

Kyte, J., and Doolittle, R. F. (1982) A simple method for displaying the hydropathic character of a protein. *J. Mol. Biol.* **157** 105–132.

Lodish, H. F. (1988) Multi-spanning membrane proteins: how accurate are the models? *Trends Biochem. Sci.* **13** 332–334.

Marqueze, B., Seagar, M., and Courand, C. (1987) Photoaffinity labelling of the K^+-channel-associated apamin-binding molecule in smooth muscle, liver and heart membranes. *Eur. J. Biochem.* **169** 295–298.

Mehraban, F., Breeze, A. L., and Dolly, J. O. (1984) Identification by cross-linking of a neuronal acceptor protein for dendrotoxin, a convulsant neurotoxin. *FEBS Lett.* **174** 116–121.

Moczydlowski, E., Lucchesi, K., and Ravindran, A. (1988) An emerging pharmacology of peptide toxins targeted against potassium channels. *J. Membr. Biol.* **105** 95–111.

Noda, M., Shimizu, S., Tanabe, T., Takai, T., Kayano, T., Ikeda, T., Takahashi, T., Nakayama, H., Kanaoka, Y., Minamino, N., Kangawa, K., Matsuo, H., Raftery, M. A., Hirose, T., Inayama, S., Hayashida, H., Miyata, T., and Numa, S. (1984) Primary structure of *Electrophorus electricus* sodium channel deduced from cDNA sequence. *Nature* **321** 121–127.

Noda, M., Ikeda, T., Kayano, T., Suzuki, H., Takeshima, H., Kurasaki, M., Takahashi, H., and Numa, S. (1986) Existence of distinct sodium channel messenger RNAs in rat brain. *Nature* **320** 188–192.

Numa, S., Noda, M., Takahashi, M., Tanabe, T., Toyosato, M., Furutani, Y., and Kikyotani, S. (1983) Molecular structure of the nicotinic acetylcholine receptor. *Cold Spring Harbor Symp. Quant. Biol.* **48** 57–69.

Papazian, D. M., Schwarz, T. L., Tempel, B. L., Jan, Y. N. J., and Jan, L. Y. (1987) Cloning of genomic and complementary DNA from *Shaker*, a putative potassium channel gene from *Drosophila. Science* **237** 749–753.

Parcej, D. N., and Dolly, J. O. (1989) Dendrotoxin acceptor from bovine synaptic plasma membranes. Binding properties, purification and subunit composition of a putative constituent of certain voltage-activated K^+ channels. *Biochem J.* **257** 899–903.

Paul, C., and Rosenbusch, J. P. (1985) Folding patterns of porin and bacteriorhodopsin. *EMBO J.* **4** 1593–1597.

Pongs, O. (1989) Molecular basis of potassium channel diversity. *Pflügers Arch.* (in press).

Pongs, O., Kecskemethy, N., Müller, R., Krah-Jentgens, I., Baumann, A., Kiltz, H. H., Canal, I., Llamazares, S., and Ferrus, A. (1988) *Shaker* encodes a family of putative potassium channel proteins in the nervous system of *Drosophila. EMBO J.* **7** 1087–1096.

Possani, L. D., Martin, B. M., and Svendsen, I. (1982) The primary structure of noxiustoxin: a K^+ channel blocking peptide, purified from the venom of the scorpion *Centruroides noxius* Hoffman. *Carlsberg Res. Comm.* **47** 285–289.

Rehm, H., and Lazdunski, M. (1988) Purification and subunit structure of a putative K^+ channel protein identified by its binding properties for dendrotoxin I. *Proc. Natl. Acad. Sci. USA* **85** 4919–4923.

Rehm, H., Bidard, J. N., Schweitz, H., and Lazdunski, M. (1988) The receptor site for bee venom mast cell degranulating peptide: affinity labelling and evidence for a common molecular target for mast cell degranulating peptide and dendrotoxin I, a snake toxin active on K^+ channels. *Biochemistry* **27** 1827–1832.

Rudy, B. (1988) Diversity and ubiquity of potassium channels. *Neuroscience* **25** 729–749.

Salkoff, L. (1983) Genetic and voltage clamp analysis of a *Drosophila* potassium channel. *Cold Spring Harbor Symp. Quant. Biol.* **48** 221–231.

Salkoff, L., and Tanouye, M. A. (1986) Genetics of ion channels. *Physiol. Rev.* **66** 301–329.

Schaeffer, W. H., Hinrichsen, R. D., Burgess-Cassler, A., Kung, C., Blair, I. A., and Watterson, D. A. (1987) A mutant *Paramecium* with a defective calcium-dependent potassium conductance has an altered calmodulin: a non-lethal selective alteration in calmodulin regulation. *Proc. Natl. Acad. Sci. USA* **84** 3931–3935.

Schmid-Antomarchi, H., Hugues, M., Norman, R., Ellroy, C., Borsotto, M., and Lazdunski, M. (1984) Molecular properties of the apamin-binding component of the Ca^{2+}-dependent K^+ channel: radiation inactivation, affinity-labelling and solubilization. *Eur. J. Biochem.* **142** 1–6.

Schmid-Antomarchi, H., Hugues, M., and Lazdunski, M. (1986) Properties of the apamin-sensitive Ca^{2+}-activated K^+ channel in PC12 phaeochromocytoma cells which hyper-produce the apamin receptor. *J. Biol. Chem.* **261** 8633–8637

Schofield, P. R., Darlison, M. G., Fujita, N., Burt, D. R., Stephenson, F. A., Rodriguez, H., Rhee, L. M., Ramachandran, J., Reale, V., Glencorse, T. A., Seeburg, P. H., and Barnard, E. A. (1987) Sequence and functional expression of the GABA receptor shows a ligand-gated receptor super-family. *Nature* **328** 221–227.

Schwarz, T. L., Tempel, B. L., Papazian, D. M., Jan, Y. N., and Jan, L. Y. (1988) Multiple potassium-channel components are produced by alternative splicing at the *Shaker* locus in *Drosophila. Nature* **331** 137–142.

Seagar, M. J., Labbe-Jullie, C., Granier, C., Van Rietschoten, J., and Couraud, F. (1985) Photoaffinity labelling of components of the apamin-sensitive K^+ channel in neuronal membranes. *J. Biol. Chem.* **260** 3895–3898.

Seagar, M. J., Labbe-Jullie, C., Granier, C., Goll, A., Glossman, H., Van Rietschoten, J., and Couraud, F. (1986) Molecular structure of rat brain apamin receptor: differential photoaffinity labelling of putative K^+ channel subunits and target size analysis. *Biochemistry* **25** 4051–4057.

Seagar, M. J., Marqueze, B., and Couraud, F. (1987) Solubilization of the apamin receptor associated with a calcium-activated potassium channel from rat brain. *J. Neurosci.* **7** 565–570.

Smith, C. D., Phillips, M., and Miller, C. (1986) Purification of charybdotoxin, a specific inhibitor of the high conductance Ca^{2+}-activated K^+ channel. *J. Biol. Chem.* **261** 14607–14613.

Strong, P. N. (1990) Potassium channel toxins. *Pharmacol. Ther.* (in press).

Stuhmer, W., Stocker, M., Sakmann, B., Seeburg, P., Baumann, A., Grupe, A., and Pongs, O. (1988) Potassium channels expressed from rat brain cDNA have delayed rectifier properties. *FEBS Lett.* **242** 199–206.

Tanabe, T., Takeshima, H., Mikami, A., Flockerzi, V., Takahashi, H., Kanagawa, K., Kojima, M., Matsuo, H., Hirose, T., and Numa, S. (1987) Primary structure of the receptor for calcium channel blockers from skeletal muscle. *Nature* **328** 313–318.

Tanouye, M. A., Ferrus, A., and Fujita, S. C. (1981) Abnormal action potentials associated with the *Shaker* complex locus of *Drosophila. Proc. Natl. Acad. Sci. USA* **78** 6548–6552.

Tempel, B. L., Papazian, D. M., Schwarz, T. L. Jan, Y. N., and Jan, J. L. Y. (1987) Sequence of a probable potassium channel component encoded at *Shaker* locus of *Drosophila. Science* **237** 770–775.

Tempel, B. L., Jan, Y. N., and Jan, L. Y. (1988) Cloning of a probable potassium channel gene from mouse brain. *Nature* **332** 837–839.

Timpe, L. C., and Jan, L. Y. (1987) Gene dosage and complementation analysis of the *Shaker* locus of *Drosophila. J. Neurosci.* **7** 1307–1317.

Timpe, L. C., Jan, Y. N., and Jan, L. Y. (1988a) Four cDNA clones from the *Shaker* locus of *Drosophila* induce kinetically distinct A-type potassium currents in *Xenopus* oocytes. *Neuron* **1** 659–667.

Timpe, L. C., Schwarz, T. L., Tempel, B. L., Papazian, D. M., Jan, Y. N., and Jan, L. Y. (1988b) Expression of functional potassium channels from *Shaker* cDNA in *Xenopus* oocytes. *Nature* **331** 143–145.

Wharton, K. A., Yedvobnick, B., Finnerty, V. G., and Artavanis-Tsakonas, S. (1985) *opa*: A novel family of transcribed repeats shared by the *Notch* locus and other developmentally regulated loci in *D. melanogaster. Cell* **40** 55–62.

Wu, C-F., and Haugland, F. N. (1985) Voltage clamp analysis of membrane currents in larval muscle fibers of *Drosophila*: alteration of potassium currents in *Shaker* mutants. *J. Neurosci.* **5** 2626–2640.

Wu, K., Carlin, R., Sachs, I., and Siekevitz, P. (1985) Existence of a calmodulin-dependent potassium channel in synaptic membrane and postsynaptic density fractions isolated from canine cerebral cortex and cerebellum, as determined by apamin binding. *Brain. Res.* **360** 183–194.

2

Voltage-dependent potassium channels: gating, ion permeation and block

D. J. Adams and **W. Nonner**

2.1 INTRODUCTION

Voltage-dependent K^+ channels form the largest and most diversified class of ion channels. First described in membranes of skeletal muscle and nerve axons, voltage-dependent K^+ conductances are ubiquitous in electrically excitable as well as inexcitable cells. Voltage-dependent K^+ conductances traditionally are classified in terms of three major families, distinguished mainly by their responses to changes in membrane voltage (Table 2.1).

Table 2.1 — Classification of voltage-dependent K^+ channels

Family	Activated by	Fast inactivation
Delayed rectifiers	Depolarization	No
Transient channels	Depolarization	Yes
Inward rectifiers	Hyperpolarization, external K^+ ion	—

(1) Current through *delayed rectifier* channels turns on with a brief delay following the onset of a membrane depolarization, and persists while the depolarization is maintained (Fig. 2.1A,B). Rectification is primarily a manifestation of this time-dependent gating process, but is enhanced by a time-independent rectification inherent to the open pores. Under physiological conditions, open-channel currents are larger in the outwaru than the inward direction, reflecting the

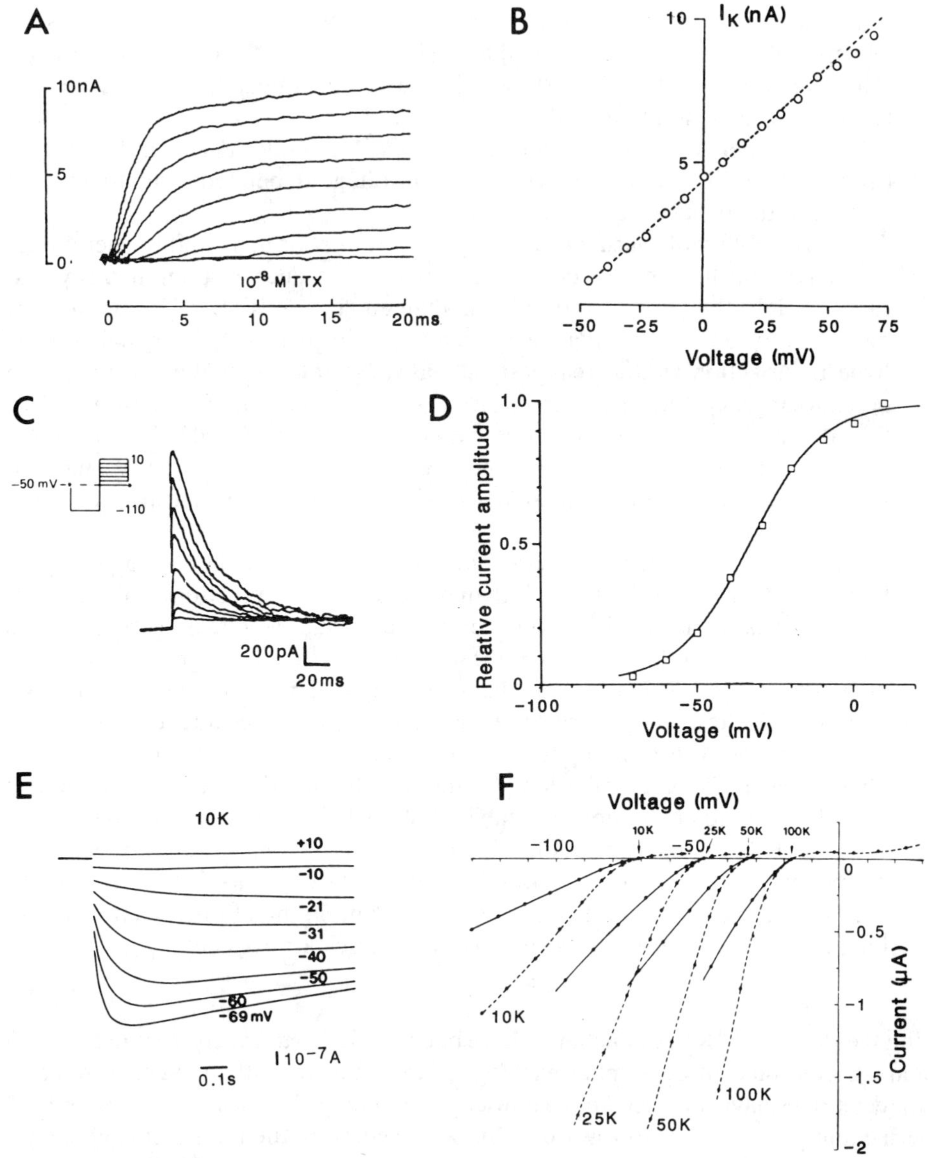

Fig. 2.1 — The three classes of voltage-dependent K^+ currents. Original records (left column) and corresponding current–voltage relations (right column). Delayed rectifier in frog node of Ranvier (from Hille, 1967): (A) depolarizing pulses between −60 and +60 mV were applied to a node treated with 10 nM tetrodotoxin to block Na^+ current; holding potential −75 mV; temperature 17°C; (B) steady-state current–voltage relation of the delayed rectifier currents shown in (A). Transient K^+ current in acutely dissociated rat hippocampal neurones (from Numann *et al.*, 1987): (C) depolarizations to the indicated levels were applied following a 150-ms hyperpolarizing conditioning pulse (see inset); temperature 22°C; (D) fraction of activated K^+conductance versus membrane potential; records shown in (C) were extrapolated to beginning of depolarizations, divided by K^+ ion driving force, and normalized. Inward rectifier in starfish egg (*Mediaster aequalis*) (from Hagiwara *et al.*, 1976): (E) membrane currents evoked under voltage clamp in 10 mM K^+, Na^+-free medium; test pulses of indicated amplitudes were applied from a holding potential of −71 mV; temperature 21–22°C; (F) steady-state (broken lines) and instantaneous (continuous lines) current–voltage relations in starfish egg bathed in four different K^+ concentrations (10, 25, 50 and 100 mM) in Na^+-free media.

different extracellular and intracellular concentrations of K^+ ion. The classical example of a delayed rectifier is the K^+ conductance of the squid giant axon (Hodgkin and Huxley, 1952a), which causes the quick repolarization that terminates the action potential.

K^+ channels regulated by intracellular Ca^{2+} ion and M-type K^+ channels (discussed in Chapters 3 and 5) are also voltage-dependent and follow the pattern of the delayed rectifier.

(2) The permeability of *transient ("A'-type) K^+ channels* is activated by membrane depolarization like that of delayed rectifiers, then decays spontaneously and rapidly while the depolarization is maintained (Fig. 2.1C,D). The amount of permeability activated by a depolarizating voltage step is maximal after a hyperpolarization and decreases steeply as the membrane voltage prior to the depolarizing step becomes less negative (Fig. 2.4B). Transient K^+ currents were first described in mollusc neurons by Hagiwara *et al.* (1961). The term 'A current' was coined by Connor and Stevens (1971); they showed that the transient K^+ current is instrumental in modulating the neuronal firing rate over a range of stimulus strength.

(3) *Inward rectifiers* mirror the characteristics of delayed rectifiers: they can pass larger K^+ currents in the inward, than in the outward, direction. This 'anomalous' rectification is due, in varying degrees, to a time-dependent gating process and/or to a time-independent rectification inherent to the permeability of the open pore (Fig. 2.1E,F). The channels are maximally opened by a large hyperpolarization and pass sustained currents. Their time-independent rectification is in the direction opposite to that expected from the asymmetry of the physiological K^+ ion distribution. As another characteristic, external K^+ ion is capable of regulating the permeability of inward rectifiers: the higher the external K^+ ion concentration, the more positive the voltage at which a permeability can be elicited. Anomalous rectification was first discovered in frog skeletal muscle (Katz, 1949); it is typcially found in membranes that generate long-lasting depolarizations, such as cardiac muscle or egg cells.

Voltage-dependent K^+ conductances have been studied extensively for more than 30 years. Electrophysiological, pharmacological and, most recently, molecular biological techniques have advanced our knowledge from empirical descriptions of the K^+ permeability of membranes towards a characterization of the underlying molecular channels and their submolecular organization. One of the major challenges in this field results from the diversity of the channels. As a rule, any particular cell membrane contains several varieties of voltage-dependent (and other) K^+ channel, and analogous channels from different cells of a species are not identical. The present classification by functional criteria, one hopes, will in the future be improved and sharpened as structural correlates for the various functional aspects become elaborated. One could envisage that a particular K^+ channel eventually will be defined by a catalogue of its molecular building blocks rather than by grouping it into one of a few families. In organizing this chapter, therefore, we will ignore family boundaries where possible and, instead, attempt to group themes by putative molecular mechanisms.

Voltage-dependent K^+ channels have been treated in recent books (Hagiwara, 1983; Hille, 1984) and reviews (Armstrong, 1975; Hille, 1975; French and Adelman, 1976; Hille and Schwarz, 1978; Adams *et al.*, 1980; Dubois, 1983; Latorre and Miller, 1983; Stanfield, 1983; Begenisich and Smith, 1984; Rogawski, 1985; Ruff, 1986; Rudy, 1988).

2.2 THE PORE

Voltage-dependent K^+ channels form highly ion-selective pores, which pass only a small range of inorganic cations. On the other hand, many inorganic and organic cations can enter and block the pore from either end, as if relatively wide vestibules narrow down to a central passsage where ion selection takes place. Several independent observations on ion interactions inside the pore have led to the picture that the narrow section can accomodate several ions at a time, but restricts them to move in a single file. Access and removal of blockers through the intracellular, but not the extracellular, mouth of the pore often interferes with the opening and closing of the channel's gate, as if the gate acts near the intracellular end of the pore. The ionic environment appears to be essential for the survival and/or proper gating of some channels.

2.2.1 Ionic selectivity

Table 2.2 gives examples of ion selectivities for one thoroughly studied channel in

Table 2.2 — Ionic permeability ratios, P_X/P_K

On	Delayed rectifier[a]	Transient channel[b]	Inward rectifier[c]
Tl^+	2.3	2.04	1.5
K^+	1	1	1
Rb^+	0.91	0.73	0.35
NH_4^+	0.13	0.18	0.035
Cs^+	<0.077	0.14	<0.03
Li^+	<0.018	0.07	—
Na^+	<0.01	0.09	<0.03
$CH_3{\cdot}NH_3^+$	<0.021	0.06	—

[a] Frog node (Hille, 1973).
[b] *Helix aspersa* neurone (Taylor, 1987).
[c] Starfish egg (Hagiwara and Takahashi, 1974).

each family. These permeability ratios were determined from current reversal potentials measured in extracellular solutions of different cations. All channels select for monovalent cations in the same order: $Tl^+ > K^+ > Rb^+ > NH_4^+ > Cs^+$. On a quantitative scale, the inward rectifier is set apart from the other channels: it discriminates more effectively among the most permanent ions. All cations with appreciable permeabilities have crystal diameters smaller than that of Cs^+ (3.38 Å); no permeant organic cations are known. A structure for this 'selectivity filter' has been proposed by Hille (reviewed by Hille, 1975). Selectivity against large cations arises here from a small pore diameter, and, against small cations (like Na^+), from an inability of the pore wall to provide to the ion electrostatic interactions equivalent to those with water.

2.2.2 Unitary conductance

The unitary conductances measured for voltage-dependent K^+ channels (Table 2.3) are considerably smaller than those of the Ca^{2+}-activated K^+ channels (reviewed by Latorre and Miller, 1983) and span about one order of magnitude. Nevertheless, these conductances translate into transfer rates of the order of 10^6 to 10^8 ions per second, values much higher than rates achieved by any known ion carrier.

Elevating the external K^+ ion concentration above physiological levels results in increase, saturation and eventual decrease of macroscopic conductance of the delayed rectifier in squid axon (Wagoner and Oxford, 1987). The unitary conductance of inward rectifiers in tunicate egg (Fukushima, 1982) and cardiac muscle (Sakmann and Trube, 1984a; Payet *et al.*, 1985) increases approximately with the square root of the external K^+ concentration; this concentration dependence is consistent with saturation as described by a Michaelis–Menten equation. On the other hand, one notes that delayed rectifiers in squid axon have conductances comparable to other channels that operate at lower salt concentrations.

2.2.3 Block by intracellular TEA and analogues

The block of the delayed rectifier and other potassium channels by intracellular tetraethylammonium (TEA) and other quaternary ammonium derivatives (QA) has been extensively reviewed by Armstrong (1975) and Stanfield (1983) and so will be discussed here with an emphasis on mechanism. Fig. 2.2 shows the effect of internally applied nonyltriethylammonium ion (C9) on the delayed rectifier current of a squid axon (Armstrong, 1971). The drug allows channels to open with an apparently normal time course, but subsequently depresses the current as if the channels had acquired the capability to 'inactivate' like transient K^+ channels. The rate and degree of this inactivation vary with drug concentration in the way expected for a one-to-one drug/channel interaction. Other QA derivatives act analogously, with potencies that increase with the size of the hydrocarbon residues.

Armstrong (Armstrong and Binstock, 1965; Armstrong, 1966, 1969, 1971, 1975) proposed the following reaction scheme to account for observations of TEA and QA block of K^+ currents in squid giant axons:

$$C_1 \leftrightarrow C_2 \leftrightarrow C_3 \leftrightarrow C_4 \underset{\beta}{\overset{\alpha}{\leftrightarrow}} O \underset{k_{-1}}{\overset{k_{+1}}{\leftrightarrow}} OB \tag{2.1}$$

where C_1–C_4 represent the four closed channel kinetic states proposed by Hodgkin and Huxley (1952b), O the open current conducting state, and OB the open but TEA- or QA-blocked non-conducting state. The rate constants (α and β) between

Table 2.3 — Unitary conductances of voltage-dependent K^+ channels

Cell	$[K^+]$out// $[K^+]$in (mM)	γ (pS)	Reference
Delayed rectifiers			
Squid axon	400//400	3	Armstrong (1971)
	10//400	12	Conti *et al.* (1975)
	460//1	18	Conti and Neher (1980)
	5//528	20, 40	Llano *et al.* (1988)
Helix roseneri	110//115	2–3	Reuter and Stevens (1980)
Frog node	(2.5//120)	4	Begenisich and Stevens (1975)
	2.5//113	2.7, 4.6	Neumcke *et al.* (1980)
	47.5//120	24	Llano *et al.* (1988)
Frog skeletal muscle	120//120	30	Standen *et al.* (1985)
	2.5//120	15	Standen *et al.* (1985)
Rabbit smooth muscle	6//126	52	Benham and Bolton (1983)
Rat hippocampal neurones	5.5//140	15	Rogawski (1986)
Rat hypothalamic neurones	5//140	48	McLarnon (1989)
Mammalian brain	2//100	9	Stühmer *et al.* (1988)
	100//100	22	Stühmer *et al.* (1988)
Chick heart	4//120	62	Clapham and De Felice (1984)
Transient K channels			
Mammalian nodose	5.4//150	23	Cooper and Shrier (1985)
neurone	150//150	38	Cooper and Shrier (1985)
Mammalian DRG neurone	5//150	20	Kasai *et al.* (1986)
Drosophila mytube	2//140	12–16	Solc *et al.* (1987)
Drosophila neurone	2//140	5–8	Solc *et al.* (1987)
Helix aspersa neurone	4//120	14	Taylor (1987)
GH3 pituitary cells	5//150	6–8	Oxford and Wagoner (1989)
Inward rectifiers			
Tunicate egg	50//400	5.5	Ohmori (1978)
	100//400	6.9	Ohmori (1978)
	100//(400)	5	Fukushima (1982)
	200//(400)	7.1	Fukushima (1982)
	400//(400)	12.3	Fukushima (1982)
Frog skeletal muscle	120//(160)	10	Schwarz *et al.* (1981)
Rat skeletal muscle	155//(150)	10.4	Ohmori *et al.* (1981)
Mammalian ventricle	150//155	22	Matsuda (1988)
	145//145	27	Sakmann and Trube (1984a)
	5.4//145	3.6	Sakmann and Trube (1984a)
	150//(150)	40	Kurachi (1985)
Bovine vascular endothelial cells	145//145	26	Olesen *et al.* (1988)

Parenthesized internal concentrations are estimated cytoplasmic K^+ ion concentrations.

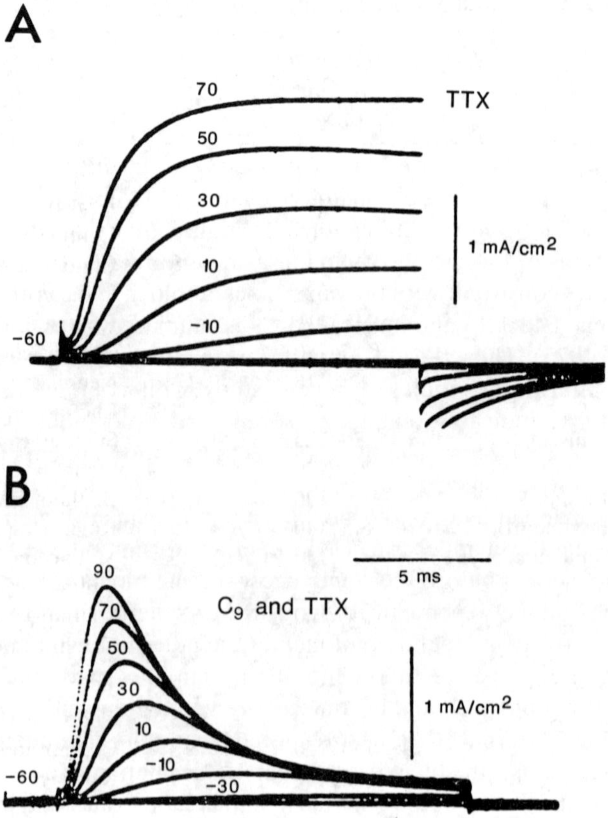

Fig. 2.2 — Block of open delayed rectifier channels by an internally applied TEA analogue, C_9. Membrane K^+ currents of a control squid giant axon (A) and an axon containing 0.11 mM C_9 (B). Depolarizing voltage clamp pulses to the indicated levels were applied from a holding potential of −50 mV (temperature 10°C) (from Armstrong, 1971).

the transition pathways of the closed states and the single open state are identical in magnitude and voltage dependence to those proposed by Hodgkin and Huxley (1952b). Thus, in the absence of drug, the model behaves exactly like their equations predict. At the holding potential of −60 mV, almost all of the channels are in the closed state C_4 because the forward rate constant, α, is small and the backward rate constant, β, is large at this potential. At the beginning of the depolarizing voltage clamp step, α instantaneously becomes large, and β small, which drives the channels from the closed towards the open conformation. Since the channel is required to pass through the several closed states, the currents do not appear instantly with the voltage step, but rather begin with a delay and rise in a sigmoid fashion to a sustained level.

With the addition of TEA or QA derivatives, transitions to the blocked state, OB, become possible, but since the channel must be in the open state for block to occur the initial onset of current is unaffected. At the start of the voltage step, the channels must still pass through the closed states to the open state in the normal manner. Only when the open-state occupancy becomes large does the blocking transition O→OB (with rate constant k_{+1}) begin to predominate, causing the currents to peak and then decay. The rate constants k_{+1} and k_{-1} then determine the rate and completeness of block.

The forward rate, k_{+1}, is proportional to the QA concentration. The reverse rate, k_{-1}, reflects the affinity of the drug to the channel: QA ions with short alkyl chains (such as TMA, TEA, and C3) bind much less tightly than does the longer chain, C9. Thus, the actions of the shorter-chain ions can be simulated by giving both k_{+1} and k_{-1} large values consistent with the rapid block/unblock and required large blocking concentrations. Block by the long-tailed C9 is simulated by making k_{-1} very small (Armstrong, 1975). Thus, the QA ions appears to attach to a hydrophobic region of a fairly large internal vestibule. A recent study with compounds larger than C9 (Swenson, 1981) indicates that the vestibule can hold molecules whose largest diameters are 1.1×1.2 nm. The hydrophobic interaction appears to be essential as introduction of hydrophilic groups increases the concentration required for block.

The voltage dependence of the blocking reactions can arise indirectly through the requirement that the channel reaches the open conformation, via a series of voltage-dependent reactions, before block can occur, or the blocking rate constants themselves may be voltage-dependent. In Armstrong's scheme the forward blocking rate constant is relatively independent of membrane potential, while the backward rate constant becomes larger as the membrane potential is made more negative. The backward rate is enhanced and becomes more voltage-dependent when the axon is bathed in a high external [K^+], as if K^+ ion entering the channel through the outer mouth can 'knock out' the blocking molecule. This phenomenon localizes the action of the drug to the interior of the pore itself, and demonstrates that more than one ion can be simultaneously present and interact in this pore.

It has also been proposed that TEA and other QA ions bind within the voltage drop across the channel and, therefore, both the forward and backward blocking rate constants are inherently voltage-dependent (French and Shoukimas, 1981, 1985; Swenson, 1981). TEA and larger compounds seem to bind at a site near the inner end of the channel, at 20% of the voltage drop across the membrane. The smaller blocker, methylammonium, appears to penetrate farther into the nodal delayed rectifier to a depth of 70% (Hille, 1975).

Internally applied TEA and other QA ions also block outward K^+ currents through delayed rectifier channels in amphibian nodes of Ranvier (Koppenhoefer and Vogel, 1969; Armstrong and Hille, 1972) and frog skeletal muscle (Fink and Wettwer, 1978) in a manner similar to the block of squid axon channels. Unlike TEA block in squid axon, where internal TEA blocks mainly outward currents, internal TEA in molluscan neurones blocks both inward and outward K^+ currents (Neher and Lux, 1972), suggesting a mechanism similar to that of the external TEA block of nodal channels.

2.2.4 Block by extracellular TEA

Delayed rectifier channels in myelinated nerve, skeletal muscle and molluscan neurons, but not in squid axon, are blocked by external TEA application (Hille, 1967; Koppenhöfer, 1967; Kao and Stanfield, 1973; Thompson, 1977; Hermann and Gorman, 1981b). The mechanism of block by external TEA and other QA ions is qualitatively different from the block by internal application. Firstly, TEA is by far the most potent blocker externally (Hille, 1967), whereas the QA ions with large hydrophobic groups are the most potent internally. Secondly, block by external QA ions is not appreciably time- or voltage-dependent. Analogous to internal block, external block appears to require one QA ion per channel (Hille, 1967). These properties suggest an external TEA receptor that is located outside the electric field of the membrane and lacks the large hydrophobic vestibule of the internal receptor.

The effects of external TEA and other QA ions on molluscan neurones have been reviewed by Adams *et al.* (1980) and Stanfield (1983). Briefly, all K^+ currents in these cells are sensitive to block by QA ions to some extent. In general, the delayed outward K^+ currents in response to depolarization are blocked by lower external TEA concentrations than are the transient (I_A) outward currents. The Ca^{2+}-activated K^+ currents in molluscan neurones can be more (Hermann and Gorman, 1981b), less (Thompson, 1977) or as (Meech and Standen, 1975) sensitive to external TEA. As with nodes of Ranvier, external TEA block of molluscan neuron delayed rectifier channels seems to be a one-to-one reaction (Meech and Standen, 1975; Hermann and Gorman, 1981b).

2.2.5 Interactions with inorganic cations

2.2.5.1 Alkali cations

Internally applied Na^+, Li^+ and Cs^+ block delayed rectifier current in squid axon (Chandler and Meves, 1965; Adelman and Senft, 1966; Benzanilla and Armstrong, 1972; French and Shoukimas, 1985) and in the node of Ranvier (Bergman, 1970; Dubois and Bergman, 1975). Externally applied Cs^+ also blocks inward K^+ currents through squid axon channels (Adelman and Senft, 1968; French and Shoukimas, 1985; Adelman and French, 1978) and in the node (Dubois and Bergman, 1975). The voltage dependence of external Cs^+ block in the squid is stronger than expected when a single blocking ion binds to a site inside the applied field and indicates multi-ion occupancy of the channel (Adelman and French, 1978).

The mechanism of block by internal application of these cations seems to be similar to that of TEA block in that the ions enter the large, relatively non-selective inner channel mouth, but their movement is arrested by a constriction through which only K^+ and Rb^+ can pass (Bezanilla and Armstrong, 1972). The block is voltage-dependent and produces a negative slope conductance at depolarizing voltages. In the case of Na^+ ion, this block of outward K^+ current can be relieved by large depolarizations (French and Wells, 1977); Na^+ ion then can become a charge carrier for outward current through the delayed rectifier channels.

Relief of internal Na^+ and Cs^+ block by raising the concentration of external permeant or even impermeant ions has been observed in delayed rectifier channels of squid axon (Wagoner and Oxford, 1987) and of node of Ranvier (Dubois and Bergman, 1977). Such relief of internal block by permeant external ions is one of the strongest arguments for Armstrongs's hypothesis that the mechanism of block by TEA, QAs and the inorganic ions involves the blocker actually entering the diffusion pathway of the permeant ions and binding somewhere within the channel's aqueous pore. The steep voltage dependence of block suggests that the binding site for these small inorganic cations is farther into the membrane electric field than that of TEA and QA derivatives (French and Shoukimas, 1981).

Block by alkali ions of delayed rectifiers is expressed in 'instantaneous' current voltage relations and, hence, lacks a measurable time dependence. Thus, the degree of block is likely to reflect a rapid equilibration between binding and unbinding of the blocking alkali ions, a situation different from that seen with QA ions. To the extent that association rates are generally diffusion-limited, the millimolar concentrations of Na^+ and Cs^+ required for block indicate that the dissociation rates for alkali cations are much larger than those for QA ions (Bezanilla and Armstrong, 1972).

Inwardly rectifying K^+ channels in egg and skeletal muscle are blocked by external Cs^+, Na^+, Li^+, and Rb^+ ion. This block, like the Cs^+ block of the delayed rectifier, is more strongly voltage-dependent than would be expected for a single monovalent cation (reviewed in Hagiwara, 1983, and Latorre and Miller, 1983). On the other hand, the block of inward rectifiers requires a significantly smaller concentration of Cs^+ ion and it exhibits a measurable time dependence (Fig. 2.3).

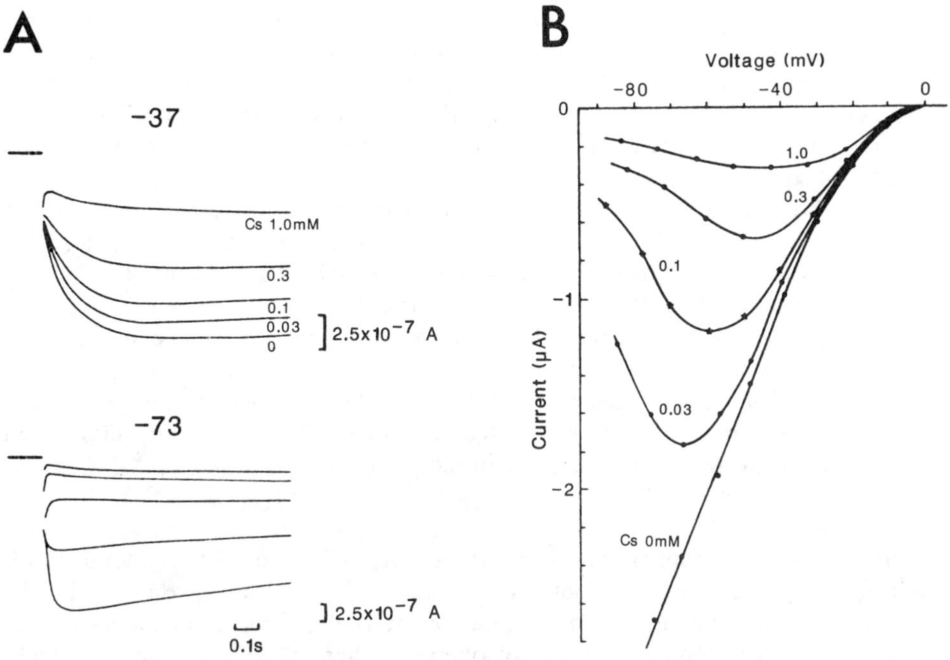

Fig. 2.3 — Block of inward rectifiers by external Cs^+ ion. (A) Effects of four different Cs^+ concentrations (0.03, 0.1, 0.3 and 1 mM) on the inward current obtained in 25 mM K^+, Na^+-free media. Each family of currents refers to one test voltage (indicated as displacement from holding level, −48 mV) and different Cs^+ concentrations. Temperature 21–22°C. (B) Current–voltage relationships resulting from Cs^+ block. Currents were measured at 0.1 s after the onset of test voltages; the latter are given as displacements from the holding level, −48 mV (from Hagiwara *et al.*, 1976).

The block by Na^+ and Rb^+ can be relieved by negative voltage, as if these relatively impermeant cations can be 'punched through' the selectivity filter by a sufficiently strong electric field (Ohmori, 1978; Standen and Stanfield, 1980). Another interesting effect of external Na^+ ion is to facilitate (macroscopic and unitary) inward current in tunicate egg (Ohmori, 1978; Fukushima, 1982) and mouse dorsal root ganglion cells (Mayer and Westbook, 1983). This phenomenon could reflect an interaction between Na^+ and K^+ ion in a multi-ion single file pore: K^+ ion bound to a channel site is more readily released into the cytoplasm due to the repulsion exerted by an Na^+ ion entering through the external mouth (Fukushima, 1982).

2.2.5.2 Hydrogen ion

Raising the concentration of external H^+ ions has two major effects on delayed rectifier currents in squid giant axon and frog node of Ranvier, both of which reduce K^+ conductance at a given voltage. One effect is to the voltage at which the gate opens (see section 2.3). By the other effect, H^+ ions reduce the conductance of open delayed rectifier channels in a manner suggesting the titration of one of more weakly acidic groups within the pore (Hille, 1968; Drouin and The, 1969; Mozhayeva and Naumov, 1970; Carbone *et al.*, 1978). The titration curve in the node of Ranvier is not that of a single acidic residue (or several residues with a single pK_a), but suggests that perhaps several groups with slightly different pK_as are involved (Hille, 1968). Wanke *et al.* (1979) reported that internal protons also block current in the delayed rectifier of squid axon.

Inward rectifier K^+ channels of frog skeletal muscle are blocked by internal, but not external, protons (Moody and Hagiwara, 1982; Blatz, 1984). This block is consistent with the simultaneous binding of H^+ ions to three equivalent sites (Moody and Hagiwara, 1982; Blatz, 1984). In an earlier study, Hagiwara *et al.*, (1978) reported that the inward rectifier of starfish egg is blocked when the external pH is lowered below 6.0 in the presence of an acetate buffer. The more recent experiments (Moody and Hagiwara, 1982; Blatz, 1984) revealed that acetate permeates cell membranes sufficiently to acidify the cytoplasm, so the effects originally interpreted as extracellular were due to changes in intracellular pH.

2.2.5.3 Barium

A divalent cation of approximately the crystal radius of K^+ ion, Ba^{2+}, interacts with delayed rectifier channels from both sides of the membrane. Internal Ba^{2+} ion blocks channels in a way reminiscent of the block by internal QA ions, mimicking a time- and voltage-dependent inactivation of open K^+ channels (Eaton and Brodwick, 1980; Armstrong and Taylor, 1980). External Ba^{2+} reduces the size of currents (Armstrong and Taylor, 1980) with little effect on time courses. Dose–effect curves suggest that one Ba^{2+} ion blocks one channel.

The blocking effects of internal and external Ba^{2+} persist for tens of seconds after the removal of Ba^{2+} from the perfusate. Washout times for internal and external Ba^{2+} are indistinguishable, and are equally shortened by depolarization, or equally increased in the external presence of some monovalent ions, e.g. NH_4^+ (Armstrong *et al.*, 1982). These results suggest that internal and external Ba^{2+} ion bind to the same site in the delayed rectifier channel and exit by the same route, the external channel mouth. The voltage dependences of Ba^{2+} block, on the other hand, cannot be explained simply by the movement of Ba^{2+} ion to a well-defined site within the membrane electric field: internal Ba^{2+} would have to bind near the outer mouth, whereas external Ba^{2+} would have to bind two-thirds of the channel depth away from the outer mouth. Armstrong *et al.* (1982) point out that interpretations of voltage-dependent block in a multi-ion channel need to consider displacements of all ions present in the pore, e.g. Ba^{2+} and K^+ ion. Apparently, the preferred location of Ba^{2+} ion in a blocked pore is affected by the electric field, other ions accessing the pore from the inner or outer mouth, and constraints exerted by the channel gate.

External Ba^{2+} (and Sr^{2+}) ion also blocks inward rectifier channels of starfish egg and frog skeletal muscle (Sperelakis *et al.*, 1967; Standen and Stanfield, 1978a; Hagiwara *et al.*, 1978). This block is time- and voltage-dependent; it is accelerated and enhanced by negative voltage. Depolarization speeds the relief of Ba^{2+} blockade (Standen and Stanfield, 1978a). Doubling the K^+ ion concentrations on both sides increases fourfold the concentration of Ba^{2+} ion required to produce a given degree of block, as if two K^+ ions compete with one Ba^{2+} ion. The voltage dependence of (external) Ba^{2+} block in inward rectifiers is tantamount to that in delayed rectifiers.

2.2.5.4 Magnesium

In inward rectifier of cardiac muscle, block of outward current by intracellular Mg^{2+} ion is involved in the instantaneous rectification. In the absence of cytoplasmic Mg^{2+} ion, the current–voltage relationship is linear, whereas, with a physiological concentration of Mg^{2+}, unitary outward current is reduced by a rapidly flickering block (Matsuda *et al.*, 1987; Vandenberg, 1987). The degree of this block varies with the internal Mg^{2+} activity in the micromolar to millimolar range, and becomes manifest at voltages more postive than E_K (Vandenberg, 1987). This rectification by an intracellular blocker is in line with a hypothesis proposed by Armstrong (1975) and elaborated by Hille and Schwarz (1978) and Standen and Stanfield (1978b). In recent experiments with internal Mg^{2+}, the inward rectifier exhibited sublevels of conductance, as if each channel comprises three pores that are gated in concert, but can be blocked individually and independently by Mg^{2+} ions (Matsuda, 1988).

2.3 THE GATE

When electrophysiologists refer to the ‘gate’ of an ion channel, they imply an intrinsic capacity by which the channel can modulate its ionic flux. Unlike the notion of a ‘pore’, the notion of a ‘gate’ does not imply a basic mechanism for the function. Indeed, one can imagine many ways of how current through a pore could be controlled (cf. Chapter 14 in Hille, 1984). Several experimental strategies are used to elucidate mechanisms and the structural correlates of gating, such as the analysis of gating kinetics, modification of gating by extrinsic agents, the measurement of gating currents and, most recently, site-directed mutagenesis.

Fluctuation analysis of macroscopic K^+ currents and recordings of single-channel currents indicate that individual K^+ channels are gated in an all-or-none manner, such that variations of the macroscopic K^+ current reflect variations in the number of open channels. The channels can exist for measurable intervals in the apparently discrete states, open or shut. The high-order kinetics expressed in the time course of macroscopic conductance changes can be attributed to transitions among multiple open or shut substates. The open/shut transition bears analogy to the behaviour of an allosteric enzyme that, triggered by the cooperative effect of several bound substrate molecules, switches among two major conformations (Monod *et al.*, 1965). From this model one may predict that gating involves a major rearrangement in the structure of the channel protein.

The gate of voltage-dependent K^+ channels must be controlled by forces that the applied electrical field exerts on charges or dipoles that are part of or closely associated with the channel. Gating currents reflect the movements of such charges of dipoles and, hence can reveal events that precede the opening or closing of the pore.

Two basic varieties of voltage-dependent gating are found in K^+ channels: activation and inactivation. By activation we will denote a process that opens channels when the membrane is depolarized. Another distinctive feature of this process is an initial delay, as if the channel proceeds through several closed conformations before it can open. By inactivation we will denote a process that shuts channels during sustained depolarization. Delayed rectifiers, thus, are gated by activation, transient K^+ channels are gated by both activation and inactivation, whilst time-dependent inward rectification is classified as inactivation.

The terms 'inactivation' and 'inward rectification' both represent perceptions of the same phenomenon, that sustained depolarization shuts off a K^+ current. Thus, one can regard channels with time-dependent inward rectification as K^+ channels in which only inactivation is functional, whereas activation is wedged in its open state. This is the opposite of the conventional nomenclature represented in Table 2.1, which describes inward rectifiers as gated by an activation process that opens channels upon hyperpolarization. However, there is at present no evidence to support the implication that time-dependent inward rectifiers are an isolated group of ion channels whose voltage-dependent gating machinery is reversed with regard to those of all other voltage-dependent (K^+, Na^+ and Ca^{2+} ion) channels.

2.3.1 Gating in delayed rectifiers

Delayed rectifiers open in response to depolarization and stay open as long as the depolarization is maintained. A formal description of this time- and voltage-dependent gating has been developed by Hodgkin and Huxley (1952b) for the classical delayed rectifier of squid axon. Here, activation is ruled by the fourth power of a first-order kinetic variable (n^4), as if four identical charged particles have to move into a permissive position to create an ionic pathway. The combined charges of these particles account for the observed steep voltage dependence of the conductance: an *e*-fold change of the K^+ conductance requires only 4.8 mV of change in voltage. This kinetic model also reproduces the observation that channels open with a sigmoidal, but close with a simple exponential, time course.

An early observation that challenged the Hodgkin–Huxley (HH) formulation was made by Cole and Moore (1960): following a strong hyperpolarization, the delay of K^+ current activation is increased to an extent such that the HH kinetic variable needs to be raised to powers n^k in which k is much larger than 4 and depends on the stimulation protocol. More recent kinetic schemes (Gilly and Armstrong, 1982; White and Bezanilla, 1985; Spires and Begensich, 1989) describe activation more generally, in terms of a sequence of closed states that lead to an open state; they no longer impose the constraints of the Hodgkin–Huxley formalism onto the number of these states and their rate constants of interconversion. Such efforts, however, have not yet converged into a generally accepted kinetic scheme.

The investigation of the gating mechanism of delayed rectifiers is complicated by a number of factors:

(1) K^+ currents of most cells cannot be interpreted as due to a single, uniform population of channels (e.g. Dubois, 1983).
(2) Kinetics of gating can be obscured by effects of diffusion polarization in a space adjacent to the membrane (Frankenhaeuser and Hodgkin, 1956).
(3) Noise measurements revealed the presence of fluctuations whose correlation times are much longer than expected from gating kinetics, indicating that channel properties are not invariant over time (Van den Berg *et al.*, 1977; Conti *et al.*, 1975, 1984); different modes of gating have been observed in single-channel recordings (Llano *et al.*, 1988; Stühmer *et al.*, 1988).
(4) Identification of gating currents associated with K^+ channel activation is difficult due to superposition with Na^+ channel contributions (Almers and Armstrong, 1980; Gilly and Armstrong, 1980; White and Bezanilla, 1985; Spires and Begenisich, 1989); furthermore, manipulation of the ionic medium, as necessary to record K^+ gating current, is likely to modify the gating of K^+ channels (see section 2.3.4).
(5) Unitary K^+ currents are quite small (see Table 2.3), limiting the temporal resolution of kinetic studies.

The rates of activation of delayed rectifiers vary over a wide range among species and tissues. Whereas the classical delayed rectifiers of nerve axons activate in a few milliseconds those of neurone somata may require tens of milliseconds, and those of skeletal and cardiac muscle hundreds of milliseconds (summarized in Table 2 in Rudy, 1988).

Whilst delayed rectifiers may sustain currents for seconds, many of them inactivate with maintained depolarization. Examples are the channels in squid axon (Ehrenstein and Gilbert, 1966), frog skeletal muscle (Adrian *et al.*, 1970), and amphibian node of Ranvier (Schwarz and Vogel, 1971). Since we cannot distinguish at the present time whether the inactivation processes of such delayed rectifiers are mechanistically different from those operating in transient K^+ channels, the distinction between inactivating delayed rectifiers and transient channels must be made with reservation.

2.3.2 Gating in transient K^+ channels ('A' currents)

The voltage-dependent activation of transient K^+ channels is analogous to that in delayed rectifiers. A sigmoidal time course of activation is observed in A currents of molluscan neurone at low temperature (Connor and Stevens, 1971; Neher, 1971) and in the soma membrane of cultured hippocampal neurones of rat at 25°C (Rizzo, 1988). 'A' currents of most cell types are activated at voltages more positive than −60 to −50 mV. The steepness of the voltage dependence ranges from 5 mV/*e*-fold change of conductance in cardiac muscle (Giles and Van Ginneken, 1985) to about 11 mV in mammalian neurones (Numann *et al.*, 1987).

The steady-state inactivation of transient K^+ conductance is significant already at the resting potential, and is complete near voltages that would activate channels. Thus, maximal A currents can be elicited only after a preceding strong hyperpolarization. The voltage dependence of this steady-state inactivation is steep; *e*-fold changes in inactivation require voltage changes of 4 mV in *Helix* neurones (Neher, 1971) to 9 mV in rat dorsal root ganglion cells (Kostyuk *et al.*, 1981). Characteristic of the inactivation kinetics is that the time constant varies little over the voltage range where significant currents are activated; time constants of inactivation and repriming are strongly voltage-dependent at more negative potentials, however. These inactivation properties have initially been described in molluscan neurones (Connor and Stevens, 1971; Neher, 1971) and have subsequently been found in vertebrate preparations (reviewed in Rogawski, 1985; Rudy, 1988).

The gating of transient K^+ channels is analogous to that of Na^+ channels. Indeed, some agents (*N*-bromoacetamide and the disulphonic stilbene derivative SITS) that are known to inhibit the inactivation of Na^+ channels similarly inhibit inactivation of transient K^+ channels in GH3 cells (Oxford and Wagoner, 1989). On the other hand, intracellular application of pronase or tannic acid was found to be ineffective on the transient K^+ current in coelenterate egg cell membrane (Hagiwara *et al.*, 1981).

2.3.3 Gating in inward rectifiers

A hyperpolarizing step in membrane voltage elicits an apparently instantaneous jump in the inward rectifier conductance, followed by a relaxation whose time course can be resolved. In one kind of inward rectifier, exemplified by that of tunicate eggs (Hagiwara and Takahashi, 1974; Hagiwara *et al.*, 1976; Ohmori, 1978), the relaxation first enhances the inward rectification during some tens of milliseconds and then slowly declines (Fig. 2.1E). In other inward rectifiers, exemplified by those in cardiac (Kurachi, 1985; Harvey and Ten Eick, 1988) and sketetal muscle (Leech and Stanfield, 1981; Hestrin, 1981), the relaxation is also biphasic, but the initial rising phase is very rapid compared to the subsequent decline. The phase of increase is faster by an order of magnitude than that observed in the tunicate egg, so it was not detected in earlier studies due to limited temporal resolution. The steady-state conductance approached after the slow decay of the biphasic conductance can exhibit a negative slope at large negative potentials.

On the premise stated above, that we refer to a process that *shuts* channels upon a positive (depolarizing) change of voltage as inactivation, and, conversely, a process that *opens* channels upon a depolarizing change of voltage as activation, then the fast increase of inward rectifier conductance during hyperpolarization is described as the removal of inactivation.

The removal of inactivation (fast relaxation) has been found to follow a simple exponential time course, which is voltage-dependent (Hagiwara *et al.*, 1976; Kurachi, 1985; Matsuda, 1988). The steady-state voltage dependence of this process is strongly modulated by variations of external, but not internal, K^+ ion concentration in egg cells (Hagiwara *et al.*, 1976; Hagiwara and Yoshii, 1979), cardiac and skeletal muscle (Harvey and Ten Eick, 1988; Kurachi, 1985; Leech and Stanfield, 1981; Hestrin, 1981). By this effect, the inward rectifier conductance appears as controlled not by membrane potential alone, but by the electrochemical gradient of K^+ ion ($E_m - E_K$). Thus, K^+ ion concentration not only determines the conductance properties of the open pore (see section 2.2.5), but is also involved in the gating. Mammalian dorsal root ganglion cells and olfactory cortex neurones (Mayer and Westbrook, 1983; Constanti and Galvan, 1983) exhibit inward rectification mediated by a non-selective cation channel, whose gating characteristics are insensitive to the external K^+ ion concentration.

The slow relaxation that reduces inward rectifier conductance has been attributed to: (1) K^+ ion depletion in traverse tubules of skeletal muscle or restricted extracellular space in cardiac muscle (Adrian and Freygang, 1962; Almers, 1972; McAllister and Noble, 1966; Carmeliet, 1982); (2) a block of open channels by an extracellular ion such as Na^+ (Standen and Stanfield, 1979; Fukushima, 1982); or (3) to a second, intrinsic gating process that deactivates the conductance as the membrane potential is changed in the hyperpolarizing direction (Sakmann and Trube, 1984b; Josephson and Brown, 1986).

2.3.4 Effects of ions on gating

2.3.4.1 Alkali cations

In delayed rectifiers of the node of Ranvier, channels open more rapidly when the external K^+ ion concentration is increased from 0 to 2.5 mM (Dubois and Bergman, 1977). A similar effect was observed in molluscan neurones (Junge, 1982). In the squid axon, the impermeant blocking ions Li^+ and Na^+ have little effect, whereas the permeant ions K^+, Rb^+ and even the marginally permeant ion Cs^+ reduce the closing, but not the opening, rate (Swenson and Armstrong, 1981; Matteson and Swenson, 1986). The strength of the effects appears to correlate with the dwell times of these ions in the pore. The permeant species, NH_4^+ and Tl^+ ion, speed opening and closing of the squid delayed rectifer (Oxford and Adams, 1981; Matteson and Swenson, 1986). Apparently, the gating of delayed rectifiers is to some degree determined by what species of ion populate the pore.

The presence of permanent ions appears to be required for maintained function of delayed rectifiers in squid axon. Chandler and Meves (1970) discovered that the K^+ conductance disappears during removal of internal K^+ ion for prolonged periods. Almers and Armstrong (1980) established conditions for survival of the conductance. Whereas in the total absence of permeant ions inside and out the conductance irreversibly decayed with a time constant of about 11 min, the decay was largely prevented in the presence of 100–400 mM external K^+, Cs^+, NH_4^+, or Rb^+ ion. The disappearance of the K^+ conductance is accompanied by the reduction of a slow component of gating current (Almers and Armstrong, 1980; Gilly and Armstrong, 1980), suggesting that the loss of K^+ channel function is associated with an immobilization of the gate.

Transient (outward) K^+ current in GH3 pituitary cells is suppressed in the absence of external K^+ ion (Oxford and Wagoner, 1989); the inhibition is reversible following brief exposure to low external K^+ media. The supportive effect of external K^+ ion begins to saturate at a concentration near 10 mM.

In inward rectifiers, gating is modulated by the external K^+ ion concentration, and external Na^+ ion can produce a time and voltage-dependent block (see section 2.3.3).

2.3.4.2 Hydrogen ion

Low external pH shifts the voltage dependence of channel activation in a depolarizing direction (Hille, 1968, 1973; Drouin and The, 1969; Mozhayeva and Naumov, 1970; Shrager, 1974; Schauf and Davis, 1976; Carbone *et al.*, 1978). This is thought to reflect the binding of H^+ ions to diffuse or aggregate negatively charged moieties near the outer mouth of K^+ channels, altering the voltage field sensed by the channel gating mechanism (Hille, 1973). Opening of delayed rectifier channels in crayfish axons is slowed by low external pH (an effect similar to the effects of Zn^{2+} and other divalent cations, which are described below). The pH dependence is consistent with the titration of a group with pK_a of around 6.0 (Shrager, 1974). This result and chemical modification experiments with diethylpyrocarbonate (a reagent that carboxymethylates histidine residues in proteins) led Shrager to propose that the imidazole group of a histidine residue was titrated.

The kinetics of inward rectifiers of skeletal muscle were not affected by reducing external pH (Blatz, 1984).

2.3.4.3 Divalent cations

Divalent cations such as Zn^{2+} (Stanfield, 1975; Arhem, 1980; Gilly and Armstrong, 1982), Ca^{2+}, Hg^{2+}, Cd^{2+}, Mn^{2+}, Ni^{2+} and Cu^{2+} (Gilly and Armstrong, 1982) modify delayed rectifier gating. Such ions slow the opening of channels to a much larger extent than they affect closing. The relative potencies in slowing the 'on' rate are: $Hg^{2+} > Zn^{2+} > Cd^{2+} > Ni^{2+} > Mn^{2+} > Cu^{2+} > Ca^{2+}$ (Gilly and Armstrong, 1982). Hg^{2+} block is irreversible and Hg^{2+}-blocked current can be further reduced by Zn^{2+} suggesting at least two divalent ion binding sites for each channel. Gilly and Armstrong (1982) proposed a mechanism for this divalent ion effect, which, unlike a surface charge effect, modifies opening and closing rates selectively. Here, divalent cations bind to sites exposed to the external environment while the channel is shut. The sites become submerged when the channel rearanges its structure to open, forcing the bound ions to leave. While a divalent cation is adsorbed to a site, it stabilizes the closed conformation and slows the rate of opening. An open channel cannot interact with a divalent cation and, hence, closes at a normal rate. Whereas the gate of delayed rectifiers channels appears to act near the internal end of the pore, these experiments with divalent ions indicate that the gating machinery may extend to the outer surface of the membrane.

Elevating the external Ca^{2+} ion concentration antagonizes the slowing effects on delayed rectifier closing that are exerted by external K^+ or Rb^+ ions (Armstrong and Matteson, 1986). High external Ca^{2+} has a depressing effect on instantaneous inward currents, as if it can enter and block the pore at negative potentials. These observations have led to the hypothesis that Ca^{2+} ion enters the pore and exerts a 'latching' effect on the gate. When Ca^{2+} is replaced by competing monovalent ions like K^+ or Rb^+, the closure of the gate is slower and less secure. Ca^{2+} ion actually appears to be an essential cofactor for the gating in delayed rectifiers (Armstrong and Lopez-Barneo, 1987). Removal of external Ca^{2+} ion eliminates delayed rectifier gating and produces a large maintained conductance (leak) comcomitantly. If Ba^{2+} ion block of the delayed rectifier precedes the removal of Ca^{2+}, the leak develops more slowly.

2.3.4.4 Anions

Substitution of internal anions affects the gating of delayed rectifier channels in squid axon: inorganic anions such as halides, sulphate or nitrate reversibly slow the opening, but not the closure, of the K^+ channels (Adams and Oxford, 1983). This effect evolves over tens of minutes and is reversed over tens of seconds. The gating remains stable with the organic anions glutamate, aspartate and citrate. Internally applied ATP selectively potentiates the magnitude of the delayed K^+ conductance in squid axons and slows its activation kinetics (Perozo *et al.*, 1989). The potentiation of K^+ current by ATP was dose-dependent (K_m=10 μM) and there was an obligatory rquirement for Mg^{2+} ion.

2.3.5 Aminopyridines

Aminopyridine compounds depress K^+ current though delayed rectifiers and transient K^+ channels. The aminopyridine action includes block of K^+ current and alteration of gating kinetics, and depends on voltage, time and stimulation pattern. In neurones with both maintained and transient K^+ currents, the transient components tend to be more sensitive to block by aminopyridines than the delayed rectifier components, making aminopyridines a useful pharmacological tool for dissection of composite K^+ currents.

4-aminopyridine (4-AP) was first shown to block delayed rectifier K^+ channels in cockroach axons (Pelhate and Pichon, 1974). Subsequently, 4-AP was found to block delayed rectifier K^+ channels in squid giant axons (Meves and Pichon, 1977; Yeh *et al.*, 1976; Schauf *et al.*, 1976), frog node of Ranvier (Ulbricht and Wagner, 1976), molluscan neurones (Hermann and Gorman, 1981a) and skeletal muscle (Gillespie and Hutter, 1975; Fink and Wettwer, 1978; Molgo, 1978). Other aminopyridines, such as 2- or 3-aminopyridine, and the diaminopyridines 3,4- and 2,3-diaminopyridine and others, also blocked delayed rectifier channels in a qualitatively similar manner, with the major difference being the concentration of drug required for block (Yeh *et al.*, 1976; Kirsch and Narahashi, 1978). Because the mechanism of block seems to be the same for all of the aminopyridines, no attempt will be made to differentiate among them.

In squid giant axons and molluscan neurones, and probably other preparations, aminopyridines are effective when applied internally or externally (Yeh *et al.*, 1976; Kirsch and Narahashi, 1978; Hermann and Gorman, 1981a, Thompson, 1982). It is still not clear whether two distinct binding sites for aminopyridines exist or whether the uncharged forms of the molecules diffuse through the membrane and bind to a single receptor site. As Ulbricht and Wagner (1976) point out, which form of these molecules, charged or uncharged, actually binds to the channels and blocks is unknown. The molecules almost certainly do not pass through the channel itself as they are too bulky to fit through the postulated 3.3 Å selectivity filter constriction (Hille, 1975). Unlike TEA and the other QA ions, 4-AP and its derivatives block inward and outward K^+ currents equally well. This suggests that the mechanism of block by 4-AP is quite different from TEA or Ba^{2+} ion block and is not affected by the ‘knock-out’ phenomenon discussed above. Although not affected by the direction of the ionic current, the block by 4-AP and derivatives is modified in a complex way by membrane potential and stimulation frequency. In general, the block is relieved by steady depolarization and increased frequency of depolarizing pulses (Yeh *et al.*, 1976; Ulbricht and Wagner, 1976; Meves and Pichon, 1977; Kirsch and Narahashi, 1978; Hermann and Gorman, 1981a).

Yeh *et al.* (1976) suggested that, like TEA, 4-AP does not alter the underlying normal kinetic scheme of the delayed rectifier channel, but channels which have the drug bound simply do not conduct. They further propose that 4-AP binds more strongly to the closed channel states than to the open (activated) state. Although there is no compelling experimental evidence for such a scheme, it does replicate many of the experimental observations, such as the relief of block by depolarization. Mechanisms that do not require exclusive binding to the open state are equally able to predict most experimental results if the blocking rate constants are given an inherent voltage dependence.

Aminopyridines also inhibit the transient K^+ current of molluscan neurones (Thompson, 1977, 1982; Adams *et al.*, 1980). Whereas they tend to slow the activation of delayed rectifier currents, aminopyridines appear to accelerate the inactivation gating of transient K^+ current as if open channels are more sensitive to the drug (Thompson, 1982). In addition to this fast time-dependent block, there is a slow time-dependent block (Fig. 2.4a). This block increases in strength during

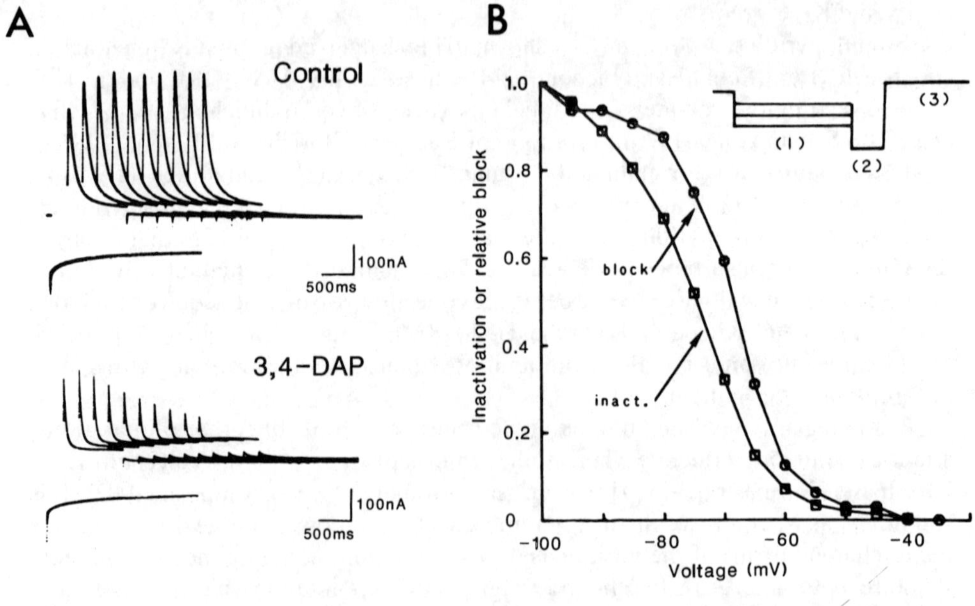

Fig. 2.4 — Effect of 3,4-diaminopyridine (3,4-DAP) on transient K^+ currents in voltage-clamped nudibranch (*Archidoris* and *Anisodoris*) neurone. (A) Progression of block during a conditioning hyperpolarization during internal exposure to 1 mM 3,4-DAP as compared to a control. The membrane potential was held at −40 mV, stepped to −100 mV for intervals between 0.25 and 2.5 s (0.25 s increment) prior to a test pulse to −30 mV. Temperature 10°C. (B) Comparison of voltage dependence of block by externally applied 1 mM 3,4-DAP ('block') with voltage dependence of normal inactivation ('inact') (from Thompson, 1982).

hyperpolarization and is alleviated by sustained depolarization; its steady-state voltage dependence approximately mimics that of the normal inactivation gating (Fig. 2.4b).

4-AP and its derivatives have been useful in the separation of components of outward K^+ currents in neurones. Thompson (1977) showed that the transient outward A current channels could be blocked by 4-AP concentrations that spared currents through delayed rectifier and Ca^{2+}-activated K^+ channels in *Tritonia* neurones. Herman and Gorman (1981a) found that delayed rectifier currents could be blocked at much lower 4-AP concentrations than were required to block Ca^{2+}-activated K^+ channels. In mammalian hippocampal neurones, 4-AP preferentially blocks transient K^+ currents (Gustafsson *et al.*, 1982; Segal *et al.*, 1984; Numann *et al.*, 1987). This block is voltage dependent and analogous to that seen with molluscan A current (Numann *et al.*, 1987).

2.4 DISTRIBUTION OF K^+ CHANNELS

Voltage-dependent K^+ channels appeared early during evolution of eukaryotic cells (Chapter 16 of Hille, 1984), and their present diversity reflects the specialization of cell function that has occurred during the subsequent evolution of multicellular organisms. The functional specialization of cells can be expressed in the varieties of K^+ channels that they possess, but may also result in a particular distribution over the cell surface. The techniques for mapping voltage-dependent K^+ channels have been restricted to electrical measurements, since specific labels for identifying functional K^+ channels have not been available. This may explain why information about voltage-gated K^+ channel distributions is only beginning to emerge (for a review see Almers and Stirling, 1984).

The nodes of Ranvier of amphibian nerve exhibit Na^+ and delayed rectifier K^+ currents comparable to those in squid axon (Dodge and Frankenhaeuser, 1958). Mammalian nodes of Ranvier appear to be largely devoid of K^+ current (Horackova *et al.*, 1968; Chiu *et al.*, 1979; Brismar, 1980). A delayed rectifier K^+ current can be evoked in these axons, however, after treatments that lift the myelin sheath from the paranodal regions of the axons (Chiu and Ritchie, 1981). The K^+ currents before and after treatment are roughly in the proportion of the exposed membrane capacitances. Hence, these observations in mammalian axons could indicate that K^+ channels actually have comparable densities in the nodal and paranodal regions. In amphibian axons, on the other hand, internodal K^+ current per unit membrane capacitance is about 2 nA/pF (Chiu and Ritchie, 1982), i.e. at least an order of magnitude smaller than that calculated for the nodal area. Thus, amphibian, but not mammalian, axons appear to have a non-uniform K^+ channel distribution, with a relatively high density at the nodes. Furthermore, the expression of nodal K^+ current in amphibian nerve depends on the diameter of axons: only the largest axons reveal relatively large K^+ currents (Smith and Schauf, 1981). In summary, K^+ channels of myelinated axons appear to be distributed in specific configurations, but the functional significance of the pattern remains a subject of speculation.

Focal recordings of action currents from mouse motor nerve terminals indicate that delayed rectifier K^+ channels are concentrated at the distal regions of these terminals, where they might serve to compensate for the depolarizing effect of the local Ca^{2+} ion influx (Brigant and Mallart, 1982).

In skeletal muscle, delayed rectifier K^+ channels appear to be preferentially located in the surface membrane; the transverse tubular density is about three times lower (Kirsch *et al.*, 1977). Focal recordings with the loose patch clamp technique have revealed that the delayed rectifier channels of the sarcolemma are non-uniformly distributed over lateral distances of 10–30 μm (Almers *et al.*, 1983). 'Hot spots' with K^+ current densities several-fold larger than those nearby were observed. Their distribution did not correlate with analogous hot spots for Na^+ current. This heterogeneous distribution of K^+ channels suggests that the lateral mobility of the channels is restricted in skeletal muscle, but its functional significance remains unclear.

2.5 CONCLUSIONS

All voltage-dependent potassium channels serve the same basic function, i.e. to create or stabilize a negative membrane potential, counteracting the depolarizing effects of channels that pass Na^+ or Ca^{2+} ions. Whilst voltage-dependent potassium channels conform in selecting their ion, they diversify in selecting their stimulus and their responses. We have attempted to summarize data and interpretations concerning the molecular properties and mechanisms of these channels.

The chapter is presented in two main sections — pore and gate — a distinction evident from several points of view. The ionic permeability mediated by the pore involves a diffusion process whose rates are usually several orders of magnitude larger than those known for channel gating; hence, permeability and gating are kinetically distinct. Interpretations of permeability properties usually include an idea of a structural correlate (such as the selectivity filter, the hydrophobic atrium of a pore, etc.), whereas interpretations of gating charactertistics are more abstract in terms of kinetic state diagrams of a protein. This reflects the fact that the electrical current through an ionic pore is a more direct indicator of physical properties of the pathway than it is of the machinery behind the gate of the channel protein. Differences between the pores of voltage-dependent K^+ channels tend to be small compared to differences in the gating characteristics. In cells containing a non-homogeneous population of channels, the permeation properties of K^+ channels can be studied unambiguously, whereas the superimposed gating characteristics are difficult to separate and analyse.

What are the future directions of research into voltage-dependent K^+ channels? More work is needed to complete the description of the diversified gating characteristics. Since the most subtle use of voltage-dependent K^+ channels is probably made by cell membranes that incorporate several classes of K^+ channels, an understanding of the function of these cells will require that their K^+ channels be characterized at the single-channel level. Channel primary structures elucidated by molecular biological techniques (see Chapter 1) will facilitate answering basic questions, for example, whether inward rectifiers are constructed from an originally different molecular blueprint than delayed rectifiers. Studies directed to elucidating mechanistic details of the gating process will be greatly facilitated when homogeneous channel populations can be created by molecular biological techniques. These techniques will also provide new means for probing the structural correlates of channel function, in particular with regard to the mechanism of the gate.

ACKNOWLEDGEMENT

This work was supported by NIH grants HL35422 and GM30377.

We dedicate this chapter to Professor Susumu Hagiwara (1922–1989) who contributed much to our understanding of voltage-dependent potassium channels.

REFERENCES

Adams, D. J., and Oxford, G. S. (1983) Interaction of internal anions with potassium channels of the squid axon. *J. Gen. Physiol.* **82**, 429–448.

Adams, D. J., Smith, S. J., and Thompson, S. H. (1980) Ionic currents in molluscan soma. *Ann. Rev. Neurosci.* **3**, 141–167.

Adelman, W. J., and French, R. J. (1978) Blocking of the squid axon potassium channel by external caesium ions. *J. Physiol.* (*Lond.*) **276** 13–25.

Adelman, W. J., and Senft, J. P. (1966) Voltage clamp studies on the effect of internal cesium ion on sodium and potassium currents in the squid giant axon. *J. Gen. Physiol.* **50** 279–293.

Adelman, W. J., and Senft, J. P. (1968) Dynamic asymmetries in the squid axon membrane. *J. Gen. Physiol.* **51** 102S–111S.

Adrian, R. H., and Freygang, W. H. (1962) The potassium and chloride conductance of frog muscle membrane. *J. Physiol.* (*Lond.*) **163** 61–103.

Adrian, R. H., Chandler, W. K., and Hodgkin, A. L. (1970) Slow changes in potassium permeability in skeletal muscle. *J. Physiol.* (*Lond.*) **208** 645–668.

Almers, W. (1972) Potassium conductance changes in skeletal muscle and the potassium concentration in the transverse tubules. *J. Physiol.* (*Lond.*) **255** 33–56.

Almers, W., and Armstrong, C. M. (1980) Survival of K^+ permeability and gating currents in squid axons perfused with K^+-free media. *J. Gen. Physiol.* **75** 61–78.

Almers, W., and Stirling, C. (1984) Distribution of transport proteins over animal cell membranes. *J. Membr. Biol.* **77** 169–186.

Almers, W., Stanfield, P. R., and Stühmer, W. (1983) Lateral distribution of sodium and potassium channels in frog skeletal muscle: measurements with a patch-clamp technique. *J. Physiol.* (*Lond.*) **336** 261–284.

Arhem, P. (1980) Effects of some hevy metal ions on the ionic currents of myelinated fibres from *Xenopus Laevis*. *J. Physiol.* (*Lond.*) **306** 219–231.

Armstrong, C. M. (1966) Time course of TEA^+ induced anomalous rectification in squid giant axons. *J. Gen. Physiol.* **50** 491–503.

Armstrong, C. M. (1969) Inactivation of the potassium conductance and related phenomena caused by quaternary ammonium ion injection in squid axons. *J. Gen. Physiol.* **54**, 553–575.

Armstrong, C. M. (1971) Interaction of tetraethylammonium ion derivatives with the potassium channels of giant axons. *J. Gen. Physiol.* **58** 413– 437.

Armstrong, C. M. (1975) Potassium pores of nerve and muscle membranes. In: *Membranes: A series of Advances*, Vol. 3, pp. 325–358, G. Eisenman (ed.). Dekker, New York.

Armstrong, C. M., and Binstock, L. (1965) Anomalous rectification in the squid giant axon injected with tetraethylammonium chloride. *J. Gen. Physiol.* **48** 859–872.

Armstrong, C. M., and Hille, B. (1972) The inner quaternary ammonium ion receptor in potassium channels of the node of Ranvier. *J. Gen. Physiol.* **59** 388–400.

Armstrong, C. M., and Lopez-Barneo, J. (1987) External calcium ions are required for potassium channel gating in squid neurons. *Science* **236** 712–714.

Armstrong, C. M., and Matteson, D. R. (1986) The role of calcium ions in the closing of K channels. *J. Gen. Physiol.* **87** 817–832.

Armstrong, C. M., and Taylor, S. R. (1980) Interaction of barium ions with potassium channels in squid giant axons. *Biophys. J.* **30** 473–488.

Armstrong, C. M., Swenson, R. P., and Taylor, S. R. (1982) Block of squid axon K channels by internally and externally applied barium ions. *J. Gen. Physiol.* **80** 663–682.

Begenisich, T., and Smith, C. (1984) Multi-ion nature of potassium channels in squid axons. *Current Topics in Membranes and Transport* **22** 353–369.

Begenisich, T., and Stevens, C. F. (1975) How many conductance states do potassium channels have? *Biophys. J.* **15** 843–846.

Benham, C. D., and Bolton, T. B. (1983) Patch-clamp studies of slow potential-sensitive potassium channels in longitudinal smooth muscle cells of rabbit jejunum. *J. Physiol. (Lond.)*, **340** 469–486.

Bergman, C. (1970) Increase of sodium concentration near the inner surface of the nodal membrane. *Pflügers Arch.* **317** 287–302.

Bezanilla, F., and Armstrong, C. M. (1972) Negative conductance caused by entry of sodium and cesium ions into the potassium channels of squid axons. *J. Gen. Physiol.* **60** 588–608.

Blatz, A. L. (1984) Asymmetric proton block of inward rectifier K channels in skeletal muscle. *Pflügers Arch.* **401** 402–407.

Brigant, J. L., and Mallart, A. (1982) Presynaptic currents in mouse motor endings. *J. Physiol. (Lond.)* **333** 619–636.

Brismar, T. (1980) Potential clamp analysis of membrane currents in rat myelinated nerve fibres. *J. Physiol. (Lond.)* **298** 171–184.

Carbone, E. R., Fioravanti, R., Prestipino, G., and Wanke, W. (1978) Action of extracellular pH on Na^+ and K^+ membrane currents in the giant axon of *Loligo vulgaris*. *J. Membr. Biol.* **43** 295–315.

Carmeliet, E. (1982) Induction of removal of inward-going rectification in sheep Purkinje fibres. *J. Physiol. (Lond.)* **327** 285–308.

Chandler, W. K., and Meves, H. (1965) Voltage clamp experiments on internally perfused giant axons. *J. Physiol. (Lond.)* **180** 798–820.

Chandler, W. K. and Meves, H. (1970) Sodium and potassium currents in squid axons perfused with fluoride solutions. *J. Physiol. (Lond.)* **211** 623–652.

Chiu, S. Y., and Ritchie, J. M. (1981) Evidence for the presence of potassium channels in the paranodal region of acutely demyelinated mammalian single nerve fibres. *J. Physiol.* (*Lond.*) **313** 415–437.

Chiu, S. Y., and Ritchie, J. M. (1982) Evidence for the presence of potassium channels in the internode of frog myelinated nerve-fibres. *J. Physiol.* (*Lond.*) **322** 485–501.

Chiu, S. Y., Ritchie, J. M., Rogart, R. B., and Stagg, D. (1979) A quantitative description of membrane currents in rabbit myelinated nerve. *J. Physiol.* (*Lond.*) **292** 149–166.

Clapham, D. E., and De Felice, L. J. (1984) Voltage-activated K channels in embryonic chick heart. *Biophys. J.* **45** 40–42.

Cole, K. S., and Moore, J. W. (1960) Ionic current measurements in the squid giant axon membrane. *J. Gen. Physiol.* **44** 123–167.

Connor, J. A., and Stevens, C. F. (1971) Voltage clamp studies of a transient outward membrane current in gastropod neural somata. *J. Physiol* (*Lond.*) **213** 21–30.

Constanti, A., and Galvan, M. (1983) Fast inward-rectifying current accounts for anomalous rectification in olfactory cortex neurones. *J. Physiol.* (*Lond.*) **335** 153–178.

Conti, F., and Neher, E. (1980) Single channel recordings of K^+ currents in squid axons. *Nature* (*Lond.*) **285** 140–143.

Conti, F., De Felice, L. J., and Wanke, E. (1975) Potassium and sodium ion current noise in the membrane of the squid giant axon. *J. Physiol.* (*Lond.*) **248** 45–82.

Conti, F., Hille, B., and Nonner, W. (1984) Non-stationary fluctations of the potassium conductance at the node of Ranvier of the frog. *J. Physiol.* (*Lond.*) **353** 199–230.

Cooper, E., and Shrier, A. (1985) Single-channel analysis of fast transient potassium currents from rat nodose neurones. *J. Physiol.* (*Lond.*) **369** 199–208.

Dodge, F. A., and Frankenhaeuser, B. (1958) Membrane currents in isolated frog nerve fibre under clamp conditions. *J. Physiol.* (*Lond.*) **143** 76–90.

Drouin, H., and The, R. (1969) The effect of reducing extracellular pH on the membrane currents of the Ranvier node. *Pflügers Arch.* **313** 80–88.

Dubois, J. M. (1983) Potassium currents in the frog node of Ranvier. *Prog. Biophys. Molec. Biol.* **42** 1–20.

Dubois, J. M., and Bergman, C. (1975) Cesium induced rectifications in frog myelinated fibers. *Pflügers Arch.* **355** 361–364.

Dubois, J. M., and Bergman, C. (1977). The steady state potassium conductance of the Ranvier node at various external K concentrations. *Pflügers Arch.* **370** 185–194.

Eaton, D. C., and Brodwick, M. S. (1980) Effects of barium on the potassium conductance of squid axon. *J. Gen. Physiol.* **75** 727–750.

Ehrenstein, G., and Gilbert, D. L. (1966) Slow changes of potassium permeability in the squid giant axon. *Biophys. J.* **6** 553–566.

Fink, R., and Wettwer, E. (1978) Modified K-channel gating by exhaustion and the block by internally applied TEA^+ and 4-aminopyridine in muscle. *Pflügers Arch.* **374** 289–292.

Frankenhaeuser, B., and Hodgkin, A. L. (1956) The after-effects of impulses in the giant nerve fibres of *Loligo*. *J. Physiol.* (*Lond.*) **131** 341–376.

French, R. J., and Adelman, W. J. (1976) Competition, saturation, and inhibition: ionic interactions shown by membrane ionic currents in nerve, muscle, and bilayer systems. *Current Topics in Membranes and Transport* **8** 161–207.

French, R. J., and Shoukimas, J. J. (1981) Blockage of squid axon potassium conductance by internal tetra-*N*-alkylammonium ions of various sizes. *Biophys. J.* **34** 271–291.

French, R. J., and Shoukimas, J. J. (1985) An ion's view of the potassium channel. The structure of the permeation pathway as sensed by a variety of blocking ions. *J. Gen. Physiol.* **85** 669–698.

French, R. J., and Wells, J. B. (1977) Sodium ions as blocking agents and charge carriers in the potassium channel of the squid giant axon. *J. Gen. Physiol.* **70** 707–724.

Fukushima, Y. (1982) Blocking kinetics of the anomalous potassium rectifier of tunicate egg studied by single channel recording. *J. Physiol.* (*Lond.*) **331** 311–331.

Giles, W. R., and Van Ginneken, A. C. G. (1985) A transient outward current in isolated cells from the crista terminalis of rabbit heart. *J. Physiol.* (*Lond.*) **368** 243–264.

Gillespie, J. I., and Hutter, O. F. (1975) The actions of 4-aminopyridine on the delayed potassium current in skeletal muscle fibres. *J. Physiol.* (*Lond.*) **252** 70–71P.

Gilly, W. F., and Armstrong, C. M. (1980) Gating current and potassium channels in the giant axon of the squid. *Biophys. J.* **29** 485–492.

Gilly, W. F., and Armstrong, C. M. (1982). Divalent cations and the activation kinetics of potassium channels in squid giant axons. *J. Gen Physiol.* **79** 965–996.

Gustafsson, B., Galvan, M., Grafe, P., and Wigstroem, H. (1982) A transient outward current in a mammalian central neurone blocked by 4-aminopyridine. *Nature* (*Lond.*) **299** 252–254.

Hagiwara, S. (1983) *Membrane potential-dependent ion channels in cell membrane: phylogenetic and developmental approaches*. Distinguished Lecture Series of the Society of General Physiologists, Vol. 3. Raven Press, New York.

Hagiwara, S., and Takahashi, K. (1974) The anomalous rectification and cation selectivity of the membrane of a starfish egg cell. *J. Membr. Biol.* **18** 61–80.

Hagiwara, S., and Yoshii, M. (1979) Effects of internal potassium and sodium on the anomalous rectification of the starfish egg as examined by internal perfusion. *J. Physiol.* (*Lond.*) **292** 251–265.

Hagiwara, S., Kusano, K., and Saito, N. (1961) Membrane changes of *Onchidium* nerve cell in potassium-rich media. *J. Physiol.* (*Lond.*) **155** 470–489.

Hagiwara, S., Miyazaki, S., and Rosenthal, N. P. (1976) Potassium current and the effect of cesium on this current during anomalous rectification in the starfish egg. *J. Physiol.* (*Lond.*) **279** 167–185.

Hagiwara, S., Miyazaki, S., Moody, W. J., and Patlak, J. (1978) Blocking effects of barium and hydrogen ions on the potassium current during anomalous rectification in the starfish egg. *J. Physiol.* (*Lond.*) **279** 167–185.

Hagiwara, S., Yoshida, S., and Yoshii, M. (1981) Transient and delayed potassium currents in the egg cell membrane of the coelenterate, *Renilla Koellikeri*. *J.Physiol.* (*Lond.*) **318** 123–141.

Harvey, R. D., and Ten Eick, R. E. (1988) Characterization of the inward-rectifying potassium current in cat ventricular myocytes. *J. Gen. Physiol.* **91** 593–615.

Hermann, A., and Gorman, A. L. F. (1981a) Effects of 4-aminopyridine on potassium currents in a molluscan neuron. *J. Gen. Physiol.* **78** 63–86.

Hermann, A., and Gorman, A. L. F. (1981b) Effects of tetraethylammonium on potassium currents in a molluscan neuron. *J. Gen. Physiol.* **78** 87–110.

Hestrin, S. (1981) The interaction of potassium with the ac.ıvation of anomalous rectification in frog muscle membrane. *J. Physiol.* (*Lond.*) **317** 497–508.

Hille, B. (1967) The selective inhibition of delayed potassium currents in nerve by teatraethylammonium ion. *J. Gen. Physiol.* **50** 1287–1302.

Hille, B. (1968) Charges and potentials at the nerve surface: divalent ions and pH. *J. Gen. Physiol.* **51** 221–236.

Hille, B. (1973) Potassium channels in myelinated nerve: selective permeability to small cations. *J. Gen. Physiol.* **61** 669–686.

Hille, B. (1975) Ionic selectivity of Na and K channels of nerve membranes. In: *Membranes: A series of Advances*, Vol. 3, pp. 255–323, G. Eisenman (ed.). Dekker, New York.

Hille, B. (1984) *Ionic channels of excitable membranes*. Sinauer Associates, Sunderland, Massachusetts.

Hille, B., and Schwarz, W. (1978) Potassium channels as multi-ion single-file pores. *J. Gen. Physiol.* **72** 409–442.

Hodgkin, A. L., and Huxley, A. F. (1952a) Currents carried by sodium and potassium ions through the membrane of the giant axon of *Loligo*. *J. Physiol.* (*Lond.*) **116** 449–472.

Hodgkin, A. L., and Huxley, A. F. (1952b) A quantitative description of membrane current and its application to conduction and excitation in nerve. *J. Physiol.* (*Lond.*) **116** 500–544.

Horackova, M., Nonner, W. and Stämpfli, R. (1968) Action potentials and voltage clamp currents of single rat Ranvier nodes. *Proc. Int. Union. Physiol. Sci.* **7** 198.

Ildefonse, M., and Rougier, O. (1971) Effet de l'ion tetraethylammonium sur les courants de cinetique lente de la fibre musculaire skelettique rapide. *J. Physiol.* (*Paris*) **63** 237A.

Josephson, I. R. and Brown, A. M. (1986) Inwardly rectifying single channel and whole cell K^+ currents in rat ventricular myocytes. *J. Membr. Biol.* **94** 19–35.

Junge, D. (1982) External K^+ ions increase rate of opening of outward current channels in snail neurons. *Pflügers Arch.* **394** 94–96.

Kao, C. Y., and Stanfield, P. R. (1970) Actions of some cations on the electrical properties and mechanical threshold of frog skeletal muscle. *J. Gen. Physiol.* **55** 620–639.

Kasai, H., Kameyama, M., Yamaguchi, Y., and Fukuda, J. (1986) Single transient K channels in mammalian central neurons. *Biophys. J.* **49** 1243–1247.

Katz, B. (1949) Les constantes electriques de la membrane du muscle. *Arch. Sci. Physiol.* **3** 285–299.

Kirsch, G. E., and Narahashi, T. (1978) 3,4-Diaminopyridine, a potent new potassium channel blocker. *Biophys. J.* **22** 507–512.

Kirsch, G. E., Nichols, R. A., and Nakajima, S. (1977) Delayed rectification in the transverse tubules: origin of the late after-potential in frog skeletal muscle. *J.*

Gen. Physiol. **70** 1–21.

Koppenhöfer, E. (1967) Die Wirkung von Tetraaethylammonium chlorid auf die Membranstroeme Ranvierscher Schnürringe von *Xenopus laevis*. *Pflügers Arch.* **293** 34–55.

Koppenhöfer, E., and Vogel, W. (1969) Wirkung von Tetrodotoxin und Tetraaethylammonium chlorid an der Innenseite der Schnüerringsmembran von *Xenopus laevis*. *Pflügers. Arch.* **313**, 361–380.

Kostyuk, P. G., Veselovsky, N. S., Fedulova, S. A., and Tsyndrenko, A. Y. (1981) Ionic currents in the somatic membrane of rat dorsal root ganglion neurons. III. Potassium currents. *Neuroscience* **6** 2439–2444.

Kurachi, Y. (1985) Voltage dependent activation of the inward-rectifier potassium channel in the ventricular cell membrane of guinea-pig heart. *J. Physiol.* (*Lond.*) **366** 365–385.

Latorre, R., and Miller, C. (1983) Conduction and selectivity in potassium channels. *J. Membr. Biol.* **71** 11–30.

Leech, C. A., and Stanfield, P. R. (1981) Inward rectification in frog skeletal muscle fibres and its dependence on membrane potential and external potassium. *J. Physiol.* (*Lond.*) **319** 295–309.

Llano, I., Webb, C. K., and Bezanilla, F. (1988) Potassium conductance of the squid giant axon: single-channel studies. *J. Gen. Physiol.* **92** 179–196.

Matsuda, H. (1988) Open-state substructure of inwardly rectifying potassium channels revealed by magnesium block in guinea-pig heart cells. *J. Physiol.* (*Lond.*) **397** 237–258.

Matsuda, H., Saigusa, A., and Irisawa, H. (1987) Ohmic conductance through the inwardly rectifying K channel and blocking by internal Mg^{++}. *Nature* (*Lond.*) **325** 156–159.

Matteson, D. R., and Swenson, R. P. (1986) External monovalent cations that impede the closing of K channels. *J. Gen. Physiol.* **87** 795–816.

Mayer, M. L., and Westbrook, G. L. (1983) A voltage-clamp analysis of inward (anomalous) rectification in mouse spinal sensory ganglion neurones. *J. Physiol.* (*Lond.*) **340** 19–45.

McAllister, R. E., and Noble, D. (1966) The time and voltage dependence of the slow outward current in cardiac Purkinje fibres. *J. Physiol.* (*Lond.*) **186** 632–662.

McLarnon, J. G. (1989) Properties of single potassium channels in hypothalamic neurons. *Pflügers Arch.* **413** 604–609.

Meech, R. W., and Standen, N. B. (1975) Potassium activation in *Helix aspersa* neurones under voltage clamp: a component mediated by Ca influx. *J. Physiol.* (*Lond.*) **249** 211–259.

Meves, H., and Pichon, Y. (1977) The effect of internal and external 4-aminopyridine on the potassium currents in intracellularly perfused squid giant axons. *J. Physiol.* (*Lond.*) **268** 511–532.

Molgo, J. (1978) Voltage-clamp analysis of the sodium and potassium currents in skeletal muscle fibres treated with 4-aminopyridine. *Experientia* **34** 1275–1276.

Monod, J., Wyman, J., and Changeux, J. P. (1965) On the nature of allosteric transitions. *J. Mol. Biol.* **12** 88–118.

Moody, W. J., and Hagiwara, S. (1982) Block of inward rectification by intracellular

H^+ in immature oocytes of the starfish *Mediaster aequalis*. *J. Gen. Physiol.* **79** 115–130.

Mozhayeva, G. N., and Naumov, A. P. (1970) Effect of surface charge on the steady-state potassium conductance of nodal membrane. *Nature* (*Lond.*) **228** 164–165.

Neher, E. (1971) Two fast transient current components during voltage clamp on snail neurons. *J. Gen. Physiol.* **58** 36–53.

Neher, E., and Lux, H. D. (1972) Differential action of TEA^+ on two K^+-current components of a molluscan neurone. *Pflügers Arch.* **336** 87–100.

Neumcke, B., Schwarz, W., and Stämpfli, R. (1980) Differences between K channels in motor and sensory nerve fibres of the frog as revealed by fluctuation analysis. *Pflügers Arch.* **387** 9–16.

Numann, R. E., Wadman, W. J., and Wong, R. K. S. (1987) Outward currents of single hippocampal cells obtained from the adult guinea-pig. *J. Physiol.* (*Lond.*) **393** 331–353.

Ohmori, H. (1978) Inactivation kinetics and steady-state current noise in the anomalous rectifier of tunicate egg cell membrane. *J. Physiol.* (*Lond.*) **281** 77–79.

Ohmori, H. (1981) Unitary current through sodium channel and anomalous rectifier channel estimated from transient current noise in the tunicate egg. *J. Physiol.* (*Lond.*) **311** 289–305.

Ohmori, H., Yoshida, S., and Hagiwara, S. (1981) Single K^+ channel currents of anomalous rectification in cultured rat myotubes. *Proc. Natl. Acad. Sci. USA* **78** 4960–4964.

Olesen, S. P., Davies, P. F., and Clapham, D. E. (1988) Muscarinic activated K^+ current in bovine aortic endothelial cells. *Circ. Res.* **62** 1059–1064.

Oxford, G. S., and Adams, D. J. (1981) Permanent cations alter K channel kinetics and permeability. *Biophys. J.* **33** 70A.

Oxford, G. S., and Wagoner, P. K. (1989) The inactivating K^+ current in GH3 pituitary cells and its modification by chemical reagents. *J. Physiol.* (*Lond.*) **410** 587–612.

Payet, M. D., Rousseau, E., and Sauve, R. (1985) Single channel analysis of a potassium inward rectifier in myocytes of newborn rat heart. *J. Membr. Biol.* **86** 79–88.

Pelhate, M., and Pichon, Y. (1974) Selective inhibition of potassium current in the giant axon of the cockroach. *J. Physiol.* (*Lond.*) **242** 90–91P.

Perozo, E., Bezanilla, F., and Dipolo, R. (1989) Modulation of K channels in dialyzed squid axons. ATP-mediated phosphorylation. *J. Gen. Physiol.* **93** 1195–1218.

Reuter, H., and Stevens, C. F. (1980) Ion conductance and ion selectivity of potassium channels in snail neurones. *J. Membr. Biol.* **57** 103–118.

Rizzo, M. A. (1988) The transient K conductance of the somatic membrane of cultured embryonic rat neurones. PhD thesis, University of Miami, Miami, Florida.

Rogawski, M. A. (1985) The A-current: how ubiquitous a feature of excitable cells is it? *Trends Neurosci.* **8** 214–219.

Rogawski, M. A. (1986) Single voltage-dependent potassium channels in cultured

rat hippocampal neurons. *J. Neurophysiol.* **56** 481–493.
Rudy, B. (1988) Diversity and ubiquity of K channels. *Neuroscience* **25** 729–749.
Ruff, R. L. (1986) Ionic channels. II. Voltage- and agonist-modified channels properties and structure. *Muscle and Nerve* **9** 767–786.
Sakmann, B., and Trube, G. (1984a) Conductance properties of single inwardly rectifying potassium channels in ventricular cells from guinea-pig heart. *J. Physiol.* (*Lond.*) **347** 641–657.
Sakmann, B., and Trube, G. (1984b) Voltage dependent inactivation of inward-rectifying single-channel currents in the guinea-pig heart cell membrane. *J. Physiol.* (*Lond.*) **347** 659–683.
Schauf, C. L., and Davis, F. A. (1976) Sensitivity of the sodium and potassium channels of *Myxicola* giant axon to changes in external pH. *J. Gen. Physiol.* **67** 185–195.
Schauf, C. L., Colton, C. A., Colton, J. S., and Davis, F. A. (1976) Aminopyridines and sparteine as inhibitors of membrane potassium conductance: effects on *Myxicola* giant axons and the lobster neuromuscular junction. *J. Pharm. Exp. Ther.* **197** 414–425.
Schwarz, J. R., and Vogel, W. (1971) Potassium inactivation in single myelinated nerve fibres of *Xenopus laevis*. *Pflügers Arch.* **330** 61–73.
Schwarz, W., Neumcke, B., and Palade, P. T. (1981) K-current fluctuations in inward-rectifying channels of frog skeletal muscle. *J. Membr. Biol.* **63** 85–92.
Segal, M., Rogawski, M. A., and Barker, J. L. (1984) A transient potassium conductance regulates the excitability of cultured hippocampal and spinal neurons. *J. Neurosci.* **4** 604–609.
Shrager, P. (1974) Ionic conductance changes in voltage clamped crayfish axons at low pH. *J. Gen. Physiol.* **64** 666–690.
Smith, K. J., and Schauf, C. L. (1981) Size dependent variation of nodal properties in myelinated nerve. *Nature* (*Lond.*) **293** 297–299.
Solc, C. K., Zagotta, W. N., and Aldrich, A. (1987) Single-channel and genetic analyses reveal two distinct A-type potassium channels in *Drosphila*. *Science* **236** 1094–1098.
Sperelakis, N., Scheider, M. F., and Harris, E. J. (1967) Decreased K^+ conductance produced by Ba^{++} in frog satorius fibres. *J. Gen. Physiol.* **50** 1565–1583.
Spires, S., and Begenisich, T. (1989) Pharmacological and kinetic analysis of K channel gating currents. *J. Gen. Physiol.* **93** 263–283.
Standen, N. B., and Stanfield, P. R. (1978a) A potential- and time-dependent blockade of inward rectification in frog skeletal muscle fibres by barium and strontium ions. *J. Physiol.* (*Lond.*) **280** 169–191.
Standen, N. B., and Stanfield, P. R. (1978b) Inward rectification in skeletal muscle: a blocking particle model. *Pflügers Arch.* **378** 173–176.
Standen, N. B., and Stanfield, P. R. (1979) Potassium depletion and sodium block of potassium currents under hyperpolarization in frog sartorius muscle. *J. Physiol.* (*Lond.*) **294** 497–520.
Standen, N. B., and Stanfield, P. R. (1980) Rubidium block and rubidium permeability of the inward rectifier of frog skeletal muscle fibres. *J. Physiol.* (*Lond.*) **304** 415–435.
Standen, N. B., Stanfield, P. R., and Ward, T. A. (1985) Properties of single

potassium channels in vesicles formed from the sarcolemma of frog skeletal muscle. *J. Physiol.* (*Lond.*) **364** 339–358.

Stanfield, P. R. (1970) The effects of tetraethylammonium ion on the delayed currents of frog skeletal muscle. *J. Physiol.* (*Lond.*) **209** 209–229.

Stanfield, P. R. (1973) The onset of the effects of zinc and tetraethylammonium ions on the action potential duration and twitch amplitude of single muscle fibres. *J. Physiol.* (*Lond.*) **235** 639–654.

Stanfield, P. R. (1975) The effect of zinc ions on the gating of the delayed potassium conductance of frog sartorius muscle. *J. Physiol.* (*Lond.*) **251** 711–735.

Stanfield, P. R. (1983) Tetraethylammonium ions and the potassium permeability of excitable cells. *Rev. Physiol. Biochem. Pharmacol.* **97** 1–67.

Stühmer, W., Stocker, M., Sakmann, B., Seeburg, P., Baumann, A., Grupe, A., and Pongs, O. (1988) Potassium channels expressed from rat brain cDNA have delayed rectifier properties. *FEBS Lett.* **242** 199–206.

Swenson, R. P. (1981) Inactivation of potassium current in squid axon by a variety of quaternary ammonium ions. *J. Gen. Physiol.* **77** 255–271.

Swenson, R. P., and Armstrong, C. M. (1981) K^+ channels close more slowly in the presence of external K^+ and Rb^+. *Nature* (*Lond.*) **291** 427–429.

Taylor, P. S. (1987) Selectivity and patch measurements of A-current channels in *Helix aspersa* neurones. *J. Physiol.* (*Lond.*) **388** 437–447.

Thompson, S. H. (1977) Three pharmacologically distinct potassium channels in molluscan neurones. *J. Physiol.* (*Lond.*) **265** 465–488.

Thompson, S. (1982) Aminopyridine block of transient potassium current. *J. Gen. Physiol.* **80** 1–18.

Ulbricht, W., and Wagner, H. H. (1976) Block of potassium channels of the nodal membrane by 4-aminopyridine and its partial removal on depolarization. *Pflügers Arch.* **367** 77–87.

Vandenberg, C. A. (1987) Inward rectification of a potassium channel in cardiac ventricular cells depends on internal magnesium ions. *Proc. Natl. Acad. Sci. USA* **84** 2560–2564.

Van den Berg, R. J., Siebenga, E., and De Bruin, G. (1977) Potassium ion noise currents and inactivation in voltage clamped node of Ranvier. *Nature* (*Lond.*) **265** 177–179.

Wagoner, P. K., and Oxford, G. S. (1987) Cation permeation through the voltage-dependent potassium channel in the squid axon: characteristics and mechanisms. *J. Gen. Physiol.* **90** 261–290.

Wanke, E., Carbone, E., and Testa, P. L. (1979) K^+ conductance modified by a titratable group accessible to protons from the intracellular side of the squid axon membrane. *Biophys. J.* **26** 319–324.

White, M. M., and Bezanilla, F. (1985) Activation of squid axon K^+ channels: ionic and gating current studies. *J. Gen. Physiol.* **85** 539–554.

Yeh, J. Z., Oxford, G. S., Wu, C. H., and Narahashi, T. (1976) Dynamics of aminopyridine block of potassium channels in squid axon membrane. *J. Gen. Physiol.* **68** 519–535.

3

Calcium-activated potassium channels

D. G. Haylett and **D. H. Jenkinson**

3.1 INTRODUCTION

The first decisive evidence for an increase in the K^+ permeability (P_K) of a cell membrane in response to a rise in cytosolic Ca^{2+} was provided by G. Gárdos in pioneering experiments with human red cells in the late 1950s. The key observation was that when red cells are ATP-depleted there is a large increase in K^+ efflux, but only if Ca^{2+} is present in the external solution. A review by Sarkadi and Gárdos (1985) summarizes this early work, and provides an excellent general survey of later developments. Other useful reviews have been compiled by Lew and Ferreira (1978: deals with Ca^{2+} transport and K^+ permeability in red cells), Meech (1978: Ca^{2+}-activated K^+ permeability in nervous tissue), Putney (1979) and Petersen (1986: role of Ca^{2+} in stimulus–permeablity coupling), Schwarz and Passow (1983: $P_{K(Ca)}$ in erythrocytes and excitable cells), Latorre and Miller (1983: conduction and selectivity in K^+ channels in general), Blatz and Magleby (1987: electrophysiology of $P_{K(Ca)}$ channels and currents), Marty (1989: Ca^{2+}-activated ion channels in general) and by Cook (1988: an excellent general survey of the pharmacology of K^+ channels).

Following the discovery of the 'Gárdos effect' in red cells, the next advance was the finding that raising cytosolic Ca^{2+} by means of intracellular injection caused an increase in the K^+ permeability of both invertebrate (Meech and Strumwasser, 1970; Meech, 1974) and vertebrate neurones (Krnjević and Lisiewicz, 1972). It has since been shown that $P_{K(Ca)}$ mechanisms of this kind occur in a wide range of excitable and inexcitable tissues.

Two recent developments have brought further rapid progress. The first is the application of single-channel recording. This makes it possible to apply known concentrations of Ca^{2+} to the inner, cytosolic face of an isolated patch of membrane, and to study the single Ca^{2+}-activated K^+ channels which open as a consequence.

Measuring the current passing through these channels allows the channel conductance to be determined. It soon became clear that by this criterion there are at least three main kinds of K_{Ca} channel. The first to be described (in chromaffin cells; see Marty, 1981) has a relatively large conductance (150–200 pS) and hence is usually described as the BK_{Ca} (Big K^+) channel. A second type of K_{Ca} channel with an intermediate conductance (hence IK_{Ca} or OK (Other K)) is found in red cells, and also in invertebrate neurones. Finally, there is a third category (SK_{Ca}: Small K) with a relatively low conductance (<15 pS) which was first analysed in embryonic skeletal muscle and in liver cells.

The second major development has been the discovery that two invertebrate toxins are able to block K_{Ca} channels with considerable potency. The first to be identified as a blocker of $P_{K(Ca)}$ mechanisms was apamin, from bee venom (Banks *et al.*, 1979; Maas *et al.*, 1980; Burgess *et al.*, 1981; Hugues *et al.*, 1982a,b; see Habermann, 1984, for an authoritative review). In these early experiments apamin was found to inhibit the K_{Ca} mechanism only in cell types (e.g. intestinal smooth muscle, hepatocytes) in which the SK_{Ca} subtype of K_{Ca} channel is now known to be involved. BK_{Ca} and IK_{Ca} channels are not affected.

The second toxin is charybdotoxin, from the scorpion *Leiurus quinquestriatus*. It was found by Miller *et al.* (1985) to block high-conductance (BK_{Ca}) channels derived from the T-tubules of skeletal muscle. It has since been shown to block IK_{Ca} channels as well (see later) although the apamin-sensitive SK_{Ca} channels are unaffected. The discovery of charybdotoxin has been especially important because it is (relatively) technically straightforward to record from the BK_{Ca} channels, and hence to begin to analyse the mechanisms of channel block (see, for example, Anderson *et al.*, 1988; McKinnon and Miller, 1988).

3.2 COMMENTS ON THE MEANS AVAILABLE FOR THE SUBCLASSIFICATION OF K_{Ca} CHANNELS

3.2.1 Single-channel conductance

Care is needed in comparing the values obtained between tissues and laboratories. For example, some channels may rectify, so that comparisons should be made at similar membrane potentials. The channel conductance will also depend on the nature and concentration of the ions on each side of the membrane. It is quite usual when studying K_{Ca} channels (particularly those of low conductance) to use symmetrical K^+-rich bathing solutions, in order to maximize current flow, and so improve the resolution of the recordings. The single-channel conductances obtained in this way may be much greater (by a factor of 3 in the case of SK_{Ca} channels in guinea-pig hepatocytes — see Capiod and Ogden, 1989b) than the value observed when the outer face of the membrane is bathed in a solution containing K^+ at the normal physiological concentration.

A related complication arises when single-channel conductances are to be compared between vertebrates and invertebrates. The solutions used in experiments with some invertebrate tissues (particularly of marine origin) are often of much greater ionic strength, and contain relatively high concentrations of Ca^{2+} and K^+ (typically 20 and 10 mM, respectively, in experiments with *Aplysia*, for example). In contrast, the K^+ concentrations employed in work with mammals are in the range 4–6 mM, and with amphibians 2–3 mM. These differences *per se* will affect single-channel conductances.

3.2.2 Channel pharmacology

The importance of toxins such as apamin and charybdotoxin has already been stressed (see also later, and the surveys by Moczydlowski *et al.*, 1988, and by Castle *et al.*, 1989). Small organic molecules such as quinine and tetraethylammonium (TEA) are also useful. More recent additions to the armamentarium include tubocurarine and cetiedil, which block SK_{Ca} and IK_{Ca} channels, respectively (see later).

These K_{Ca} channel blockers differ markedly in their selectivity. Probably the most selective is apamin, which blocks the SK_{Ca} channel at nanomolar concentrations that have no effect on other K^+ channels. It has been reported, however, to have some action on Ca^{2+} channels in chick heart muscle (Bkaily *et al.*, 1985) and in frog auricles (Filippov *et al.*, 1988) at low concentrations. Charybdotoxin is also highly potent but has a wider spectrum of action. Half-maximal block of BK_{Ca} channels is observed with 3 nM, whereas even less is needed to inhibit the K_{Ca} channel in erythrocytes (Wolff *et al.*, 1988) and the voltage-dependent K^+ channel in lymphocytes (Sands *et al.*, 1988). Volume-activated K^+ channels are also affected (Grinstein *et al.*, 1988), as are voltage-sensitive IK_{Ca} channels (see later).

The blockers of lower molecular weight (e.g. TEA, quinine, tubocurarine and cetiedil) are less selective than the toxins. Quinine in particular is active at many channels, not only those for K^+ (including all K_{Ca} channels) but also for Na^+ and Ca^{2+} (see Chapter 8). Nevertheless quinine and TEA continue to be useful, especially if the concentration dependence of the block is taken into account. For example, BK_{Ca} channels are blocked by TEA in the 0.1–1 mM range, whereas much higher concentrations (often by a factor of 10 or more) are needed for the SK_{Ca} subtype (see, for example, Romey and Lazdunski, 1984). Another example is the differential sensitivity to TEA and quinidine of K_{Ca} channels in *Aplysia* neurones (p. 82).

Experiments by Blatz and Magleby (1986) provide a particularly clear illustration of the value of blockers (in this instance, TEA and apamin) in discriminating between channel types (Fig. 3.1). Note that the use of these agents reveals a small current component which is resistant to the blockers (record *e*). This may imply the existence of an additional K_{Ca} channel with a very low conductance and high sensitivity to Ca^{2+} (see the discussions by Blatz and Magleby, 1986, 1987).

3.2.3 Other approaches

3.2.3.1 Kinetic studies

In principle, K_{Ca} channels, like those that underlie I_A and I_K, should be categorizable in terms of the rate constants for the transitions between the open and closed states of the channel, and of the voltage and calcium dependence of these constants. However, detailed information of this kind is as yet available only for the BK_{Ca} channels (see section 3.3.1).

The ability of other ions to substitute for Ca^{2+} in activating the channel may also

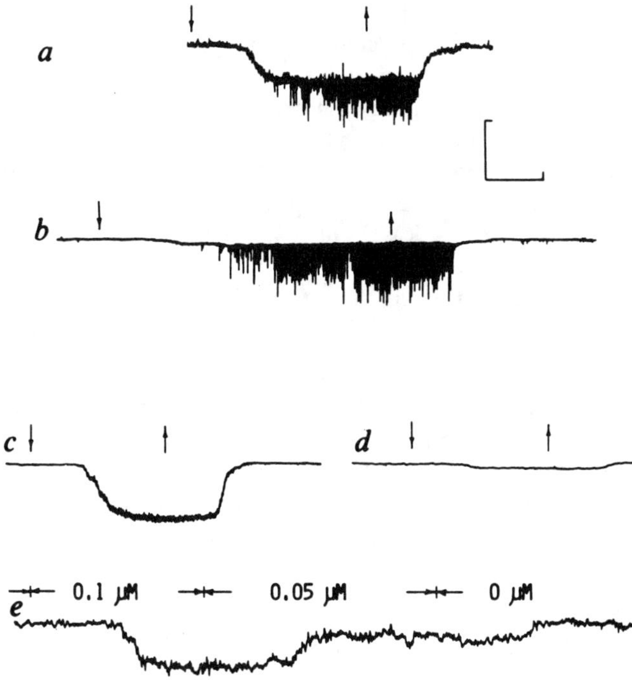

Fig. 3.1 — Differential blockade by TEA and apamin of BK_{Ca} and SK_{Ca} channels respectively. The records are of current flow through an inside-out membrane patch which was excised from cultured rat skeletal muscle, and held at a potential of −40 mV. At the first arrow in each trace, the Ca^{2+} concentration of the solution bathing the 'inside' of the membrane was raised from 0 to 1 μM, and back again at the second arrow. The increase in Ca^{2+} first caused the SK_{Ca} channels to open, as shown by the downward deflection in *a*. As Ca^{2+} rose further, the BK_{Ca} channels also present in the patch began to be activated giving rise to the transient downward spikes. The sequence was then repeated, but with apamin in the patch pipette, to block the SK_{Ca} channels. Record *b* shows that raising Ca^{2+} now activated mainly the BK_{Ca} channels. The converse is illustrated in *c*. The pipette contained TEA (5 mM), to block the BK_{Ca} channels, so that the current is largely attributable to the SK_{Ca} channels. Accordingly, it was greatly reduced when apamin was also present (*d*). Some residual current still flowed, however, and it is illustrated at a higher gain in record *e*, in response to an increase in Ca^{2+} from 0 to 0.1 μM, followed by a reduction first to 0.05 μM and finally to 0 μM. Vertical bar: *a* to *d*, 10 pA; *e*, 1 pA. Horizontal bar: *a* to *d*, 20 s; *e*, 10 s. From Blatz and Magleby (1986), with permission.

aid classification. In most cases Sr^{2+} can activate K_{Ca} channels but Pb^{2+} appears to be more selective, activating the channels in red blood cells (see later) but having little effect on BK_{Ca} channels (Oberhauser *et al.*, 1988).

3.2.3.2 Ionic selectivity

Although each K_{Ca} channel subtype might be expected to have a characteristic selectivity in its relative permeability to different ions, in practice such information as is available (mainly for the BK_{Ca} and IK_{Ca} channels — see later) shows only small differences. The SK_{Ca} channels have yet to be studied in any detail.

3.2.3.3 Selective chemical modification of the channels

Just as with drug receptors, ion channels can be expected to differ in their susceptibility to chemical modification. Although this approach is in its infancy, a promising starting point is the finding that bromoacetamide causes striking changes in the function of BK_{Ca} channels, which lose their Ca^{2+} sensitivity, while remaining reponsive to voltage (Pallotta, 1985; Cornejo *et al.*, 1987). High external Ca^{2+} protects against this action (Cornejo *et al.*, 1987).

Covalent modification of the ion channel at the inner surface of the membrane is also possible, and indeed occurs physiologically. For example, some K_{Ca} channels are modified by phosphorylation via the action of cAMP-dependent protein kinase.

3.2.3.4 Classification by channel structure

In due course this will provide an unequivocal basis for categorization. The isolation of K_{Ca} channel proteins by using selective ligands such as apamin, calmodulin and probably also charybdotoxin should allow the identification of the genes that code for these proteins, and make possible the determination of their primary structure (as has already been achieved for A-channel components — see Chapter 1).

3.3 THE MAIN KINDS OF K_{Ca} CHANNEL

It has to be stressed that the scheme for K_{Ca} channel classification that we have used is tentative, and may well require revision, or indeed replacement, as the structures and amino acid sequences of the channels become known. Already it is clear that the IK_{Ca} category contains subtypes and, as already discussed in relation to Fig. 3.1, there is evidence for an additional SK_{Ca} channel with an even lower conductance (Blatz and Magleby, 1986, 1987).

Note also that the expression of a particular channel subtype in a given tissue can vary greatly between species. For example, rat hepatocytes lack the apamin-sensitive $P_{K(Ca)}$ that has been found in the liver cells of all the other species that have been studied (see also section 3.3. and Fig. 3.3). Again, some animal red cells do not display the $P_{K(Ca)}$ mechanism that underlies the Gárdos effect (reviewed by Schwarz and Passow, 1983, and by Sarkadi and Gárdos, 1985).

3.3.1 Large-conductance K_{Ca} channels (BK_{Ca})

Since their discovery in chromaffin cells (Marty, 1981), skeletal muscle myotubes (Pallotta *et al.* (1981) and a pituitary cell line (Wong *et al.*, 1982), these channels have been identified in a very wide range of excitable and inexcitable cells. They have been studied in particular detail in T-tubule membranes, reconstitued into artificial lipid bilayers, by Latorre and Miller and their colleagues. In neurones they underlie I_C, a current which can contribute to the fast phase of action potential repolarization (Adams *et al.*, 1982); in exocrine gland cells they are closely involved in ion secretion (Petersen, 1986) and in the β-cells of pancreatic islets they play a key role in generating the bursting pattern of electrical activity which leads to the pulsatile release of insulin (see Chay, 1986, for a theoretical treatment, and Sherman *et al.*, 1988).

Gating
Both $[Ca^{2+}]_i$ and voltage control the gating of these channels. The effects of Ca^{2+} and voltage on the open state probability, p_o, have been investigated in several types of cell using single-channel recording methods, and are illustrated in Fig. 3.2 for

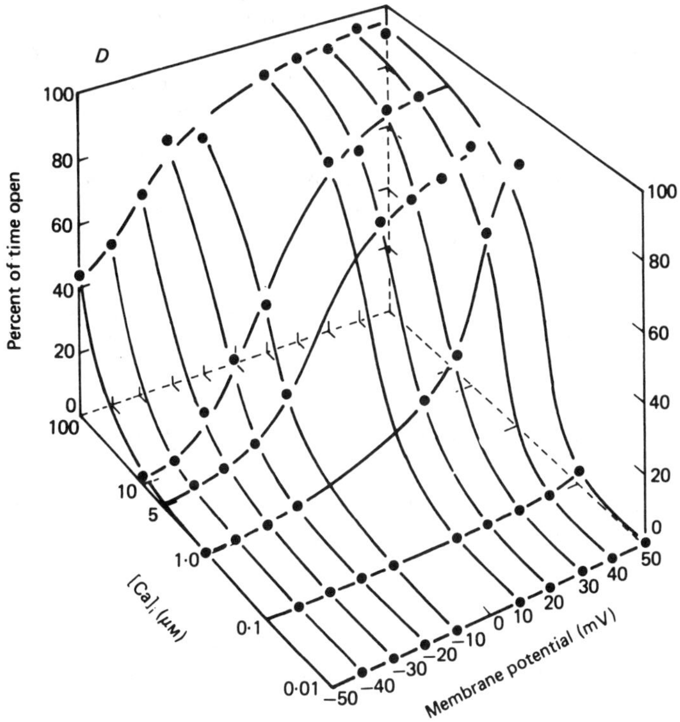

Fig. 3.2 — Channel open time of BK_{Ca} channels as a function of $[Ca^{2+}]_i$ and membrane potential. Single-channel activity was measured in excised membrane patches from rat myotubes. The probability that the channel is open is enhanced by increases in both $[Ca^{2+}]_i$ and membrane positivity. From Barrett *et al.* (1982), with permission.

embryonic skeletal muscle. Making the membrane potential more positive increases p_o, the effect of raising $[Ca^{2+}]_i$ being to shift the sigmoidal curve, relating p_o to membrane potential, to less positive potentials. The sensitivity of the channels to Ca^{2+} varies markedly from one cell type to another. The channels in smooth muscle (Benham *et al.*, 1986) and pancreatic acinar cells (Maruyama *et al.*, 1983) are very sensitive, maximum activation being achieved with $[Ca^{2+}]_i$ as low as 0.2 μM, whereas in myotubes (Barrett *et al.*, 1982) and T-tubules (Latorre *et al.*, 1982) 10 μM or more is required. There is a steep relationship between p_o and $[Ca^{2+}]_i$, the Hill coefficient varying between tissues. In rat muscle plasma membranes a Hill coefficient between 1 and 2 is interpreted as a need for two Ca^{2+} to bind for maximal channel activation (Moczydlowski and Latorre, 1983), whereas in myotubes a Hill coefficient of 3.6

(McManus and Magleby, 1985) suggests that as many as four Ca^{2+} must bind. A later analysis by McManus and Magleby (1988) in rat myotubes indicates that there may be three or four open and six or more closed states of the channel, each open state correponding perhaps to the binding of different numbers of Ca^{2+}. A further complication is revealed by studies in which $[Mg^{2+}]_i$ is found to modify gating (Golowasch *et al.*, 1986). In the presence of 1–10 μM Mg^{2+} there is an apparent increase in the affinity of the channel for Ca^{2+} and an increase in the Hill coefficient, which can reach values close to 6. The authors suggest that Mg^{2+} may bring into play Ca^{2+}-binding sites which are not operational in the absence of Mg^{2+}. The steepening of the relationship between $[Ca^{2+}]_i$ and p_o would make the cell more responsive to small changes in $[Ca^{2+}]_i$. The detailed analysis of single-channel records by McManus and Magleby referred to above has also identified 'modes' of channel gating. In the most common, 'normal' mode, all states of the channel are accessible, whereas in the other modes only subsets of the channel states can be entered. These channels can also be activated by other divalent ions in addition to Ca^{2+}, including Cd^{2+} and Sr^{2+} (Oberhauser et al., 1988). An intermediary role of calmodulin in the activation of the channels by Ca^{2+} would seem to be ruled out by the activation seen in isolated membrane patches or after channel incorporation in artificial lipid bilayers, where calmodulin might be expected to be absent. Counter to this, however, are the findings of Klærke *et al.* (1987), who reported that solubilized BK_{Ca} channels could be purified using a calmodulin affinity column, and that the Ca^{2+} sensitivity of the purified channel, when reconstituted into vesicles, could be increased by exogenous calmodulin.

Selectivity

The relative ionic selectivity of BK_{Ca} channels has been studied in several cell types and the *permeability* sequence for myotubes (Blatz and Magleby, 1984) is typical: Tl^+ (1.2)>K^+ (1)>Rb^+ (0.67)>NH_4^+ (0.11)≫Cs^+, Na^+, Li^+. It should be noted that *conductance* for Rb^+ is relatively much lower, a finding attributed to strong binding to a site in the channel (see Gallacher *et al.*, 1984). Na^+ ions, which hardly permeate, can in fact block BK_{Ca} channels. In single-channel records the block produced by Na^+ is described as 'flickery' (Yellen, 1984), reflecting the speed at which blocking and unblocking occurs. This effect of Na^+ can be reversed by raising $[K^+]_o$ and this observation, plus the finding that channel conductance for mixtures of K^+ and Rb^+ shows 'anomalous mole fraction' behaviour, reflecting interactions between ions in the channel (Eisenman *et al.*, 1986), suggests that BK_{Ca} channels are not single-ion channels as once believed but are multi-ion pores. Ca^{2+} ions do not penetrate the channel and indeed at higher concentrations and at positive potentials can have a blocking action. Such channel block has slow kinetics, resulting in long shut periods (e.g. Vergara and Latorre, 1983).

Channel conductance

The single-channel conductance of BK_{Ca} channels is characteristically very high. For channels exposed on both sides to ~140 mM K^+ it ranges from 100 to 300 pS in different cell types, 250 pS being typical. Under a physiological K^+ gradient the conductance is lower, ~100 pS. Intermediate conductance substates are occasionally seen (e.g. Barrett *et al.*, 1982; Tabares *et al.*, 1985). The increase in conductance as $[K^+]$ is elevated is not linear but tends towards a maximum of about 640 pS as K^+ (in and out) is raised above 600 mM (Blatz and Magleby, 1984). A simple Michaelis–Menten relationship does not, however, apply (Latorre, 1986).

Blockers

As already discussed, non-permeant inorganic cations such as Na^+ and Ca^{2+} can block the channel from the inside. Ba^{2+} is particularly effective, acting at concentrations of 1 μM or less (Vergara and Latorre, 1983). The block, which resembles that to Ca^{2+}, is characterized by long shut periods and is voltage-dependent, indicating binding to a site within the channel. Channel closure can trap Ba^{2+} in the channel (Miller, 1987). As noted previously, TEA shows some selectivity in its blockade of BK_{Ca} channels and is more effective from the outside (K_D=0.29 mM) than the inside (K_D=45 mM) of the membrane (Latorre, 1986). The kinetics of TEA block are so rapid that individual blocking events are not resolved in single-channel recording. Rather the single-channel conductance appears to be reduced. The internal TEA binding site also binds other quaternary ammonium ions with an affinity which increases as hydrophobic groups are added, nonyltriethylammonium ion yielding a K_D of 3 μM. The bisquaternary decamethonium is also an effective blocker from the inside, demonstrating a greater voltage dependence than the monoquaternary ions, probably as a result of its double charge (see Latorre, 1986).

Two scorpion toxins are known to block BK_{Ca} channels. Charybdotoxin blocks from the outside with high affinity (K_D~10 nM) and promises to be valuable in the further study and isolation of BK_{Ca} channels (Miller *et al.*, 1985). As already noted (section 3.2.2), however, its selectivity is limited. BK_{Ca} channels are also blocked by noxiustoxin (better known for its action on voltage-activated channels) but with a much lower affinity (K_D~450 nM) than charybdotoxin (Valdivia *et al.*, 1988). Quinine produces 'flickery block' of the open channel and for this reason is a poor blocker unless p_o is high (>0.85) (e.g. Glavinović and Trifaró, 1988). Neuroleptic drugs have also been shown to block open BK_{Ca} channels from the internal side and in a voltage-dependent fashion. The relative potencies of haloperidol, trifluoperazine, thioridazine and chlorpromazine, all of which have K_Ds in the range of 1–5 μM, differ from those seen for inhibition of calmodulin and it is concluded that the action is at the channel level rather than due to interference with a calmodulin-like Ca^{2+}-binding protein (McCann and Welsh, 1987).

Channel structure

The large conductance of BK_{Ca} channels coupled with a high selectivity for K^+ imposes severe constraints on possible models of their structure. Considerable attention has been focused on the problem, especially by Miller and Latorre and their colleagues. Yellen (1987) provides a recent review. Selectivity for K^+ demands a narrow channel region which, to allow for the high conductance, cannot be very long. The models devised thus have large 'mouths' and/or antechambers on either side of the membrane which allow easy movement of K^+ towards the narrow 'selectivity filter'. The models also incorporate negative charges or dipoles within the channel to provide binding sites for permeant or blocking cations.

Modulation

In pancreatic β-cells lowering pH_i moves the curve relating p_o to membrane potential to more positive potentials (Cook *et al.*, 1984). This is consistent with observations in intact β-cells where raising pH_i by application of NH_4Cl can increase membrane conductance (Rosario and Rojas, 1986). Whether this effect is important in the regulation of insulin release is not yet known. Phosphorylation of BK_{Ca} channels in plasma membranes from rat brain (but not those from T-tubules) by application of the catalytic subunit of cAMP-dependent protein kinase can result in a marked increase in channel activity (Reinhart and Levitan, 1987). An increase in p_o by cAMP-initiated phosphorylation of BK_{Ca} channel proteins has also been proposed for chick kidney cells (Guggino *et al.*, 1985). In contrast, the effect of intracellular ATP to reduce p_o in *Amphiuma* distal tubules is thought not to involve channel phosphorylation (Hunter and Giebisch, 1988). Finally a modulatory role for protein kinase C is suggested by the inhibitory effect of a phorbol ester on BK_{Ca} channel activity in an insulin-secreting cell line (Ribalet *et al.*, 1988).

Channel number

Since these channels have such a large unitary conductance only a few need to be open in a small cell to have a substantial effect on its input conductance. This problem has been addressed for pancreatic acinar cells by Petersen and Findlay (1987). These cells have about 50 BK_{Ca} channels each so that, with a p_o of only 0.002 at rest, for most of the time no channels can be open. The occasional opening of one or two channels might then lead to marked changes in membrane resistance and potential. Instability of the membrane potential is, however, avoided by electrical coupling of the acinar cells. Within the syncytium any individual channel opening will have a proportionately smaller effect on membrane potential. Coupling of pancreatic β-cells also has important consequences for their electrical activity (Sherman *et al.*, 1988).

3.3.2 Intermediate-conductance K_{Ca} channels (IK_{Ca})

3.3.2.1 Voltage-insensitive IK_{Ca} channels

The best-studied channel of this type is the one found in the red blood cells of most animals and which underlies the 'Gárdos effect' (see section 3.1). In red blood cells potassium permeability may be increased by up to 1000× (Lew and Ferreira, 1976). Extensive reviews of K_{Ca} channels in erythrocytes are available (Lew and Ferreira, 1978; Sarkadi and Gárdos, 1985). Similar channels may be present in lymphocytes (Grinstein *et al.*, 1982), macrophages (Oliveira Castro and Dos Reis, 1981) and Ehrlich ascites cells (Valdeolmillos *et al.*, 1982; Hoffmann *et al.*, 1984) but their characteristics are much less well established and no direct measurements have been made of their unitary conductances. Their classification as IK_{Ca} channels is thus circumstantial. It is possible that the K_{Ca} channels in these other cells may in fact be identical with the K^+ channels responsible for the regulatory volume decrease (RVD) seen after the cells swell in hypotonic media. In Ehrlich cells there are clear parallels between volume and Ca^{2+}-induced increases in K^+ permeability (Hoffmann, 1987), and in lymphocytes depletion of cell Ca^{2+} can inhibit the K^+ permeability increase induced by swelling (Grinstein *et al.*, 1984). Hazama and Okada (1988) have also obtained evidence that the K^+ channels activated by swelling in intestinal epithelial cells are K_{Ca} channels. Direct measurements of $[Ca^{2+}]_i$ during swelling have, however, produced conflicting findings: in lymphocytes Rink *et al.* (1983) were unable to detect any increase in $[Ca^{2+}]_i$, in contrast to findings with toad urinary bladder (Wong and Chase, 1986). Rather more decisive evidence for separate volume- and Ca^{2+}-activated channels comes from single-channel recording in frog red cells, where Hamill (1983) found K^+ channels with unitary conductances of 50 pS and 18 pS (in high-K^+ media). The larger one exhibited clear Ca^{2+} dependence whereas the smaller channel had a much weaker dependence on $[Ca^{2+}]_i$ but was activated by cell swelling. Further single-channel studies are required in other cells which show an RVD in order to clarify the possible role of K_{Ca} channels.

Channel gating

The prime stimulus for channel opening is an increase in $[Ca^{2+}]_i$, which in the 'Gárdos effect' is a consequence of the depletion of cellular ATP. Activation can also be induced by the calcium ionophore A23187, when added in a Ca^{2+}-containing solution (Reed, 1976). Activation normally occurs in the range 0.1–10 μM as shown by Simons (1976) and Heinz and Passow (1980) using Ca^{2+} buffers in red cell ghosts, and by Hamill (1983) and Grygorczyk and Schwarz (1983) using direct measurements of the open time of the channels in membrane patches. The relationship betwen p_o and $[Ca^{2+}]_i$ is steep, suggesting that opening involves the binding of two Ca^{2+} ions (Grygorczyk and Schwarz, 1983). The finding that Ca^{2+} can open channels in cell-free membrane patches (Grygorczyk *et al.*, 1984) argues against a requirement for calmodulin; the same conclusion was reached by Plishker (1984) in studies with red cell ghosts. Some role for calmodulin was, however, suggested by the results of Pape and Kristensen (1984), who found that exogenous calmodulin could increase Rb^+ uptake into inside-out vesicles and that this could be inhibited by calmodulin antagonists.

In addition to Ca^{2+} the channels are also activated by Sr^{2+} (Simons, 1976). Lead also has long been known to cause K^+ loss from red blood cells (see Passow, 1981) and single-channel recording has now shown this to be a direct effect of Pb^{2+} on channel opening (Grygorzyk and Schwarz, 1983; Shields *et al.*, 1985). Pb^{2+} activates at lower concentrations than Ca^{2+} (Simons, 1985). Mg^{2+} inhibits channel opening (García-Sancho *et al.*, 1982).

The sensitivity of the channel to Ca^{2+} can also be modified. For example, the electron donor system ascorbate+phenazine methosulphate increases the apparent affinity for Ca^{2+} in inside-out red cell vesicles by about 5× (Alvarez *et al.*, 1984).

Open channels do not show inactivation, so that the increase in membrane permeability can be maintained for more than an hour (Lew and Ferreira, 1978). Channel opening is not significantly voltage-sensitive (e.g. Hamill, 1981, 1983).

There is an interesting dependence of channel opening on extracellular $[K^+]$. Prolonged exposure to K^+-free solution renders the channels refractory to activation (e.g. Grygorczyk *et al.*, 1984). Channel opening is also impaired with high external K^+, maximal activation being seen with a $[K^+]_o$ of ~1 mM (Blum and Hoffmann, 1971). Intracellular pH also has a marked effect on the Ca^{2+}-activated permeability. Between pH 6 and 6.5 the change in P_K has a Hill coefficient of ~2.5, with the relationship becoming even steeper at lower pH (Stampe and Vestergaard-Bogind, 1985).

Conductance

Despite the small size of human erythrocytes, successful single-channel recordings have been obtained; first by Hamill (1981), who found with ~100 mM K^+ on either side of the membrane patch a unitary conductance of 18 pS which was voltage-independent over the range studied. Studies by Grygorczyk and Schwarz (1983) and by Grygorczyk *et al.* (1984) over a wider range of voltage demonstrated inward rectification of the single-channel currents. With 140 mM K^+ outside, the unitary conductance varied between 40 pS at −280 mV to 10 pS at +160 mV. At 0 mV the conductance of 25 pS was close to Hamill's figure. More recently, Bennekou and Christophersen (1988) have reported a higher single-channel conductance of ~40 pS (130 mM K^+ outside). Grygorczyk and Schwarz (1983) also found (although less frequently) Ca^{2+}-activated channels with a smaller conductance (~40% of the main level). In the larger frog red cell Hamill (1983) found a K_{Ca} channel with a conductance of 50 pS (high external K^+). Patch recordings from HeLa cells (Sauvé *et al.*, 1986) revealed K_{Ca} channels with similar characteristics to those of human red cells, namely an open probability which is insensitive to voltage in the physiological range, and inward rectification with conductances of 50 pS at negative potentials and 10 pS at positive potentials (symmetrical 200 mM K^+ solutions).

Selectivity

Electrophysiological measurements (Grygorczyk *et al.*, 1984) have suggested that P_K/P_{Na} for K_{Ca} channels in red cells is 17, whereas flux studies (Simons, 1976) indicate a value of at least 40. Lew and Ferreira (1978) have suggested that P_K/P_{Na} might exceed 1000. High concentrations of intracellular sodium can, in fact, reduce potassium permeability (e.g. Simons, 1976; Yingst and Hoffman, 1984). The behaviour of Rb^+ is reminiscent of that seen in BK_{Ca} channels in that its permeability (measured by tracer flux) is similar to that for K^+ (Simons, 1976; Grygorczyk *et al.*, 1984), whereas its conductance is very low when it is present as the major charge carrier (Hamill, 1983).

Channel structure

In comparison with BK_{Ca} or SK_{Ca} channels, rather little is known of the structure of red cell IK_{Ca} channels. Measurements of undirectional fluxes (Vestergaard-Bogind *et al.*, 1985) revealed deviations from Ussing's relationship consistent with single-file movement in channels holding three ions. The dependence of channel opening on external K^+, and the K^+ dependence of block by several drugs, suggests the presence of a binding site for K^+ on the external side of the channel.

Channel numbers

Two approaches have been used to estimate the number of K^+ channels in red blood cells. The first involves inside-out vesicles. These can be prepared from red blood cells in numbers which exceed the number of K_{Ca} channels likely to be present, so that some vesicles will be devoid of channels, and elevating $[Ca^{2+}]$ will not increase their rate of K^+ exchange. Determination of the proportion of vesicles lacking channels thus allows an estimate of the number of channels per cell. In this way Lew *et al.* (1982) estimated each cell to have 100–200 channels; Alvarez and García-Sancho (1987) reported a value of 142. These figures are higher than those obtained by comparing single-channel conductance with the permeability of the cell as a whole. Using the latter method Grygorczyk and Schwarz (1983) and Grygorczyk *et al.* 1984) obtained a mean value of only 10 per cell, but with evidence for subpopulations of cells containing different numbers of channels. For Ehrlich cells Hoffmann *et al.* (1986) estimated 100 K_{Ca} channels per cell (assuming a unitary conductance of 20 pS).

Blockers

Agents which block K_{Ca} channels in erythrocytes at submillimolar concentrations include quinine and quinidine (Armando-Hardy *et al.*, 1975), carbocyanine dyes (Simons, 1979), the anti-sickling agent cetiedil (Berkowitz and Orringer, 1981) and oligomycin A (Riordan and Passow, 1971)). Reichstein and Rothstein (1981) have shown that the action of quinine is inhibited by external K^+. The same may well be true of the carbocyanine dyes and cetiedil, whose action to inhibit the RVD in lymphocytes is also reduced by external K^+ (Sarkadi *et al.*, 1985).

More recently it has been found that the scorpion toxin charybdotoxin (CTX) can not only block BK_{Ca} channels (see above) but also, at low concentrations, the IK_{Ca} channel in red blood cells. (Wolf *et al.*, 1988; see also Abia *et al.*, 1986, and Castle and Strong, 1986). It may be noted that CTX will also block the ionomycin-induced hyperpolarization and RVD in lymphocytes (Grinstein *et al.*, 1988).

3.3.2.2 Voltage-sensitive IK_{Ca} channels

The presence of a K_{Ca} mechanism in nerve cells was first demonstrated in studies with the mollusc *Aplysia* (Meech and Strumwasser, 1970) and soon after with *Helix* (Meech, 1974). In contrast to the findings with red cells, the proportion of the channels that are open at a given Ca^{2+} concentration increases with depolarization. Their physiological role is thought to lie in the control of the bursts of action potentials that characterize the activity of these molluscan neurones. The entry of Ca^{2+} during the action potentials causes intracellular Ca^{2+} to increase, as has been demonstrated using the Ca^{2+} indicator arsenazo III (Gorman and Thomas, 1978; Gorman *et al.*, 1982a). The resulting activation of $P_{K(Ca)}$ will reduce excitability, so tending to decrease the frequency of action potentials towards the end of the burst. (The marked hyperpolarization that occurs between bursts has also been attributed to the activation of $P_{K(Ca)}$ although this is now in question: a more important factor is likely to be the inhibition by intracellular Ca^{2+} of a resting Ca^{2+} current. For references and discussion, see Ewald and Levitan, 1987).

Conductance

The single-channel conductance in *Helix* neurones is 40–60 pS when both faces of the membrane are bathed in K^+-rich solution (Ewald *et al.*, 1985), and 20 pS when the external K^+ is low (Lux *et al.*, 1981; Ewald *et al.*, 1985). In *Aplysia californica* neurones a similar voltage-dependent K_{Ca} channel with a conductance of 35 pS in asymmetrical K^+ solution has been described (Hermann and Erxleben, 1987). When the open-channel probability of the *Aplysia* channel is low, it increases *e*-fold for a 50-mV change in membrane potential (inside-out patch: bath $[Ca^{2+}]=0.1$ μM) and the current–voltage relationship for single channels shows outward rectification of the magnitude to be expected on the basis of the constant field equation.

Note that in the *Aplysia* and *Helix* literature, the IK_{Ca} channels are often described as *small*. This is in relation to the BK_{Ca} variety already discussed, and the 'small' molluscan IK_{Ca} channel should not be confused with the SK_{Ca} channels of lower conductance dealt with in the following section. The difference is underlined by the insensitivity of the molluscan channel to apamin (see, for example, Hermann and Hartung, 1983; Deitmer and Eckert, 1985) and by its voltage dependence.

Selectivity
The ionic selectivity of the IK_{Ca} channel in *Aplysia* has been studied by Gorman *et al.* (1982b), who reported the permeability sequence for monovalent cations to be $Tl^+ \geqslant K^+ > Rb^+ \geqslant NH_4^+ > Cs^+ > Li^+$, Na^+. This is very similar to that of the BK_{Ca} channel, and also of the delayed rectifier in various tissues including *Helix* neurones (see the discussion by Hermann and Hartung, 1983).

Blockers
Quinidine and TEA but not apamin are highly effective in *Helix* neurones (see Hermann and Hartung, 1983). The 35-pS channel in *Aplysia californica* is blocked by charybdotoxin at nanomolar concentrations (Hermann and Erxleben, 1987). The apparent K_D is 30 nM at a membrane potential of −30 mV, and the affinity decreases with depolarization.

Work by Deitmer and Eckert (1985) suggests that there may be two distinct K_{Ca} currents in the IK_{Ca} category in certain *Aplysia* neurones, the proportion of each varying with the type of neurone. One of the currents is blocked by TEA at relatively low concentrations (K_D<1 mM) and by quinidine (0.2–0.3 mM). The other is much less sensitive to both agents, the apparent K_D for TEA being at least 100 mM. It seems that charybdotoxin distinguishes between these channels, blocking only the subtype sensitive to TEA and quinidine (Hermann and Erxleben, 1987).

Modulation, and possible occurrence in vertebrates
A striking feature of the molluscan IK_{Ca} channels is the alteration in function produced by phosphorylation due to the action of cAMP-dependent protein kinase: the outcome is an increase in current at a given $[Ca^{2+}]_i$ and membrane potential. Application of the catalytic subunit of the enzyme is equally effective (e.g. DePeyer *et al.* 1982). The analysis by Ewald *et al.* (1985) suggests that phosphorylation increases the Ca^{2+} sensitivity of the channel (see also the discussion by Ewald and Levitan, 1987). These findings are particularly interesting in relation to the possiblity that IK_{Ca} channels also occur in vertebrate neurones. Farley and Rudy (1988) have inserted K^+ channels from mammalian brain synaptosomal membranes into planar phospholipid bilayers, and observed unitary currents corresponding to a number of distinct K^+ channels. Among them was a charybdotoxin-sensitive K_{Ca} channel with a conductance of 75–80 pS in symmetrical K^+-rich (200 mM) solutions. The open-time probability of this channel was increased by the catalytic subunit of cAMP-dependent protein kinase. In discussing these findings, Farley and Rudy raise the possibility that this K_{Ca} subtype in mammalian brain may correspond to the molluscan channel.

Another vertebrate voltage-dependent K_{Ca} channel with a unitary conductance (30 pS in symmetrical K^+-rich solution) just within the IK_{Ca} range has been detected in guinea-pig parotid acinar cells by Suzuki and Petersen (1988). It can be blocked by TEA, albeit at a high concentration (10 mM). The relationship of this channel to the invertebrate neurone types just described is not yet known.

3.3.3 Small-conductance K_{Ca} channels (SK_{Ca})

Although single SK_{Ca} channel recordings have only recently been obtained, the existence of a $P_{K(Ca)}$ mechanism with the properties now ascribed to the SK_{Ca} subtype has been known for many years. For example, α_1-adrenoceptor activation causes a Ca^{2+}-mediated increase in the K^+ permeability of the longitudinal smooth muscle of the intestine, and hence inhibits the spontaneous electrical and mechanical activity of the tissue (see the review by Bülbring and Tomita, 1987). The effect was found to be potently and selectively blocked by apamin (Banks *et al.*, 1979; Maas *et al.*, 1980), which has become a valuable tool in identifying the presence of the SK_{Ca} channel. A similar apamin-sensitive $P_{K(Ca)}$ mechanism was described in hepatocytes (Banks *et al.*, 1979; Burgess *et al.*, 1981) and in neuroblastoma cells (Hugues *et al.*, 1982a). It is now known to be widely distributed.

Conductance

An electrophysiological study by Romey and Lazdunski (1984) indicated that the neuroblastoma channels were of very low conductance, certainly much lower than the other K_{Ca} channels which had been analysed at the time. Spectral analysis of current noise in voltage-clamped guinea-pig hepatocytes also suggested a low conductance (Capiod and Ogden, 1985, 1988, 1989a). Direct confirmation has come from the single-channel recordings of Blatz and Magleby (1986), who found the conductance to be 10–14 pS (measured in symmetrical K^+, 140 mM) in cultured rat skeletal muscle, and of Capiod and Ogden (1987, 1988, 1989b), who reported a value of 6 pS in asymmetrical K^+ (cf. 20 pS in symmetrical K^+) in membrane patches excised from guinea-pig hepatocytes. These single-channel studies have also provided the first direct information on the Ca^{2+} dependence of the SK_{Ca} channels. In cultured skeletal muscle, activation was half-maximal with Ca^{2+} at 200–500 nM (Blatz and Magleby, 1986), and in hepatocytes at 600 nM (Capiod and Ogden, 1988, 1989b). There is little dependence on membrane potential (Capiod and Ogden 1989a).

Blockers

Apamin is by far the most active and selective compound available. Structure–activity studies with apamin congeners have shown that two arginines (or lysines substituted for them) at positions 13 and 14 are essential for activity (Vincent *et al.*, 1975; Habermann and Fischer, 1979a; Habermann, 1984). These adjoining amino acids carry positive charges, suggesting that smaller bis-charged molecules, such as the neuromuscular blocking agent tubocurarine, might also be effective. This has been found to be so, first in hepatocytes (Jenkinson *et al.*, 1983; Cook and Haylett, 1983, 1985) and later in neurones (Nohmi and Kuba, 1984). Tubocurarine has been studied in some detail, and shown to inhibit apamin binding to hepatocytes (implying a common site of action) with a K_D of 7.5 μM. This is similar to the concentration (3 μM) which causes a half-maximal inhibition of Ca^{2+}-mediated K^+ release from the same cells (Cook and Haylett, 1985). Tubocurarine and related nueromuscular blockers are also active in intestinal muscle although somewhat larger concentrations are needed (Gater *et al.*, 1985), as in neurones (e.g. Dun *et al.*, 1986; Bourque and Brown, 1987). These agents are less selective than apamin (see, for example, Smart, 1987).

The binding of apamin (and presumably also its action), is reduced when K^+ is lowered (Habermann and Fischer, 1979b; Hugues *et al.*, 1982b; Seagar *et al.*, 1984). This suggests that K^+ can modulate the channel in some way. Interestingly, the single-channel conductance of the SK_{Ca} channels is unexpectedly sensitive to *external* K^+ (Capiod and Ogden, 1989b), which is clearly in keeping with the idea that K^+ acts at a modulatory site at the external surface (see also the discussions by Lazdunski *et al.*, 1988, and by Moczydlowski *et al.*, 1988).

The venom of *Leiurus quinquestriatus hebraeus* is now known to contain a peptide which blocks the SK_{Ca} channel (Abia *et al.*, 1986; Castle and Strong, 1986), and is distinct from both apamin and charybdotoxin (a major constituent of this venom). Chicchi *et al.* (1988) have purified a peptide component (leiurotoxin I) with these properties, and have determined its amino acid sequence, which bears little obvious relationship to that of apamin. Nevertheless, it is able to inhibit apamin binding to rat brain synaptosomal membranes.

Channel structure

Radiation inactivation analysis suggests that the functional apamin receptor protein has a relative molecular mass in the range ~100 000 (Seagar *et al.*, 1986) to ~250 000 (Schmid-Antomarchi *et al.*, 1984). The application of covalent labelling techniques, using apamin derivatives, has allowed the isolation of polypeptide components of the apamin binding site. These have M_rs ranging from 22 000 to 86 000 (Schmid-Antomarchi *et al.*, 1984; Seagar *et al.*, 1986; Marquèze *et al.*, 1987). From this it seems likely that the SK_{Ca} channel is made up of at least three and possibly as many as eight subunits (see the discussions by Marquèze *et al.*, 1987, and by Lazdunski *et al.*, 1988).

Channel numbers

The availability of labelled apamin allows the number of apamin binding sites to be estimated. Fig. 3.3 illustrates the binding to isolated guinea-pig and rat hepatocytes under physiological conditions. With the guinea-pig cells, the maximum inhitable binding corresponds to ~2000 sites per hepatocyte, with a K_D of 400 pM (Cook and Haylett, 1985). The rat cells, in contrast, show little binding. This is in keeping with other evidence that in this species the hepatocytes lack the apamin-sensitive K_{Ca} mechanism (Burgess *et al.*, 1981; see also Jenkinson *et al.*, 1988). Returning to the guinea-pig results, if each site is tentatively assumed to correspond to a single K^+ channel, multiplying the number of sites by the single-channel conductance, 6 pS, in this tissue (Capiod and Ogden, 1987, 1988, 1989b) provides an estimate of 12 nS for the maximum conductance these channels could contribute in a single cell, were they all to open. Maximal hormonal stimulation activates a K^+ conductance of from 6 nS (whole-cell recording by Capiod and Ogden, 1985, 1988, 1989a) to 10 nS (intracellular recording, in which larger cells are favoured; Field and Jenkinson, 1987). This suggests that under these conditions p_o is high.

The binding of apamin has also been studied in primary neuronal cultures, with

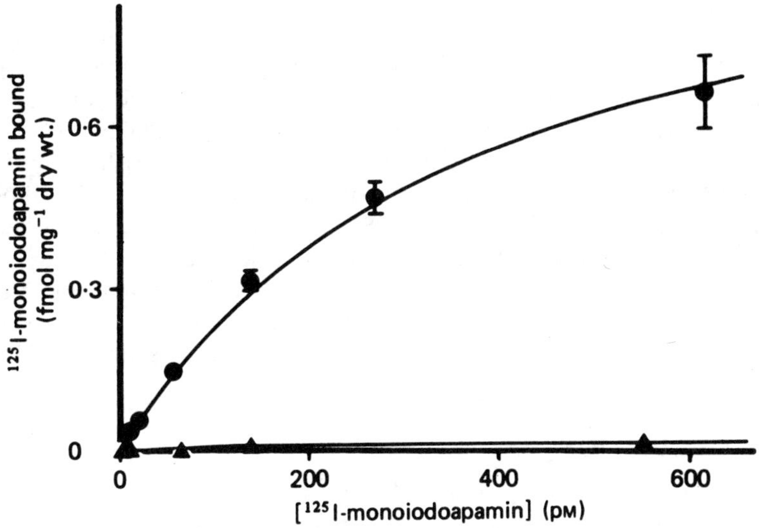

Fig. 3.3 — Inhibitable binding of [^{125}I]monoiodoapamin to guinea-pig and rat hepatocytes. Experiments were done at 37°C in physiological medium for an incubation period of 2 min. Points show the inhibitable binding component in each species, obtained by subtraction of the non-inhibitable binding (which was linear over the concentration range tested in each case), measured in the presence of 10^{-6} M unlabelled apamin, from the total binding. Points are means ± s.e. from five (guinea-pig; ●) or six (rat; ▲) experiments, four to eight observations per experiments at each concentration of [^{125}I]monoiodoapamin. From Cook and Haylett (1985), with permission.

synaptosomes and with membrane preparations from a variety of cells (Habermann and Fischer, 1979b; Hugues *et al.*, 1982b; Seagar *et al.*, 1984, 1985; Strong and Evans, 1987; for reviews see Habermann, 1984; Lazdunski *et al.*, 1988). The results show that the number of binding sites (and presumably of channels) is low, certainly in relation to tetrodotoxin-sensitive Na^+ channels measured under the same conditions in excitable cells. An important exception occurs with the PC12 phaeochromocytoma cell line, which binds at least 50 times more labelled apamin than other cells which possess the apamin-sensitive K_{Ca} channel (Schmid-Antomarchi *et al.*, 1986).

The number of apamin-binding sites changes during development, and under pathophysiological conditions. For example, apamin-binding sites, and SK_{Ca} channels, do not normally occur in adult mammalian skeletal muscle, but appear in response to chronic denervation (Schmid-Antomarchi *et al.*, 1985). They are also present in fetal muscle, and in the muscles of patients with myotonic muscular dystrophy (Renaud *et al.*, 1986).

Role

The physiological role of the SK_{Ca} channel is known in some instances at least. In intestinal smooth muscle, its activation underlies the inhibitory action of α_1-adrenoceptors, and of the receptors for neurotensin and ATP (P_2-subtype). Its presence in hepatocytes can account for adrenaline hyperkalaemia (Coats, 1983, 1985). In brown fat cells, activation of apamin-sensitive K_{Ca} channels holds the membrane potential at more negative values than would otherwise obtain during the action of noradrenaline: the outcome is a greater and more sustained influx of Na^+, which is considered to contribute to the mobilization of intracellular Ca^{2+} in this tissue (Nånberg *et al.*, 1985). Opening of SK_{Ca} channels underlies the maintained hyperpolarization that follows action potentials in many neurones, including those of the sympathetic ganglia (Nohmi and Kuba, 1984; Pennefather *et al.*, 1985), where its function has been studied in detail (see Chapter 14). Indeed, an even more general role in nervous tissue is suggested by the finding that the brain contains an endogenous apamin-like peptide (for references see Lazdunski *et al.*, 1988). The wide distribution of apamin-binding sites in rat brain has been described by Mourre *et al.* (1986).

3.4 FUTURE DIRECTIONS

The most important outstanding task is to identify and sequence the genes that code for the main kinds of K_{Ca} channels. As recent work on the molecular biology of the A channel in *Drosophila* neurones illustrates (see Chapter 1), relatively simple organisms can provide an appropriate starting point, and work of this kind is now under way. It is already known that a single-gene mutation in *Paramecium* results in a large increase in a Ca^{2+}-activated K^+ current (Hennessey and Kung, 1987). The control of the expression of these genes has also to be understood, and Lazdunski's group has provided valuable information on alterations in expression in skeletal muscle during development, and in disease (see above).

A related problem is to establish how the synthesis and degradation of K_{Ca} channel proteins are regulated. It has already been shown that the apamin-binding protein (presumably part of the SK_{Ca} channel) can be taken into the cell interior (Strong and Evans, 1987).

Other outstanding problems include the nature of the relationship between the K_{Ca} channels we have discussed and the less well understood K^+ permeabilities that are activated by changes in cell volume (see section 3.3.2.1) or by shear, as in endothelial cells (see Olesen *et al.* (1988). The pharmacological analysis of K_{Ca} channels also needs further refinement, not least because the development of blockers (or activators) which are selective for the various subtypes may open the way to new therapeutic agents.

ACKNOWLEDGEMENTS

Our work is supported by the Medical Research Council, and by the Wellcome Trust. We are most grateful to Miss V. Grant and to Mrs I. Kapadia for their assistance with the preparation of the manuscript.

REFERENCES

Abia, A., Lobaton, C. D., Moreno, A., and García-Sancho, J. (1986) *Leiurus quinquestriatus* venom inhibits different kinds of Ca^{2+}-dependent K^+ channels. *Biochim. Biophys. Acta* **856** 403–407.

Adams, P. R., Constanti, A., Brown, D. A., and Clark, R. B. (1982) Intracellular

Ca^{2+} activates a fast voltage-sensitive K^+ current in vertebrate sympathetic neurones. *Nature* **296** 746–749.

Alvarez, J., and García-Sancho, J. (1987) An estimate of the number of Ca^{2+}-dependent K^+ channels in the human red cell. *Biochim. Biophys. Acta* **903** 543–546.

Alvarez, J., García-Sancho, J., and Herreros, B. (1984) Effects of electron donors on Ca^{2+}-dependent K^+ transport in one-step inside-out vesicles from the human erythrocyte membrane. *Biochim. Biophys. Acta* **771** 23–27.

Anderson, C. S., McKinnon, R., Smith, C., and Miller, C. (1988) Charybdotoxin block of single Ca^{2+}-activated K^+ channels: effects of channel gating, voltage and ionic strength. *J. Gen. Physiol.* **91** 317–333.

Armando-Hardy, M., Ellory, J. C., Ferreira, H. G., Fleminger, S., and Lew, V. L. (1975) Inhibition of the calcium-induced increase in the potassium permeability of human red blood cells by quinine. *J. Physiol.* **250** 32–33P.

Banks, B. E. C., Brown, C., Burgess, G. M., Burnstock, G., Claret, M., Cocks, T. M., and Jenkinson, D. H. (1979) Apamin blocks certain neurotransmitter-induced increases in potassium permeability. *Nature* **282** 415–417.

Barrett, J. N., Magleby, K. L., and Pallotta, B. S. (1982) Properties of single calcium-activated potassium channels in cultured rat muscle. *J. Physiol.* **331** 211–230.

Benham, C. D., Bolton, T. B., Lang, R. J. and Takewaki, T. (1986) Calcium-activated potassium channels in single smooth muscle cells of rabbit jejunum and guinea-pig mesenteric artery. *J. Physiol.* **371** 45–67.

Bennekou, P., and Christophersen, P. (1988) The occurrence of a low conductance channel in the human red cell. *Acta Physiol. Scand.* **134** (Suppl. 575) 73.

Berkowitz, L. R., and Orringer, E. P. (1981) Effect of cetiedil, an *in vitro* antisickling agent, on erythrocyte membrane cation permeability. *J. Clin. Invest.* **68** 1215–1220.

Bkaily, G., Sperelakis, N., Renaud, J.-N., and Payet, M. D. (1985) Apamin, a highly-specific Ca^{2+} blocking agent in heart muscle. *Am. J. Physiol.* **248** H961–H965.

Blatz, A. L., and Magleby, K. L. (1984) Ion conductance and selectivity of single calcium-activated potassium channels in cultured rat muscle. *J. Gen. Physiol.* **84** 1–23.

Blatz, A. L., and Magleby, K. L. (1986) Single apamin-blocked Ca-activated K^+ channels of small conductance in rat cultured skeletal muscle. *Nature* **323** 718–720.

Blatz, A. L., and Magleby, K. L. (1987) Calcium-activated potassium channels. *Trends Neurosci* **10** 463–467.

Blum, R. M., and Hoffman, J. F. (1971) The membrane locus of Ca-stimulated K transport in energy depleted human red blood cells. *J. Membr. Biol.* **6** 315–328.

Bourque, C. W., and Brown, D. A. (1987) Apamin and *d*-tubocurarine block the after-hyperpolarization of rat supraoptic neurosecretory neurons. *Neurosci. Lett.* **82** 185–190.

Bülbring, E., and Tomita, T. (1987) Catecholamine action on smooth muscle. *Pharmacol. Rev.* **39** 49–96.

Burgess, G. M., Claret, M., and Jenkinson, D. H. (1981) Effects of quinine and apamin on the calcium-dependent potassium permeability of mammalian hepatocytes and red cells. *J. Physiol.* **317** 67–90.

Capiod, T., and Ogden, D. C. (1985) Noise analysis of α-adrenergic activated K-conductance in isolated guinea-pig hepatocytes. *J. Physiol.* **369** 107P.

Capiod, T., and Ogden, D. C. (1987) The properties of single apamin-sensitive potassium ion channels in guinea-pig hepatocytes. *J. Physiol.* **394** 46P.

Capiod, T., and Ogden, D. C. (1988) Analysis by voltage clamp and single channel recording of noradrenaline action on isolated guinea-pig hepatocytes. In: *Gastrointestinal and hepatic secretions: mechanism and control*, pp. 74–79, J. S. Davison and E. A. Shaffer (eds). University of Calgary Press, Calgary.

Capiod, T., and Ogden, D. C. (1989a) Properties of membrane ion conductances evoked by hormonal stimulation of guinea-pig and rabbit isolated hepatocytes. *Proc. Roy. Soc. B* **236** 187–201.

Capiod, T., and Ogden, D. C. (1989b) The properties of calcium-activated potassium ion channels in guinea-pig isolated hepatocytes. *J. Physiol.* **409** 285–295.

Castle, N. A., and Strong, P. N. (1986) Identification of two toxins from scorpion (*Leiurus quinquestriatus*) venom which block distinct classes of calcium-activated potassium channel. *FEBS Lett.* **209** 117–121.

Castle, N. A., Haylett, D. G., and Jenkinson, D. H. (1989) Toxins in the characterization of potassium channels. *Trends Neurosci.* **12** 59–65.

Chay, T. R. (1986) On the effect of the intracellular calcium-sensitive K^+ channel in the bursting pancreatic β-cell. *Biophys. J.* **50** 765–777.

Chicchi, G. G., Gimenez-Gallego, G., Ber, E., Garcia, M. L., Winquist, R., and Cascieri, M. A. (1988) Purification and characterization of a unique, potent inhibitor of apamin binding from *Leiurus quinquestriatus hebraeus* venom. *J. Biol. Chem.* **263** 10192–10197.

Coats, R. A. (1983) Effects of apamin on α-adrenoceptor-mediated changes in plasma potassium in guinea-pigs. *Br. J. Pharmacol.* **80** 573–580.

Coats, R. A. (1985). The effects of adrenoceptor agonists and antagonists on plasma potassium concentration in anaesthetized guinea-pigs, rabbits and rats. *Br. J. Pharmacol.* **86** 827–836.

Cook, D. L., Ikeuchi, M., and Fujimoto, W. Y. (1984) Lowering of pH_i inhibits Ca^{2+}-activated K^+ channels in pancreatic β-cells. *Nature* **311** 269–271.

Cook, N. S. (1988) The pharmacology of potassium channels and their therapeutic potential. *Trends Pharmacol. Sci.* **9** 21–28.

Cook, N. S., and Haylett, D. G. (1983) Effects of tubocurarine and other neuromuscular blocking agents on Ca-activated K-channels in guinea-pig liver cells. *Br. J. Pharmacol.* **80** 546P.

Cook, N. S., and Haylett, D. G. (1985) Effects of apamin, quinine and neuromuscular blockers on calcium-activated potassium channels in guinea-pig hepatocytes. *J. Physiol.* **358** 373–394.

Cornejo, M., Guggino, S. E., and Guggino, W. B. (1987) Modification of Ca^{2+}-activated K^+ channels in cultured medullary thick ascending limb cells by *N*-bromoacetamide. *J. Membr. Biol.* **99** 147–155.

Deitmer, J. W., and Eckert, R. (1985) Two components of Ca-dependent potassium current in identified neurones of *Aplysia californica*. *Pflügers Arch.* **403** 353–359.

DePeyer, J. E., Cachelin, A. B., Levitan, I. B., and Reuter, H. (1982) Ca^{2+}-activated K^+ conductance in internally perfused snail neurons is enhanced by protein phosphorylation. *Proc. Natl. Acad. Sci. USA* **79** 4207–4211.

Dun, N. J., Jiang, Z. G., and Mo, N. (1986) Tubocurarine suppresses slow calcium-dependent after-hyperpolarization in guinea-pig inferior mesenteric ganglion cells. *J. Physiol.* **375** 499–514.

Eisenman, G., Latorre, R., and Miller, C. (1986) Multi-ion conduction and selectivity in the high-conductance Ca^{++}-activated K^+ channel from skeletal muscle. *Biophys. J.* **50** 1025–1034.

Ewald, D. A., and Levitan, I. B. (1987) Ion channels regulated by calcium. In: *Neuromodulation: the biochemical control of neuronal excitability,* pp. 138–158. L. K. Kaczmarek and I. B. Levitan (eds). Oxford University Press, New York.

Ewald, D. A., Williams, A., and Levitan, I. B. (1985) Modulation of single Ca^{2+}-dependent K^+-channel activity by protein phosphorylation. *Nature* **315** 503–506.

Farley, J., and Rudy, B. (1988) Multiple types of voltage-dependent Ca^{2+}-activated K^+ channels of large conductance in rat brain synaptosomal membranes. *Biophys. J.* **53** 919–934.

Field, A. C., and Jenkinson, D. H. (1987) The effect of noradrenaline on the ion permeability of isolated mammalian hepatocytes, studied by intracellular recording. *J. Physiol.* **392** 493–512.

Filippov, A. K., Skatova, G. E., Porotikov, V. I., Orlov, B. N., and Asafova, N. N. (1988) Effect of the neurotoxin apamin on ion currents in membranes of heart fibres. *Biofizika* **33** 109–112.

Gallacher, D. V., Maruyama, Y., and Petersen, O. H. (1984) Patch-clamp study of rubidium and potassium conductances in single cation channels from mammalian exocrine acini. *Pflügers Arch.* **401** 361–367.

García-Sancho, J., Sanchez, A., and Herreros, B. (1982) All-or-none response of the Ca^{2+}-dependent K^+ channel in inside-out vesicles. *Nature* **296** 744–746.

Gater, P. R., Haylett, D. G., and Jenkinson, D. H. (1985) Neuromuscular blocking agents inhibit receptor-mediated increases in the potassium permeability of intestinal smooth muscle. *Br. J. Pharmacol.* **86** 861–868.

Glavinović, M. I., and Trifaró, J. M. (1988) Quinine blockade of currents through Ca^{2+}-activated K^+ channels in bovine chromaffin cells. *J. Physiol.* **399** 139–152.

Golowasch, J., Kirkwood, A., and Miller, C. (1986) Allosteric effects of Mg^{2+} on the gating of Ca^{2+}-activated K^+ channels from mammalian skeletal muscle. *J. Exp. Biol.* **124** 5–13.

Gorman, A. L. F., and Thomas, M. V. (1978) Changes in the intracellular concentration of calcium ions in a pace-maker neurone, measured with the metallochromic indicator dye arsenazo III. *J. Physiol.* **275** 357–376.

Gorman, A. L. F., Hermann, A., and Thomas, M. V. (1982a) Ionic requirements for membrane oscillations and their dependence on calcium concentration in a molluscan pace-maker neuron. *J. Physiol.* **327** 185–217.

Gorman, A. L. F., Woolum, J. C., and Cornwall, M. C. (1982b) Selectivity of the Ca^{2+}-activated and light-dependent K^+ channels for monovalent cations. *Biophys. J.* **38** 318–322.

Grinstein, S., Dupre, A., and Rothstein, A. (1982) Volume regulation by human lymphocytes: role of calcium. *J. Gen. Physiol.* **79** 849–868.

Grinstein, S., Rothstein, A., Sarkadi, B., and Gelfand, E. W. (1984) Reponses of lymphocytes to anisotonic media: volume-regulating behavior. *Am. J. Physiol.* **246** C204–C215.

Grinstein, S., Smith, J. D., and Miller, C. (1988) Charybdotoxin inhibits regulatory volume decrease and mitogen- or Ca^{2+} ionophore-induced hyperpolarization in lymphocytes. *Biophys. J.* **53** 148a.

Grygorczyk, R., and Schwarz, W. (1983) Properties of the Ca^{2+}-activated K^+ conductance of human red cells as revealed by the patch-clamp technique. *Cell Calcium* **4** 499–510.

Grygorczyk, R., Schwarz, W., and Passow, H. (1984) Ca^{2+}-activated K^+ channels in human red cells. Comparison of single-channel currents with ion fluxes. *Biophys. J.* **45** 693–698.

Guggino, S. E., Suarez-Isla, B. A., Guggino, W. B., and Sacktor, B. (1985) Forskolin and antidiuretic hormone stimulate a Ca^{2+}-activated K^+ channel in cultured kidney cells. *Am. J. Physiol.* **249** F448–F455.

Habermann, E. (1984) Apamin. *Pharmacol. Ther.* **25** 255–270.

Habermann, E., and Fischer, K. (1979a) Bee venom neurotoxin (apamin): iodine labeling and characterization of binding sites. *Eur. J. Biochem.* **94** 355–364.

Habermann, E., and Fischer, K. (1979b) Apamin, a centrally acting neurotoxic peptide: binding and actions. In: *Advances in cytopharmacology,* Vol. 3, pp. 387–394, B. Ceccarelli and F. Clementi (eds). Raven Press, New York.

Hamill, O. P. (1981) Potassium channel currents in human red blood cells. *J. Physiol.* **319** 97–98P.

Hamill, O. P. (1983) Potassium and chloride channels in red blood cells. In: *Single channel recording,* pp. 451–471, B. Sakmann and E. Neher (eds). Plenum Press New York.

Hazama, A., and Okada, Y. (1988) Ca^{2+}-sensitivity of volume-regulatory K^+ and Cl^--channels in cultured human epithelial cells. *J. Physiol.* **402** 687–702.

Heinz, A., and Passow, H. (1980) Role of external potassium in the calcium-induced potassium efflux from human red blood cell ghosts. *J. Membr. Biol.* **57** 119–131.

Hennessey, T. M., and Kung, C. (1987) A calcium-dependent potassium current is increased by a single-gene mutation in *Paramecium. J. Membr. Biol.* **98** 145–155.

Hermann, A., and Hartung, K. (1983) Ca^{2+} activated K^+ conductance in molluscan neurones. *Cell Calcium* **4** 387–405.

Hermann, A., and Erxleben, C. (1987) Charybdotoxin selectively blocks small Ca-activated K channels in *Aplysia* neurons. *J. Gen. Physiol.* **90** 27–47.

Hoffmann, E. K. (1987) Volume regulation in cultured cells. *Current Topics in Membranes and Transport* **30** 125–180.

Hoffmann, E. K., Simonsen, L. O., and Lambert, I. H. (1984) Volume-induced increase of K^+ and Cl^- permeabilities in Ehrlich ascites tumor cells: role of internal Ca^{2+}. *J. Membr. Biol.* **78** 211–222.

Hoffmann, E. K., Lambert, I. H., and Simonsen, L. O. (1986) Separate, Ca^{2+}-activated, K^- and Cl^- transport pathways in Ehrlich ascites tumor cells. *J. Membr. Biol.* **91** 227–244.

Hugues, M., Romey, G., Duval, D., Vincent, J.-P., and Lazdunski, M. (1982a) Apamin as a selective blocker of the calcium-dependent potassium channel in neuroblastoma cells: voltage clamp and biochemical characterization of the toxin receptor. *Proc. Nat. Acad. Sci. USA* **79** 1308–1312.

Hugues, M., Duval, D., Kitabgi, P., Lazdunski, M., and Vincent, J.-P. (1982b) Preparation of a pure monoiodo derivative of the bee venom neurotoxin apamin and its binding properties to rat brain synaptosomes. *J. Biol. Chem.* **257** 2762–2769.

Hunter, M., and Giebisch, G. (1988) Calcium-activated K^+ channels of *Amphiuma* early distal tubule: inhibition by ATP. *Pflügers Arch.* **412** 331–333.

Jenkinson, D. H., Haylett, D. G., and Cook, N. S. (1983) Calcium-activated potassium channels in liver cells. *Cell Calcium* **4** 429–437.

Jenkinson, D. H., Capiod, T., and Field, A. C. (1988) Effects of hormones on ion movements in hepatocytes. In: *Gastrointestinal and hepatic secretions: mechanisms and control,* pp. 30–36, J. S. Davison and E. A. Shaffer (eds). University of Calgary Press, Calgary.

Klærke, D. A., Petersen, J., and Jørgensen, P. L. (1987) Purification of Ca^{2+}-activated K^+ channel protein on calmodulin affinity columns after detergent solubilization of luminal membranes from outer renal medulla. *FEBS Lett.* **216** 211–216.

Krnjević, K., and Lisiewicz, A. (1972) Injections of calcium ions into spinal motoneurones. *J. Physiol.* **225** 363–390.

Latorre, R. (1986) The large calcium-activated potassium channel. In: *Ion channel reconstitution,* pp. 431–467, C. Miller (ed.). Plenum Press, New York.

Latorre, R., and Miller, C. (1983) Conduction and selectivity in potassium channels. *J. Membr. Biol.* **71** 11–30.

Latorre, R., Vergara, C., and Hidalgo, C. (1982) Reconstitution in planar lipid bilayers of a Ca^{2+}-dependent K^+ channel from transverse tubule membranes isolated from rabbit skeletal muscle. *Proc. Natl Acad. Sic. USA* **79** 805–809.

Lazdunski, M., Romey, G., Schmid-Antomarchi, H., Renaud, J. F., Mourre, C., Hugues, M., and Fosset, M. (1988) The apamin-sensitive Ca^{2+}-dependent K^+ channel: molecular properties, differentiation, involvement in muscle disease, and endogenous ligands in mammalian brain. *Hbk. Exp. Pharmacol.* **83** 135–145.

Lew, V. L., and Ferreira, H. G. (1976) Variable Ca sensitivity of a K^+-selective channel in intact red cell membranes. *Nature* **263** 336–338.

Lew, V. L., and Ferreira, H. G. (1978) Calcium transport and the properties of a calcium-activated potassium channel in red cell membranes. *Current Topics in Membranes and Transport* **10** 217–277.

Lew, V. L., Muallem, S., and Seymour, C. A. (1982) Properties of the Ca^{2+}-activated K^+ channel in one-step inside-out vesicles from human red cell membranes. *Nature* **296** 742–744.

Lux, H. D., Neher, E., and Marty, A. (1981) Single channel activity associated with the calcium dependent outward current in *Helix pomatia. Pflügers Arch.* **389** 293–295.

Maas, A. J. J., Den Hertog, A., Ras, R., and Van Den Akker, J. (1980) The action of apamin on guinea-pig Taenia caeci. *Eur. J. Pharmac.* **67** 265–274.

Marquèze, B., Seagar, M. J., and Couraud, F. (1987) Photoaffinity labeling of the K^+-channel associated apamin-binding molecule in smooth muscle, liver and heart membranes. *Eur. J. Biochem.* **169** 295–298.

Marty, A. (1981) Ca-dependent K channels with large unitary conductance in chromaffin cell membranes. *Nature* **291** 497–500.

Marty, A. (1989) The physiological role of calcium-dependent channels. *Trends Neurosci.* **12** 420–424.

Maruyama, Y., Petersen, O. H., Flanagan, P., and Pearson, G. T. (1983) Quantification of Ca^{2+}-activated K^+ channels under hormonal control in pig pancreas acinar cells. *Nature* **305** 228–232.

McCann, J. D., and Welsh, M. J. (1987) Neuroleptics antagonise a calcium-activated potassium channel in airway smooth muscle. *J. Gen. Physiol.* **89** 339–352.

McKinnon, R., and Miller, C. (1988) Mechanism of charybdotoxin block of the high conductance, Ca^{2+}-activated K^+ channel. *J. Gen. Physiol.* **91** 335–349.

McManus, O. B., and Magleby, K. L. (1985) The large conductance calcium-activated potassium channel in cultured rat muscle has at least three open states of similiar conductance and six shut states. *Biophys. J.* **47** 137a.

McManus, O. B., and Magleby, K. L. (1988) Kinetic states and modes of single large-conductance calcium-activated potassium channels in cultured rat skeletal muscle. *J. Physiol.* **402** 79–120.

Meech, R. W. (1974) The sensitivity of *Helix aspersa* neurones to injected calcium ions. *J. Physiol.* **237** 259–277.

Meech, R. W. (1978) Calcium-dependent potassium activation in nervous tissues. *Ann. Rev. Biophys. Bioeng.* **7** 1–18.

Meech, R. W., and Strumwasser, F. (1970) Intracellular calcium injection activates potassium conductance in *Aplysia* nerve cells. *Fed. Proc.* **29** 834.

Miller, C. (1987) Trapping single ions inside single ion channels. *Biophys. J.* **52** 123–126.

Miller, C., Moczydlowski, E., Latorre, R., and Phillips, M. (1985) Charybdotoxin, a protein inhibitor of single Ca^{2+}-activated K^+ channels from mammalian skeletal muscle. *Nature* **313** 316–318.

Moczydlowski, E., and Latorre, R. (1983) Gating kinetics of Ca^{2+}-activated K^+ channels from rat muscle incorporated into planar lipid bilayers: evidence for two voltage-dependent Ca^{2+} binding reactions. *J. Gen. Physiol.* **82** 511–542.

Moczydlowski, E., Luchesi, K., and Ravindran, A. (1988) An emerging pharmacology of peptide toxins targeted against potassium channels. *J. Membr. Biol.* **105** 95–111.

Mourre, C., Hugues, M., and Lazdunski, M. (1986) Quantitative autoradiographic mapping in rat brain of the receptor of apamin, a polypeptide toxin specific for one class of Ca^{2+}-dependent K^+ channels. *Brain Res.* **382** 239–249.

Nånberg, E., Connolly, E., and Nedergaard, J. (1985) Presence of a Ca^{2+}-dependent K^+ channel in brown adipocytes: possible role in maintenance of α_1-adrenergic stimulation. *Biochim. Biophys. Acta* **844** 42–49.

Nohmi, M., and Kuba, K. (1984) (+)-Tubocurarine blocks Ca^{2+}-dependent K^+ channel of the bullfrog sympathetic ganglion cell. *Brain Res.* **301** 146–148.

Oberhauser, A., Alvarez, O., and Latorre, R. (1988) Activation by divalent cations of a Ca^{2+}-activated K^+ channel from skeletal muscle membranes. *J. Gen.*

Physiol. **92** 67–86.

Olesen, S.-P., Clapham, D. E., and Davies, P. F. (1988) Haemodynamic shear stress activates a K^+ current in vascular endothelial cells. *Nature* **331** 168–170.

Oliveira-Castro, G. M., and Dos Reis, G. A. (1981) Electrophysiology of phagocytic membranes III. Evidence for a calcium-dependent potassium permeability change during slow hyperpolarizations of activated macrophages. *Biochim. Biophys. Acta* **640** 500–511.

Pallotta, B. S. (1985) *N*-Bromoacetamide removes a calcium-dependent component of channel opening from calcium-activated potassium channels in rat skeletal muscle. *J. Gen. Physiol.* **86** 601–611.

Pallotta, B. S., Magleby, K. L., and Barrett, J. N. (1981) Single channel recordings of Ca^{2+}-activated K^+ currents in rat muscle cell culture. *Nature* **293** 471–474.

Pape, L., and Kristensen, B. I. (1984) A calmodulin activated Ca^{2+}-dependent K^+ channel in human erythrocyte membrane inside-out vesicles. *Biochim. Biophys. Acta* **770** 1–6.

Passow, H. (1981) Selective enhancement of potassium efflux from red blood cells by lead. In: *The function of red blood cells: erythrocyte pathobiology,* pp. 79–104, D. F. H. Wallach (ed.). Liss, New York.

Pennefather, P., Lancaster, B., Adams, P. R., and Nicoll, R. A. (1985) Two distinct Ca-activated K currents in bullfrog sympathetic ganglion cells. *Proc. Natl Acad. Sci. USA* **82** 3040–3044.

Petersen, O. H. (1986) Calcium-activated potassium channels and fluid secretion by exocrine glands. *Am. J. Physiol.* **251** G1–G13.

Petersen, O. H., and Findlay, I. (1987) Electrophysiology of the pancreas. *Physiol. Rev.* **67** 1054–1116.

Plishker, G. A. (1984) Phenothiazine inhibition of calmodulin stimulates calcium-dependent potassium efflux in human red blood cells. *Cell Calcium* **5** 177–185.

Putney, J. W. (1979) Stimulus-permeability coupling: role of calcium in the regulation of membrane permeability. *Pharmacol. Rev.* **30** 209–245.

Reed, P. W. (1976) Effects of the divalent cation ionophore A23187 on potassium permeability of rat erythrocytes. *J. Biol. Chem.* **251** 3489–3494.

Reichstein, E., and Rothstein, A. (1981) Effects of quinine on Ca^{++}-induced K^+ efflux from human red blood cells. *J. Membr. Biol.* **59** 57–63.

Reinhart, P. H., and Levitan, I. B. (1987) Modulation of charybdotoxin-sensitive calcium-dependent potassium channels by protein phosphorylation. *Soc. Neurosci. Abs* **13** 1439.

Renaud, J.-F., Desnuelle, C., Schmid-Antomarchi, H., Hugues, M., Serratrice, G., and Lazdunski, M. (1986). Expression of the apamin receptor in muscles of patients with myotonic muscular dystrophy. *Nature* **319** 678–680.

Ribalet, B., Eddlestone, G. T., and Ciani, S. (1988) Metabolic regulation of the K(ATP) and a maxi K(V) channel in the insulin-secreting RINm5F cell. *J. Gen. Physiol.* **92** 219–237.

Rink, T. J., Sanchez, A., Grinstein, S., and Rothstein, A. (1983) Volume restoration in osmotically swollen lymphocytes does not involve changes in free Ca^{2+} concentration. *Biochim. Biophys. Acta* **762** 593–596.

Riordan, J. R., and Passow, H,. (1971) Effects of calcium and lead on potassium permeability of human erythrocyte ghosts. *Biochim. Biophys. Acta* **249** 601–605.

Romey, G., and Lazdunski, M. (1984) The coexistence in rat muscle cells of two distinct classes of Ca^{2+}-dependent K^+ channels with different pharmacological properties and different physiological functions. *Biochem. Biophys. Res. Commun.* **118** 669–674.

Rosario, L. M., and Rojas, E. (1986) Modulation of K^+ conductance by intracellular pH in pancreatic β-cells. *FEBS Lett* **200** 203–209.

Sands, S. B., Lewis, R. S., and Cahalan, M. D. (1988) Charybdotoxin blocks voltage-gated K^+ channels in T lymphocytes. *Biophys. J.* **53** 260a.

Sarkadi, B., and Gárdos, G. (1985) Calcium-induced potassium transport in cell membranes. In: *The enzymes of biological membranes* (2nd edn), Vol. 3, pp. 193–234, A. N. Martonosi (ed.). Plenum Press, New York.

Sarkadi, B., Cheung, R., Mack, E., Grinstein, S., Gelfand, E. W., and Rothstein, A. (1985) Cation and anion transport pathways in volume regulatory response of human lymphocytes to hypoosmotic media. *Am. J. Physiol.* **248** C480–C487.

Sauvé, R., Simoneau, C., Monette, R., and Roy, G. (1986) Single-channel analysis of the potassium permeability in HeLa cancer cells: evidence for a calcium-activated potassium channel of small unitary conductance. *J. Membr. Biol.* **92** 269–282.

Schmid-Antomarchi, H., Hugues, M., Norman, R., Ellory, C., Borsotto, M., and Lazdunski, M. (1984) Molecular properties of the apamin-binding component of the Ca^{2+}-dependent K^+ channel: radiation-inactivation, affinity labelling and solubilization. *Eur. J. Biochem.* **142** 1–6.

Schmid-Antomarchi, H., Renaud, J.-F., Romey, G., Hugues, M., Schmid, A., and Lazdunski, M. (1985) The all-or-none role of innervation in expression of apamin receptor and of apamin-sensitive Ca^{2+}-activated K^+ channel in mammalian skeletal muscle. *Proc. Natl. Acad. Sci. USA* **82** 2188–2191.

Schmid-Antomarchi, H., Hugues, M., and Lazdunski, M. (1986) Properties of the apamin-sensitive Ca^{2+}-activated K^+ channel in PC12 pheochromocytoma cells which hyper-produce the apamin receptor. *J. Biol. Chem.* **261** 8633–8637.

Schwarz, W., and Passow, H. (1983) Ca^{2+}-activated K^+ channels in erythrocytes and excitable cells. *Ann. Rev. Physiol.* **45** 359–374.

Seagar, M. J., Granier, C., and Couraud, F. (1984) Interaction of the neurotoxin apamin with a Ca^{2+}-activated K^+ channel in primary neuronal cultures. *J. Biol. Chem.* **259** 1491–1495.

Seagar, M. J., Labbé-Juillié, C., Granier, C., Van Reitschoten, J., and Couraud, F. (1985) Photoaffinity labelling of components of the apamin-sensitive K^+ channel in neuronal membranes. *J. Biol. Chem.* **260** 3895–3898.

Seagar, M. J., Labbé-Juillié, C., Granier, C., Goll, A., Glossmann, H., Van Rietschoten, J., and Couraud, F. G. (1986) Molecular structure of rat brain apamin receptor: differential photoaffinity labelling of putative K^+ channel subunits and target size analysis. *Biochemistry* **25** 4051–4057.

Sherman, A., Rinzel, J., and Keizer, J. (1988) Emergence of organized bursting in clusters of pancreatic β-cells by channel sharing. *Biophys. J.* **54** 411–425.

Shields, M., Grycorczyk, R., Fuhrmann, G. F., Schwarz, W., and Passow, H. (1985) Lead-induced activation and inhibition of potassium-selective channels in the human red blood cell. *Biochim. Biophys. Acta* **815** 223–232.

Simons, T. J. B. (1976) Calcium-dependent potassium exchange in human red cell ghosts. *J. Physiol.* **256** 227–244.

Simons, T. J. B. (1979) Actions of a carbocyanine dye on calcium-dependent potassium transport in human red cell ghosts. *J. Physiol.* **288** 481–507.

Simons, T. J. B. (1985) Influence of lead ions on cation permeability in human red cell ghosts. *J. Membr. Biol.* **84** 61–71.

Smart, T. G. (1987) Single calcium-activated potassium channels recorded from cultured rat sympathetic neurones. *J. Physiol.* **389** 337–360.

Stampe, P., and Vestergaard-Bogind, B. (1985) The Ca^{2+}-sensitive K^+-conductance of the human red cell membrane is strongly dependent on cellular pH. *Biochim. Biophys. Acta* **815** 313–321.

Strong, P. N., and Evans, W. H. (1987) Receptor-mediated endocytosis of apamin by liver cells. *Eur. J. Biochem.* **163** 267–273.

Suzuki, K., and Petersen, O. H. (1988) Patch-clamp study of single-channel and whole-cell K^+ currents in guinea-pig pancreatic acinar cells. *Am. J. Physiol.* **255** G275–G285.

Tabares, L., Lopez-Barneo, J., and de Miguel, C. (1985) Calcium- and voltage-activated potassium channels in adrenocortical cell membranes. *Biochim. Biophys. Acta* **814** 96–102.

Valdeolmillos, M., García-Sancho, J., and Herreros, B. (1982) Ca^{2+}-dependent K^+ transport in the Ehrlich ascites tumor cell. *Biochim. Biophys. Acta* **685** 273–278.

Valdivia, H. H., Smith, J. S., Martin, B. M., Coronado, R., and Possani, L. D. (1988) Charybdotoxin and noxiustoxin, two homologous peptide inhibitors of the K^+ (Ca^{2+}) channel. *FEBS Lett.* **226** 280–284.

Vergara, C., and Latorre, R. (1983) Kinetics of Ca^{2+}-activated K^+ channels from rabbit muscle incorporated into planar bilayers. Evidence for a Ca^{2+} and Ba^{2+} blockade. *J. Gen. Physiol.* **82** 543–568.

Vestergaard-Bogind, B., Stampe, P., and Christophersen, P. (1985) Single-file diffusion through the Ca^{2+}-activated K^+ channel of human red cells. *J. Membr. Biol.* **88** 67–75.

Vincent, J.-P., Schweitz, H., and Lazdunski, M. (1975) Structure–function relationships and site of action of apamin, a neurotoxic polypeptide of bee venom with an action on the central nervous system. *Biochemistry* **14** 2521–2525.

Wong, B. S., Lecar, H., and Adler, M. (1982) Single calcium-dependent potassium channels in clonal anterior pituitary cells. *Biophys. J.* **39** 313–317.

Wong, S. M. E., and Chase, H. S. (1986) Role of intracellular calcium in cellular volume regulation. *Am. J. Physiol.* **250** C841–C852.

Yellen, G. (1984) Ionic permeation and blockade in Ca^{2+}-activated K^+ channels of bovine chromaffin cells. *J. Gen. Physiol.* **84** 157–186.

Yellen, G. (1987) Permeation in potassium channels: implications for channel structure. *Ann. Rev. Biophys. Biophys. Chem.* **16** 227–246.

Yingst, D. R., and Hoffman, J. F. (1984) Ca-induced K transport in human red blood cell ghosts containing arsenazo III: transmembrane interactions of Na, K, and Ca and the relationship to the functioning Na–K pump. *J. Gen. Physiol.* **83** 19–45.

Wolff, D., Cecchi, X., Spalvins, A., and Canessa, M. (1988) Charybdotoxin blocks with high affinity the Ca^{2+}-activated K^+ channel of HbA and HbS red cells: individual differences in the number of channels. *J. Membr. Biol.* **106** 243–252.

4

Biophysics and physiology of ATP-regulated K^+ channels (K_{ATP})

P. Rorsman and **G. Trube**

4.1 INTRODUCTION

When pancreatic β-cells are stimulated by glucose, they depolarize and give action potentials. Today there is substantial evidence that the initial depolarization is mediated by the closure of a type of K^+ channel which is inhibited by intracellular ATP — the ATP-regulated K^+ channels (K_{ATP}) (Ashcroft *et al.*, 1984, Cook and Hales, 1984). Channels with similar properties have also been demonstrated to exist in myocardial cells (Noma, 1983; Trube and Hescheler, 1984), skeletal muscle cells (Spruce *et al.*, 1985) and neurones (Ashford *et al.*, 1988), but it is only in the β-cells where the physiological importance (regulation of insulin secretion) of these channels is obvious (Petersen and Findlay, 1987; Ashcroft, 1988). This review will therefore concentrate on K_{ATP} in insulin-secreting cells when discussing the physiological role(s) of the channels, and possible functions in other cell types will be mentioned only briefly. The biophysical properties of K_{ATP} are similar in the various cells and are considered in general. The pharmacology of K_{ATP} in insulin-secreting cells is dealt with in Chapter 11.

4.2 DISTRIBUTION AND DENSITY OF K_{ATP}

K_{ATP} have been observed in β-cells from several species (rat: Cook and Hales, 1984; Ashcroft *et al.*, 1984; mouse: Rorsman and Trube, 1985; guinea-pig: Rorsman and Hellman, 1988) including man (Ashcroft *et al.*, 1987b; Misler *et al.*, 1989a). A convenient way of obtaining cells is to use insulin-secreting cell lines such as RINm5F cells. These cell lines contain K_{ATP} in numbers similar to those encountered in normal pancreatic β-cells. Cell lines may therefore be an acceptable choice for studies concerning the basic properties of the channel. When exploring the mechanisms underlying normal glucose-induced stimulus–secretion coupling, cell lines are less suitable as most of them (if not all) do not respond to glucose stimulation with an enhancement of insulin secretion. Patch-clamp results obtained from normal β-cells also allow a comparison to be made with membrane potential recordings from β-cells in intact islets (review: Henquin and Meissner, 1984).

The activity of K_{ATP} in cell-attached patches on β-cells is very variable. Whereas some patches contain several active channels (seen as superpositions of several simultaneous openings), others display no or very little activity. From our own experience there are two ways of increasing the chances of finding channel openings in a cell-attached patch. First, more channel activity is observed when recordings are made from multicellular clusters than from single β-cells. Secondly, channel activity is higher when the cells have been cultured at 5 mM of glucose (i.e. normal serum glucose) instead of 11 mM, the standard glucose concentration in many tissue media (e.g. RPMI 1640).

To evaluate the possibility of purifying the K_{ATP} protein it is necessary to estimate the number of K_{ATP} per cell. This can be derived by measuring the increase of the cell conductance (G_c) following washout of intracellular ATP. Knowing the single-channel conductance (γ) and the open probability of the channel (P_{open}) the number of channels (N) in a β-cell can be estimated from the equation:

$$N = G_c/(\gamma \times P_{open})$$

With values of G_c of maximally 20 nS (Rorsman and Trube, 1985), γ of 20 pS (Trube *et al.*, 1986) and P_{open} of 0.1–0.2 (own unpublished results; Cook *et al.*, 1988), a value of N corresponding to 5000–10 000 channels/cell is derived. Taking the β-cell to be a sphere with a diameter of 13 μm this would correspond to 10–20 channels/μm^2. This is comparable to the density in skeletal muscle (Spruce *et al.*, 1985) but higher than in heart cells (0.5/μm^2; Noma and Shibasaki, 1985).

4.3 BIOPHYSICAL PROPERTIES

4.3.1 Single-channel conductance

The single-channel conductance (γ) is most reliably measured in excised patches where full control of the membrane potential and the composition of the solutions on both sides of the membrane are ensured. When measured in symmetrical solutions containing ≈150 mM K^+, current flow is inward at negative potential. Under these circumstances, values of γ of 50–65 pS have been reported (Cook and Hales, 1984; Rorsman and Trube, 1985; Misler *et al.*, 1986). Similar values for the inward current have been obtained in cell-attached patches (Ashcroft *et al.*, 1984; Rorsman and Trube, 1985). In the cell-attached patches the channel shows pronounced inward rectification and current amplitudes increase very little at voltages above +20 mV (Ashcroft *et al.*, 1984; Arkhammar *et al.*, 1987). This behaviour results from internal Mg^{2+}, Na^+ and Ca^{2+} blocking outward currents (Findlay, 1987a, Horie *et al.*, 1987). In fact, the degree of rectification has been used to estimate the cytosolic concentrations of Na^+ and Mg^{2+} (Ciani and Ribalet, 1988). The blocking action of Mg^{2+} and Na^+ is due to the binding of these ions within the channel (see section 4.3.2). With normal ionic gradients ($[K^+]_i$=150 mM; $[K^+]_o$≈5 mM) the current flow is *outward* within the physiological range of membrane potentials and a conductance of 27 pS is observed in the absence of internal Mg^{2+} and Na^+ (Findlay, 1987a). Again the conductance is lower in cell-attached patches and amounts to 13–15 pS (Misler *et al.*, 1986; Arkhammar *et al.*, 1987). This figure is similar to the value of γ observed in outside-out patches with an $[Na^+]_i$ of 10 mM and an $[Mg^{2+}]_i$ of 1.1 mM (Findlay, 1987a; Ciani and Ribalet, 1988). Conductance states other than those presented above, which is by far the most dominant, have been observed (Kakei *et al.*, 1985).

4.3.2 K_{ATP} permeability does not obey independence

K_{ATP} discriminates strongly against external Na^+, and in physiological solutions containing 140 mM Na^+ the reversal potential approximates the K^+ equilibrium potential (Rorsman and Trube, 1985, Trube *et al.*, 1986). The relative permeability (P_{Na}/P_K) is similar in the various cell types and ranges between 0.007 in β-cells (Ashcroft *et al.*, 1989) and 0.015 in frog sarcolemma (Spruce *et al.*, 1987). Within a limited voltage range and at a particular $[K^+]_o$, the single-channel current–voltage relationships can be reasonably approximated to the Goldman–Hodgkin–Katz (GHK) equation. When the recordings are made with pipettes containing 140 mM K^+, values for the single-channel potassium permeability (P_K) of $0.86–1.05\times10^{-13}$ cm^3s^{-1} are obtained in β-cells (Ashcroft *et al.*, 1988a; Ashcroft *et al.*, 1989). A similar figure (1.4×10^{-13} cm^3s^{-1} at 30–36°C) has been reported for P_K in cardiac ventricular myocytes (Kakei *et al.*, 1985). However, higher values of P_K are obtained when it is measured with low external $[K^+]$. For instance, in skeletal muscle cells a value for P_K of 2.6×10^{-13} cm^3 s^{-1} and a single-channel conductance of 15 pS were obtained in $[K^+]_o$ of 2.5 mM (Spruce *et al.*, 1987). If K_{ATP} obeyed independence this value would predict a single-channel conductance of 82.6 pS in a $[K^+]_o$ of 60 mM, which is about twice the conductance recorded experimentally (42.3 pS; Spruce *et al.*, 1987). This indicates that K_{ATP} saturates with increasing $[K^+]_o$. Similar observations have been made in β-cells (Ashcroft *et al.*, 1988a). The properties of saturation and voltage-dependent block are inconsistent with the GHK equation, suggesting that ion flow through the channel does not obey independence, but are to be expected if ion permeation through the channel involves the binding to site(s) within the channel protein. To account for the experimental findings models with three energy barriers and two energy wells (binding sites) have been proposed (Spruce *et al.*, 1987; Ciani and Ribalet, 1988). Experiments on the block of K_{ATP} by extracellular Cs^+ and Ba^{2+} suggest that K_{ATP} is a multi-ion pore (Quayle *et al.*, 1988). The binding sites have affinities to K^+ of 290 mM and 85 mM, the lower value corresponding to the site closer to the 'cytoplasm'. The inner (high-affinity) site is postulated to bind Na^+ and Mg^{2+} more strongly than K^+. Affinity constants for Na^+ and Mg^{2+} of 46 mM and 34 mM, respectively, have been calculated (Ciani and Ribalet, 1988; Horie *et al.*, 1987). In addition, at least one of the energy barriers (probably the outer) must be very high for Na^+ and Mg^{2+} to account for the low permeabilities of these ions. In this context it is of interest that the Na^+ permeability, relative to that of potassium, is ≈50 times higher for internal Na^+ than for external (Ashcroft *et al.*, 1989). K_{ATP} is also permeable to Rb^+ and the relative permeability (P_{Rb}/P_K), determined from the shift in reversal potential, is close to 0.7 in skeletal muscle and β-cells (Spruce *et al.*, 1987; Ashcroft *et al.*, 1989). However, the single-channel currents recorded using Rb^+ as the charge carrier are much smaller than those observed with K^+. The fact that the P_{Rb} is similar to P_K indicates that the barrier heights for Rb^+ can only be slightly higher than they are for K^+. To account for the small unitary inward currents and the block of the outward current, the dominant energy well has to be deeper for Rb^+ than K^+ (1.64 times; Spruce *et al.*, 1987). When exposed to mixtures of Rb^+ and K^+, K_{ATP} shows anomalous mole-fraction effects (Spruce *et al.*, 1986). It should be understood that as ^{86}Rb binds more tightly to the channel than K^+, it is an acceptable probe only for estimating relative changes of P_K (Ashcroft *et al.*, 1989). Previous attempts to derive the magnitude of the β-cell membrane K^+ conductance from ^{86}Rb flux measurements probably have yielded falsely low values (Dawson *et al.*, 1983, but cf. Dawson *et al.*, 1986).

4.3.3 Voltage dependence

$I_{K(ATP)}$ observed in whole-cell experiments is not voltage-dependent as indicated by linear current–voltage relationships and time-independent current responses to voltage steps (Rorsman and Trube 1985; Trube and Rorsman, 1987). Recordings of the single-channel currents show that openings of the channel are grouped in bursts (see Fig. 4.1C). The lifetimes of the bursts and the shut periods between the bursts are quite variable from patch to patch. Most studies have therefore been limited to analyses of the openings and the shut intervals within the bursts. For high extracellular $[K^+]$ ($\geq$60 mM) and at potentials around −70 mV, the mean lifetime of the closures within bursts (τ_c) ranges between 0.3 and 0.7 ms irrespective of the cell type and patch configuration (Trube and Hescheler, 1984; Rorsman and Trube, 1985; Spruce *et al.*, 1985; Ashcroft *et al.*, 1988a, Zilberter *et al.*, 1988). Because of the short duration of τ_c, part of the closures within the bursts are too short to be detected (see Fig. 4.1C), with resultant apparent lengthening of the duration of the channel openings. The observed distribution of the open times will therefore depend on the frequency resolution of the analysis. Mean open times (τ_o) between 1.8 and 2.8 ms have been reported for the conditions specified above. The dependence of K_{ATP} kinetics on membrane voltage and K^+ concentration has been studied most extensively in inside-out patches of the heart cell membrane (Zilberter *et al.*, 1988). The results indicate that the gating of the channel depends on the electromotive force rather than the absolute membrane potential. τ_o approaches a maximum at the K^+ equilibrium (E_K) and decreases again for more positive voltages. τ_c shows the reverse relationship with a minimum at E_K.

4.4 EFFECTS OF ATP ON K_{ATP}

4.4.1 Inhibition of K_{ATP} does not require ATP hydrolysis

The open-state probability of K_{ATP} is reduced in the presence of ATP at the intracellular face of the membrane. Half maximal inhibition is attained with 15–50 μM ATP in experiments on inside-out patches excised from β-cells (Cook and Hales, 1984; Ohno-Shosaku *et al.*, 1987), 135 μM ATP applied to skeletal muscle membrane patches (Spruce *et al.*, 1985) and 100 μM added to patches from heart cells (Noma, 1983). Millimolar concentrations of ATP cause complete channel inhibition (Fig. 4.1B). It is unclear whether several molecules of ATP bind to the channel cooperatively. Concentration–inhibition relationships with Hill coefficient close to 1 have been reported for β-cells (Cook and Hales, 1984; Kakei *et al.*, 1986) and skeletal muscle (Spruce *et al.*, 1987) but values of 1.5–1.8 for β-cells (Ohno-Shosaku

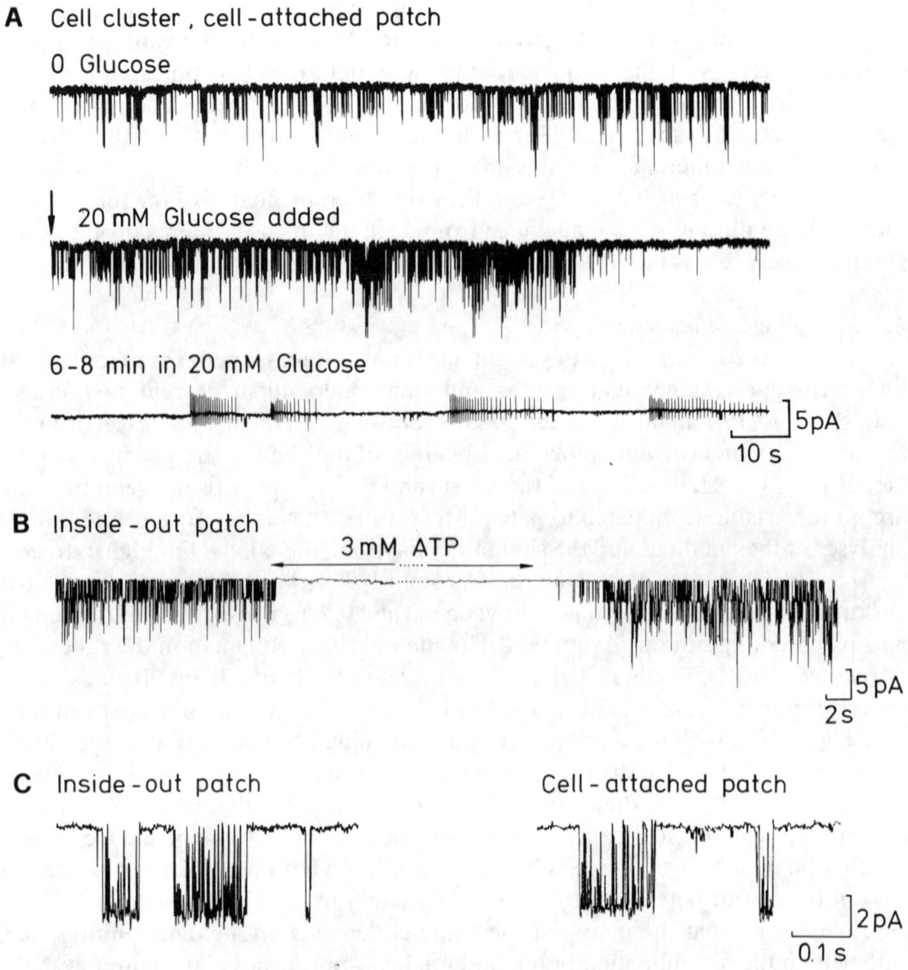

Fig. 4.1 — (A) Channel activity seen in a cell-attached patch on a β-cell on small cluster in the absence of glucose (1st trace), immediately after addition of 20 mM glucose (2nd trace) and 6–8 min after glucose addition (3rd trace). Note increase in channel activity immediately after addition of glucose (see section 4.6.1) and inhibition of channel activity and appearance of bursts of biphasic action 'potentials' after 6 min. (B) Effect of 3 mM Mg–ATP on channel activity in an inside-out patch. Channel activity is completely inhibited. Note that more activity is obtained after removal of ATP ('refreshment'; see section 4.4.4). (C) Examples of channel openings observed in a cell-attached patch in the absence of glucose (right) and in an inside-out patch in the absence of ATP (left). Note similarity of channel openings.

et al., 1987; Ribalet and Ciani, 1987) and up to 4 for cardiac myocytes (Kakei *et al.*, 1985) have also been obtained. The inhibitory effect of ATP does not require ATP to be metabolized, and non-hydrolysable analogues such as AMP–PCP, AMP–PNP and ATPγS are about as effective as ATP itself (Cook and Hales, 1984; Kakei *et al.*, 1985; Spruce *et al.*, 1985; Misler *et al.*, 1986; Ohno-Shosaku *et al.*, 1987). ADP has weak inhibitory effects with 50% block at 0.44 mM in skeletal muscle (Spruce *et al.*, 1987) and 0.8–2 mM in heart and insulin-secreting cells (Noma, 1983; Kakei *et al.*, 1986; Ribalet and Ciani, 1987; see Findlay, 1988b for Mg dependence of ADP inhibition). Inhibition by AMP is hardly detectable except in skeletal muscel (Spruce *et al.*, 1987). Other purines (GTP, ITP, UTP, etc.) have also been reported to reduce K_{ATP} activity in heart (Kakei *et al.*, 1985) and skeletal muscle (Spruce *et al.*, 1985).

4.4.2 Effects of ATP on K_{ATP} kinetics

The influence of ATP on K_{ATP} kinetics has been studied most extensively in inside-out patches from skeletal muscle cells (Spruce *et al.*, 1985, 1987). Attempts to carry out similar studies with isolated patches of heart and β-cell membrane were hampered by the difficulty that a decrease of ATP often causes only transient activation of the channel ('run-down', see section 4.4.4). There is qualitative agreement, however, that the channel openings are grouped in bursts in all cell types and that the dominant effect of ATP is a decrease in the number of openings per burst. The duration of the long (≈100 ms) closed periods between the bursts is also increased by ATP (Spruce *et al.*, 1985, 1987; Isenberg *et al.*, 1988). A similar effect was seen when the bath glucose concentration was increased while recording the current from a β-cell in the cell-attached configuration (Ashcroft *et al.*, 1988a).

Histograms of the channel closed times have to be fitted by sums of at least three exponentials, and a fourth exponential is sometimes needed to account for the very long closed times (>1 s). The fastest exponential (time constant $\tau_c \approx 0.3$ ms at −60 mV and $[K^+]_o \geq 60$ mM) accounts for about 95% of all events (Spruce *et al.*, 1985, 1987; Isenberg *et al.*, 1988). These are the gaps within the bursts, which are not or only weakly ATP-dependent. Histograms of channel open times in the absence of ATP sometimes can be fitted reasonably well by a single exponential. In the presence of ATP, however, sums of two exponentials are needed to obtain a good fit (Spruce *et al.*, 1985, 1987; Isenberg *et al.*, 1988). The main effect of ATP on the open times is an increase in the percentage of events belonging to the more rapid component. The relative increase is probably due to the fact that ATP suppresses the open state of longer average lifetime more strongly than the state of short lifetime (Isenberg *et al.*, 1988). The results above indicate that K_{ATP} is able to occupy at least two open and three to four closed states already in the absence of ATP. More work has to be done to elaborate a reaction scheme showing how these states may be interconnected and where ATP interferes with the transitions between the channel's states.

4.4.3 Modulation of K_{ATP} ATP sensitivity by ADP

The concentrations necessary to produce 50% inhibition of channel activity in inside-out patches are much lower than the concentration of ATP supposed to exist in the cytoplasm. The *total* β-cell content ranges from 3–4 mM in the absence of glucose to 5–6 mM in the presence of >5 mM glucose (Ashcroft *et al.*, 1973). However, it is known that much of the ATP is stored in organelles. The *cytosolic* concentration has been estimated using a digitonin permeabilization technique, and a value of 0.5 mM was obtained in the absence of glucose (Malaisse and Sener, 1987). Cell permeabilization by saponin or digitonin has also been used to control the intracellular ATP concentration while recording channel activity from cell-attached patches.

This technique avoids the disruption of the investigated piece of membrane from the rest of the cell, which may destroy functionally important links between membrane proteins and the cytoskeleton (Noma and Shibasaki, 1985). Using permeabilized heart cells, Kakei *et al.* (1985) found 50% inhibition of K_{ATP} by 0.5 mM ATP instead of the 0.1 mM observed in excised inside-out patches. A similar shift of the ATP sensitivity has also been reported in insulinoma cells (Dunne *et al.*, 1986a). The blocking potency of ATP also appears to be reduced when the resting K^+ conductance is recorded from a whole pancreatic β-cell. When comparing currents measured with different ATP concentrations in the pipette, 50% inhibition is again observed at about 0.5 mM (Panten *et al.*, 1989 and own unpublished observations). A similar figure was derived from simultaneous measurements of $^{86}Rb^+$ efflux and intracellular ATP levels in RINm5F cells during metabolic inhibition (Schmid-Antomarchi *et al.*, 1987). The different sensitivities of K_{ATP} to ATP obtained using the various patch-clamp configurations may reflect the presence of intracellular modulatory factors which are washed away in the inside-out patch configuration. One such factor has been identified. When both the ratio and concentrations of ATP and ADP approximating those encountered intracellularly in β-cells are used, channel activity can be recorded from inside-out patches even in the presence of millimolar concentrations of ATP (Kakei *et al.*, 1986; Dunne and Petersen, 1986a; Dunne *et al.*, 1988b). Changes in cytosolic ATP/ADP ratio are more pronounced than changes in the ATP concentration (Kakei *et al.*, 1986; Malaisse and Sener, 1987). The site involved in the activation by ADP is probably distinct from that of inhibition (Bokvist *et al.*, 1988; Findlay, 1988a). A model has been proposed in which oscillations in glycolysis, which are seen in permeabilized RINm5F cells and cell-free muscle extracts, result in variations in the ATP/ADP ratio. This in turn is suggested to modulate K_{ATP} activity and thus the bursts of electrical activity (Corkey *et al.*, 1988a, b).

4.4.4 ATP must be hydrolysed for K_{ATP} function

In most inside-out patches channel activity decays rapidly ('run-down') after patch excision if ATP is not present (Trube and Hescheler, 1984; Findlay and Dunne, 1986; Misler *et al.*, 1986; Ohno-Shosaku *et al.*, 1987). Most of the initial channel activity reappears after the channel has been blocked by ATP again for a few minutes ('refreshment effect'; Misler *et al.*, 1986). This action of ATP requires the presence of Mg^{2+} and non-hydrolysable ATP analogues are unable to mimick the 'refreshing' action of ATP (Ohno-Shosaku *et al.*, 1987; Findlay, 1987a). This finding suggests that ATP must be hydrolysed (channel protein phosphorylation?) in order to keep the channel is a functional state. This may explain why more channel activity is observed when the channel is exposed to a very low concentration of ATP (0.1–1 μM) than in its absence (F. M. Ashcroft, personal communication).

4.5 OTHER INTRACELLULAR MODULATORS OF K_{ATP}

4.5.1 Pyridine nucleotides

During glucose stimulation of β-cells not only the ATP/ADP ratio but also the redox ratio is increased (Panten and Ishida, 1975). Both the reduced and the oxidized forms of the pyridine nucleotides NADP and NAD lead to channel activation when applied at low concentrations (10–100 μM) and inhibition at higher concentrations (>500 μM; Dunne *et al.*, 1988a). The reduced forms produce somewhat larger effects, but the importance of this difference is unclear since the cytoplasmic concentrations of ATP, ADP and the pyridine nucleotides are not known with sufficient accuracy. It has been suggested that NAD contributes to the tonic inhibition of the ATP-regulated K^+ channel in the absence of glucose (Dunne *et al.*, 1988a).

4.5.2 GTP and GDP

Low concentrations of intracellular GDP, GTP and other guanosine base nucleotides lead to an increase in channel activity (Dunne and Petersen, 1986b; Findlay, 1987a; Zünkler *et al.*, 1988b). This may provide a link between the K_{ATP} and the G-proteins (see also section 4.7). The effects of GTP and GDP (inhibitory or activating) depend on the Mg^{2+} concentration (Findlay, 1987a). At the millimolar Mg^{2+} concentrations which are believed to exist in the β-cell only the activating properties will be expressed.

4.5.3 Cations

Although not being glucose metabolites in the true sense, the concentrations of the cations H^+, Ca^{2+} and possibly Mg^{2+} also change during glucose stimulation (Rorsman *et al.*, 1984; Arkhammar *et al.*, 1986). The reduction of outwardly directed single-channel currents by intracellular Mg^{2+} and Ca^{2+} has been described above (section 4.3.1). In addition, these alkali earth cations reduce the open-state probability of K_{ATP} in the absence of ATP (Findlay 1987a, b). The effects of Mg^{2+} and Sr^{2+} are immediately reversible, but the action of Ca^{2+} probably has to be explained by an acceleration of channel run-down, since it can be reversed only by the application of Mg–ATP (Findlay, 1987b; see section 4.4.4). Mg^{2+} also reduces the efficacy whereby ATP blocks the channel in insulin-secreting cells. Since Mg^{2+} binds to ATP^{4-}, the latter effect suggests that the channel is closed only by the tetravalent nucleotide, but not by the Mg^{2+} salt (Ashcroft and Kakei, 1989; Dunne *et al.*, 1987). ATP^{4-} at a concentration of 4 μM was found to inhibit K_{ATP} activity by 50%, whereas a 7 times higher concentration was necessary to produce a comparable effect in the presence of Mg^{2+} (Ashcroft and Kakei, 1989). However, in heart cells, both the free acid and the Mg^{2+} salt seem to block K_{ATP} (Findlay, 1988b).

Although originally reported not to be influenced by cytoplasmic pH (Cook and Hales, 1984), more recent evidence suggests that the channel in fact is sensitive to variations in intracellular pH. Addition of NH_4Cl (which transiently makes the cytoplasm more alkaline) leads to a transient two- to fourfold increase of channel activity. Upon removal of NH_4Cl (which transiently makes cytoplasm more acidic) channel activity drops below basal values (Misler *et al.*, 1988). Whereas glucose simulation results in a slow, sustained and moderate increase in the cytosolic pH in pancreatic β-cells, the glucose analogue D-glyceraldehyde leads to a rapid decrease in pH (Arkhammar *et al.*, 1986). It is possible that part of the inhibitory effect of D-glyceraldehyde on K_{ATP} is attributable to its effect on cytoplasmic pH. Such a notion would be compatible with the rapid onset of the action of D-glyceraldehyde (see section 4.6.3).

4.5.4 Bicarbonate and internal Cl^-

To date almost all the patch-clamp experiments on pancreatic β-cells have been performed in the presence of HEPES-buffered media lacking bicarbonate. It has recently been demonstrated in cell-attached patches that K_{ATP} is blocked in the presence of bicarbonate (Carroll *et al.*, 1988). Addition of HCO_3^- is known to induce a transient acidification of the cytoplasm and, as discussed above (section 4.5.3), a reduction in pH_i reduces K_{ATP} activity. Therefore, it seems possible that the cells were not allowed enough time to reach a new steady state in the presence of HCO_3^-. The fact that long recovery periods were necessary (10–20 min; Carroll *et al.*, 1988) upon removal of HCO_3^- for K_{ATP} to reappear supports this interpretation.

Up to now experiments on K_{ATP} in inside-out patches have always been carried out with solutions containing Cl^- as the main anion. Recent experiments using glutamate$^-$ or aspartate$^-$ suggest that the open-state probability is doubled upon removal of Cl^-, whereas ATP sensitivity and single-channel conductance are unaffected (Trube *et al.*, 1989).

4.5.5 Protein kinases

It has recently been demonstrated that D-glyceraldehyde induces the closure of the K_{ATP} in RINm5F cells before there is a significant increase of the ATP content, and the role of ATP as the signalling substance in fuel-stimulated insulin secretion has therefore been questioned (Wollheim *et al.*, 1988). During the same period of time the cell content of 1,2-diacylglycerol (an activator of protein kinase C) was increased. It was therefore suggested that phosphorylation of K_{ATP} via diacylglycerol-induced activation of protein kinase C causes the closure of the channel. In support of this idea, application of the exogenous protein kinases C activators PMA (12-phorbol-myristate 13-acetate) and DC_{10} (1,2-didecanoylglycerol) also inhibited K_{ATP}. It is an open question whether these findings can be extended to explain the effect of glucose in normal β-cells. In these cells, ATP does change with sufficient speed to account for the inhibition of K_{ATP} (Malaisse *et al.*, 1979; Malaisse and Herchuelz, 1982). Furthermore, in a different study on the same cell line, the protein kinase C activator TPA was found to *increase* the activity of K_{ATP} (Ribalet *et al.*, 1988).

4.6 EFFECTS OF FUEL METABOLITES ON CHANNEL ACTIVITY IN INSULIN-SECRETING CELLS

4.6.1 Effects of glucose

In the absence of glucose, K_{ATP} dominates the resting conductance of the β-cell. When the input conductance is measured *immediately* after disrupting the patch membrane in a cell not exposed to glucose, values of up to 2 nS (at +20°C) are obtained†. This is comparable to the 4 nS obtained in the absence of glucose and at +37°C in input conductance measurements from intact islets using intracellular electrodes (Smith, 1988). Radioisotipic measurements have revealed that glucose reduces ^{86}Rb efflux in a concentration-dependent manner with an IC_{50} of about

† This current reverses at about −70 mV and rapidly decreases to <100 pS when the cell is dialysed with an intracellular solution containing 3 mM ATP, excluding the possibility that the above value for the input conductance primarily reflects leak.

3mM (Malaisse and Herchuelz, 1982). Fig. 4.2 shows the effects of glucose on the activity of the ATP-regulated K^+ channel in a cell-attached patch on a mouse β-cell. Fifty per cent inhibition is again obtained at about 2 mM (cf. Misler *et al.*, 1986; Ashcroft *et al.*, 1988a). At a physiological glucose concentration (5 mM), channel activity is reduced by >90%. However, the remaining active channels are sufficient to keep the β-cell hyperpolarized and action potentials are not elicited†. This result suggests that a substantial reduction of channel activity (≈99%) is required to produce a significant depolarization (Trube *et al.*, 1986; Arkhammar *et al.*, 1987). An increase of the glucose concentration to >10 mM results in almost complete inhibition of the channel, leading to depolarization and initiation of spiking (Ashcroft *et al.*, 1984).

It is generally accepted that the glucose-induced inhibition of K_{ATP} in β-cells is a consequence of metabolism of the sugar. However, the identity of the 'signal' still remains to be established. Circumstantial evidence that ATP is the intracellular regulator is provided by the following observations. (i) Activity of the channel in cell-attached patches increases in the presence of metabolic inhibitors which lower $[ATP]_i$ (Ribalet and Ciani, 1987). (ii) In normal pancreatic β-cells changes in the activity of K_{ATP} and $[ATP]_i$ occur in the same concentration range of glucose and with a similar time course (Ashcroft *et al.*, 1988a; Malaisse *et al.*, 1979). (iii) Immediately after addition of glucose there is sometimes rather an increase than a decrease of channel activity. (Arkhammer *et al.*, 1987; Ribalat *et al.*, 1988; see Fig. 4.1A). This is an agreement with previous membrane potential measurements from intact islets which have demonstrated that the input resistance is transiently reduced after an increase of the glucose concentration (Atwater *et al.*, 1978). The increase in channel activity is coincident with a reduction in the islet ATP content, possibly reflecting the consumption of ATP in the early glycolytic steps (Malaisse *et al.*, 1979). Given the evidence suggesting that ATP may indeed be the intracellular regulator, how can this role be reconciled with the discrepancy between the ATP sensitivity of K_{ATP} and the intracellular ATP concentration? It has been pointed out that the density of K_{ATP} is high in the β-cell membrane (section 4.2) and that only a fraction of these channels has to be active to account for the low K^+ permeability of the intact cell (Rorsman and Trube, 1985). Starting from this conclusion, Cook *et al.* (1988) recently estimated the expected relationship between membrane potential and intracellular ATP concentration. Their model predicts an approximately 300-fold shift to higher ATP concentrations of the membrane potential relationship relative to the dose–response curve for channel inhibition. A system of membrane potential regulation based on a large number of channels operating near the fully closed state ensures maximal sensitivity to (metabolically induced) changes in channel activity at a minimal expenditure of energy to maintain the ionic gradients (Cook *et al.*, 1988).

4.6.2 Effects of α-ketoisocaproate

α-Ketoisocaproate (KIC) is an interesting substance as it is primarily metabolized by the mitochondria (Panten *et al.*, 1974). When this substrate (20 mM) was added to

† Inhibition of channel activity in pancreatic β-cells by glucose concentrations of 5 mM or less is not associated with a decrease in the single-channel amplitude, which would be expected if the cell depolarized.

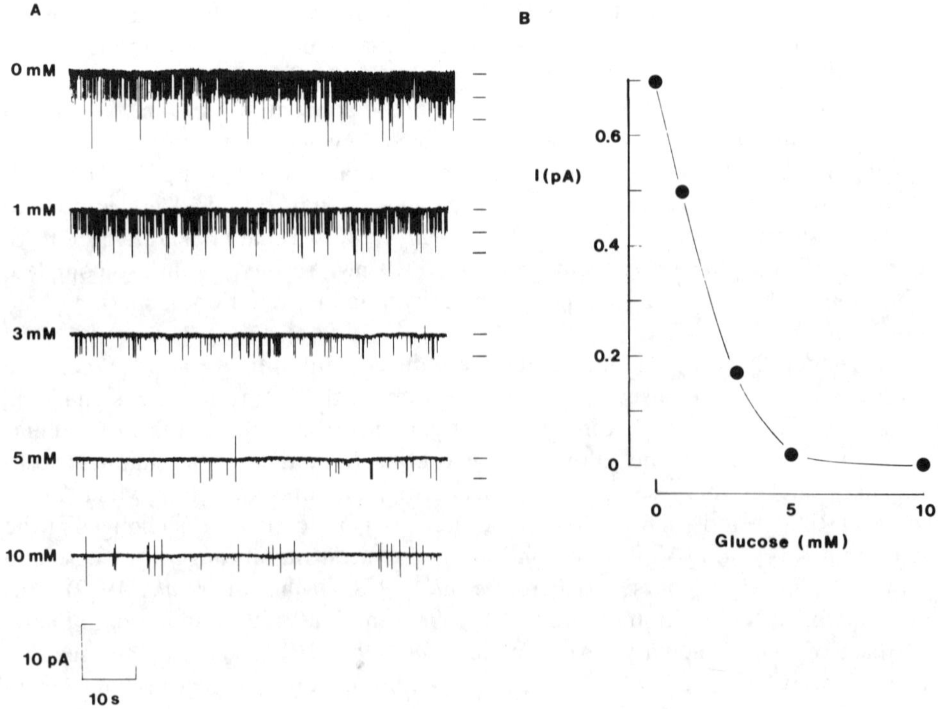

Fig. 4.2 — (A) Channel activity observed in a cell-attached patch on a β-cell in a small cluster the absence of glucose and 7 min after the respective addition of 1, 3, 5 and 10 mM glucose. All records from same cell. (B) Channel activity plotted against the respective glucose concentration. Channel activity was estimated by integration of the current during the last 2 min at each glucose concentration. Half maximal inhibition was attained at about 2 mM. In the presence of 10 mM glucose the cell is producing biphasic current deflections which correspond to action potentials.

the bath while recording currents from cell-attached patches, K_{ATP} was strongly but not completely inhibited (Ashcroft *et al.*, 1987a). This would appear to be in agreement with observations made in heart cells, where glycolysis was more effective than oxidative phosphorylation in promoting the closure of K_{ATP} (Weiss and Lamp, 1987). The effects of KIC on channel kinetics are similar to those of glucose and are manifest as a decrease in the burst duration and increase in the interburst closures (Ashcroft *et al.*, 1987a).

4.6.3. Effects of D-glyceraldehyde

D-Glyceraldehyde at a concentration of 5 mM blocks K_{ATP} completely in cell-attached patches on rat β-cells (Misler *et al.*, 1986). In RINm5F cells, 10 mM glyceraldehyde reduced channel activity by ≈90% and initiated action potentials. The effects in the RINm5F cells developed within 20 s (Dunne *et al.*, 1986), which is much faster than the 2–6 min required for the effects of glyceraldehyde (Misler *et al.*, 1986; own unpublished results) and glucose to develop (Ashcroft *et al.*, 1984; Rorsman and Trube, 1985; Misler *et al.*, 1986) in normal β-cells. This suggests that D-glyceraldehyde in RINm5F cells may produce the closure of K_{ATP} by a pathway unrelated to that utilized by glucose in normal β-cells (see sections 4.5.3 and 4.5.5).

4.7 EFFECTS OF HORMONES AND TRANSMITTERS ON K_{ATP}

Galanin, somatostain and adrenaline inhibit insulin release, hyperpolarize the β-cell and lower cytosolic free Ca^{2+} (Cook and Perara, 1982; Ahrén *et al.*, 1986; Nilsson *et al.*, 1988, 1989). The actions of these compounds, which are normally present within the islet, are strikingly similar (Nilsson *et al.*, 1989) and it is therefore reasonable to speculate that they share a common intracellular mechanism. Recent studies have demonstrated that the effects of these compounds are abolished subsequent to pretreatment of the cells with pertussis toxin, indicating the involvement of GTP-binding protein(s). It is therefore worthy of note that K_{ATP} can be activated by intracellular GTP (Dunne and Petersen, 1986b). The hyperglycaemic sulphonamide diazoxide affects membrane potential and $[Ca^{2+}]_i$ in a way reminiscent of clonidine, somatostatin and galanin. The action of diazoxide is due to activation of K_{ATP} (Trube *et al.*, 1986; Dunne *et al.* 1987; Sturgess *et al.*, 1988). It is therefore attractive to postulate that inhibition of insulin secretion by hormones and neurotransmitters also involves the opening of K_{ATP}. However, such a notion is difficult to reconcile with observations that they do not seem to stimulate ^{86}Rb from pancreatic islets (Ismail *et al.*, 1983; Tamagawa and Henquin, 1983; Ahrén *et al.*, 1988)†. It was recently demonstrated that galanin, applied extracellularly during inside-out patch recordings, maintains K_{ATP} in an open state even in the presence of ATP (De Weille *et al.*, 1988). A similar effect has been observed in cell-attached patches on mouse β-cells, and galanin promoted the opening of K_{ATP}, which had previously been inhibited by 20 mM glucose (Ahrén *et al.*, 1989) but the effect was small and difficult to reproduce. However, a small change in the open probability of K_{ATP} may well be sufficient to explain the repolarization induced by galanin. If galanin produces an *increase* in K_{ATP} open probability comparable to the *decrease* obtained when raising the glucose from the subthreshold concentration of 5 mM (channel block: 95%) to the stimulatory 10 mM (channel block: 99%), then the change will not be easily observable and hardly noticeable in radioisotopic flux measurements using $^{86}Rb^+$.

4.8 K_{ATP} AND POSSIBLE IMPLICATIONS FOR DIABETES

The fact that sulphonylureas, which are selective blockers of the K_{ATP} (Trube *et al.*, 1986; Schmid-Antomarchi *et al.*, 1987; Sturgess *et al.*, 1988; Zünkler *et al.*, 1988a), are effective in the treatment of certain types of human diabetes suggests that either the channel protein itself of its regulation may be defective in diabetic individuals. Human β-cells obtained from non-diabetic organ donors are equipped with K_{ATP} similar to that observed in rodent β-cells (Ashcroft *et al.*, 1987b; Misler *et al.*, 1989). However, pancreatic islets can rarely be obtained from diabetic patients. A way of circumventing the difficulty in obtaining diabetic β-cells is to study animal or tissue culture models. Fetal β-cells respond with an attenutated insulin secretion when

† cf. De Weille *et al.* (1988) for effects of galanin on ^{86}Rb efflux from RINm5F cells.

stimulated by glucose (Freinkel *et al.*, 1984) but secrete when $[Ca^{2+}]_i$ is raised by increasing external K^+ (Hole *et al.*, 1988; Rorsman *et al.*, 1989). As in other types of insulin-secreting cells, openings of K_{ATP} are seen in fetal β-cells in the absence of glucose. However, in contrast to normal adult cells, K_{ATP} is not inhibited by the sugar. The fetal β-cells are consequently not depolarized by glucose action potentials as well as secretion are not initiated. The channels are rapidly blocked by the sulphonylurea tolbutamide and display the same sensitivity to intracellular ATP as do their adult counterparts (Rorsman *et al.*, 1989). In the fetal β-cells the lack of glucose sensitivity can probably be explained by an immaturity of glucose metabolism (Rorsman *et al.*, 1989). One can speculate that a disturbance of the energy metabolism in the β-cell is a possible cause also of human type II diabetes. It is of interest to note that ATP-consuming reactions (i.e. futile glucose cycles) are enhanced in diabetic liver cells (Efendic *et al.*, 1985). It should be emphasized that as an almost complete inhibition of channel activity is required before a marked depolarization is obtained (Arkhammar *et al.*, 1987; Cook *et al.*, 1988), even small disturbances of the cellular ATP metabolism may have profound effects on the electrical and secretory activities of the insulin-secreting β-cell.

4.9 K_{ATP} IN OTHER CELLS

4.9.1 Heart cells

The cardiac action potential loses its long plateau phase during hypoxia or inhibition of the energy metabolism by cyanide or uncouplers of oxidative phosphorylation (Isenberg *et al.*, 1983; see references therein for previous work). Normal cells from heart atrium and ventricle have a high K^+ conductance at the resting potential (≈ -80 mV), but outward currents around 0 mV, i.e. in the range of the action potential plateau, are quite small due to an N-shaped curvature of the cells' current–voltage relationship ('inward rectification'). Much larger outward currents develop, however, during metabolic inhibition and the current–voltage relationship becomes almost linear under extreme experimental conditions (Fig. 6 in Isenberg *et al.*, 1983; Fig. 3 in Noma and Shibasaki, 1985). The increase in outward current shortens the action potential and decreases influx of Ca^{2+}, leading to a reduced contraction. Activation of an additional outward K^+ current may consequently provide an energy-sparing mechanism in heart cells during conditions of metabolic insufficiency. It has been shown that K_{ATP} produces the additional outward current observed in the presence of metabolic poisons (Noma, 1983; Trube and Hescheler, 1984; Kakei *et al.*, 1985; Noma and Shibasaki, 1985) but the evidence that these channels are also activated during hypoxia and ischaemia is less direct. However, the sulphonylurea glibenclamide† was recently found to reduce the K^+ efflux from whole heart during ischaemia (Kantor *et al.*, 1987). Furthermore, the arrhythmias, which are frequently seen in ischaemic hearts, are prevented by glibenclamide (Kantor *et al.*, 1987).

Biochemical assays of ATP in dissociated cardiac cells and analyses of whole

† Sulphonylureas have been shown to block K_{ATP} in insulin-secreting cells as well as heart cells (Trube *et al.*, 1986; Belles *et al.*, 1987; Fosset *et al.*, 1988).

hearts by NMR spectroscopy suggest that total intracellular ATP decreases only very slowly during hypoxia (≈10% with the first 10 min; Piper *et al.*, 1984; Allen *et al.*, 1985). As 50% inhibition of K_{ATP} was already found with 0.1 mM ATP applied to inside-out patches (Noma, 1983; Trube and Hescheler, 1984) the slow decrease in ATP content seems insufficient to explain the activation of K_{ATP}. However, it was possible to resolve this apparent discrepancy by single-channel recordings from permeabilized cells and whole-cell experiments showing that activation of K_{ATP} starts already at ⩽2 mM ATP when the patch is not disrupted from the cell (Kakei *et al.*, 1985; Noma and Shibasaki, 1985; see section 4.4.3). Part of the lower ATP sensitivity in these experiments may be due to the generation of ADP and the channel-activating effects of this nucleotide (Findlay, 1988a). Other possible reasons for the differences between inside-out patches have already been discussed (Noma and Shibasaki, 1985; see section 4.4.3). In this context it is interesting to note that permeabilized cells are still able to metabolize glycolytic and mitochondrial substrates. Using this preparation, it was recently shown that K_{ATP} in heart cells is more tightly controlled by glycolytic than mitochondrial ATP production (Weiss and Lamp, 1987). This agrees with earlier findings that high concentrations of glucose prevent the activation of outward K^+ current even in the presence of inhibitors of mitochondrial metabolism (Isenberg *et al.*, 1983; see therein for older literature).

4.9.2 Skeletal muscle

The K^+ permeability of skeletal muscle fibres also increases during metabolic fatigue (Fink *et al.*, 1983; see therein for older literature) and excitability is consequently reduced. It has been suggested that the increased membrane conductance is caused by activation of Ca^{2+}-dependent K^+ channels, but the change in intracellular $[Ca^{2+}]_i$ during metabolic inhibition appeared insufficient to explain all of the effect. An additional modulatory influence by the fall of intracellular ATP was therefore postulated (Fink *et al.*, 1983). K_{ATP} was later discovered in sarcolemma of adult frog and mouse skeletal muscle (Spruce *et al.*, 1985, 1987; Woll *et al.*, 1988) and it seems likely that these channels mediate at least part of the K^+ conductance increase seen during metabolic inhibition.

4.9.3 Neurones

The existence of K_{ATP} in the brain is suggested by the finding that, in addition to membranes from insulin-secreting cells, brain membranes are the only other preparation binding hypoglycaemic sulphonylureas (see section 4.8) with high affinity and significant capacity (Geisen *et al.*, 1985; Lupo and Bataille, 1987). Recently, the activation of K^+ currents during the removal of intracellular ATP has been directly shown by patch-clamp experiments on tissue-cultured neurones from embryonal and neonatal brain of mouse and rat (Trussel and Jackson, 1987, Fig. 9; Ashford *et al.*, 1988). The single-channel recordings indicate, however, that the properties of these channels (conductance, ATP sensitivity) are different from those in heart, pancreatic β-cells and skeletal muscle (Ashford *et al.*, 1988). Further studies showing the distribution of K_{ATP} in adult brain have to be awaited for understanding of the possible roles of these channels in neurones.

4.10 PERSPECTIVES

Only six years have elapsed since the discovery of K_{ATP} (Noma, 1983). With the aid of the patch-clamp techniques and its different recording modes a detailed biophysical and electrophysiological characterization has been possible. Taking advantage of the selective ligands provided by the hypoglycaemic sulphonylureas (Trube *et al.*, 1986; Niki *et al.*, 1989; Zünkler *et al.*, 1988a), the molecular characterization of the ATP channel has recently begun. Using glibenclamide it has been possible to identify of sulphonylurea-binding protein with a molecular weight of 140 kDa which may correspond to K_{ATP} (Kramer *et al.*, 1988). It has also been shown that a tolbutamide-sensitive K^+ channel is expressed in *Xenopus laevis* oocytes after injection with poly(A^+) mRNA from the insulin-secreting cell line HIT T15. (Ashcroft *et al.*, 1988b). Several groups have embarked on the project of cloning K_{ATP}. However, the density of K_{ATP} is too low to permit a conservative strategy for purification and sequencing of the protein. Recent data suggest that there may exist a superfamily of K^+ channels with great structural homology (Tempel *et al.*, 1988; Pongs *et al.*, 1988). Using a cDNA probe for an already sequenced K^+ channel may consequently represent a more feasible way for isolation of the DNA coding for K_{ATP}.

ACKNOWLEDGEMENTS

We thank Drs F. M. Ashcroft and P. A. Smith for reading and commenting on the manuscript.

REFERENCES

Ahrén, B., Arkhammar, P., Berggren, P.-O., and Nilsson, T. (1986) Galanin inhibits glucose-stimulated insulin release by a mechanism involving hyperpolarization and lowering of cytosolic free Ca^{2+} concentration. *Biochem. Biophys. Res. Commun.* **140** 1059–1063.

Ahrén, B., Rorsman, P., and Berggren, P.-O. (1988) Galanin and the endocrine pancreas. *FEBS Lett.* **229** 233–237.

Ahrén, B., Berggren, P.-O., Bokvist, K., and Rorsman, P. (1989) Does galanin inhibit insulin secretion by opening the ATP-regulated K^+ channels in the β-cell? *Peptides* **10** 453–457.

Allen, D. G., Morris, P. G., Orchard, C. H., and Pirolo, J. S. (1985) A nuclear magnetic resonance study of metabolism in the ferret heart during hypoxia and inhibition of glycolysis. *J. Physiol. (Lond.)* **361** 185–204.

Arkhammar, P., Berggren, P.-O., and Rorsman, P. (1986) Direct evidence for opposite effects of D-glucose and D-glyceraldehyde on cytoplasmic pH of mouse pancreatic β-cells. *Biosci. Rep.* **6** 355–361.

Arkhammar, P., Nilsson, T., Rorsman, P., and Berggren, P.-O. (1987) Inhibition of ATP-regulated K^+ channels precedes depolarization-induced increase in cytoplasmic free Ca^{2+} concentration. *J. Biol. Chem.* **262** 5448–5454.

Ashcroft, F. M. (1988) Adenosine-5′-triphosphate-sensitive potassium channels. *Ann. Rev. Neurosci.* **11** 97–118.

Ashcroft, F. M., and Kakei, M. (1989) ATP-sensitive K^+ channels in rat pancreatic β-cells: modulation by ATP and Mg^{2+} ions. *J. Physiol.* **416** 349–367.

Ashcroft, F. M., Harrison, D. E., and Ashcroft, S. J. H. (1984) Glucose induces closure of single potassium channels in isolated rat pancreatic β-cells. *Nature* **312** 446–448.

Ashcroft, F. M., Ashcroft, S. J. H., and Harrison, D. E. (1987a) Effects of 2-ketoisocaproate on insulin release and single channel activity in dispersed rat pancreatic β-cells. *J. Physiol. (Lond)* **385** 517–529.

Ashcroft, F. M., Kakei, M., Kelly, R. P., and Sutton, R. (1987b) ATP-sensitive K^+ channels in human pancreatic B-cells. *FEBS Lett.* **215** 9–12.

Ashcroft, F. M., Ashcroft, S. J. H., and Harrison, D. E. (1988a) Properties of single potassium channels modulated by glucose in rat pancreatic β-cells. *J. Physiol. (Lond.)* **400** 501–527.

Ashcroft, F. M., Ashcroft, S. J. H., Berggren, P.-O., Betszholz, C., Rorsman, P., Trube, G., and Welsh, M. (1988b) Expression of K channels in *Xenopus laevis* oocytes injected with poly(A^+)mRNA from the insulin-secreting β-cell line, HIT T15. *FEBS Lett.* **239** 185–189.

Ashcroft, F. M., Kakei, M., and Kelly, R. P. (1989) Rubidium and sodium permeability of the ATP-sensitive K^+ channel in single rat pancreatic β-cells. *J. Physiol. (Lond.)* **408** 413–430.

Ashcroft, S. J. H., Weerasinghe, L. C. C., and Randle, P. J. (1973) Interrelationship of islet metabolism, adenosine triphosphate content and insulin release. *Biochem. J.* **134** 223–231.

Ashford, M. L. J., Sturgess, N. C., Trout, N. J., Gardner, N. J., and Hales, C. N. (1988) Adenosine-5′-triphosphate-sensitive ion channels in neonatal rat cultured central neurones. *Pflügers Arch.* **412** 297–304.

Atwater, I., Ribalet, B., and Rojas, E. (1978) Cyclic changes in potential and resistance of the B-cell membrane induced by glucose in islets of Langerhans from mouse. *J. Physiol. (Lond.)* **278** 117–139.

Belles, B., Hescheler, J., and Trube, G. (1987) Changes in membrane currents in cardiac cells induced by long whole-cell recordings and tolbutamide. *Pflügers Arch.* **409** 582–588.

Bokvist, K. Berggren, P.-O., and Rorsman, P. (1988) Evidence that ADP and ATP bind at different sites at the G-channel. *Diabetologia* **30** 473A.

Carroll, P. B., Li, M. X., Rojas, E., and Atwater, I. (1988) The ATP-sensitive potassium channel in pancreatic B-cells is inhibited in physiological bicarbonate buffer. *FEBS Lett.* **234** 208–212.

Ciani, S., and Ribalet, B. (1988) Ion permeation and rectification in ATP-sensitive channels from insulin-secreting cells (RINm5F): Effects of K^+, Na^+ and Mg^{2+}. *J. Membr. Biol.* **103** 171–180.

Cook, D. L., and Hales, C. N. (1984) Intracellular ATP directly blocks K^+ channels in pancreatic B-cells. *Nature* **311** 271–273.

Cook, D. L., and Ikeuchi, M. (1988) Tolbutamide as mimic of glucose on beta-cell electrical activity — ATP-sensitive K^+-channels as common pathway for both stimuli. *Diabetes* **38** 416–421.

Cook, D. L., and Perara, E. (1982) Islet electrical pacemaker response to alpha-adrenergic stimulation. *Diabetes* **31** 985–990.

Cook, D. L., Satin, L. S., Ashford, M. L. J., and Hales, C. N. (1988) ATP-sensitive K^+ channels in pancreatic β-cells. Spare-channel hypothesis. *Diabetes* **37** 495–498.

Corkey, B. E., Deeney, J. T., Glennon, M. C., Matschinsky, F. M., and Prentki,

M. (1988a) Regulation of steady-state free Ca^{2+} levels by the ATP/ADP ratio and orthophosphate in permeabilized RINm5F insulinoma cells. *J. Biol. Chem.* **263** 4247–4253.

Corkey, B. E., Tornheim, K., Deeney, J. T., Glennon, M. C., Parker, J. C., Matschinsky, F. M., Ruderman, N. B., and Prentki, M. (1988b) Linked oscillations of free Ca^{2+} and the ATP/ADP ratio in permeabilized RINm5F insulinoma cells supplemented with a glycolyzing cell-free muscle extract. *J. Biol. Chem.* **263** 4254–4258.

Dawson, C. M., Croghan, P. C., Atwater, I., and Rojas, E. (1983) Estimation of potassium permeability in mouse islets of Langerhans. *Biomed. Res* **4** 389–392.

Dawson, C. M., Croghan, P. C., Scott, A. M., and Bangham, J. A. (1986) Potassium and rubidium permeability and potassium conductance of the β-cell membrane in mouse islet of Langerhans. *Quarterly J. Exp. Physiol.* **71** 205–222.

De Weille, J., Schmid-Antomarchi, H., Fosset, M., and Lazdunski, M. (1988) ATP-sensitive K^+ channels that are blocked by hypoglycemia-inducing sulfonylureas in insulin-secreting cells are activated by galanin, a hyperglycemia-inducing hormone. *Proc. Natl. Acad. Sci. USA* **85** 1312–1316.

Dunne, M. J., and Petersen, O. H. (1986a) Intracellular ADP activates K^+ channels that are inhibited by ATP in an insulin-secreting cell line. *FEBS Lett.* **208** 58–62.

Dunne, M. J. and Petersen, O. H. (1986b) GTP and GDP activation of K^+ channels that can be inhibited by ATP. *Pflügers Arch.* **407** 564–565.

Dunne, M. J., Findlay, I., Petersen, O. H., and Wollheim, C. B. (1986) ATP-sensitive K^+ channels in an insulin-secreting cell line are inhibited by D-glyceraldehyde and activated by membrane permeabilization. *J. Membr. Biol.* **93** 271–279.

Dunne, M. J., Illot, M. C., and Petersen, O. H. (1987) Interaction of diazoxide, tolbutamide and ATP^{4-} on nucleotide-dependent K^+-channels in an insulin-secreting cell line *J. Membr. Biol.* **99** 215–224.

Dunne, M. J., Findlay, I., and Petersen, O. H. (1988a) Effects of pyridine nucleotides on the gating of ATP-sensitive potassium channels in insulin-secreting cells. *J. Membr. Biol.* **102** 205–216.

Dunne, M. J., West-Jordan, J. A., Abraham, R. J., Edwards, R. H. T., and Petersen, O. H. (1988b) The gating of nucleotide-sensitive K^+ channels in insulin-secreting cells can be modulated by changes in the ratio ATP^{4-}/ADP^{3-} and by nonhydrolyzable derivatives of both ATP and ADP. *J. Membr. Biol.* **104** 165–177.

Efendic, S., Wajngot, A., and Vranic, M. (1985) Increased activity of the glucose cycle in the liver: Early characteristic of type 2 diabetes. *Proc. Natl. Acad. Sci. USA* **82** 2965–2969.

Findlay, I. (1987a) The effects of magnesium upon adenosine triphosphate-sensitive potassium channels in a rat insulin-secreting cell line. *J. Physiol. (Lond.)* **391** 611–629.

Findlay, I. (1987b) ATP-sensitive K^+ channels in rat ventricular myocytes are blocked and inactivated by internal divalent cations. *Pflügers Arch.* **410** 313–320.

Findlay, I. (1988a) Effects of ADP upon the ATP-sensitive K^+ channel in rat ventricular myocytes. *J. Membr. Biol.* **101** 83–92.

Findlay, I. (1988b) ATP^{4-} and ATP.Mg inhibit the ATP-sensitive K^+ channel of rat ventricular myocytes. *Pflügers Arch.* **412** 37–41.

Findlay, I., and Dunne, M. J. (1986) ATP maintains ATP-inhibited K^+ channels in an operational state. *Pflügers Arch.* **407** 238–240.

Fink, R., Hase, S., Lüttgau, H. C., and Wettwer, E. (1983) The effects of cellular energy reserves and internal calcium ions on the potassium conductance in skeletal muscle of the frog. *J. Physiol. (Lond.)* **336** 211–218.

Fosset, M., De Weille, J. R., Green, R. D., Schmid-Antomarchi, H., and Lazdunski, M. (1988) Antidiabetic sulfonylureas control action potential properties in heart cells via high affinity receptors that are linked to ATP-dependent K^+ channels. *J. Biol. Chem.* **263** 7933–7936.

Freinkel, N., Lewis, N. J., Johnson, R., Swenne, I., Bone, A., and Hellerström, C. (1984) Differential effects of age versus glycemic stimulation on the maturation of insulin stimulus-secretion coupling during culture of fetal rat islets. *Diabetes* **33** 1028–1038.

Geisen, K., Hitzel, V., Oekomonopoulos, R., Pünter, J., Weyer, R., and Summ, H. D. (1985) Inhibition of ^{3}H-glibencamide binding to sulfonylurea receptors by oral antidiabetics. *Arzneimittelforsch./Drug Res.* **35** 707–712.

Henquin, J. C., and Meissner, H. P. (1984) Significance of ionic fluxes and changes in membrane potential for stimulus–secretion coupling in pancreatic B-cells. *Experientia* **40** 1043–1054.

Hole, R. L., Pian-Smith, M. C., and Sharp, G. W. G. (1988) Development of the biphasic response to glucose in fetal and neonatal rat pancreas. *Am. J. Physiol.* **254** E167–E178.

Horie, M., Irisawa, H., and Noma, A. (1987) Voltage-dependent magnesium block of adenosine-triphosphate-sensitive potassium channel in guinea-pig ventricular cells. *J. Physiol. (Lond.)* **387** 251–272.

Isenberg, G., Vereecke, J., van der Heyden, G., and Carmeliet, E. (1983) The shortening of the action potential by DNP in guinea-pig ventricular myocytes is mediated by an increase of time-independent K conductance. *Pflügers Arch.* **397** 251–259.

Isenberg, C., Schröter, K., and Trube, G. (1988) Kinetics of single ATP-dependent K^+-channels in the membrane of murine pancreatic β-cells. *Pflügers Arch.* **411** R111.

Ismail, N. A., El-Denshary, E. S. M., Idahl, L.-Å., Lindström, P., Sehlin, J., and Täljedal, I.-B. (1983) Effects of alpha-adrenoreceptor agonists and antagonists on insulin secretion, calcium uptake, and rubidium efflux in mouse pancreatic islets. *Acta Physiol. Scand.* **118** 167–174.

Kakei, M., Noma, A., and Shibasaki, T. (1985) Properties of adenosine-triphosphate-regulated potassium channels in guinea-pig ventricular cells. *J. Physiol. (Lond.)* **363** 441–462.

Kakei, M., Kelly, R. P., Ashcroft, S. J. H., and Ashcroft, F. M. (1986) The ATP-sensitivity of K^+ channels in rat pancreatic B-cells is modulated by ADP. *FEBS Lett.* **208** 63–66.

Kantor, P. F., Coetzee, W. A., Dennis, S. C., and Opie, L. H. (1987) Effects of glibenclamide on ischemic arrythmias. *Circulation* **76** (Suppl. IV) 17.

Kramer, W., Oekonomopulos, R., Pünter, J., and Summ, H. D. (1988) Direct photoaffinity labeling of the putative sulfonylurea receptor in rat β-cell tumor membranes by [^{3}H]glibenclamide. *FEBS Lett.* **229** 355–359.

Lupo, B., and Bataille, D. (1987) A binding site for [^{3}H] glipizide in the rat cerebral cortex. *Eur. J. Pharmacol.* **140** 157–169.

Malaisse, W. J., and Herchuelz, A. (1982) Nutritional regulation of K^+ conductance: An unsettled aspect of pancreatic B-cell physiology. In: *Biochemical actions of hormones*, Vol. IX, pp. 69–92, G. Litwack (ed.). Academic Press, New York.

Malaisse, W. J., and Sener, A. (1987) Glucose-induced changes in cytosolic ATP content in pancreatic islets. *Biochim. Biophys. Acta* **927** 190–195.

Malaisse, W. J., Hutton, J. C., Kawazu, S., Herchuelz, A., Valverde, I., and Sener, A. (1979) The stimulus secretion coupling of glucose-induced insulin release. XXXV. The links between metabolic and cationic events. *Diabetologia* **16** 331–341.

Misler, S., Falke, L. C., Gillkis, K., and McDaniel, M. L. (1986) A metabolite-regulated potassium channel in rat pancreatic B-Cells. *Proc. Natl. Acad. Sci. USA* **83** 7119–7123.

Misler, S., Gee, W. M., Gillis, K. D., Scharp, D. W., and Falke, L. C. (1989a) A metabolite-regulated potassium channel in human pancreatic islet cells. *Diabetes* **38** 422–427.

Misler, S., Gillis, K., and Tabcharani, J. (1989b) Modulation of gating of a metabolically regulated, ATP-dependent K^+ channel by intracellular pH in B-cells of the pancreatic islet. *J. Membrane Biol.* **109** 135–143.

Niki, I., Kelly, R. P., Ashcroft, S. J. H., and Ashcroft, F. M. (1989) ATP-sensitive K-channels in HITT15 β-cells studied by patch clamp methods, ^{85}Rb efflux and glibenclanide binding. *Pflügers Arch.* **415** 47–55.

Nilsson, T., Arkhammar, P., Rorsman, P., and Berggren, P.-O. (1988) Inhibition of glucose-stimulated insulin release by α_2-adrenoceptor activation is paralleled by both a repolarization and a reduction in cytoplasmic free Ca^{2+} concentration. *J. Biol. Chem.* **263** 1855–1860.

Nilsson, T., Arkhammar, P., Rorsman, P., and Berggren, P.-O. (1989) Suppression of insulin release by galanin and somatostatin is mediated by G-protein. An effect involving repolarization and reduction in cytoplasmic free Ca^{2+} concentration. *J. Biol. Chem.* **264** 973–980.

Noma, A. (1983) ATP-regulated K^+ channels in cardiac muscle. *Nature* **305** 147–148.

Noma, A., and Shibasaki, T. (1985) Membrane currents through adenosine-triphosphate-regulated potassium channels in guinea-pig ventricular cells. *J. Physiol. (Lond.)* **363** 463–480.

Ohno-Shosaku, T., Zünkler, B. J., and Trube, G. (1987) Dual effects of ATP on K^+ currents of mouse pancreatic β-cells. *Pflügers Arch.* **408** 133–138.

Panten, U., and Ishida, H. (1975) Fluorescence of oxidized flavoproteins from perifused isolated pancreatic islets. *Diabetologia* **11**, 569–573.

Panten, U., Christians, J., Kriegstein, E., Poser, W., and Hasselblatt, A. (1974)

Studies on the mechanisms of L-leucine and α-ketoisocaproic acid induced insulin release from perifused isolated pancreatic islets. *Diabetologia* **10** 149–154.

Panten, U., Heipel, C., Lins, S., and Zünkler, B. J. (1989) Activity of ATP-sensitive K^+ channels in B-cells is regulated by ATP utilization in the vicinity of the plasma membrane. *Diabetologia* **32** 527A.

Petersen, O. H., and Findlay, I. (1987) Electrophysiology of the pancreas. *Physiol. Rev.* **67** 1054–1116.

Piper, H. M., Hütter, J. F., and Spieckermann, P. G. (1984) Relation between enzyme release and metabolic changes in reversible anoxic injury of myocardial cells. *Life Sci.* **35** 127–134.

Pongs, O., Kecskemethy, N., Müller, R., Krah-Jentgenf, I., Baumann, A., Kiltz, A. A., Canal, I., Llamazares, S., and Ferrus, A. (1988) Shaker encodes a family of putative potassium channel proteins in the nervous system of *Drosphilia. Embo J.* **7** 1087–1096.

Quayle, J. M., Standen, N. B., and Stanfield, P. R. (1988) The voltage-dependent block of ATP-sensitive potassium channels of frog skeletal muscle by caesium and barium ions. *J. Physiol. (Lond.)* **405** 677–697.

Ribalet, B. and Ciani, S. (1987) Regulation by cell metabolism and adenine nucleotides of a K channel in insulin-secreting B cells (RINm5F). *Proc. Natl. Acad. Sci. USA* **84** 1721–1725.

Ribalet, B., Eddlestone, G. T., and Ciani, S. (1988) Metabolic regulation of the K(ATP) and maxi-K(V) channel in the insulin-secreting RINm5F-cell. *J. Gen. Physiol.* **92** 219–237.

Rorsman, P., and Hellman, B. (1988) Voltage-activated currents in guinea pig pancreatic α_2 cells. *J. Gen. Physiol.* **91** 223–242.

Rorsman, P., and Trube, G. (1985) Glucose-dependent K^+-channels in pancreatic β-cells are regulated by intracellular ATP. *Pflügers Arch* **405** 305–309.

Rorsman, P., Abrahamsson, H., Gylfe, E., and Hellman, B. (1984) Dual effects of glucose on cytosolic Ca^{2+} activity of mouse pancreatic β-cells. *FEBS Lett.* **170** 196–200.

Rorsman, P., Arkhammar, P., Bokvist, K., Hellerström, C., Nilsson, T., Welsh, M., Welsh, N., and Berggren, P.-O. (1989) Failure of glucose to elicit a normal secretory response in fetal pancreatic beta-cells results from glucose-insensitivity of the ATP-regulated K^+ channels. *Proc. Natl. Acad. Sci. USA* **86** 4505–4509.

Schmid-Antomarchi, H., De Weille, K., Fosset, M., and Lazdunski, M. (1987) The receptor for antidiabetic sulfonylureas controls the activity of the ATP-modulated K^+ channel in insulin-secreting cells. *J. Biol. Chem.* **262** 15840–15844.

Smith, P. A. (1988) Electrophysiology of β-cells from the islets of Langerhans. PhD thesis, University of East Anglia, Norwich, UK.

Spruce, A. E., Standen, N. B., and Stanfield, P. R. (1985) Voltage-dependent. ATP-sensitive potassium channels of skeletal muscle membrane. *Nature* **316** 736–738.

Spruce, A. E., Standen, N. B., and Stanfield, P. R. (1986) Rubidium permeability of ATP-dependent potassium channels of frog skeletal muscle. *J. Physiol. (Lond.,)* **378** 105P.

Spruce, A. E., Standen, N. B., and Stanfield, P. R. (1987) Studies of the unitary properties of adenosine-5′-triphosphate-regulated potassium channels of frog skeletal muscle. *J. Physiol. (Lond.)* **382** 213–236.

Sturgess, N. C., Kozlowski, R. Z., Carrington, C. A., Hales, C. N., and Ashford, M. L. J. (1988) Effects of sulphonylureas and diazoxide on insulin secretion and nucleotide-sensitive channels in an insulin-secreting cell line. *Br. J. Pharmacol.* **95** 83–94.

Tamagawa, T., and Henquin, J. C. (1983) Epinephrine modifications of insulin release and of $^{86}Rb^+$ or $^{45}Ca^{2+}$ fluxes in rat islets. *Am. J. Physiol.* **244** E245–252.

Tempel, B., Jan, Y. N., and Jan, L. Y. (1988) Cloning of a probable potassium channel gene from mouse brain. *Nature* **332** 837–839.

Trube, G., and Hescheler, J. (1984) Inward-rectifying channels in isolated patches of the heart cell membrane: ATP-dependence and comparison with cell-attached patches. *Pflugers Arch.* **401** 178–184.

Trube, G., and Rorsman, P. (1987) Calcium and potassium currents recorded from pancreatic β-cells under voltage clamp control. In: *Biophysics of the pancreatic β-cell*, pp. 167–175, I. Atwater, E. Rojas, and B. Soria (eds). Plenum, New York.

Trube, G., Rorsman, P., and Ohno-Shosaku, T. (1986) Opposite effects of tolbutamide and diazoxide on the ATP-dependent K^+ channel in mouse pancreatic β-cells. *Pflügers Arch.* **407** 493–499.

Trube, G., Hescheler, J., and Schröter, K. (1989) Regulation and function of ATP-dependent K^+ channels in pancreatic B-cells. In: *Secretion and its control*, Society of General Physiologists Series, Vol. 44, G. S. Oxford and C. M. Armstrong (eds). Rockefeller University Press, New York, pp. 83–95.

Trussel, L. D., and Jackson, M. B. (1987) Dependence of an adenosine-activated potassium current on GTP-binding protein in mammalian central neurons. *J. Neurosci.* **7** 3306–3316.

Weiss, J. N., and Lamp, S. T. (1987) Glycolysis preferentially inhibits ATP-sensitive K^+ channels in isolated guinea pig cardiac myocytes. *Science* **238** 67–69.

Woll, K. H., Lönnendonker, U., and Neumcke, B. (1988) ATP-sensitive K^+ channels in adult mouse skeletal muscle: different modes of blockage by internal cations, ATP and tolbutamide. *Pflügers Arch.* **414** 622–628.

Wollheim, C. B., Dunne, M. J., Peter-Riesch, B., Bruzzone, R., Pozzan, T., and Petersen, O. H. (1988) Activators of protein kinase C depolarize insulin-secreting cells by closing K^+ channels. *EMBO J.* **7** 2443–2449.

Zilberter, Y., Burnashev, N., Papin, A., Portnov, V., and Khodorov, B. (1988) Gating kinetics of ATP-sensitive single potassium channels in myocardial cells depends on electromotive force. *Pflügers Arch.* **411** 584–589.

Zünkler, B. J., Lenzen, S., Männer, K., Panten, U., and Trube, G. (1988a) Concentration-dependent effects of tolbutamide, meglitidine, glipizide, glibenclamide and diazoxide on ATP-regulated K^+ currents in pancreatic B-cells. *Naunyn-Schmiedeberg's Arch. Pharmacol.* **337** 225–230.

Zünkler, B. J., Lins, S., Ohno-Shosaku, T., Trube, G., and Panten, U. (1988b) Cytosolic ADP enhances the sensitivity to tolbutamide of ATP-dependent K^+ channels from pancreatic B-cells. *FEBS Lett.* **239** 241–244.

5

K^+ channel modulation by G proteins and second messengers

Paul J. Pfaffinger and **Steven A. Siegelbaum**

5.1 INTRODUCTION

Potassium channels are prime targets for modulatory actions of hormones and transmitters. These modulatory actions can involve alterations in potassium current magnitude and/or kinetics and can lead to a number of changes in the electrical properties of nerve and muscle cells. Thus, potassium channel modulation has been associated with changes in resting potential and input conductance, action potential duration, threshold, accommodation, spontaneous bursting activity and neurotransmitter release.

Neurotransmitter and neurohormone actions are thought to be mediated by two receptor gene superfamilies. One superfamily comprises the directly gated transmitter receptors, where the receptor and ion channel are part of a single macromolecular protein complex (see Miller, 1989). Here, the binding of transmitter to its receptor directly opens (or gates) a channel. This type of transmitter receptor is exemplified by the nicotinic acetylcholine receptor–cation channel at the neuromuscular junction (Fig. 5.1A) (Karlin, 1980; Noda *et al.*, 1983; Popot and Changeux, 1984). The inhibitory amino acid receptors to GABA and glycine, which both gate a chloride channel, also fall into this class of directly gated receptor channels (Grenningloh *et al.*, 1987; Schofield *et al.*, 1987). The directly gated receptor channels are composed of four or five homologous subunits, where each subunit is an integral membrane protein that contains four membrane-spanning domains.

In contrast to the directly gated transmitter receptors, the second receptor gene superfamily comprises the G protein-mediated transmitter receptors, where the receptor and channel are two distinct macromolecules (Fig. 5.1B). Here, binding of transmitter to its receptor does not directly lead to a change in channel function. Rather, the activated receptor first activates a guanine nucleotide binding protein or G protein (Gilman, 1987) which then interacts with one (or more) effector proteins, leading to a change in channel activity. These G protein-mediated receptors include the α- and β-adrenergic receptors, the muscarinic receptors, the 5-HT_{1c} receptor and various peptide receptors. The receptors are homologous to the bacterial photopigment bacteriorhodopsin and all share a common structural motif in that they are composed of a single subunit with seven membrane-spanning regions (Hall, 1987).

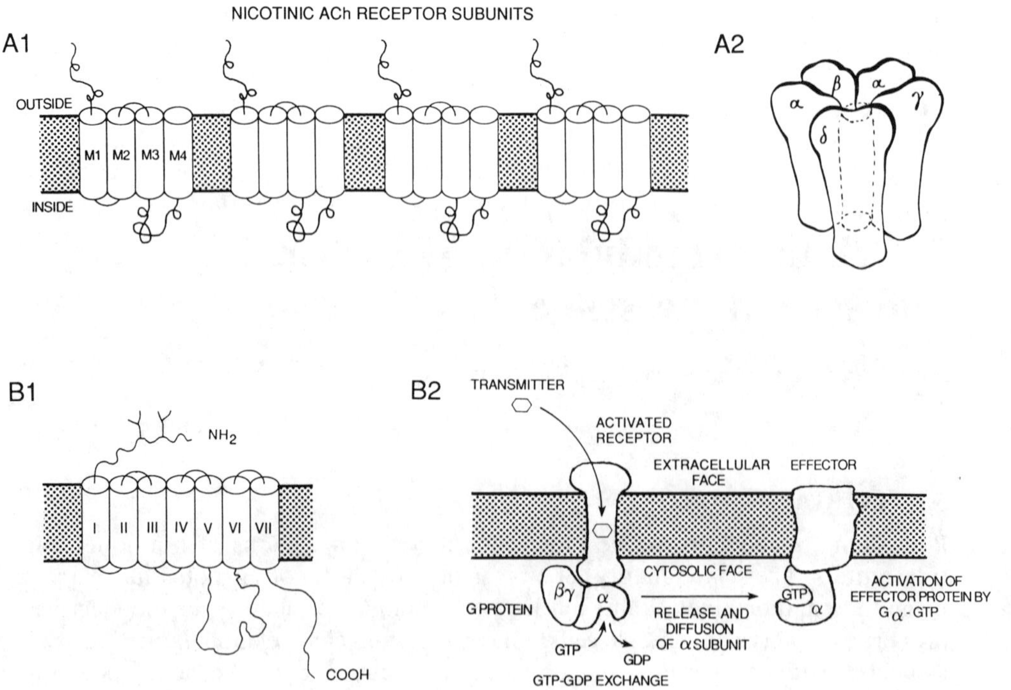

Fig. 5.1 — Neurotransmitters affect ion channels through two classes of receptor mechanisms. (A) Directly gated receptor channel, exemplified by nicotine ACh receptor. The receptor is a hetero-multimer composed of 2 α, 1 β, 1γ and 1 δ subunit. Each α subunit binds one molecule of ACh. Each subunit contains four membrane-spanning α-helical regions (M1–M4). A channel is formed from the M2 membrane spanning regions of all five subunits. (Modified from Miller, 1989.) (B) G protein-mediated receptor actions. The transmitter receptor interacts with trimeric G protein. The activated α subunit of G protein (GαGTP) activates the effector protein. (Modified from Stryer, 1988.)

The two receptor families are not only structurally distinct, but they also have very different physiological functions. Thus, the directly gated transmitter actions are very rapid, producing conductance changes that last only a few milliseconds. The function of these transmitter actions is to provide for rapid synaptic transmission. They act as simple on or off switches, either eliciting or inhibiting an action potential. In contrast, the G protein-mediated receptor actions tend to be much slower. They can produce conductance changes that last for seconds or even minutes. Their function is to produce longer-lasting modulatory changes in channel and cellular activity. So far, all known examples of K^+ channel modulation occur via the G protein-mediated class of transmitter actions. In this chapter, we will discuss the biochemical mechanisms involved in K^+ channel modulation and review some selected examples of transmitter actions on specific types of K^+ channels.

5.2 STRUCTURE AND FUNCTION OF G PROTEINS

GTP-binding proteins (G proteins) are a family of transducing proteins that couple receptors that are members of the opsin gene superfamily to their effectors (Gilman,

1984, 1987; Stryer and Bourne, 1986). G proteins are related to the RAS oncogenes and more distantly to Ef-Tu (a bacterial transcription elongation factor), although neither of these proteins are thought to interact with opsin superfamily receptors (Masters *et al.*, 1986). G protein structure and the basic mechanism of G protein function is conserved among all members of the gene family. The G protein is a heterotrimer consisting of a hydrophilic α subunit (M_r=39–45 kDa), and hydrophobic β (M_r=39 kDa) and γ subunits (M_r=11 kDa) (Fig. 5.1B). When the G protein is inactive (in the absence of agonist), a molecule of guanosine diphosphate (GDP) is bound to the α subunit, which forms a membrane-associated complex with the β and γ subunits. In the presence of an agonist, activated receptors interact with the G protein and induce an exchange of GTP for GDP on the α subunit. Upon binding GTP, the α subunit dissociates from the $\beta\gamma$ subunits and becomes activated. The free α subunit remains active until it hydrolyses the GTP back to GDP, leading to its reassociation with the $\beta\gamma$ subunits.

G proteins are traditionally classified according to their α subunits; the $\beta\gamma$ subunits are thought to be structurally similar and interchangeable. Originally only the α subunit was thought to be active, although recent experiments suggest roles for the $\beta\gamma$ subunits in the activation of phospholipase A_2 (Jelsema and Axelrod, 1987) and possibly inhibition of adenylate cyclase (Gillman, 1984; Katada *et al.*, 1986). The best-characterized G proteins are G_t (or transducin) (α_t; M_r=39 kDa), G_s (α_s; M_r=42–52 kDa), G_i(α_i; M_r=40–41 kDa), G_o (α_o; M_r=39 kDa), and G_p (α_p; M_r=?). Recent molecular cloning studies suggest that G_i and G_o are in fact two related members of a G_i multigene family. To date, four genes have been identified: α_{i1}, α_{i2}, α_{i3} and α_o (Graziano and Gilman, 1987).

Two bacterial toxins are known to modify different G proteins through an ADP ribosylation of the α subunit. Cholera toxin acts on two different G proteins to inhibit the hydrolysis of GTP, thus locking the G protein in its active form (Cassel and Selinger, 1977; Abood *et al.*, 1982). Pertussis toxin (PTX, or islet-activating protein, IAP) ADP ribosylates certain G proteins at a different site, preventing the interaction of receptor with the G protein and thus blocking G protein activation (Ui, 1984).

G_t is only found in photoreceptor cells and is involved in activating cyclic GMP phosphodiesterase during phototransduction (see Hurley, 1987). Separate G_t gene products serve the same transduction function in rods and cones (Lerea *et al.*, 1986). G_t is sensitive to both cholera toxin and pertussis toxin.

G_s couples receptors to stimulation of adenylate cyclase (Rodbell, 1980; Gilman, 1987). The two different sized forms of G_s are produced by alternative splicing of a single gene (Robishaw *et al.*, 1986). Recent experiments in the heart suggest that G_s may have a second function in the direct activation of L-type calcium channels (Yatani *et al.*, 1987b). G_s is irreversibly activated by cholera toxin.

Members of the G_i family appear to couple receptors to a number of different effectors. The initially described function of G_i was the inhibition of adenylate cyclase (Ui *et al.*, 1984; Gilman, 1984). In addition, G_i and G_o have been reported to couple to activation of phospholipases in some tissues (Gilman, 1987). Recently, different forms of G_i have been shown to lead to activation of potassium channels in different tissues. In the heart, α_i activates a potassium channel similar to the ACh-activated potassium channel (Codina *et al.*, 1987), while in brain α_o activates one or more types of potassium channels (VanDongen *et al.*, 1988). Other reported functions for the G_i family of proteins include the inhibition of neuronal Ca^{2+} channels by G_o (but not by G_i) (Hescheler *et al.*, 1987). Both G_i and G_o are inhibited by pertussis toxin.

One additional G protein, G_p, has not yet been purified or cloned but has been functionally identified due to its proposed role in activation of phospholipase C (Aub *et al.*, 1986). G_p is distinguished by its insensitivity to both pertussis toxin and cholera toxin. In addition to activation of phospholipase C, G_p is proposed to be involved in suppression of M-current in sympathetic neurones (Pfaffinger, 1988) and activation of potassium channels in some brain neurones (Nakajima *et al.*, 1988).

5.3 SECOND MESSENGER CASCADES INVOLVED IN K^+ CHANNEL MODULATION

In addition to direct actions on ion channel activity, G proteins also influence ion channel function by activating various second messenger cascades (see Kaczmarek and Levitan, 1987). Many (but not all) of the actions of these second messengers are mediated by protein phosphorylation through activation of various protein kinases (Nairn *et al.*, 1985; Levitan, 1985, 1988). Of the many second messenger-mediated cascades, the best characterized involves cyclic AMP-mediated protein phosphorylation, described by Sutherland, Krebs and their colleagues in the 1950s and 1960s (see Krebs and Beavo, 1979; Cohen, 1982; Levitzki, 1988, for reviews). In addition to cyclic AMP, a number of other second messengers are known to modulate K^+ channel activity, including intracellular Ca^{2+}; the breakdown products of phosphatidylinositol — IP_3 and diacylglycerol; and arachidonic acid (another phospholipid metabolite). We shall first summarize the biochemical properties of these cascades and then discuss selected examples of K^+ channel modulation involving these second messenger systems.

5.3.1 The cyclic AMP cascade

The first step in the cyclic AMP cascade is the activation of the stimulatory G protein, G_s, through an interaction with its associated receptor as described above. Receptors known to be coupled to G_s include the β-adrenergic receptors, a 5-HT receptor in molluscs, the dopamine D_1 receptor, as well as receptors for various peptides and prostaglandins (see Drummond, 1983; Cooper *et al.*, 1986). The activated α subunit of G_s stimulates the activity of adenylate cyclase, an integral membrane protein, to catalyse the conversion of ATP to cyclic AMP (Fig. 5.2A), which activates cyclic AMP-dependent protein kinase (PKA) (for reviews see Krebs and Beavo, 1979; Edelman *et al.*, 1987; Blackshear *et al.*, 1988). The holoenzyme of PKA is a tetramer consisting of two identical regulatory subunits (R) and two identical catalytic subunits (C) (Fig. 5.2B). The two R subunits are linked by disulphide bonds. In the absence of cyclic AMP, the R subunits bind to and inactivate the C subunits with a very high affinity (around 1 nM). In the presence of cyclic AMP, each R subunit binds two molecules of cyclic AMP (K_D around 10 nM), causing a 10^4-fold decrease in affinity of the regulatory subunits for the C subunits. As a result, the catalytic subunit dissociates from the regulatory subunit, leading to the phosphorylation of a variety of substrate proteins on serine or threonine residues.

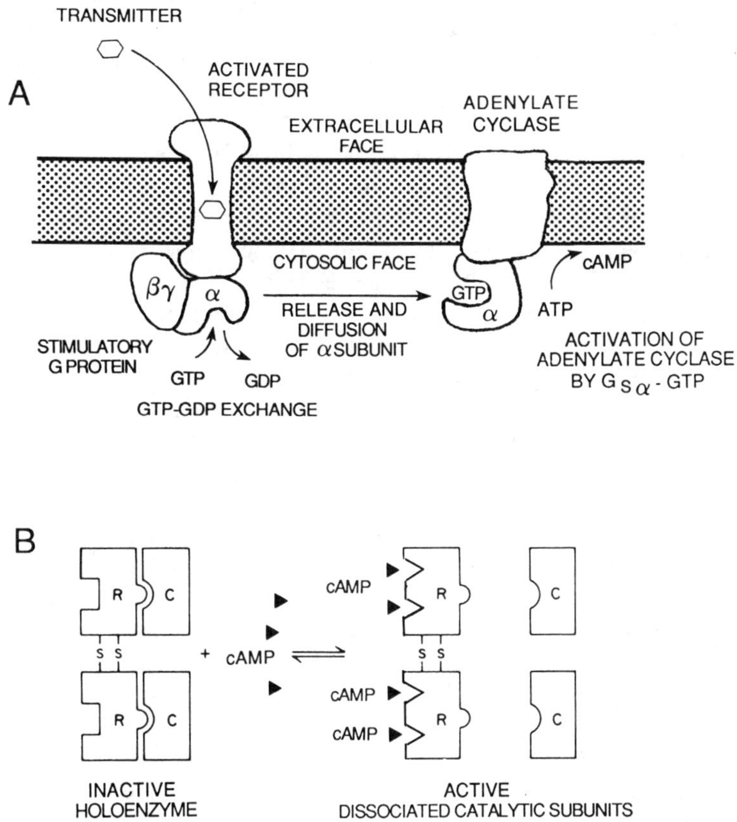

Fig. 5.2 — Cyclic AMP cascade. (A) Activation of adenylate cyclase by Gα-S GTP leads to production of cyclic AMP from ATP. (B) Cyclic AMP-dependent protein kinase activation due to binding of cyclic AMP to R (regulatory) subunit, leading to dissociation of holoenzyme into free R and C (catalytic) subunits.

Two major isoforms of the regulatory subunit, R_I (M_r=42 800) and R_{II} (M_r=45 000), have been identified. There may be several different species of the R_{II} subunit, accounting for tissue heterogeneity (Schwartz and Greenberg, 1987; Nairn *et al.*, 1985; Krebs and Beavo, 1979). The R_{II} subunit can also be auto-phosphorylated, decreasing the rate at which R reassociates with C. In addition to the R subunit, there is a small heat-stable inhibitory protein (PKI) isolated from skeletal muscle that binds to free catalytic subunit with a very high affinity (K_D around 2 nM), blocking the activity of free C subunit (see Krebs and Beavo, 1979).

Free C subunit phosphorylates the hydroxyl group of serine or threonine residues that occur within the context of the following consensus sequence: B–B–X–S; where B is a basic residue (lys or arg), X is any amino acid and S is the target serine or threonine. Both R_I and R_{II} subunits contain sequences similar to this consensus phosphorylation sequence at the site of interaction between the R and C subunits. In R_{II}, a serine residue in this sequence is the site of autophosphorylation. In R_I, however, the serine is replaced by an alanine residue. Thus R_I is thought to inhibit the C subunit by acting as a pseudosubstrate (e.g. see Blackshear *et al.*, 1988).

The phosphorylation event is terminated through the action of a phosphoprotein phosphatase. Several classes of phosphoprotein phosphatases have been identified in cells, including types 1, 2a, 2b (the Ca^{2+} calmodulin-dependent enzyme, calcineurin), and type 2C (a Mg^{2+}-dependent enzyme) (Ingebritsen and Cohen, 1983). Protein phosphatase 1 is inhibited by a specific protein called Inhibitor 1 that is only active after it has been phosphorylated by PKA. There is some evidence that certain agents (e.g. insulin) act to increase phosphatase activity.

Muscarinic and α-adrenergic agonists, among others, inhibit adenylate cyclase activity by activating an inhibitory G protein, G_i (Gilman, 1984). This may be due to a direct inhibitory action of α_i on adenylate cyclase and/or may be due to the generation of excess free βγ subunits which can then reassociate with any active α_s.

Cyclic AMP-dependent transmitter actions were the first modulatory mechanism to be studied in detail and a number of such actions on K^+ channels have been reported (Siegelbaum and Tsien, 1983; Kaczmarek and Levitan, 1987; Levitan, 1988). In all cases so far examined, the effects of cyclic AMP are thought to be mediated by PKA. However, with the identification of cationic channels in photoreceptors (Fesenko *et al.*, 1985) and olfactory neurones (Nakamura and Gold, 1987) that are directly gated by cyclic nucleotides, it is possible that some cyclic AMP-dependent modulatory actions on K^+ channels may be independent of phosphorylation.

A number of different experimental approaches have been used to study the role of the cyclic AMP cascade in modulating K^+ channel activity. One approach often used is to increase artificially the intracellular level of cyclic AMP by intracellular injection of cyclic AMP through microelectrodes, extracellular application of membrane-permeant analogues of cyclic AMP such as dibutyryl cyclic AMP or 8-bromo cyclic AMP, use of phosphodiesterase inhibitors, and use of the diterpene compound forskolin, a direct activator of adenylate cyclase (Seamon *et al.*, 1981). However, forskolin has now been shown to block directly potassium channels (Coombs and Thompson, 1987; Hoshi *et al.*, 1988; Krause *et al.*, 1988), independent of its action to increase cyclic AMP. Thus, results with this drug must be treated with caution.

It is also possible to bypass cyclic AMP and directly introduce the C subunit of PKA into cells, by pressure injection from microelectrodes (Castellucci *et al.*, 1980; Kaczmarek *et al.*, 1980), by internal perfusion through patch pipettes during whole-cell recording (DePeyer *et al.*, 1982), or by bath application to inside-out membrane patches (Shuster *et al.*, 1985; Ewald *et al.*, 1985; Avenet *et al.*, 1988). One stringent test as to whether a transmitter or hormone action is mediated by cyclic AMP-dependent phosphorylation involves intracellular injection or perfusion of PKI (Castellucci *et al.*, 1982; Adams and Levitan, 1982). Finally, at the biochemical level, the neurotransmitter should lead to elevated levels of cyclic AMP and protein phosphorylation.

5.3.2 The protein kinase C cascade

Other receptors act through the unidentified G_p protein to activate phospholipase C, leading to the hydrolysis of phosphatidylinositol and release of diacylglycerol (DAG) and inositol trisphosphate (IP_3) (Nishizuka, 1984; Berridge and Irvine, 1984) (Fig. 5.3). IP_3 releases calcium from intracellular stores (Berridge and Irvine, 1984) and thus can lead to the opening of calcium-activated K$^+$ channels (Trautmann and

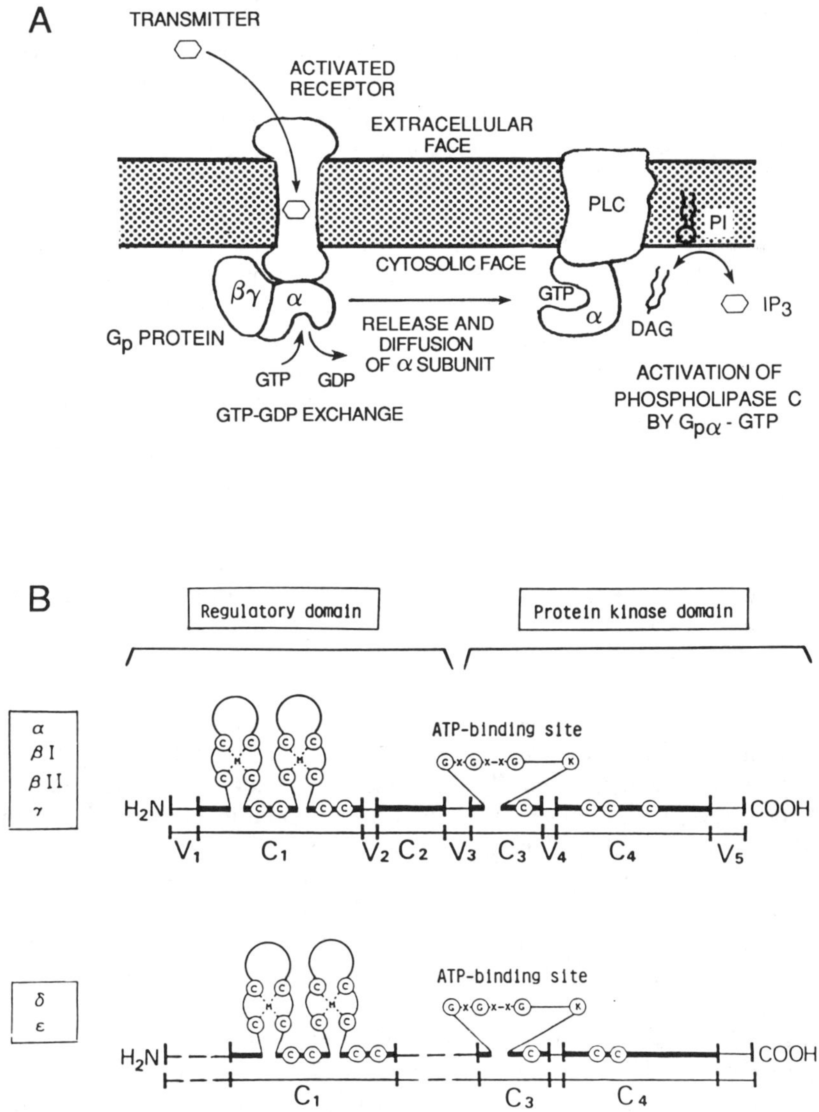

Fig. 5.3 — Phosphatidylinositol cascade. (A) Receptor coupled to G_p leads to activation of phospholipase C (PLC), leading to hydrolysis of phosphatidylinositol (PI) and production of diacylglycerol (DAG) and inositol trisphosphate (IP_3). (B) Structure of different forms of protein kinase C (PKC). The different forms of PKC are made up of both constant domains (C1–C4) and variable domains (V1–V5). Each kinase molecule contains a regulatory domain with cysteine-rich (C) metal (m) binding sites and a catalytic domain with an ATP binding site. (From Nishizuka, 1988.)

Marty, 1984). In addition, Dutar and Nicoll (1988) have reported that IP_3 leads to a decrease in the M potassium current of hippocampal neurones through a Ca^{2+}-independent mechanism (see below). Recent results suggest that IP_3 can be metabolized to other active compounds including IP_4, IP_5, IP_6 and cyclic IP_3, which may have a synergistic role in potentiating the actions of IP_3 (Berridge, 1988).

DAG leads to activation of protein kinase C (PKC), a calcium- and phospholipid-dependent enzyme that was first described by Nishizuka and his colleagues in 1977 as a proteolytically activated kinase (see Nishizuka, 1984, 1986, 1988). Protein kinase C has a molecular weight of 80 kDa and contains both a catalytic domain (50 kDa) and a regulatory domain (30 kDa). Treatment with a calcium-dependent protease activates PKC by cleaving a free catalytic domain from the enzyme. In the absence of DAG, the enzyme requires very high unphysiological concentrations of calcium (100 μM) for activity. However, in the presence of DAG, the enzyme can be activated by much lower concentrations of calcium (<1 μM). PKC can also be activated by various tumour-promoting agents such as the phorbol esters. Activation of PKC is often accompanied by a translocation of the enzyme from the cytoplasm to the membrane. In addition, a low molecular weight protein inhibitor of PKC has been identified (McDonald and Walsh, 1985).

At present, seven different forms of PKC have been cloned and sequenced (Nishizuka, 1988). These species show significant homology but differ in their exact sequences, tissue distribution and calcium and lipid sensitivity. The βI and βII species are very sensitive to activation by DAG and do not require Ca^{2+}. In contrast, the γ species, which is expressed solely in brain and spinal cord, is rather insensitive to DAG but is efficiently activated by arachidonic acid.

One note of caution concerning effects of phorbol esters on ion channels comes from a recent report that these compounds may have direct blocking effects on calcium channels (Hockberger *et al.*, 1989). Whether similar effects occur with K^+ channels is not yet known.

5.3.3 The arachidonate cascade

A third pathway implicated in K^+ channel modulation involves the lipoxygenase metabolites of arachidonic acid (Fig. 5.4) (see Needleman *et al.*, 1986, for review). Arachidonic acid is a 20-carbon fatty acid with four double bonds and is esterified to membrane phospholipids at the 2-position of the glycerol backbone. Arachidonic acid can be released from membranes, either through the direct action of phospholipase A_2 or through the combined action of phospholipase C and diacylglycerol lipase. Arachidonic acid can then be metabolized through a cyclooxygenase enzyme to produce the prostaglandins, prostacyclins and thromboxanes. Alternatively, arachidonic acid can be metabolized through various lipoxygenase enzymes, which insert a hydroperoxy group (–OOH) at different positions of arachidonic acid, to form hydroperoxyeicosatetraenoic acids (HPETEs). These unstable hydroperoxy acids are rapidly reduced to corresponding hydroxyeicosatetraenoic acids (HETEs). In addition, the 5-lipoxygenase product, 5-HPETE, is metabolized to the leukotrienes, compounds that are potent vasoconstrictors. The 12-lipoxygenase product, 12-HPETE, is also subject to further metabolism to epoxy alcohols and other derivatives (Pace-Asciak *et al.*, 1983; Piomelli *et al.*, 1989).

The best-characterized means by which arachidonate metabolites produce their

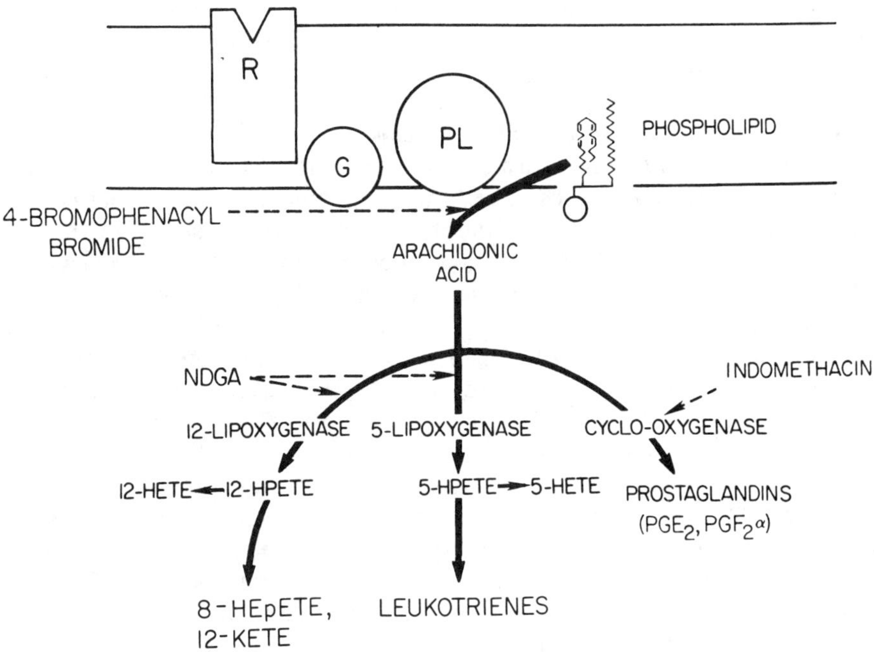

Fig. 5.4 — Arachidonic acid cascade. Receptor (R) activates G protein (G), leading to phospholipase (PL) activation and release of arachidonic acid. 12-Lipoxygenase enzyme leads to production of 12-hydroperoxyeicosatetraenoic acid (12-HPETE). 12-HPETE is reduced to 12-hydroxyeicosatetraenoic acid (12-HETE) and metabolized to 8-hydroxy-11,12,epoxy-5,9,14-eicosatrienoic acid (8-HEpETE) and 12-keto-5,8,10,14-eicosatetraenoic acid (12-KETE). 5-Lipoxygenase enzyme produces 5-hydroperoxyeicosatetraenoic acid (5-HPETE), which is metabolized to 5-HETE or leukotrienes. A cyclooxygenase leads to production of prostaglandins. 4-Bromophenacyl bromide inhibits PL activity. NDGA inhibits lipoxygenase activity. Indomethacin inhibits cyclooxygenase activity.

physiological actions is by acting on external membrane surface receptors after being released from the cells in which they are produced (see Halushka *et al.*, 1989). Thus different eicosanoids have been reported to interact with receptors to activate or inhibit adenylate cyclase and/or to activate phospholipase C through a G protein-dependent mechanism. In addition, lipoxygenase metabolites have been reported to activate guanylate cyclase in neuroblastoma (Snider *et al.*, 1984). Recent evidence suggests that ion channels may also be a site of action for arachidonic acid and other fatty acids (Kim and Clapham, 1989; Ordway *et al.*, 1989) or for more specific arachidonic acid metabolites (Belardetti *et al.*, 1989; Buttner *et al.*, 1989).

5.4 EXAMPLES OF MODULATORY ACTIONS ON K$^+$ CHANNELS

5.4.1 The cardiac muscarinic-activated K$^+$ channel

Until a few years ago, all G protein-mediated receptor actions on ion channel function were thought to be mediated by one or more intracellular second messengers. However, a number of recent experiments suggest that G proteins may directly modulate channel activity, without involvement of a freely diffusible second messenger. The first described and most thoroughly studied example of this proposed direct G protein channel gating is the muscarinic ACh-activated K^+ channel (K_{ACh}) in cardiac atrial myocytes.

5.4.1.1 Direct activation of K_{ACh} by G proteins

Soejima and Noma (1984) investigated the activation of single K_{ACh} channels by ACh using the patch-clamp technique. Their experiments revealed that channels isolated in a cell-attached patch could only be activated if ACh was applied inside the pipette. Application of ACh to the bath solution had no effect, arguing against the involvement of a diffusible second messenger (Fig. 5.5A). However, if patches were excised from the cells to the inside-out configuration, channel activity was rapidly lost, suggesting that some cytosolic factor was essential for channel activity.

Pfaffinger *et al.* (1985) demonstrated by whole-cell dialysis of atrial cells that the activation of K_{ACh} by ACh required the presence of GTP in the pipette solution. Furthermore, pretreatment of the cells with PTX completely blocked channel activation with ACh (Fig. 5.5B). The only known targets for this toxin are G proteins in the 39–41-kDa size range. Breitweiser and Szabo (1985) reported independent evidence for a G protein based on experiments showing that dialysis of atrial cells with the non-hydrolysable GTP analogue GPP(NH)P transformed the normally reversible activation of the K^+ current by ACh into an irreversible response.

These results were tied together by the observation of Kurachi *et al.* (1986a) that addition of GTP to the inside face of inside-out patches restored K^+ channel activation with ACh (ACh was present in the pipette). Furthermore, PTX blocked channel activation by GTP, while the non-hydrolysable GTP analogue GTP-γ-S produced an irreversible activation of K_{ACh} in the inside-out patches (Kurachi *et al.*, 1986b) (Fig. 5.6). Finally, Yatani *et al.* (1987b) demonstrated that a purified activated G protein from red blood cells could also activate the K_{ACh} channel in inside-out patches. They named this protein G_k (a terminology originally proposed by Breitweiser and Szabo, 1985) because it interacted with the K^+ channel with a very high affinity, in the picomolar range. These several findings, indicating that G protein activation in cell-free patches opens the K_{ACh} channel, provide strong evidence for tight G protein–K^+ channel coupling without the involvement of second messenger systems.

One surprising development in the study of K_{Ach} channel activation was the report by Logothetis *et al.* (1987) that purified βγ subunits, but not activated α39 or α41 subunits, of G proteins from bovine brain activated K_{ACh} in inside-out patches. In contrast, Codina *et al.* (1987) reported a few months later that the α subunit of G_k, but not the βγ subunits, activated the K_{ACh} channels in picomolar concentrations.

The lively controversy that these apparently conflicting reports generated has now been largely resolved. There is now general agreement that it is the α subunit of a G_i-like protein that mediates the actions of ACh (e.g. Cerbai *et al.*, 1988; Logothetis *et al.*, 1988). First, α subunits activate the channel at concentrations 100 times lower than those required by the βγ subunits (Codina *et al.*, 1987). Second, an antibody that binds to α subunits but not to βγ subunits blocks the activation of K_{ACh}(Yatani *et al.*, 1988a). Third, antibodies to phospholipase A_2 block activation of K_{ACh} by the βγ subunits but do not block activation by ACh or GTP-γ-S (Kim *et*

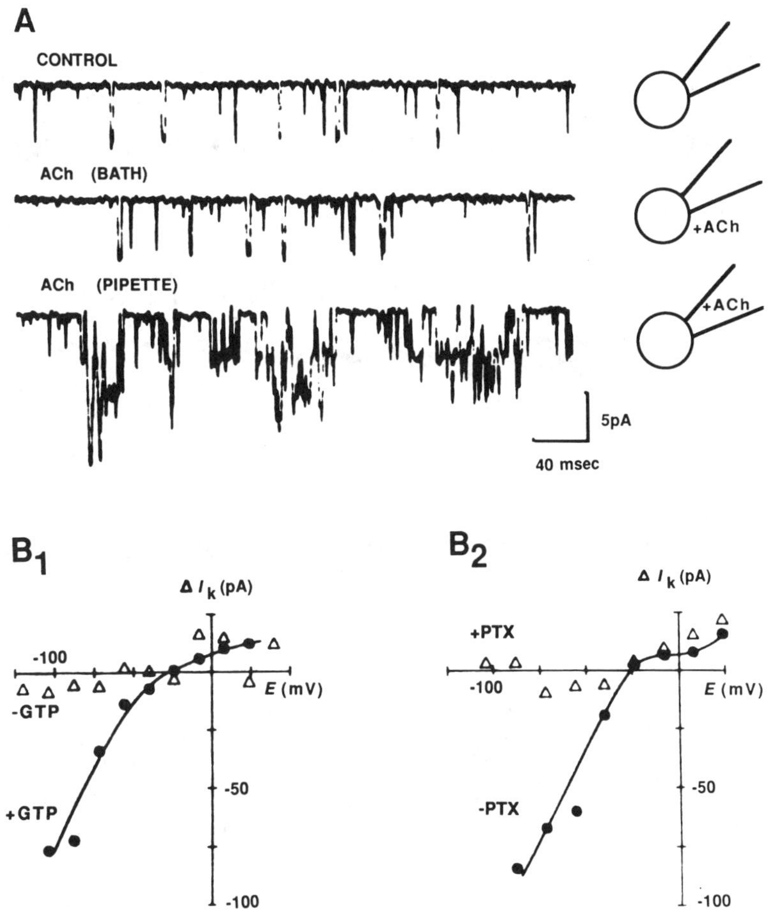

Fig. 5.5 — Muscarinic ACh-activated K^+ channel of cardiac cells. (A) ACh activates the K^+ channel only when applied inside pipette. (From Soejima and Noma, 1984.) (B) Activation of K_{ACh} requires PTX-sensitive G protein. Current–voltage plots of ACh-induced currents from whole cell voltage-clamp of embryonic chick atrial cells. B_1 K_{ACh} is GTP-dependent. Triangles, no GTP in pipette; filled circles, 100 μM GTP in pipette. (B_2) K_{ACh} is blocked by PTX. Open triangles, with PTX pretreatment; filled circles, no PTX pretreatment. (From Pfaffinger *et al.*, 1985.)

al., 1989). Later we will discuss the possible involvement of the βγ subunits in the regulation of K^+ channel activation in the heart through the arachadonic acid cascade.

One unexplained finding in the initial studies was the relationship between the G_k protein purified from red blood cells with the G protein responsible for the physiological control of K_{ACh} in the heart. Subsequent cDNA cloning and sequence analysis has shown that the red blood cell α_k was in fact encoded by an α_i cDNA clone, α_{i3} — one of three clones coding for α_i subunits (Codina *et al.*, 1988). Moreover, α_{i3} produced through recombinant DNA expression in bacteria was

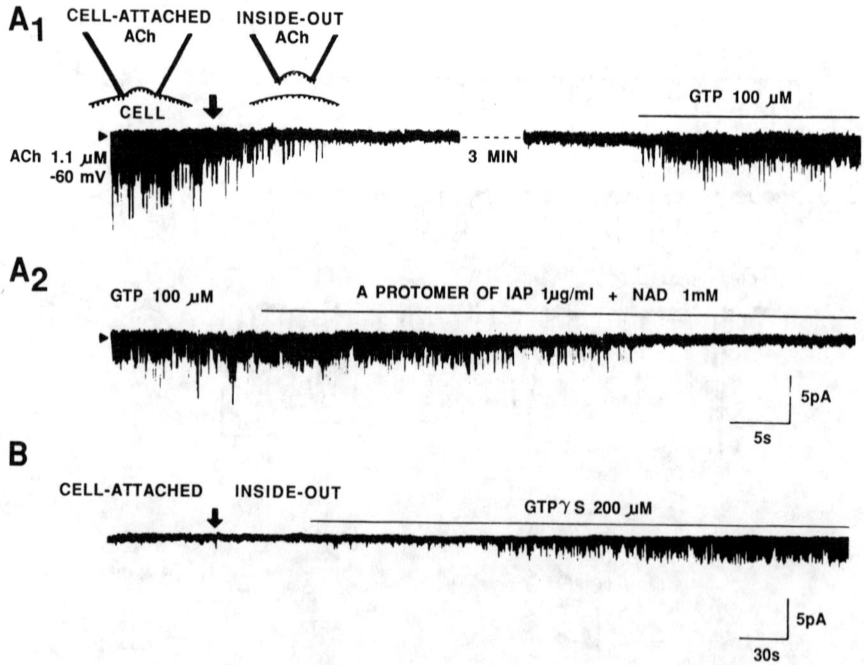

Fig. 5.6 — G protein activation opens K_{ACh} channels in cell-free conditions. In the cell-attached patch, K_{ACh} channels open due to ACh in pipette. Upon formation of inside-out patch, channel opening ceases. Addition of 100 μM GTP to bath causes channel opening to resume. (A_2) In the presence of GTP (in bath) and ACh (in pipette), K_{ACh} channel opening is inhibited by active A protomer of IAP (PTX). (B) GTP-γ-S, a non-hydrolysable GTP analogue that irreversibly activates G proteins, leads to K_{ACh} channel opening in the absence of ACh (in pipette) From Kurachi *et al.*, 1986a,b.)

capable of activating K_{ACh} (Yatani *et al.*, 1988b). However, any hopes for a specific targeting of α_{i3} to K_{ACh} channel activation were dashed by the finding that all three α_i clones produced in bacterial expression systems are equipotent in activating K_{ACh} (Yatani *et al.*, 1988b). It is not yet known if recombinant α_o can activate the K_{ACh} channel, although G_o purified from brain does activate this channel, but only at rather high concentrations (Yatani *et al.*, 1987a). Since all of the α_i proteins both inhibit adenylate cyclase and activate K_{ACh}, Yatani *et al.* (1988b) concluded that G proteins must be multifunctional, interacting with more than one effector. One remaining question is why three different α_i proteins are required to do the same job. Is it possible that their function in intact cells may show more specificity than presently observed using *in vitro* systems?

5.4.1.2 Other G protein-gated K^+ channels

Channel activation by G proteins appears to be a widely utilized channel regulatory scheme. In excised patches from GH3 cells, G_i activates a K^+ channel that is indistinguishable from channels opened by somatostatin (Yatani *et al.*, 1987c). The other PTX-sensitive G protein, G_o, activates three different classes of K^+ channels from central neurones (VanDongen *et al.*, 1988), at least one of which is similar to channels opened by 5-HT. It is unclear whether G proteins other than G_i and G_o directly regulate potassium channels. However, G_s has been shown to directly activate a Ca^{2+} channel in heart, a response previously attributed solely to cyclic AMP-dependent protein kinase (Yatani *et al.*, 1987b).

In a number of other systems, a direct control of K^+ channel activity by G proteins has been suggested, based on the evidence that G protein activation is necessary for K^+ channel modulation combined with negative results that fail to implicate known second messenger pathways (Andrade *et al.*, 1986; Sasaki and Sato, 1987; Trussell and Jackson, 1987; Nakajima *et al.*, 1988; Pfaffinger, 1988).

Although the list of examples of G protein gating of K^+ channel activity in cell-free membrane patches is now quite impressive (and still growing), the cell-free patch is likely to be a complex biochemical unit. Thus, until a purified G protein can be shown to regulate a purified K^+ channel reconstituted in a lipid bilayer, the question as to whether the regulatory actions of these G proteins is indeed direct or is mediated by some second messenger contained in the patch membrane must remain open.

5.4.1.3 The role of arachidonic acid metabolites in activation of K_{ACh}

Recent studies by Kurachi *et al.* (1989a,b) and Kim *et al.* (1989) demonstrate that the 5-lipoxygenase metabolites of arachidonic acid also activate the cardiac K_{ACh} channel. Thus, both groups report that arachidonic acid activates a K^+ channel in cell-attached patches from cardiac atrial cells that is indistinguishable from the K_{ACh} channel. In addition, both groups find that NDGA, a lipoxygenase blocker, blocks the response to arachidonic acid, whereas indomethacin, a cyclooxygenase blocker, is ineffective. Kurachi *et al.* (1989a) also report that AA-861, a 5-lipoxygenase inhibitor, blocks the action of arachidonate, while baicalein, a 12-lipoxygenase inhibitor, does not. Experiments applying specific metabolites are generally in agreement with the blocker studies, as Kurachi *et al.* (1989a) find that various 5-HPETE metabolites activate K_{ACh} (including 5-HPETE and the leukotrienes LTA4 and LTC4 but not LTB4), whereas 12-HPETE, 12-HETE and 5-HETE are ineffective. Kim *et al.* (1989) report similar results except they find that 12-HETE and LTB4 are also active.

One important aspect of these studies is that they help resolve the controversy as to which G protein subunit is responsible for K_{ACh} activation. Thus, Kim *et al.* (1989) find that the effects of the $\beta\gamma$ subunits on K_{ACh} channel activation can be blocked by an anti-phospholipase A_2 antibody and by NDGA. This led these authors to conclude that the $\beta\gamma$ subunits act by stimulating PLA_2, in agreement with the earlier findings of Jelsema and Axelrod (1987). Although not directly tested, the action of the α subunit does not appear to involve the PLA_2 pathway as K_{ACh} channel activation by ACh or by GTP-γ-S is not affected by the antibody or by NDGA. Thus the K_{ACh} channel appears to be under dual regulation by both the α subunits, perhaps acting directly on the channel, and the $\beta\gamma$ subunits, acting through the arachidonic acid cascade. The mechanism by which the 5-lipoxygenase metabolites activate K_{ACh} is unclear, however. Kurachi *et al.* (1989a) report that activation of K_{ACh} by the lipoxygenase metabolites requires GTP and propose that the metabolites activate G proteins, possibly directly. However, no such GTP dependence was seen by Kim *et al.* (1989). Finally, since neither NDGA nor the anti-PLA_2 antibodies block channel activation by ACh, the arachadonic acid pathway is unlikely to mediate the effects of this transmitter. However, Kurachi *et al.*, (1989b) do find that activation of $I_{K(ACh)}$ by α-adrenergic agonists is mediated by the 5-lipoxygenase pathway.

5.4.2 The S potassium current

In *Aplysia* mechanoreceptor sensory neurones, 5-HT and other excitatory modulatory inputs produce a slow decreased-conductance EPSP, an increase in action potential duration, and an increase in transmitter release from the sensory neurone terminals (Klein and Kandel, 1980; Kandel and Schwartz, 1982; Kandel *et al.*, 1987). This presynaptic facilitation is thought to underlie sensitization of the gill-withdrawal reflex, a simple form of learning. Using the voltage clamp technique, 5-HT was found to decrease current flow through a class of novel K^+ channels termed the serotonin-sensitive or S–K^+ channels (Fig. 5.7A), which contribute to the resting conductance of the membrane and action potential repolarization (Klein *et al.*, 1982; Pollock *et al.*, 1985). The action of serotonin is mediated by cAMP-dependent protein phosphorylation. Thus, the effects of 5-HT are mimicked by cyclic AMP and intracellular injection of the catalytic subunit of PKA (Castellucci *et al.*, 1980) and inhibited by the PKA inhibitor protein (Castellucci *et al.*, 1982). Moreover, biochemical studies confirm that 5-HT increases cylic AMP levels (Bernier *et al.*, 1982) and protein phosphorylation in the sensory neurones (Sweatt and Kandel, 1989).

Using single channel recording, the decrease in the S current by 5-HT was shown to result from prolonged all-or-none closure of about half of the active S channels in the membrane (Fig. 5.7A) (Siegelbaum *et al.*, 1982; Brezina *et al.*, 1987a; Pollock and Camardo, 1987). This appears as a decrease in the number of functional channels in the patch (N) with no change in open probability (p_o) of those channels that remain active. This effect of 5-HT is reproduced by application of the purified catalytic subunit of PKA to cell-free inside-out patches (Fig. 5.13c; Shuster *et al.*, 1985). A very similar cyclic AMP-dependent K^+ current decrease by 5-HT has been characterized by Gerschenfeld and his colleagues in *Helix* neurones (Deterre *et al.*, 1981, 1982). In addition to S channel modulation, 5-HT also decreases a Ca^{2+}-activated K^+ current in the sensory neurones through a cyclic AMP-dependent mechanism (Walsh and Byrne, 1989) and alters a delayed rectifier current through a cyclic AMP-independent mechanism (Baxter and Byrne, 1987).

Aside from the facilitatory modulatory inputs, the sensory neurones are also subject to inhibitory modulation by the peptide FMRFamide, which produces presynaptic inhibition and inhibition of the gill and siphon-withdrawal reflex (Abrams *et al.*, 1984; Mackey *et al.*, 1987). At the molecular level of single S channel function, FMRFamide also produces an opposing modulatory action to 5-HT, leading to an increase in S channel activity (Belardetti *et al.*, 1987; Brezina *et al.*, 1987a) (Fig. 5.7). However, the action of FMRFamide is not the mirror-image of that of 5-HT. Fig. 5.7 shows that FMRFamide does not increase the number of active S channels — rather it increases channel open probability (see also Belardetti *et al.*,

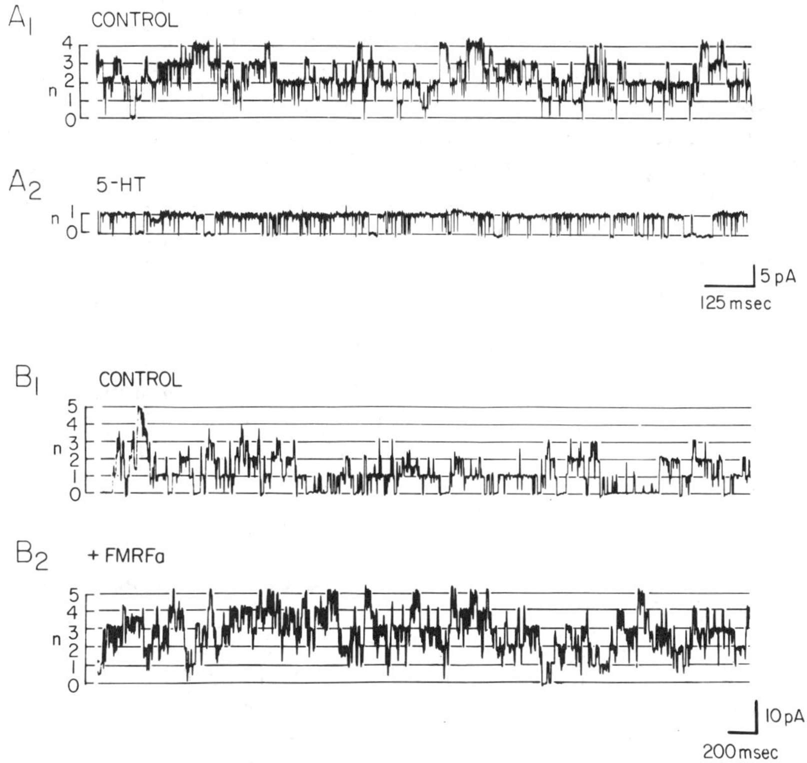

Fig. 5.7 — S–K^+ channel modulation in cell-attached patches. (A) 5-HT (applied to bath) causes all-or-none closure of three of four active S channels in the patch. (From Siegelbaum and Tsien, 1983). (B) FMRFamide (applied to bath) causes an increase in S channel open probability with no change in number of active channels. (From Belardetti *et al.*, 1987.)

1987). In addition to S channel modulation, FMRFamide also acts to decrease a calcium current (Brezina *et al.*, 1987b; Edmonds *et al.*, 1988). The synergistic effects of increased K^+ current and decreased Ca^{2+} current are thought to underlie the presynaptic inhibitory actions of FMRFamide.

As FMRFamide application to the bath was able to modulate S channel activity under the patch, the transmitter appeared to act through an intracellular second messenger. A role for arachidonic acid metabolites in the action of FMRFamide was first suggested by the finding that histamine, another presynaptic inhibitory transmitter, caused release of 5- and 12-HETE from *Aplysia* nervous tissue (Piomelli *et al.*, 1987a). In a combined biophysical and biochemical study, FMRFamide was also shown to release 12- and 5-HETE from *Aplysia* sensory neurones (Piomelli *et al.*, 1987b). Moreover, arachidonic acid and 12-HPETE simulate the actions of FMRFamide in the sensory neurones, producing presynaptic inhibition and an increase in S channel opening. Both the phospholipase blocker bromophenacyl bromide and the lipoxygenase blocker NDGA block the action of FMRFamide (but importantly have little or no effect on the cyclic AMP-dependent response to 5-HT) (Fig. 5.8). Finally, the action of FMRFamide depends on a perussis toxin-sensitive G protein that probably serves to couple the FMRFamide receptor to phospholipase activation (Volterra and Siegelbaum, 1988).

How does 12-HPETE lead to K^+ channel modulation? Recent data indicate that 12-HPETE may require further metabolic processing in *Aplysia* to the epoxy-alcohol 8-hydroxy- 11,12,epoxy-5,9,14-eicosatrienoic acid (8-HEpETE) and/or the keto derivative 12-keto-5,8,10,14-eicosatetraenoic acid (12-KETE) (Piomelli *et al.*, 1989). In the *Aplysia* neurone L14, these compounds are physiologically active and mimic different components of a biphasic (depolarizing–hyperpolarizing) response to histamine, with 12-KETE producing a depolarization (Piomelli *et al.*, 1988) and 8-HEpETE producing a hyperpolarization (Piomelli *et al.*, 1989). The action of the arachidonic acid metabolites on S channel gating does not appear to require phosphorylation, dephosphorylation or G protein activation since they produce a reversible increase in S channel activity in cell-free membrane patches that lack ATP or GTP (Belardetti *et al.*, 1989; Buttner *et al.*, 1989).

Thus, the S K^+ channel is a target for dual antagonistic modulation through independent second messenger systems. However, as illustrated in Fig. 5.9, the 5-HT and FMRFamide pathways do interact. Thus, during simultaneous application of these transmitters, the FMRFamide pathway overrides or antagonizes the 5-HT pathway (Fig. 5.9A). At the single-channel level, this override appears as a reopening by FMRFamide of S channels closed by 5-HT or cyclic AMP (Fig. 5.9B). Biochemical experiments suggest that this override is due to a decrease in protein phosphorylation by FMRFamide that does not involve a change in cAMP levels (Volterra *et al.*, 1988; Occor and Byrne, 1985).

5.4.3 Mechanism of M current modulation

The M current was originally identified in frog sympathetic neurones as a potassium current that is suppressed by the cholinergic agonist muscarine (Brown and Adams, 1980). Later the M current was also shown to be suppressed by a variety of other transmitters including LHRH, substance P, and nucleotide triphosphates, all acting on different receptors (Fig. 5.10). M currents have now been found in many cell types under the control of a variety of receptors (see Brown, 1988, for review). The most extensively studied M currents are in sympathetic neurones, smooth muscle cells, hippocampal neurones, and a neuroblastoma–glioma cell line, NG108–15. The biophysical characteristics of M current, summarized in Fig. 5.11, appear to be nearly identical in these disparate cells. The M current is a time- and voltage-dependent current that is activated at potentials positive to −60 or −70 mV. M current gating is slow, with time constants of activation and deactivation ranging from tens to hundreds of milliseconds.

There is also general agreement that M current suppression by agonists falls into the G protein-dependent class of transmitter actions (Fig. 5.12). Thus, in frog sympathetic neurones, M current suppression is linked to the activation of a PTX-insensitive G protein (Pfaffinger, 1988). A similar PTX-insensitive G protein is involved in M current suppression in mammalian sympathetic neurones (Marion, 1987) and probably also in hippocompal neurones (Dutar and Nicoll, 1988). In NG108–15 cells, M current modulation by bradykinin appears to involve a PTX-

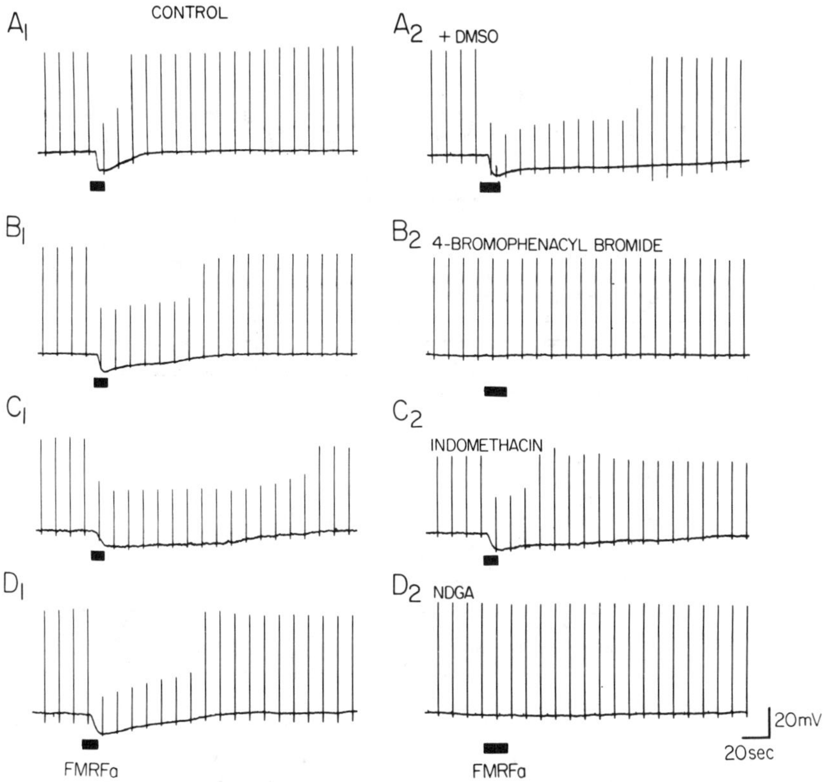

Fig. 5.8 — Action of arachidonate inhibitors on hyperpolarizing response to FMRFamide (2 μM). (A_1–D_i) Control responses. (A_2–D_2) Responses in presence of inhibitors. (A) The solvent DMSO has no effect on FMRFamide response. (B) 10 μM bromophenacyl bromide blocks response to FMRFamide (but not depolarizing response to 5-HT, not shown). (C) Indomethacin (5 μM) has no effect on hyperpolarizing response to FMRFamide. (D) NDGA (5 μM) blocks response to FMRFamide (but not response to 5-HT, not shown). (From Piomelli *et al.*, 1987b.)

sensitive G protein (Higashida *et al.*, 1986; Higashida and Brown, 1986; Brown and Higashida, 1988).

However, the mechanism by which G protein activation is coupled to M current suppression remains controversial. One consistent finding is that receptors that couple with the phospholipase C second messenger system are able to suppress M current. Two possible interpretations of this result have been proposed: first, the phospholipase C second messengers IP_3 and/or DAG cause the suppression of M current; second, the G protein that couples receptors to activation of phospholipase C also couples in an independent manner, to suppression of M current.

A role for the phospholipase C second messengers in the control of M current was first suggested by Higashida and Brown (1986), based on their studies of the suppression of M current in NG108–15 cells by bradykinin. They were able to show that protein kinase C activation by phorbol esters causes suppression of M current,

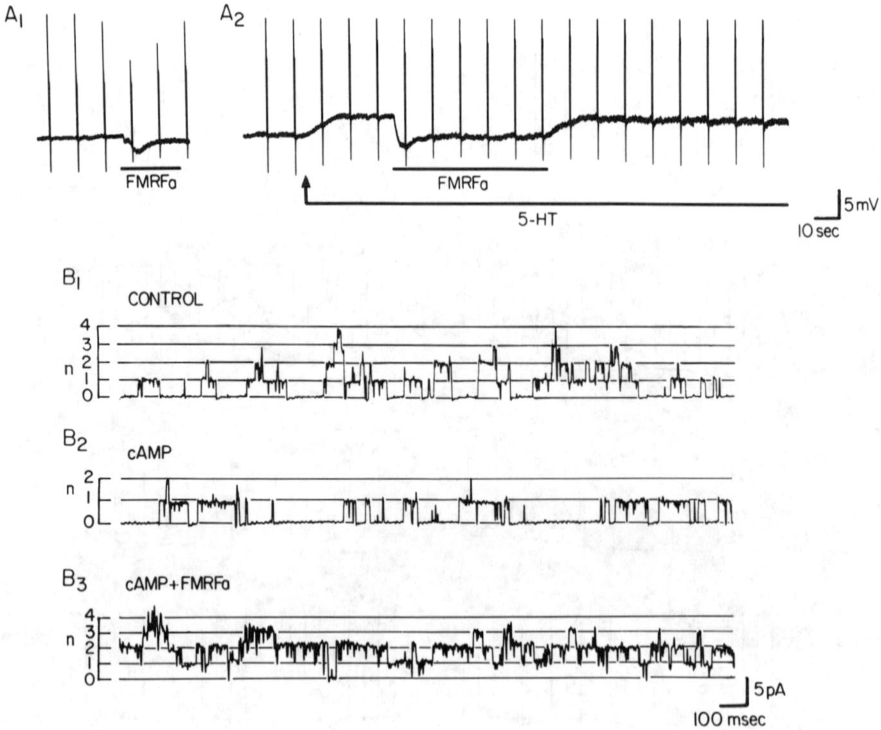

Fig. 5.9 — Interaction between 5-HT, cyclic AMP and FMRFamide. (A) Hyperpolarization to FMRFamide alone (A_1) or in presence of 5-HT (A_2). (B) FMRFamide reopens channels closed by cyclic AMP. (B_1) Control. (B_2) Dibutyryl cAMP closes two of four active S channels. (B_3) FMRFamide (in maintained presence of dibutyryl cyclicAMP) reopens the two closed channels and produces a normal increase in open probability in all four active channels. (From Belardetti *et al.*, 1987.)

mimicking the bradykinin effect. Since an earlier study suggested that bradykinin receptors couple to phospholipase C via a PTX-sensitive G protein (Higashida *et al.*, 1986) they proposed that this was the likely mechanism for M current suppression. Studies on M current regulation in smooth muscle cells (Clapp *et al.*, 1988) and some experiments in sympathetic neurones (Brown and Adams, 1987) have concluded that a similar mechanism is involved in M current inhibition. However, in hippocampal neurones, Dutar and Nicoll (1988) have concluded that IP_3, the other phospholipase C messenger, is responsible for M current suppression. In these neurones, agonists that activate phospholipase C also suppress M current but, unlike NG108–15 cells, phorbol ester activation of protein kinase C has no effect (Malenka *et al.*, 1986). In contrast, intracellular injection of IP_3 results in a loss of M current in an apparently Ca^{2+}-independent manner, suggesting a direct effect of IP_3 (or an IP_3 metabolite) on M current. Whether this IP_3 effect is the physiological mechanism for agonist suppression of M current remains to be shown.

Studies by Hille and his colleagues have provided strong evidence that the

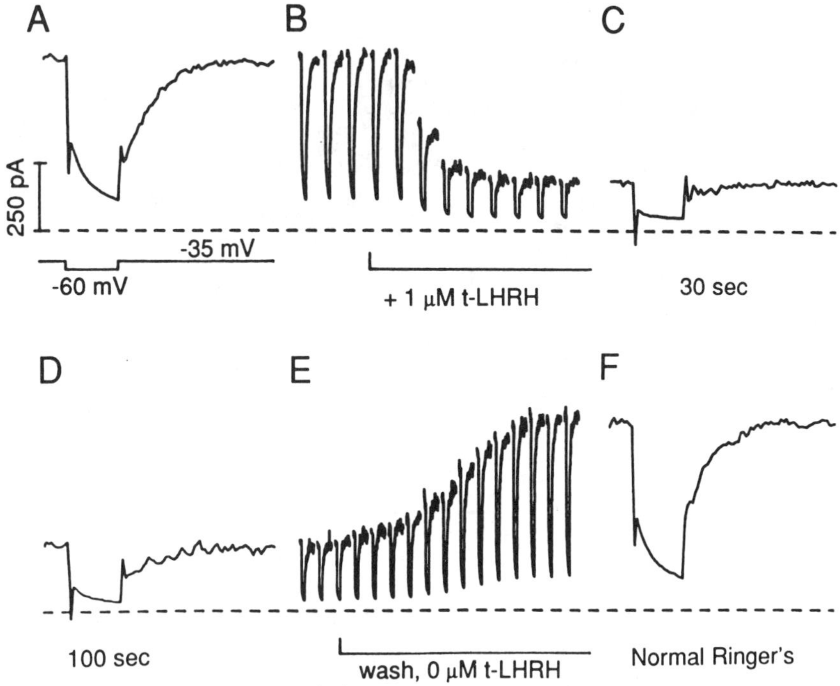

Fig. 5.10 — LHRH suppresses M current in frog sympathetic neurones. (A) Control M current in response to hyperpolarizing voltage-clamp step. (B) M current recorded in response to voltage-clamp hyperpolarization once every 5 s as *t*-LHRH is applied. Note rapid suppression of M current. (C and D) Residual current at 30 sec (C) and 100 sec (D) in the presence of *t*-LHRH lacks characteristic M current kinetics. (E) Reversal of *t*-LHRH effect upon washing requires tens of seconds. (F) M current recovers to a peak amplitude 20% greater than initial control size. (From Pfaffinger, 1988.)

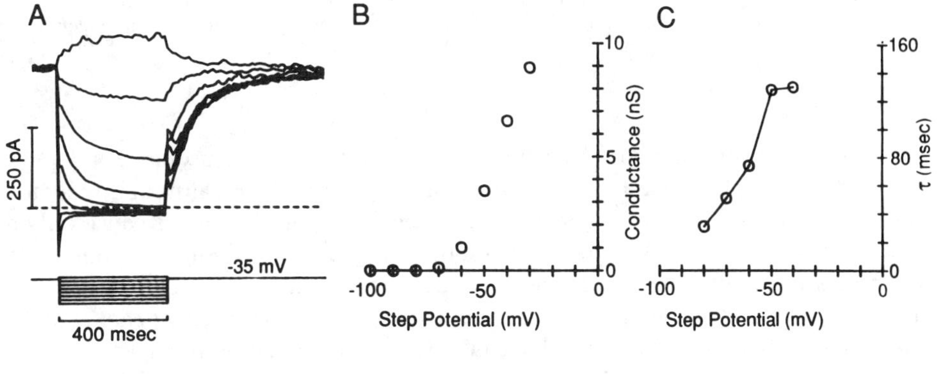

Fig. 5.11 — Biophysical properties of M current. Whole-cell voltage-clamp of M current from frog sympathetic neurone. Cell held at −35 mV where M current is dominant. (A) M current gating in response to 10-mV steps between −30 and −100 mV. Reversal potential is −90 mV. (B) Plot of conductance at end of test step as a function of test potential. (C) M current time constant of deactivation as a function of test potential. (From Pfaffinger, 1988.)

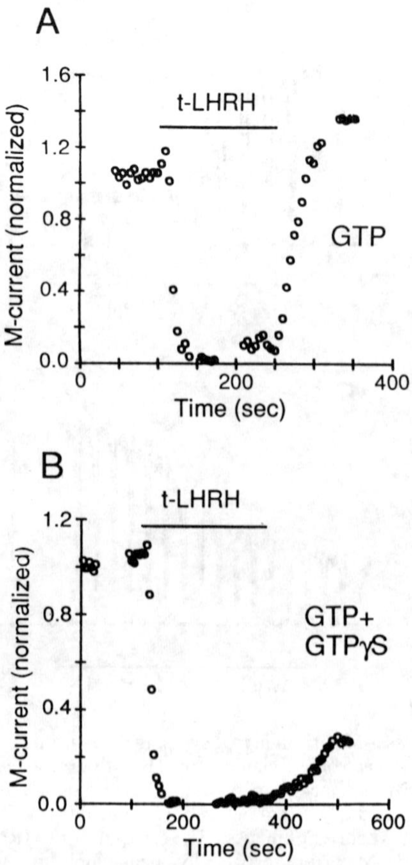

Fig. 5.12 — M current suppression depends on G protein activation. Amplitude of M current plotted as a function of time during response to *t*-LHRH. (A) With 100 μM GTP in pipette, the M current is completely suppressed and current 'over-recovers' following washout. (B) With 100 μM GTP plus 1 μM GTP-γ-S in pipette M current suppression is almost completely irreversible, indicating stimulation of GTP-γ-S incorporation by *t*-LHRH. (From Pfaffinger, 1988.)

agonist-induced suppression of M current in frog sympathetic neurones is independent of the phospholipase C second messenger cascade. Thus, although receptor activation causes PI turnover in sympathetic neurones, phorbol esters cause only a partial suppression of M current. Addition of LHRH following phorbol ester suppresses the remaining current normally, suggesting that LHRH receptors can inhibit M current in a protein kinase C-independent manner (Pfaffinger *et al.*, 1988). Furthermore, IP_3 or internal Ca^{2+} have no effect on M current or its suppression by agonists. Finally Bosma and Hille (1989) have shown that the protein kinase C blockers H-7 and staurosporine have no effect on agonist suppression of M current, although they do block the inhibition of M current by phorbol esters. Therefore, it appears that, at least in sympathetic neurones, the PTX-insensitive G protein may couple both to activation of phospholipase C and to a separate, possibly direct, suppression of M current.

In addition to suppression by certain agonists, M current can also be up-regulated by other agonists. In smooth muscle, this activation is seen with β-adrenergic agonists (Sims *et al.*, 1988). The response is also mimicked by the cell permeant analogue 8Br-cAMP, suggesting that protein kinase A phosphorylation may activate the M channel. Interestingly, ACh antagonizes the increase in M current with adrenaline or cyclic AMP, similar to the override of the cyclic AMP-dependent 5-HT response on S channel activity with FMRFamide in *Aplysia.* A similar increase in M current is produced in hippocampal neurones by somatostatin, although it is not known if an increase in cyclic AMP is the mechanism of action there (Moore *et al.*, 1988). While a sensitivity of M current to cyclic AMP has not been seen in sympathetic neurones, an up-regulation of M current termed 'over-recovery' occurs following the removal of an agonist that suppresses M current (Pfaffinger, 1988; Pfaffinger *et al.*, 1988; Bosma and Hille, 1989). It is not known if this over-recovery is a physiologically important response.

5.4.4 Modulation of other K^+ currents

5.4.4.1 Cardiac delayed rectifier channel

One of the first examples of cyclic AMP-dependent channel modulation to be described was the increase in the delayed rectifier current (I_x or I_K) in cardiac muscle cells in response to β-adrenergic agonists (Tsien *et al.*, 1972). Thus, Tsien *et al.* (1972) found that mono-butyryl cyclic AMP as well as the phosphodiesterase inhibitors theophylline and RO 7–2956 mimic the actions of adrenaline in increasing the delayed K^+ current in cardiac Purkinje fibers. Brum *et al.* (1983) found that the effects of adrenaline were also mimicked by intracellular injection of the catalytic subunit of PKA in ventricular muscle cells. Using a kinetic analysis of the action of noradrenaline on voltage-clamped cardiac Purkinje fibers, Bennett *et al.* (1986) proposed that noradrenaline selectively increases a single rate constant for channel activation. The increase in K^+ current acts to offset a simultaneous increase in calcium current with β-adrenergic stimulation and serves to keep the action potential duration relatively constant.

More recently, it has been shown that the cardiac delayed rectifier is also modulated by PKC activation (Walsh and Kass, 1988; Tohse *et al.*, 1987). Interestingly, PKC and PKA activation produce different voltage-dependent increases in I_K, with PKA producing a larger increase at negative voltages and PKC producing a larger increase at positive voltages (Walsh and Kass, 1989). This result argues for separate modulatory control sites on the I_K channel.

5.4.4.2 Neuronal delayed rectifier channels

PKC activation also modulates a delayed rectifier K^+ current in hippocampal pyramidal neurones; however, here the effect is to decrease the magnitude of the current (Doerner *et al.*, 1988). In mouse dorsal root ganglion neurones, phorbol esters and forskolin decrease the same delayed voltage-dependent potassium current (Grega *et al.*, 1987), thus exhibiting a convergent modulation similar to that seen with the cardiac delayed K^+ current.

5.4.4.3 *Inward rectifier channel*

In the spontaneously bursting neurone R15 of *Aplysia*, 5-HT increases an inwardly rectifying K^+ conductance (Benson and Levitan, 1983) by increasing the apparent number of active inward rectifier channels (Gunning, 1987). 5-HT also causes an increase in phosphorylation in these cells (Lemos *et al.*, 1985), and its physiological action is blocked by the injection of PKI (Adams and Levitan, 1982). The increase in resting potassium conductance leads to a hyperpolarization of R15 and an increase in the inter-burst interval. Recently, Kramer *et al.* (1988) have reported that Ca^{2+} influx during bursting activity down-regulates the inward rectifier current, possibly by activating a cyclic AMP phosphodiesterase.

The inward rectifier channel is also the target for modulation in magnocellular cholinergic neurones of the globus pallidus in rats (Stanfield *et al.*, 1985). Here substance P increases neuronal excitability by inhibiting the inward rectifier, an effect that depends on the activation of a PTX-insensitive G protein (Nakajima *et al.*, 1988).

5.4.4.4 *Calcium-activated K^+ channels*

Voltage-dependent calcium-activated K^+ channels are also regulated by cyclic AMP-dependent protein phosphorylation. In *Helix* neurones, intracellular perfusion of the catalytic subunit of PKA leads to an increase in calcium-activated K^+ current (DePeyer *et al.*, 1982). Using the patch-clamp technique, Ewald *et al.* (1985) showed that the opening of a 20-pS Ca^{2+}-activated K^+ channel from *Helix*, first identified by Lux *et al.* (1981), was increased by purified catalytic subunit of PKA in cell-free inside-out patches. Larger-conductance 100–200-pS Ca^{2+}-activated K^+ channels from mammalian brain reconstituted in planar lipid bilayers have also been shown to respond to the C subunit of PKA with an increase in channel open probability (Farley and Rudy, 1988). So far, a physiological role for modulation of these channels by neurotransmitters has not been identified, although Kehoe (1985b) has reported the synaptic modulation of a Ca^{2+}-activated K^+ channel in *Aplysia* through a cyclic AMP-dependent mechanism.

A second distinct type of Ca^{2+}-activated K^+ channel underlies a slow afterhyperpolarization in a variety of nerve and muscle cells (Pennefather *et al.*, 1985; Blatz and Magleby, 1986). This AHP channel is relatively insensitive to TEA, is blocked by the bee venom apamin at nanomolar concentrations (Pennefather *et al.*, 1985) and has a smaller single-channel conductance of around 10–14 pS (Blatz and Magleby, 1986). In hippocampal pyramidal neurones, noradrenaline, acting through cyclic AMP, appears to decrease the magnitude of this AHP current (Madison and Nicoll, 1982). As a result of the decrease in this K^+ current, noradrenaline leads to a decrease in spike accommodation in these cells. This same AHP current also appears to be regulated by PKC activation, as phorbol esters also lead to a decrease in the slow AHP in pyramidal neurones (Baraban *et al.*, 1985; Malenka *et al.*, 1986). Thus, as seen above, PKA and PKC appear to converge to produce similar modulatory actions on a common ion conductance.

5.4.4.5 *K^+ currents of* Aplysia *bag cells*

In the neuroendocrine bag cells of *Aplysia*, forskolin and/or cyclic AMP analogues modulate the activity of a variety of potassium currents (Strong and Kaczmarek,

1987). Thus forskolin increases the rate of inactivation of a transient outward current (I_A) (Strong, 1984). In addition, forskolin reduces the amplitute of a non-inactivating delayed rectifier K^+ current and increases the rate of inactivation of a second inactivating delayed rectifier K^+ current (Strong and Kaczmarek, 1986). The inhibition of these K^+ currents is thought to account partly for the increase in action potential duration and amplitude during the afterdischarge of the bag cells, a prolonged phase of spontaneous repetitive firing that is associated with release of bag cell peptides during egg-laying (see Strong and Kaczmarek, 1987). In these neurones, PKC activation with phorbol esters has no effect on K^+ currents, although it does cause an increase in inward calcium current (DeRiemer *et al.*, 1985).

5.4.4.6 Modulation of transient and calcium-activated K^+ currents in Hermissenda *photoreceptors*

In photoreceptors of the mollusc *Hermissenda,* classical conditioning of a photoresponse is associated with a long-lasting decrease in a transient K^+ current (I_A) and a calcium-activated K^+ current ($I_{K(Ca)}$) (Alkon *et al.*, 1982, 1985). These effects can be mimicked by intracellular injection of calcium/calmodulin-dependent protein kinase (Sakakibura *et al.*, 1986) and/or by injection of PKC (Farley and Auerbach, 1986). Alkon *et al.* (1986) found similar decreases in I_A and $I_{K(Ca)}$ in these cells in response to phorbol esters. However, the phorbol esters were only active if the photoreceptor had been first loaded with calcium (for example using a train of action potentials). In a later study, Alkon *et al.* (1988) obtained the unexpected result that injection of purified PKC caused an *increase* in the calcium-activated K^+ current. However, if the cell was pretreated with phorbol ester, PKC injection now *decreased* both I_A and $I_{K(Ca)}$, simulating the previously reported effects of phorbol ester on the calcium-loaded cells. Alkon *et al.* (1988) attribute these different effects to differences between the actions of cytoplasmic PKC (increasing $I_{K(Ca)}$) and membrane-bound PKC (decreasing I_A and $I_{K(Ca)}$), with the phorbol ester treatment causing translocation of cytoplasmic PKC to the membrane. The reason for the discrepancy between the results of Farley and Auerbach (1986) and Alkon *et al.* (1988) regarding the action of PKC in non-phorbol ester-treated cells is unclear (although cf. Alkon *et al.*, 1988).

5.4.4.7 Synaptically activated and inactivated K^+ currents

Most of the K^+ currents considered above function in the absence of neurotransmitter or hormone; their activity is either up- or down-modulated in response to agonists. However, other K^+ currents (including K_{ACh}) are only active in the presence of agonist. These transmitter-activated K^+ conductances can be distinguished from the directly gated class of receptor channels in that a G protein or second messenger system has been implicated. Thus, North and his colleagues have described an inwardly rectifying K^+ current in rat locus ceruleus neurones (Williams *et al.*, 1988) and in myenteric neurones (Surprenant and North, 1988) that is activated by *u*-opioid and α2-adrenergic agonists through a G protein-dependent mechanism. This K^+ channel is only activated by agonists inside the patch pipette in cell-attached patches, suggesting a tight coupling between the agonist receptor and K^+ channel (Miyake *et al.*, 1989). In hippocampal neurones, 5-HT and $GABA_B$

receptor agonists activate a K$^+$ current through a PTX-sensitive G protein (Andrade *et al.*, 1986).

Furthermore, synaptically activated K$^+$ conductances can, in turn, be modulated by other second messenger dependent transmitter actions. Thus, in many *Aplysia* neurons, ACh evokes an increase in K$^+$ current via a PTX-sensitive G protein (Sasaki and Sato, 1987). In the medial neurons of the pleural ganglion, Kehoe (1985a) has shown that the ACh-induced K$^+$ current is inhibited by stimulation of three interneurons through a cAMP-dependent mechanism.

5.4.4.8 Fatty acid-activated K$^+$ channels

Arachidonic acid, as well as other fatty acids, appears to have a second mode of action on ion channel activity that does not depend on metabolic processing. Thus, in cardiac atrial cells, Kim and Clapham (1989) have found that arachidonic acid activates a 160-pS K$^+$ channel in cell-attached or inside-out patches. This effect is not blocked by NDGA or indomethacin, nor is it mimicked by any lipoxygenase metabolite; however, it is also produced by linoleic acid. Finally, phosphatidylcholine and phosphatidylserine, but not arachidonic acid, activate a 68-pS K$^+$ channel in these same cells. Both of these channels are distinct from the K_{ACh} channel. In smooth muscle cells, Ordway *et al.* (1989) have described a 25-pS K$^+$ channel that is activated by a variety of fatty acids, including arachidonic acid and oleic acid.

5.5 HOW DOES PHOSPHORYLATION MODULATE K$^+$ CHANNEL ACTIVITY?

One of the major unsolved puzzles regarding channel modulation is how protein phosphorylation alters K$^+$ channel function. At present, we do not know for any K$^+$ channel whether its modulation is due to direct phosphorylation of the channel or whether the modulation depends on the phosphorylation of some intermediate protein. For example, PKA could phosphorylate a membrane ion pump or transport mechanism, leading to an alteration in intracellular ionic activity (e.g. of Ca^{2+} or pH) which might in turn alter K$^+$ channel activity. Alternatively, the PKA could phosphorylate some other enzyme, perhaps another kinase, which could then alter channel activity. Such an indirect mechanism has been well studied in the case of glycogen metabolism. Here, epinephrine increases glycogen metabolism by first activating PKA, which phosphorylates and activates phosphorylase kinase, which then phosphorylates and activates glycogen phosphorylase (Cohen, 1982).

Although there is no direct evidence that K$^+$ channels are phosphorylated, several lines of indirect evidence suggest that K$^+$ channel modulation may indeed be often associated with direct K$^+$ channel phosphorylation. First, other ion channels that have been biochemically purified have been found to be good substrates for protein kinases, including PKA and PKC. Thus Huganir, Greengard and their colleagues have shown that the nicotinic AChR is phosphorylated on several sites by PKA, PKC and tyrosine kinase (see Miles and Huganir, 1988). Recently, several groups have demonstrated that these phosphorylation events are associated with an increase in receptor desensitization (Huganir *et al.*, 1986; Middleton *et al.*, 1986; Downing and Role, 1987). Both the voltage-dependent sodium channel (Costa and Catterall, 1984) and the voltage-dependent L-type calcium channel (Curtis and Catterall, 1985; Hofmann *et al.*, 1987) are also substrates for PKA. For the calcium channel, this phosphorylation leads to an increase in calcium channel opening, while for the sodium channel phosphorylation enhances the inactivation process (see Catterall, 1988). Recently, several K^+ channels have been cloned and sequenced from *Drosophila*, mouse and rat that are homologous to the Shaker I_A K^+ current and shown to contain a putative cAPK phosphorylation site (Tempel *et al.*, 1987, 1988; Baumann *et al.*, 1987). Whether this site is indeed phosphorylated and what functional effect it has on A channel activity remains to be determined.

A second line of indirect evidence for K^+ channel phosphorylation comes from experiments using cell-free membrane patches and channels reconstituted in bilayers. Several groups have shown that application of purified C subunit of PKA to cell-free inside-out patches can modulate K^+ channel activity. Thus Shuster *et al.* (1985) showed that the C subunit of PKA produced all-or-none closures of the S-type potassium channel from *Aplysia* sensory neurones in inside-out patches (Fig. 5.13C). Ewald *et al.* (1985) showed that the C subunit caused an increase in opening of a calcium-activated K^+ channel from *Helix* neurones in cell-free patches (Fig. 5.13B), consistent with the effects of cyclic AMP on whole-cell currents (DePeyer *et al.*, 1982). In frog taste receptor cells, the C subunit of PKA also closed a K^+ channel in inside-out patches (Avenet *et al.*, 1988), simulating the action of exogenous cyclic AMP on intact taste cells (Fig. 5.13A). These cell-free experiments argue against a role for some cytoplasmic intermediate or a change in ionic activity as the mode of PKA action in controlling channel function. However, since cell-free patches are likely to contain a number of peripheral and integral membrane proteins as well as some cytoskeletal elements, these types of experiments do not provide definitive proof that the channel itself is phosphorylated.

In some cases, it has been possible to reconstitute ion channels from membrane vesicles into membranes made from purified lipids. Thus Ewald *et al.* (1985) also showed that the calcium-activated K^+ channel from *Helix* could be modulated by PKA under these conditions. Farley and Rudy (1988) showed similar modulation of calcium-activated K^+ channels from rat brain synaptosomal membranes. However, while reconstituted channels in bilayers represent a 'cleaner' preparation than the cell-free patch, one still cannot rule out the possibility that some intermediary protein remains associated with the channel during the reconstitution. Thus, the final answer to the question as to the role that K^+ channel phosphorylation plays in K^+ channel modulation must await the purification and/or cloning and expression of these different types of K^+ channels.

5.6 SUMMARY

Potassium channel modulation is clearly a widespread phenomenon in both nerve and muscle cells and produces marked changes in cellular electrical activity. Much progress has been made in characterizing the actions of modulatory transmitters on ionic channels and currents and in defining the roles that different G proteins and second messengers play in channel modulation. However, several major questions remain. In view of the increasing number of second messenger pathways, how do different second messenger pathways interact? How does FMRFamide manage to antagonize the action of 5-HT in *Aplysia* sensory neurones and ACh override the effects of adrenaline in smooth muscle? A second major area of ignorance concerns

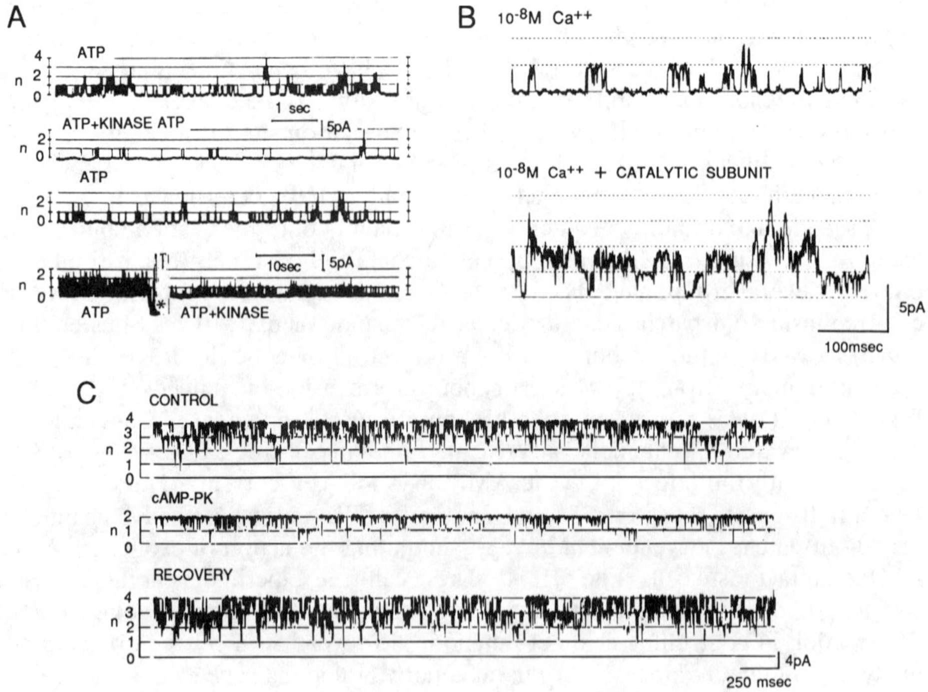

Fig. 5.13 — K$^+$ channel regulation by catalytic subunit of PKA in cell-free inside-out patches. (A) PKA closes a 44-pS taste receptor K$^+$ channel. (From Avenet *et al.*, 1988.) (B) PKA increases opening of a 20-pS Ca-activated K$^+$ channel in *Helix* neurones (From Ewald *et al.*, 1985.) (C) PKA closes two out of four active S channels from *Aplysia* sensory neurone. (From Shuster *et al.*, 1985.)

the molecular mechanism of channel modulation. How does phosphorylation modulate channel activity? Is there a molecular logic behind the vast array of modulatory actions that phosphorylation produces? Do G proteins actually bind to channels and directly modulate them or are intermediary membrane proteins involved? With the recent progress in cloning of potassium channels, protein kinases and G proteins, the answers to such questions may soon be at hand.

REFERENCES

Abood, M. E., Hurley, J. B., Pappone, M. C., Bourne, H. R., and Stryer, L. (1982) Functional homology between signal-coupling proteins. Cholera toxin inactivates the GTPase activity of transducin. *J. Biol. Chem.* **257** 10540–10543.

Abrams, T. W., Castellucci, V. F., Camardo, J. S., Kandel, E. R., and Lloyd, P. E. (1984) Two endogenous neuropeptides modulate the gill and siphon withdrawal reflex in *Aplysia* by presynaptic facilitation involving cAMP-dependent closure of a serotonin-sensitive potassium channel. *Proc. Natl. Acad. Sci. USA* **81** 7956–7960.

Adams, W. B., and Levitan, I. B. (1982) Intracellular injection of protein kinase inhibitor blocks the serotonin-induced increase in K^+ conductance in *Aplysia* neuron R15. *Proc. Natl Acad. Sci. USA* **79** 3877–3880.

Alkon, D. L., Lederhendler, I., and Shoukimas, J. J. (1982) Primary changes of membrane currents during retention of associative learning. *Science* **226** 1037–1045.

Alkon, D. L., Sakakibara, M., Forman, R., Harrigan, J., Lederhendeler, I., and Farley, J. (1985) Reduction of two voltage-dependent K^+ currents mediates retention of a learned association. *Behav. Neural Biol.* **44** 278–300.

Alkon, D. L., Naito, S., Kubota, M., Chen, C., Bank, B., Smallwood, J., Gallant, P., and Rasmussen, H. (1988) Regulation of *Hermissenda* K^+ channels by cytoplasmic and membrane-associated protein kinase C. *J. Neurochem.* **51** 903–917.

Alkon, D., Kubota, M., Neary, J. T., Naito, S., Coulter, D., and Rasmussen, H. (1986) C-kinase activation prolongs Ca^{2+}-dependent inactivation of K^+ currents. *Biochem. Biophys. Res. Commin.* **134** 1215–1222.

Andrade, R., Malenka, R. C., and Nicoll, R. A. (1986) A G protein couples serotonin and $GABA_B$ receptors to the same channels in hippocampus. *Science* **234** 1261–1265.

Aub, D. L., Frey, E. A., Sekura, R. D., and Cote, T. E. (1986) Coupling of the thyrotropin-releasing hormone receptor to phospholipase C by a GTP-binding protein distinct from the inhibitory or stimulatory GTP-binding protein. *J. Biol. Chem.* **261** 9333–9340.

Avenet, P., Hofmann, F., and Lindemann, B. (1988) Transduction in taste receptor cells requires cAMP-dependent protein kinase. *Nature* **331** 351–354.

Baraban, J. M., Snyder, S. H., and Alger, B. E. (1985) Protein kinase C regulates ionic conductance in hippocampal pyramidal neurons: Electrophysiological effects of phorbol esters. *Proc. Natl Acad. Sci. USA* **82** 2538–2542.

Baumann, A., Krah-Jentgens, I., Mueller, R., Mueller-Holtkamp, F., Seidel, R., Kecskemethy, N., Casal, J., Ferrus, A., and Pongs, O. (1987) Molecular organization of the Shaker complex of *Drosophila:* characterization of an I_A channel transcript with homology to vertebrate Na^+ channel. *EMBO J.* **6** 3419–3429.

Baster, D. A., and Byrne, J. H. (1987) Modulation of membrane currents and excitability by serotonin and cAMP in pleural sensory neurons of *Aplysia. Soc. Neurosci. Abstr.* **13** 1440.

Belardetti, F., Kandel, E. R., and Siegelbaum, S. A. (1987) Neuronal inhibition by the peptide FMRFamide involves opening of S K^+ channels. *Nature* **325** 153–156.

Belardetti, F., Campbell, W., Flack, J. R., Demontis, G., and Rosolowsky, M., (1988) Products of heme-catalysed transormation of the arachidonate derivative 12-HPETE open S-type K^+ channels in *Aplysia. Neuron* **3** 497–505.

Bennett, P., McKinney, L., Begenisich, T., and Kass, R. S. (1986) Adrenergic modulation of the delayed rectifier potassium channel in calf cardiac Purkinje fibers. *Biophys. J.* **49** 839–848.

Benson, J. A., and Levitan, I. B. (1983) Serotonin increases an anomalously rectifying K^+ current in *Aplysia* neuron R15. *Proc. Natl Acad. Sci. USA* **80** 3522–3525.

Bernier, L., Castellucci, V. F., Kandel, E. R., and Schwartz, J. H. (1982) Facilatory transmitter causes a selective and prolonged increase in adenosine 3′:5′-monophosphate in sensory neurons mediating the gill and siphon withdrawal reflex in *Aplysia. J. Neurosci.* **2** 1682–1691.

Berridge, M. J. (1988) Inositol lipids and calcium signalling. *Proc. Roy. Soc. Lond. B* **234**, 359–378.

Berridge, M. J., and Irvine, R. F. (1984) Inositol triphosphate, a novel second messenger in cellular signal transduction. *Nature* **312** 315–321.

Blackshear, P. J., Nairn, A. C., and Kuo, J. F. (1988) Protein kinases 1988: a current perspective. *FASEB J.* **2** 2957–2969.

Blatz, A. L., and Magleby, K. L. (1986) Single apamin-blocked Ca-activated K^+ channels of small conductance in cultured rat skeletal muscle. *Nature* **323** 718–720.

Bosma, M. M., and Hille, B. (1989) Protein kinase C is not necessary for peptide-induced suppression of M-current or for desensitization of the peptide receptors. *Proc. Natl Acad. Sci. USA* **86** 2943–2947.

Breitweiser, G. E., and Szabo, G. (1985) Uncoupling of cardiac muscarinic and β-adrenergic receptors from ion channels by a guanine nucleotide analogue. *Nature* **317** 538–540.

Brezina, V., Eckert, R., and Erxleben, C. (1987a) Modulation of potassium conductances by an endogenous neuropeptide in neurones of *Aplysia californica. J. Physiol. (Lond.)* **382** 267–290.

Brezina, V., Eckert, R., and Erxleben, C. (1987b) Suppression of calcium current by an endogenous neuropeptide in neurones of *Aplysia californica. J. Physiol. (Lond.)* **388** 565–595.

Brown, D. A. (1988) M currents: an update. *Trends Neurosci.* **11** 294–299.

Brown, D. A., and Adams, P. R. (1980) Muscarinic suppression of a novel voltage-sensitive K^+ current in a verterbrate neurone. *Nature* **283** 673–676.

Brown, D. A., and Adams, P. R. (1987) Effects of phorbol dibutyrate on M currents and M current inhibition in bullfrog sympathetic neurons. *Cell. Mol. Neurobiol.* **7** 255–269.

Brown, D. A., and Higashida, H. (1988) Inositol 1,4,5-trisphosphate and diacylglycerol mimic bradykinin effects on mouse neuroblastoma × rat glioma hybrid cells. *J. Physiol. (Lond.)* **397** 185–207.

Brum, G., Flockerzi, V., Hofmann, F., Osterrieder, W., and Trautwein, W. (1983) Injection of catalytic subunit of cAMP-dependent protein kinase into isolated cardiac myocytes. *Pflügers Arch.* **398** 147–154.

Buttner, N., Siegelbaum, S. A., and Volterra, A. (1989) Direct modulation of *Aplysia* S-K^+ channels by a lipoxygenase metabolite of arachidonic acid. *Nature* (in press).

Cassel, D., and Selinger, Z. (1977) Mechanism of adenylate cyclase activation by cholera toxin: Inhibition of GTP hydrolysis at the regulatory site. *Proc. Natl Acad. Sci. USA* **74** 3307–3311.

Catellucci, V. F., Kandel, E. R., Schwartz, J. H., Wilson, F. D., Nairn, A. C., and Greengard, P. (1980) Intracellular injection of the catalytic subunit of cyclic AMP-dependent protein kinase simulates facilitation of transmitter release underlying behavioral sensitization in *Aplysia*. *Proc. Natl Acad. Sci. USA* **77** 7492–7496.

Castellucci, V. F., Nairn, A., Greengard, P., Schwartz, J. H., and Kandel, E. R. (1982) Inhibitor of adenosine 3′:5′-monophosphate-dependent protein kinase blocks presynaptic facilitation in *Aplysia*. *J. Neurosci.* **2** 1673–1681.

Catterall, W. (1988) Structure and function of voltage-sensitive ion channels. *Science* **242** 50–61.

Cerbai, E., Klockner, U., and Isenberg, G. (1988) The α subunit of the GTP binding protein activates muscarinic potassium channels of the atrium. *Science* **240** 1782–1783.

Clapp, L. H., Sims, S. M., Singer, J. J., and Walsh, J. V. (1988) Cholinergic suppression of both endogenous and isoproterenol-induced M current in isolated smooth muscle cells is mimicked by a diacylglycerol analog. *Soc. Neurosci. Abstr.* **14** 1088.

Codina, J., Olate, J., Abramowitz, J., Mattera, R., Cook, R. G., and Birnbaumer, L. (1988) A i-3 cDNA encodes the α subunit of Gk, the stimulatory G protein of receptor-regulated K^+ channels. *J. Biol. Chem.* **263** 6746–6750.

Codina, J., Yatani, A., Grenet, D., Brown, A. M., and Birnbaumer, L. (1987) The α subunit of the GTP binding protein Gk opens atrial potassium channels. *Science* **236** 442–445.

Cohen, P. (1982) The role of protein phosphorylation in neural and hormonal control of cellular activity. *Nature* **296** 613–620.

Costa, M. R., and Catterall, W. A. (1984) Phosphorylation of the α subunit of the sodium channel by protein kinase C. *Cell. Mol. Neurobiol.* **4** 291–297.

Coombs, J., and Thompson, S. (1987) Forskolin's effect on transient K^+ currents in nudibranch neurons is not reproduced by cAMP. *J. Neurosci.* **7** 443–452.

Cooper, J. R., Bloom, F. E., and Roth, R. H. (1986) *The Biochemical Basis of Neuropharmacology*. Oxford University Press, New York.

Curtis, B. M., and Catterall, W. A. (1985). Phosphorylation of the Ca antagonist receptor of the voltage-sensitive calcium channel by cAMP-dependent protein kinase. *Proc. Natl. Acad. Sci. USA* **82** 2528–2532.

DePeyer, J. E., Cachelin, A. B., Levitan, I. B., and Reuter, H. (1982) Ca^{2+}-activated K^+ conductance in internally perfused snail neurons is enhanced by protein phosphorylation. *Proc. Natl Acad. Si. USA* **79** 4207–4211.

DeRiemer, S. A., Strong, J. A., Albert, K. A., Greengard, P., and Kaczmarek, L. K. (1985) Enhancement of calcium current in *Aplysia* neurones by phorbol ester and protein kinase C. *Nature* **313** 313–316.

Deterre, P., Paupardin-Tritsch, D., Bockaert, J., and Gerschenfeld, H. M. (1981) Role of cyclic AMP in a serotonin-evoked slow inward current in snail neurones. *Nature* **290** 783–785.

Deterre, P., Paupardin-Tritsch, D., Bockaert, J., and Gerschenfeld, H. M. (1982) cAMP-mediated decrease in K^+ conductance evoked by serotonin and dopamine in the same neuron: a biochemical and physiological single-cell study.

Proc. Natl. Acad. Sci. USA **79** 7934–7938.

Doerner, D., Pitler, T. A., and Alger, B. E. (1988) Protein kinase C activators block specific calcium and potassium current components in isolated hippocampal neurons. *J. Nerosci.* **8** 4069–4078.

Downing, J. E., and Role, L. W. (1987) Activators of protein kinase C enhance acetylcholine receptor desensitization in sympathetic ganglion neurons. *Proc. Natl Acad. Sci. USA* **84** 7739–7743.

Drummond, G. I. (1983) Cyclic nucleotides in the nervous system. *Adv. Cyclic Nucleotide Res.* **15** 373–494.

Dutar, P., and Nicoll, R. A. (1988) Stimulation of phosphatidylinositol (PI) turnover may mediate the muscarinic suppression of the M-current in hippocampal pyramidal cells. *Neursoci. Lett.* **85** 89–94.

Edelman, A. M., Blumenthal, D. K., and Krebs, E. G. (1987) Protein serine/threonine kinases. *Ann. Rev. Biochem.* **56** 567–613.

Edmonds, B. W., Kandel, E. R., and Klein, M. (1988) Different functional roles for two components of calcium current in *Aplysia* sensory neurons. *Soc. Neurosci. Abstr.* **14** 1206.

Ewald, D. A., Williams, A., and Levitan, I. B. (1985) Modulation of single Ca^{2+}-dependent K^+-channel activity by protein phosphorylation. *Nature* **315** 503–506.

Farley, J., and Auerbach, S. (1986) Protein kinase C activation induces conductance changes in *Hermissenda* photoreceptors like those seen in associative learning. *Nature* **319** 220–223.

Farley, J., and Rudy, B. (1988) Multiple types of voltage-dependent Ca^{2+}-activated K^+ channels of large conductance in rat brain synaptosomal membranes. *Biophys. J.* **53** 919–934.

Fesenko, E. E., Kolesnikov, S. S., and Lyubarsky, A. L. (1985). Induction by cyclic GMP of cationic conductance in plasma membrane of retinal rod outer segment. *Nature* **313** 310–313.

Gilman, A. G. (1984) G proteins and dual control of adenylate cyclase. *Cell* **36** 577–579.

Gilman, A. G. (1987) G proteins: transducers of receptor generated signals. *Ann. Rev. Biochem.* **56** 615–649.

Graziano, M. P., and Gilman, A. G. (1987) Guanine nucleotide regulatory proteins: mediators of transmembrane signalling. *Trends Pharm. Sci.* **8** 478–481.

Grega, D. S., Werz, M. A., and MacDonald, R. L. (1987) Forskolin and phorbol esters reduce the same potassium conductance of mouse neurons in culture. *Science* **235** 345–348.

Grenningloh, G., Rienitz, A., Schmitt, B., Methfessel, C., Zensen, M., Beyreuther, K., Gundelfinger, E. D., and Betz, H. (1987) The strychnine-binding subunit of the glycine receptor shows homology with nicotinic acetylcholine receptors. *Nature* **328** 215–220.

Gunning, R. (1987) Increased numbers of ion channels promoted by an intracellular second messenger. *Science* **235** 80–82.

Hall, Z. W. (1987) Three of a kind: the β-adrenergic receptor, the muscarinic acetylcholine receptor and rhodoposin. *Trends Neurosci.* **10** 99–101.

Halushka, P. V., Mais, D. E., Mayeux, P. R., and Morinelli, T. A. (1989)

Thromboxane, prostaglandin and leukotriene receptors. *Ann. Rev. Pharm. Tox.* **10** 213–239.

Hescheler, J., Rosenthal, W., Trautwein, W., and Schultz, G. (1987) The GTP-binding protein, G_o, regulates neuronal calcium channels. *Nature* **325** 445–447.

Higashida, H., and Brown, D. A. (1986) Two polyphosphatidylinositide metabolites control two K^+ currents in a neuronal cell. *Nature* **323** 333-335.

Higashida, H., Streaty, R. A., Klee, W., and Nirenberg, M. (1986) Bradykinin-activated transmembrane signals are coupled via N_o or N_i to production of inositol 1,4,5-trisphosphate, a second messenger in NG108–15 neuroblastoma-glioma hybrid cells. *Proc. Natl Acad. Sci. USA* **83** 942–946.

Hochner, B., Klein, M., Schacher, S., and Kandel, E. R. (1986) Additional component in the cellular mechanism of presynaptic facilation contributes to behavioral distribution in *Aplysia. Proc. Natl Acad. Sci. USA* **83** 8794–8798.

Hockberger, P., Toselli, M., Swandulla, D., and Lux, H. D. (1989) A diacylglycerol analogue reduces neuronal calcium currents independently of protein kinase C activation. *Nature* **338** 340–342.

Hoffman, F., Nastainczyk, W., Rohrkasten, A., Schneider, T., and Sieber, M. (1987) Regulation of the L-type calcium channel. *Trends Pharm. Sci.* **8** 393–398.

Hoshi, T., Garber, S. S., and Aldrich, R. W. (1988) Effect of forskolin on voltage-gated K^+ channels is independent of adenylate cyclase activation. *Science* **240** 1652–1655.

Huganir, R. L., Delcour, A. H., Greengard, P., and Hess, G. P. (1986) Phosphorylation of the nicotinic acetylcholine receptor regulates its rate of desensitization. *Nature* **321** 774–776.

Hurley, J. B. (1987) Molecular properties of the cGMP cascade of vertebrate photoreceptors. *Ann. Rev. Physiol.* **49** 793–812.

Ingebritsen, T. S., and Cohen, P. (1983) Protein phosphatases: Properties and role in cellular regulation. *Science* **221** 331–338.

Jelsema, C. L., and Axelrod, J. (1987) Stimulation of phospholipase A2 activity in bovine rod outer segments by the βγ subunits of transducin and its inhibition by the α subunit. *Proc. Natl Acad. Sci. USA* **84** 3623–3627.

Kaczmarek, L. K., Jennings, K. R., Strumwasser, F., Nairn, A. C., Walter, U., Wilson, F. D., and Greengard, P. (1980) Microinjection of catalytic subunit of cyclic AMP-dependent protein kinase enhances calcium action potentials of bag cell neurons in cell culture. *Proc. Natl Acad. Sci. USA* **77** 7487–7491.

Kaczmarek, L. K., and Levitan, I. B. (1987) *Neuromodulation.* Oxford University Press, New York.

Kandel, E. R., Klein, M., Hochner, B., Shuster, M., Siegelbaum, S. A., Hawkins, R. D., Glanzman, D. L., Castellucci, V. F., and Abrams, T. W. (1987) Synaptic modulation and learning: New insights into synaptic transmission from the study of behavior. In: *Synaptic function,* G. M. Edelman, E. Gall and M. W. Cowan (eds). Wiley, New York.

Kandel, E. R., and Schwartz, J. H. (1982) Molecular biology of learning: modulation of transmitter release. *Science* **218** 433–443.

Karlin, A. (1980) Molecular properties of nicotinic acetylcholine receptors. In: *The cell surface and neuronal function,* pp. 191–260. C. W. Cotman, G. Poste and G. L. Nicholson (eds), Elsevier/North-Holland, New York.

Katada, T. Oinuma, M., and Ui, M. (1986) Mechanisms for inhibition of the catalytic activity of adenylate cyclase by the guanine nucleotide-binding proteins serving as the substrate of islet-activating protein, pertussis toxin. *J. Biol. Chem.* **261** 5215–5221.

Kehoe, J. (1985a) Synaptic block of a transmitter-induced potassium conductance in *Aplysia* neurones. *J. Physiol. (Lond.)* **369** 399–437.

Kehoe, J. (1985b) Synaptic block of a calcium-activated potassium conductance in *Aplysia* neurones. *J. Physiol. (Lond.)* **369** 439–474.

Kim, D., and Clapham, D. E. (1989) Potassium channels in cardiac cells activated by arachidonic acid and phospholipids. *Science* **244** 1174–1176.

Kim, D., Lewis, D. L., Graziadei, L., Neer, E. J., Bar-Sagi, D., and Clapham, D. E. (1989) G-protein βγ subunits activate the cardiac muscarinic K^+ channel via phospholipase A_2. *Nature* **337** 557–560.

Klein, M., Camardo, J. S., and Kandel, E. R. (1982) Serotonin modulates a specific potassium current in the sensory neurons that show presynaptic facilitation in *Aplysia. Proc. Natl Acad. Sci. USA* **79** 5713–5717.

Klein, M., and Kandel, E. R. (1980) Mechanism of calcium current modulation underlying presynaptic facilitation and behavioral sensitization in *Aplysia. Proc. Natl Acad. Sci. USA* **7** 6912–6916.

Kramer, R. H., Levitan, E. S., Wilson, M. P., and Levitan, I. B. (1988) Mechanism of calcium-dependent inactivation of a potassium current in *Aplysia* neuron R15: interaction between calcium and cyclic AMP. *J. Neurosci.* **8** 1804–1813.

Krause, D., Lee, S. C., and Deutsch, C. (1988) Forskolin effects on the voltage-gated K^+ conductance of human T cells. *Pflugers Arch.* **412** 133–140.

Krebs, E. G., and Beavo, J. A. (1979) Phosphorylation–dephosphorylation of enzymes. *Ann. Rev. Biochem.* **48** 923–959.

Kurachi, Y., Nakajima, T., and Sugimoto, T. (1986a) Acetylcholine activation of K^+ channels in cell-free membrane of atrial cells. *Am. J. Physiol.* **251** H681–684.

Kurachi, Y., Nakajima, T., and Sugimoto, T. (1986b) On the mechanism of activation of muscarinic K^+ channels by adenosine in isolated atrial cells: involvement of GTP-binding proteins. *Pflügers Arch.* **407** 264–274.

Kurachi, Y., Ito, H., Sugimoto, T., Shimuzu, T., Miki, I., and Ui, M. (1989a) Arachidonic acid metabolites as intracellular modulators of the G protein-gated cardiac K^+ channel. *Nature* **337** 555–557.

Kurachi, Y., Ito, H., Sugimoto, T., Shimizu, T., Miki, I., and Ui, M. (1989b) α-Adrenergic activation of the muscarinic K^+ channel is mediated by arachidonic acid metabolites. *Pflügers Arch.* **414** 102–104.

Lemos, J. R., Novak-Hofer, I., and Levitan, I. B. (1985) Phosphoproteins associated with the regulation of a specific potassium channel in the identified *Aplysia* neuron R15. *J. Biol. Chem.* **260** 3207–3214.

Lerea, C. L., Somers, D. E., Hurley, J. B., Klock, I. B., and Bunt-Milan, A. H. (1986) Identification of specific transducin α subunits in retinal rod and cone photoreceptors. *Science* **234** 77–80.

Levitan, I. B. (1985) Phosphorylation of ion channels. *J. Membr. Biol.* **87** 177–190.

Levitan, I. B. (1988) Modulation of ion channels in neurons and other cells. *Ann. Rev. Neurosci.* **11** 119–136.

Levitzki, A. (1988) From epinephrine to cyclic AMP. *Science* **241** 800–806.

Logothetis, D. E., Kim, D. H., Northup, J. K., Neer, E. J., and Clapham, D. E. (1988) Specificity of action of guanine nucleotide-binding regulatory protein subunits on the cardiac muscarinic K^+ channel. *Proc. Natl Acad. Sci. USA* **85** 5814–5818.

Logothetis, D. E., Kurachi, Y., Galper, J., Neer, E. J., and Clapham, D. E. (1987) The βγ subunits of GTP-binding proteins activate the muscarinic K^+ channel in heart. *Nature* **325** 321–326.

Lux, H. D., Neher, E., and Marty, A. (1981) Single channel activity associated with the calcium-dependent outward current in *Helix pomatia. Pflügers Arch.* **389** 293–295.

McDonald, J. R., and Walsh, M. P. (1985) Inhibition of the Ca^{2+}- and phospholipid-dependent protein kinase by a novel M_r 17,000 Ca^{2+}-binding protein. *Biochem. Biophys. Res. Commun.* **129** 603–610.

Mackey, S. L., Glanzman, D. L., Small, S. A., Dyke, A. M., Kandel, E. R., and Hawkins, R. D. (1987) Tail shock produces inhibition as well as sensitization of the siphon-withdrawal reflex of *Aplysia:* Possible behavioral role for presynaptic inhibition mediated by the peptide Phe-Met-Arg-Phe-NH_2. *Proc. Natl Acad. Sci. USA* **84** 8730–8734.

Madison, D. V., and Nicoll, R. A. (1982) Noradrenaline blocks accommodation of pyramidal cell discharge in the hippocampus. *Nature* **299** 636–638.

Malenka, R. C., Madison, C. V., Andrade, R., and Nicoll, R. A. (1986) Phorbol esters mimic some cholinergic actions in hippocampal pyramidal neurons. *J. Neurosci.* **6** 475–480.

Marion, N. V. (1987) Probable role of a GTP-binding protein in mediating M current inhibition by muscarine in rat sympathetic neurones. *J. Physiol. (Lond.)* **396**, 87p.

Masters, S. B., Stroud, R. M., and Bourne, H. R. (1986) Family of G-protein α-chains — amphipathic analysis and predicted structure of functional domains. *Protein Eng.* **1** 47–54.

Middleton, P., Jaramillo, F., and Schuetze, S. M. (1986) Forskolin increases the rate of acetylcholine receptor desentization at rat soleus endplates. *Proc. Natl Acad. Sci. USA* **83** 4967–4971.

Miles, K., and Huganir, R. L. (1988) Regulation of nicotinic acetylcholine receptors by protein phosphorylation. *Mol. Neurobiol.* **2** 91–124.

Miller, C. (1989) Genetic manipulation of ion channels: A new approach to structure and mechanism. *Neuron* **2** 1195–1205.

Miyake, M., Christie, M. J., and North, R. A. (1989) Single potassium channels opened by opioids in rat locus ceruleus neurons., *Proc. Natl Acad. Sci. USA* **86** 3419–3422.

Moore, S. D., Madamba, S. G., Joels, M., and Siggins, G. R. (1988) Somatostatin augments the M- current in hippocampal neurons. *Science* **239** 278–280.

Nairn, A. C., Hemmings, H. C., Jr, and Greengard, P. (1985) Protein kinases in the brain. *Ann. Rev. Biochem.* **54** 931–976.

Nakajima, Y., Nakajima, S., and Inoue, M. (1988) Pertussis toxin-insensitive G protein mediates substance P-induced inhibition of potassium channels in brain neurons. *Proc. Natl Acad. Sci. USA* **85** 3643–3647.

Nakamura, T., and Gold, G. H. (1987) A cyclic nucleotide-gated conductance in olfactory receptor cilia. *Nature* **325** 442–444.

Needleman, P., Turk, J., Jakschik, B. A., Morrison, A. R., and Lefkowitz, J. B. (1986) Arachidonic acid metabolism. *Ann. Rev. Biochem.* **55** 69–102.

Nishizuka, Y. (1984) The role of protein kinase C in cell surface signal transduction and tumour promotion. *Nature* **308** 693–698.

Nishizuka, Y. (1986) Studies and perspectives of protein kinase C. *Science* **233** 305–312.

Nishizuka, Y. (1988) The molecular heterogeneity of protein kinase C and its implications for cellular regulation. *Nature* **334** 661–665.

Noda, M., Takahashi, H., Tanabe, T., Toyosato, M., Kikyotani, S., Furatani, Y., Hirose, T., Takashima, H., Inayama, S., Miyata, T., and Numa, S. (1983) Structural homology of *Torpedo californica* acetylcholine receptor subunits. *Nature* **302** 528–532.

Ocorr, K. A., and Byrne, J. H. (1985) Membrane responses and changes in cAMP levels in *Aplysia* sensory neurons produced by serotonin, tryptamine, FMRFamide, and Small Cardioactive Peptide B (SCPB). *Neurosci. Lett.* **55** 113–118.

Ordway, R. W., Walsh, J. V. Jr., and Singer, J. J. (1989) Arachidonic acid and other fatty acids directly activate potassium channels in smooth muscle cells. *Science* **244** 1176–1179.

Pace-Asciak, C. R., Granstrom, E., and Samuelsson, B. (1983) Arachidonic acid epoxides: Isolation and structure of two hydroxy epoxide intermediates in the formation of 8,11,12- and 10,11,12-trihydroxyeicosatrienoic acid. *J. Biol. Chem.* **258** 6835–6840.

Pennefather, P., Lancastar, B., Adams, P. R., and Nicoll, R. A. (1985) Two distinct Ca-dependent K$^+$ currents in bullfrog sympathetic ganglion cells. *Proc. Natl Acad. Sci. USA* **82** 3040–3044.

Pfaffinger, P. J. (1988) Muscarine and *t*-LHRH suppress M-current by activating an IAP-insensitive G-protein. *J. Neurosci.* **8** 3343–3353.

Pfaffinger, P. J., Martin, J. M., Hunter, D. D., Nathanson, N. M., and Hille, B. (1985) GTP-binding proteins couple cardiac muscarinic receptors to a K$^+$ channel. *Nature* **317** 536–538.

Pfaffinger, P. J., Leibowitz, M. D., Subers, E. M., Nathanson, N. M., Almers, W., and Hille, B. (1988) Agonists that suppress M-current elicit phosphoinositide turnover and Ca transients, but these events do not explain M-current suppression. *Neuron* **1** 477–484.

Piomelli, D., Feinmark, S. J., Shapiro, E., and Schwartz, J. H. (1988) Formation and biological activity of 12-ketoeicosatetraenoic acid in the nervous system of *Aplysia. J. Biol. Chem.* **263** 16591–16596.

Piomelli, D., Shapiro, E., Feinmark, S. J., and Schwartz, J. H. (1987a) Metabolites of arachidonic acid in the nervous system of *Aplysia:* Possible mediators of synaptic modulation. *J. Neurosci.* **7** 3675–3686.

Piomelli, D., Shapiro, E., Zipkin, R., Schwartz, J. H., and Feinmark, S. J. (1989) Formation and action of 8-hydroxy-11,12-epoxy-5,9,14-eicosatrienoic acid in *Aplysia;* A possible second messenger in neurons. *Proc. Natl Acid. Sci. USA* (in press).

Piomelli, D., Volterra, A., Dale, N., Siegelbaum, S. A., Kandel, E. R., Schwartz, J.

H., and Belardetti, F. (1987b) Lipoxygenase metabolites of arachidonic acid as second messengers for presynaptic inhibition of *Aplysia* sensory cells. *Nature* **328** 38–43.

Pollock, J. D., and Camardo, J. S. (1987) Regulation of single potassium channels by serotonin in the cell bodies of the tail mechansosensory neurons of *Aplysia californica*. *Brain Res.* **410** 367–370.

Pollock, J. D., Camardo, J. S., and Bernier, L. (1985) Serotonin and cyclic adenosine 3′:5′-monophosphate modulate the S potassium current in tail sensory neurons in the pleural ganglion of *Aplysia*. *J. Neurosci.* **7** 1862–1871.

Popot, J. L., and Changeux, J. P. (1984) The nicotinic receptor of acetylcholine: structure of an oligomeric integral membrane protein. *Physiol. Rev.* **64** 1162–1239.

Robishaw, J. D., Smigel, M. D., and Gilman, A. G. (1986) Molecular basis for two forms of the G protein that stimulates adenylate cyclase. *J. Biol. Chem.* **261** 9587–9590.

Rodbell, M. (1980) The role of hormone receptors and GTP-regulatory proteins in membrane transduction. *Nature* **284** 17–22.

Sakakibura, M., Alkon, D. L., DeLorenzo, R., Goldenring, J. R., Neary, J. T., and Heldman, E. (1986) Modulation of calcium-mediated inactivation of ionic currents by Ca^{2+}/Calmodulin-dependent protein kinase II. *Biophys. J.* **50** 319–327.

Sasaki, K., and Sato, M. (1987) A single GTP-binding protein regulates K^+-channels coupled with dopamine, histamine and acetylcholine receptors. *Nature* **325** 259–262.

Schofield, P. R., Darlison, M. G., Fujita, N., Burt, D. R., Stephenson, F. A., Rodriguez, H., Rhee, L. M., Ramachandran, J., Reale, V., Glencorse, T. A., Seeburg, P. H., and Barnard, E. A. (1987) Sequence and functional expression of the $GABA_A$ receptor shows a ligand-gated receptor super-family. *Nature* **328** 221–227.

Schwartz, J. H., and Greenberg, S. M. (1987) Molecular mechanisms for memory: Second-messenger induced modifications of protein kinases in nerve cells. *Ann. Rev. Neurosci.* **10** 459–476.

Seamon, K. B., Padgett, W., and Daly, J. W. (1981) Forskolin: unique diterpene activator of adenylate cyclase in membranes and intact cells. *Proc. Natl Acad. Sci. USA* **78** 3363–3367.

Shuster, M. J., Camardo, J. S., Siegelbaum, S. A., and Kandel, E. R. (1985) Cyclic AMP-dependent protein kinase closes the serotonin-sensitive K^+ channels of *Aplysia* sensory neurones in cell-free membrane patches. *Nature* **313** 392–395.

Siegelbaum, S. A., Camardo, J. S., and Kandel, E. R. (1982) Serotonin and cyclic AMP close single K^+ channels in *Aplysia* sensory neurones. *Nature* **299** 413–417.

Siegelbaum, S. A., and Tsien, R. W. (1983) Modulation of gated ion channels as a mode of neurotransmitter action. *Trends Neurosci.* **6** 307–313.

Sims, S. M., Singer, J. J., and Walsh, J. V. Jr. (1988) Antagonistic adrenergic–muscarinic regulation of M current in smooth muscle cells. *Science* **239** 190–193.

Snider, R. M., McKinney, M., Forray, C., and Richelson, E. (1984) Neurotransmitter receptors mediate cGMP formation by involvement of arachidonic acid and lipoxygenase. *Proc. Natl Acad. Sci. USA* **81** 3905–3909.

Soejima, M., and Noma, A. (1984) Mode of regulation of the ACh-sensitive K-channel by the muscarinic receptor in rabbit atrial cells. *Pflügers Arch.* **400** 424–431.

Stanfield, P. R., Nakajima, U., and Yamaguchi, K. (1985) Substance P raises neuronal membrane excitability by reducing inward rectification. *Nature* **315** 498–501.

Strong, J. A., and Kaczmarek, L. K. (1986) Multiple components of delayed potassium current in peptidergic neurons of *Aplysia:* modulation by an activator of adenylate cyclase. *J. Neurosci.* **6** 814–822.

Strong, J. A., and Kaczmarek, L. (1987) Potassium currents that regulate action potentials and repetitive firing. In: *Neuromodulation,* L. K. Kaczmarek and I. B. Levitan (eds). Oxford University Press, New York.

Strong, J. A. (1984) Modulation of potassium current kinetics in bag cell neurons of *Aplysia* by an activator of adenylate cyclase. *J. Neurosci.* **4** 2772–2783.

Stryer, L. (1988) *Biochemistry*. Freeman, New York.

Stryer, L., and Bourne, H. R. (1986) G-proteins: A family of signal transducers. *Ann. Rev. Cell. Biol.* **2** 391–419.

Surprenant, A., and North, R. A. (1988) Mechanism of synaptic inhibition by noradrenaline acting at α_2-adrenoceptors. *Proc. Roy. Soc. Lond. B* **234** 85–114.

Sweatt, J. D., and Kandel, E. R. (1989) 5-HT causes a persistent increase in protein phosphorylation that is dependent of transcription: A molecular mechanism contributing to long-term facilitation in *Aplysia* sensory neurons. *Nature* **339** 51–54.

Tempel, B. L., Jan, Y. N., and Jan, L. Y. (1988) Cloning of a probable potassium channel gene from mouse brain. *Nature* **332** 837–839.

Tempel, B. L., Papazian, D. M., Schwarz, T. L., Jan, Y. N., and Jan, L. Y. (1987) Sequence of a probable potassium channel component encoded at Shaker locus of *Drosophila. Science* **237** 770–775.

Tohse, N., Kameyama, M., and Irisawa, H. (1987) Intracellular Ca^{2+} and protein kinase C modulate K^+ current in guinea pig heart cells. *Am. J. Physiol.* **253** H1321–H1324.

Trautmann, A., and Marty, A. (1984) Activation of Ca-dependent K^+ channels by carbamoylcholine in rat lacrimal glands. *Proc. Natl Acad. Sci. USA* **81** 611–615.

Trussel, L. O., and Jackson, M. B. (1987) Dependence of an adenosine-activated potassium current on a GTP-binding protein in mammalian central neurons. *J. Neurosci.* **7** 3306–3316.

Tsien, R. W., Giles, W., and Greengard, P. (1972) Cyclic AMP mediates effects of adrenaline in cardiac purkinje fibers. *Nature New Biol.* **240** 181–183.

Ui, M. (1984) Islet-activating protein, perussis toxin: a probe for functions of the inhibitory guanine nucleotide regulatory component of adenylate cyclase *Trends Pharmacol Sci.* **5** 277–279.

Ui, M., Katada, T., Murayama, T., Kurose, H., Yajima, M., Tamura, M., Nakamura, T., and Nogomori, K. (1984) Islet-activating protein, pertussis toxin: a specific uncoupler of receptor mediated inhibition of adenylate cyclase. *Adv. Cyclic Nucleotide Protein Phosphorylation Res.* **17** 145–151.

VanDongen, A. M. J., Codina, J., Olate, J., Mattera, R., Joho, R., Birnbaumer, L., and Brown, A. M. (1988) Newly identified brain potassium channels gated by the guanine nucleotide binding protein G_o *Science* **242** 1433–1437.

Volterra, A., and Siegelbaum, S. A. (1988) Role of two different guanine nucleotide-binding proteins in the antagonistic modulation of the S-type K^+ channel by cAMP and arachidonic acid metabolites in *Aplysia* sensory neurons. *Proc. Natl Acad. Sci. USA* **85** 7810–7814.

Volterra, A., Siegelbaum, S. A. Kandel, E. R., and Sweatt, J. D. (1988) FMRFamide antagonizes S current modulation and protein phosphorylation produced by 5-HT and cAMP. *Soc. Neurosci. Abstr.* **14** 152.

Walsh, J. P., and Byrne, J. H. (1984) Forskolin mimics and blocks a serotonin-sensitive decreased K^+ conductance in tail sensory neurons of *Aplysia*. *Neurosci. Lett.* **52** 7–11.

Walsh, J. P., and Byrne, J. H. (1989) Modulation of a steady-state Ca^{2+}-activated, K^+ current in tail sensory neurons: Role of serotonin and cAMP. *J. Neurophysiol.* **61** 32–44.

Walsh, K. B., and Kass, R. S. (1988) Regulation of a heart potassium channel by protein kinase A and C. *Science* **242** 67–69.

Walsh, K. B., and Kass, R. S. (1989) Modulation of heart delayed rectification by protein kinase A and C: Evidence for different mechanisms of action. *Biophys. J.* **55** 588a.

Williams, J. T., North, R. A., and Tokimasa, T. (1988) Inward rectification of resting and opiate-activated potassium currents in rat locus coeruleus neurons. *J. Neurosci.* **8** 4299–4306.

Yatani, A., Codina, J., Brown, A, M., and Birnbaumer, L. (1987a) Direct activation of mammalian atrial muscarinic potassium channels by GTP regulatory protein G_k. *Science* **235** 207–211.

Yatani, A., Codina, J., Imoto, Y., Reeves, J. P., Birnbaumer, L., and Brown, A. M. (1987b) A G protein directly regulates mammalian cardiac calcium channels. *Science* **238** 1288–1292.

Yatani, A., Codina, J., Sekura, R. D., Birnbaumer, L., and Brown, A. M. (1987c) Reconstitution of somatostatin and muscarinic receptor mediated stimulation of K^+ channels by isolated GK protein in clonal rat anterior pituitary cell membranes. *Mol. Endocrinol.* **1** 283–289.

Yatani, A. Hamm, H., Codina, J., Mazzoni, M. R., Birnbaumer, L., and Brown, A. M. (1988a) A monoclonal antibody to the α subunit of G_k blocks muscarinic activation of atrial K^+ channels. *Science* **241** 828–831.

Yatani, A., Mattera, R., Codina, J., Graf, R., Okabe, K., Padrell, E., Iyengar, R., Brown, A. M., and Birnbaumer, L. (1988b) The G protein-gated atrial K^+ channel is stimulated by three distinct G_i α-subunits. *Nature* **336** 680–682.

6

Sodium-activated potassium currents

C. R. Bader, L. Bernheim, D. Bertrand and **C. Haimann**

6.1 INTRODUCTION

In many excitable cells the opening of sodium channels is responsible for the depolarizing phase of action potentials. In addition to the inward movement of sodium through these channels, an inward flux of sodium can occur through several other voltage and chemically gated membrane channels. These sodium fluxes can lead to modifications of the intracellular sodium concentration near the cell membrane and this, in turn, may activate potassium channels. Potassium channels activated by intracellular sodium could provide excitable cells with a feedback mechanism, permitting faster repolarization during spiking activity. They might also contribute to maintain the membrane potential near the potassium equilibrium potential when cells are faced with intracellular sodium accumulation. To date, there are few reports describing sodium-activated potassium currents (Kameyama *et al.*, 1984; Hartung, 1985; Bader *et al.*, 1985; Dryer *et al.*, 1989; Haimann *et al.*, 1990; Schwindt *et al.*, 1989). This chapter will review the evidence for the existence of sodium-activated potassium currents (K_{Na}) both in whole cells and membrane patches.

6.2 PRESENCE OF K_{Na} IN WHOLE-CELL RECORDINGS

Results obtained in whole-cell recording from trigeminal sensory neurones in culture will first be described; recordings were made from freshly dissociated neurones of the trigeminal ganglion. The ganglia were dissected from quail embryos at the 7th day of embryonic development. After dissociation, the suspension of neurones was transferred into 'modified' Petri dishes (Bader *et al.*, 1987) in which neurones were allowed to attach to for 2 hours before the electrophysiological experiments were started. During the first hours in culture, freshly dissociated neurones have essentially no neurites, and this favours the control of the membrane potential during voltage-clamp experiments.

An indication that phasic influxes of sodium ions in the neuronal cytoplasm may activate K_{Na} is suggested by the experiment illustrated in Fig. 6.1. The membrane potential was held in voltage clamp at −100 mV and was stepped every 5 s to −10 mV for 30 ms. This protocol was applied first in the absence (Fig. 6.1A, continuous line) and then in the presence of tetrodotoxin (Fig. 6.1A, dotted line). Subtraction of the dotted line from the continuous line yields the current suppressed by tetrodotoxin (Fig. 6.1B). It can be seen that this current is made up of an initial inward current, the sodium current, followed by an outward current. In this experiment, calcium currents were blocked by 3 mM cobalt. The same result as illustrated in Fig. 6.1 is obtained when the influx of sodium ions is prevented by superfusing a sodium-free solution (rather than a solution containing tetrodotoxin). The outward current suppressible by tetrodotoxin reversed its polarity at E_K and it could be blocked by a mixture of TEA (20 mM) and 4AP (4 mM) (Bader *et al.*, 1985), indicating that it is a potassium current. Recently, Dryer *et al.* (1989) reported that a similar current in chick brainstem neurons is blocked by 4AP (4 mM) but not by a low concentration of TEA (2 mM).

It is important to mention that an outward current such as illustrated in Fig. 6.1 might be generated as a consequence of an inadequate voltage clamp. Indeed, if the potential of part of the membrane was not adequately controlled, an action potential (partial or complete) might be triggered in that part of the membrane during a depolarizing step and this action potential could, in turn, activate additional voltage-gated potassium currents. Obviously, ‘unwanted’ sodium action potentials will not occur in the presence of tetrodotoxin; therefore, in the case of inadequate voltage clamp, a difference between outward currents recorded in the absence and presence of tetrodotoxin (cf. Fig. 6.1.B) may simply be due to additional voltage-gated potassium currents activated by the ‘unwanted’ action potential.

The quality of the voltage clamp can be assessed by examining the ability to control the inward sodium current in a range of membrane potentials between the threshold for the activation of the sodium current and the voltage at which this current reaches its maximum amplitude. It is in this range that escape from adequate clamping would be likely to occur (i.e. in the region where the fast regenerative process of the action potential develops). Fig. 6.2A is an example of an adequate voltage clamp in a trigeminal neurone. It can be seen that the amplitude of the sodium current (triangles) is well controlled at all voltages above threshold. Fig. 6.2B also shows that K_{Na} (circles) is present and that the peak amplitude of the potassium conductance increases in parallel with the inward sodium current (continuous line) and drops progressively when the sodium current decreases at more depolarized voltages (the driving force for sodium decreases). It is important to note that the drop of the K_{Na} conductance at voltages more depolarized than −20 mV occurs in spite of the fact that the driving force for potassium continues to increase with the depolarization. Therefore, the size of K_{Na} is correlated primarily with the size of the sodium current, as one would expect from a current activated by the entry of sodium into the cell.

Another useful experiment to exclude the possibility that K_{Na} measured in whole-cell recording is a voltage-clamp artefact is to replace sodium with an ion capable of carrying the inward current upon depolarization but unable to activate K_{Na}. In brainstem neurones, Dryer *et al.* (1989) found that during superfusion with lithium large inward currents during depolarizing steps did not activate K_{Na}.

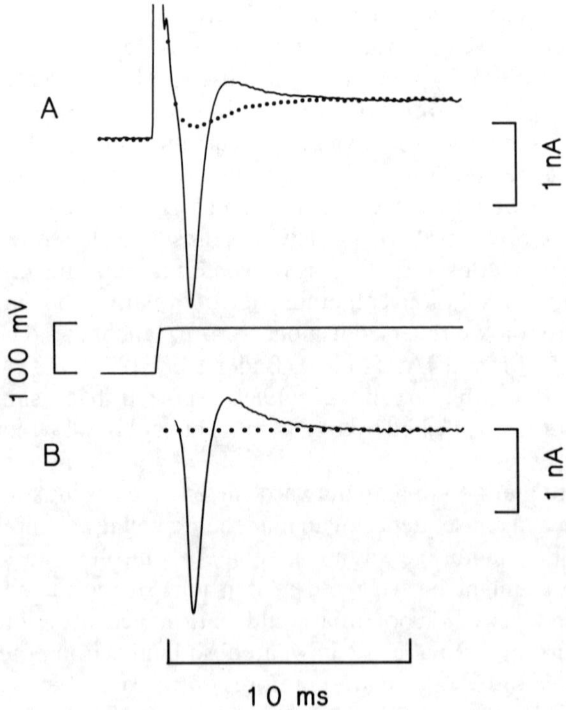

Fig. 6.1 — Currents suppressed by tetrodotoxin. Trigeminal neurone, 3–4 hours in culture. (A) when the neurone is depolarized from 100 mV to −10 mV, the current represented by the continuous line is recorded in the absence of tetrodotoxin. In the presence of tetrodotoxin (1 μM), the current during the same voltage protocol is represented by the dotted line. (B) Difference between the two current traces in (A). This difference represents the current suppressed by tetrodotoxin. The early downward deflection is the inward sodium current and the upward deflection is the sodium-activated potassium current.

Methods: The methods for preparing dissociated neurones from quail trigeminal ganglia were described previously (Bader *et al.*, 1987). For recording, patch pipettes (borosilicate glass) were pulled on a BB-CH puller (Mecanex). The fine-polishing which occurred during the symmetrical pulling procedure of this puller (Bertrand *et al.*, 1983) was sufficient to provide effective seals in excess of 5 GΩ between the electrode and the cell membrane. The solution inside the pipette was (mM): NaCl (10), KCl (20), K-acetate (130), EGTA (20), HEPES (5). The extracellular solution contained: NaCl (150), KCl (5.4), $CoCl_2$ (3), and $MgSO_4$ (0.8), HEPES (5). The pH of both solutions was adjusted to 7.4. Current injection and voltage recording were done with a single electrode, using a chopping technique (Brennecke and Lindemann, 1974) at a frequency of 27.7 kHz. Voltage and current records were digitized (12-bit resolution) and stored on disk using an IBM PC-AT computer. Current traces were filtered at 3 kHz with a low-pass filter before digitization. Acquisition, analysis and graphical representation of data was performed with the program DATAC (Bertrand and Bader, 1986).

Furthermore, these authors demonstrated that a neurone which initially, in the presence of sodium, possessed K_{Na} lost part of this current in the presence of lithium; the amplitude of the inward current itself was essentially the same with sodium and lithium. The authors do not indicate whether the effect was reversible. Our own experiments with sodium substitution by lithium in whole-cell recordings were inconclusive; they will be discussed in section 6.7.

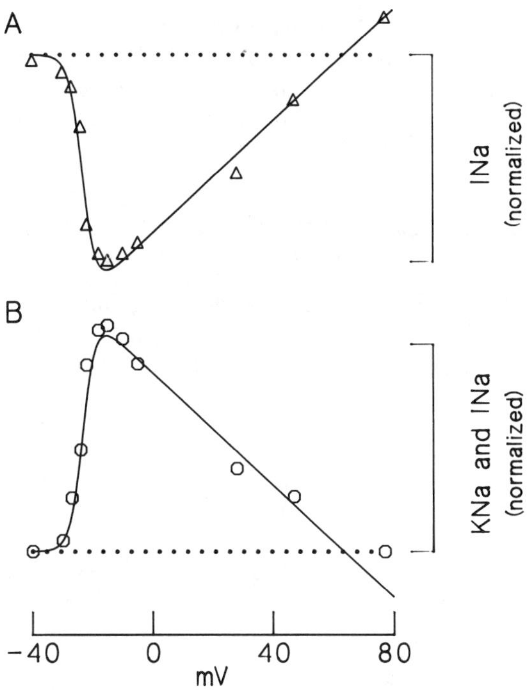

Fig. 6.2 — Correlation between the size of the sodium current and the conductance of K_{Na}. Trigeminal neurone, 4 hours in culture. (A) Peak amplitude of the sodium current (triangles) as a function of the voltage during a depolarizing step from −100 mV. The peak inward current suppressed by tetrodotoxin at each voltage was normalized to the maximum value of the current. The continuous line was fitted by eye to the data. (B) Relationship between the normalized conductance of K_{Na} (circles) and the potential during a step. The continuous line is the same curve as in (A) and facilitates the comparison between normalized sodium current and K_{Na} conductance.

Another feature that might be expected from K_{Na} is that its time course should be prolonged when the sodium current is prolonged. To this effect, we added an agent known to remove sodium inactivation, *N*-bromoacetamide (Oxford *et al.*, 1978; Patlak and Horn, 1982), to the solution in the patch pipette. This increased the duration of the sodium current and, in parallel, the duration of K_{Na} (data not shown).

Taken together, these results suggest that K_{Na} can be activated by a phasic influx of sodium ions such as occurs during an action potential. It must be emphasized, however, that special attention must be given to the adequateness of the voltage clamp before one accepts the existence of K_{Na} in whole-cell recording.

In quail sensory neurones, whole-cell recording with patch pipettes containing 50 mM sodium indicates that a potassium current can be steadily activated by intracellular sodium (Bader *et al.*, 1985). The current was activated between −130 mV and −60 mV, in a range where voltage-gated potassium currents, such as the delayed rectifier (Hodgkin and Huxley, 1952) and the A current (Connor and Stevens, 1971) are normally not activated in trigeminal neurones (Bader, Bernheim,

Bertrand and Distasi, unpublished results). The sodium-activated potassium current could be blocked by a mixture of TEA and 4AP. This supports the notion that K_{Na} can be activated not only by phasic changes in sodium concentration but also by a steady elevation in the intracellular sodium concentration.

6.3 SODIUM-ACTIVATED SINGLE-CHANNEL CURRENTS IN QUAIL TRIGEMINAL NEURONES

Single-channel activity could be evoked in cultured trigeminal neurones by continuous application of sodium to the cytoplasmic face of an inside-out membrane patch. It can be seen on the chart recording illustrated in Fig. 6.3A that there is

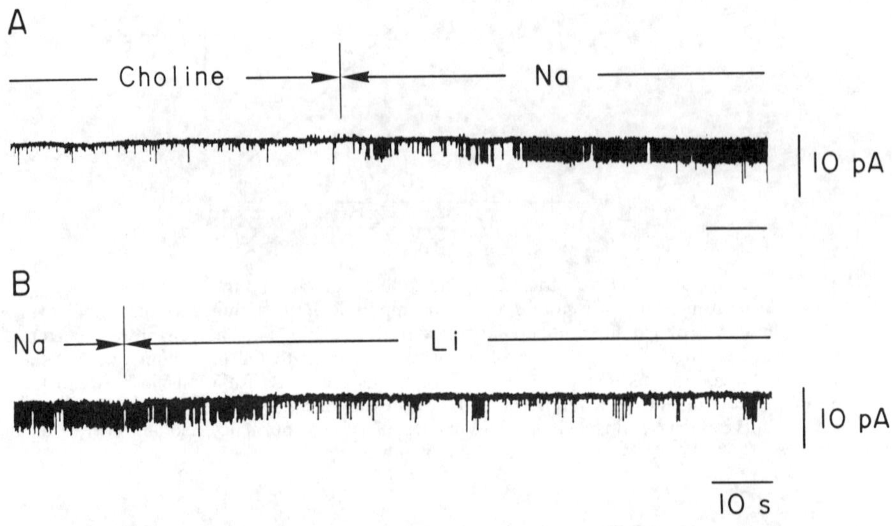

Fig. 6.3—Effect of sodium, choline and lithium on K_{Na}. An inside-out patch of membrane was excised from a trigeminal neuron. These patches were obtained by realizing first a cell-attached patch (Hamill *et al.*, 1981) in which unidentified channels were usually present. The superfusion solution was then switched to a solution without calcium (see composition below). The pipette was slowly pulled away and all channel activity disappeared at that stage. The tip of the pipette was then removed from the solution for approximately 500 ms and dipped again in the superfusion solution. The patch was maintained at a potential of −50 mV. (A) The intracellular face of the patch is exposed to a sodium-free solution (choline), and then at the time indicated by a vertical bar, to a solution containing sodium (Na). (B) Same patch as in (A). At the time indicated by the vertical bar, lithium (Li) replaced sodium in the superfusion solution. The solution inside the pipette was (mM): NaCl (75), KCl (70), KF (5), HEPES (5), EGTA (5), tetrodotoxin (1 μM). The superfusion solution was : NaCl (75), KF (75), HEPES (5), EGTA (20). In the superfusion solution, 75 mM choline or lithium replaced the corresponding concentration of sodium during sodium substitution experiments.

little single-channel activity in the presence of 75 mM choline and that immediately after switching to 75 mM sodium the activity increases progressively, presumably due to the progressive increase in sodium concentration. In the same patch, 75 mM lithium (Fig. 6.3B) suppressed the single-channel activity. The effect of lithium was reversible. However, in the presence of lithium, occasional bursts of activity were observed, suggesting that this cation is a weak agonist. This possibility has to be tested more extensively.

In patches with K_{Na} channels, steady application of a solution containing 10 μM Ca^{2+} in the absence of sodium failed to evoke any single-channel activity. This indicates that the single-channel activity elicited by sodium is not due to activation of calcium-activated channels. Overall, these results indicate that the single-channel activity we report here is indeed induced by the presence of sodium ions near the cytoplasmic face of the patch membrane.

At a higher time resolution, Fig. 6.4A shows that when the channel is open the

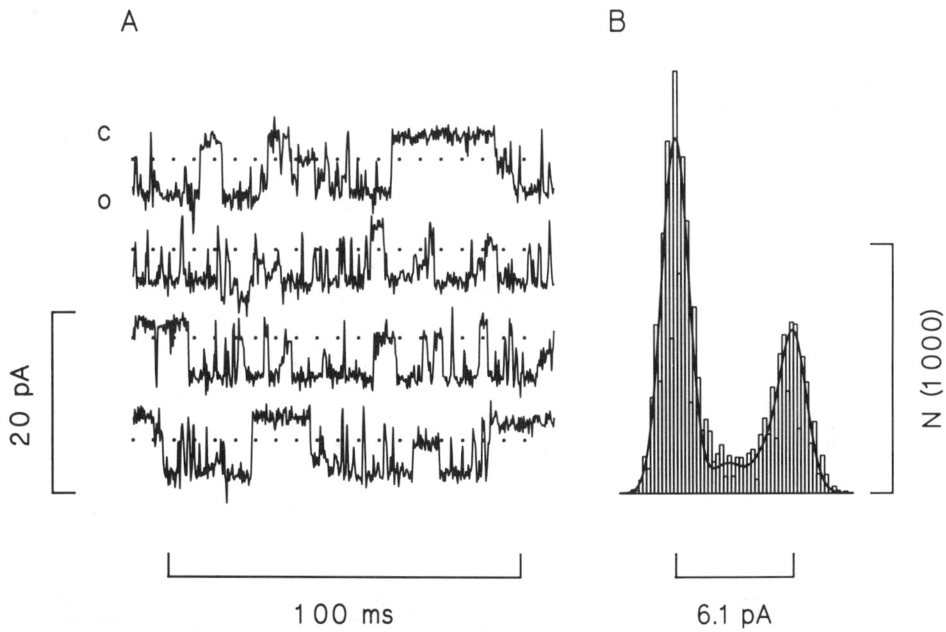

Fig. 6.4 — K_{Na} single channel in an inside-out patch exposed to a steady application of sodium. An inside-out patch of membrane was excised from a trigeminal neurone. The patch was maintained at a potential of −50 mV during exposure to 75 mM sodium. (A) Four traces are represented and the letters c and o indicate the level of the current when the channel is closed or open, respectively. The dotted line indicates the level of opening of a smaller-amplitude current. There is another, less well-defined level of current near 4.7 pA. (B) Amplitude histogram of 40 current traces similar to those illustrated in (A), except that each trace lasted 200 ms. The histogram was fitted by eye with a sum of 4 Gaussian functions, centred at 0, 2.7, 4.7 and 6.1 pA. The solution inside the pipette was as in Fig. 6.3. The superfusion solution was (mM): NaCl (75), KF (75), HEPES (5), EGTA (20).

amplitudes of the current have at least two values, 6.1 pA and 2.7 pA (holding potential, −50 mV). Generally, the small current is seen when the channel is already open and when the current decays from the large current level. Both the small and the large currents are activated by sodium. On an amplitude histogram (Fig. 6.4B), the small current lies between the noise level (left peak, 0 pA) and the large current level (right peak, 6.1 pA). There was no additional current peak to the right of the large current level. Thus, it is possible that the two currents seen in Fig. 6.4 are determined by different substates of the same channel. As judged from the width of the Gaussian distribution at 0 and 6.1 pA, the noise level was approximately the same when the channel was open or closed.

6.4 NATURE OF THE CHARGE CARRIER

The current-voltage relationship of this channel was investigated in different inside-out patches exposed to 75 mM intra- and extracellular potassium (Fig. 6.5). The

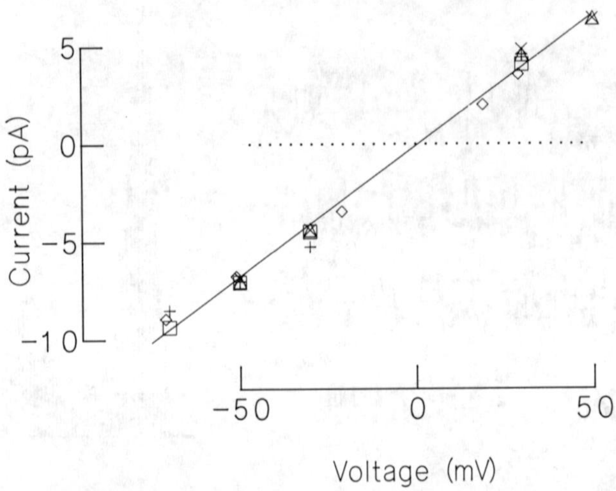

Fig. 6.5 — Current–voltage relationship of K_{Na}. Five inside-out patches of membrane were excised from trigeminal neurones and their cytoplasmic face was exposed to 75 mM sodium. The voltage was changed stepwise and the amplitude of the current measured as a function of voltage. The line through the data is a linear regression line. The reversal potential was 0.7 mV. The solution in the pipette was as in Fig. 6.3 (75 mM K^+). The superfusion solution contained (mM): NaCl (75), KF (75), EGTA (20), HEPES (5). Different symbols represent different patches.

relationship was linear, the conductance was 134 pS and the reversal potential was at the potassium equilibrium potential (E_K). Using different potassium concentrations either in the patch pipette (i.e. in the solution facing the extracellular face of the patch membrane) or in the solution facing the inside face of the membrane, we observed that the reversal potential of the single-channel activity was always in the vicinity of E_K. The effects of changes in potassium concentration on the current–voltage relationship of the sodium-activated current were reversible. Modifications of the intracellular sodium concentration had no influence on the reversal potential of K_{Na}. These results suggest that the major charge carriers of the sodium-activated current are potassium ions and justify the name K_{Na}.

6.5 EFFECT OF DIFFERENT INTRACELLULAR SODIUM CONCENTRATIONS

The results presented so far suggest that the opening of K_{Na} channels in trigeminal neurones is modulated by sodium ions near the cytoplasmic face of the membrane. However, one might object that concentrations of sodium as high as those used in our experiments may never exist during the physiological functioning of a neurone. It was therefore of interest to evaluate the relationship between the opening of the K_{Na} channel and the sodium concentration. In most patches, a clear increase in activity only occurred for sodium concentrations greater than 12 mM. Fig. 6.6 illustrates two

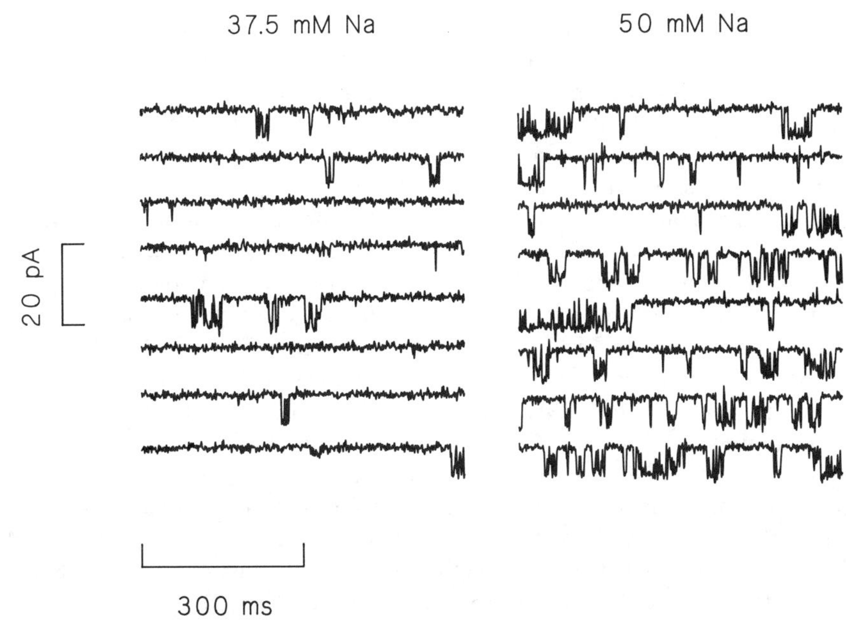

Fig. 6.6 — Effect of different intracellular sodium concentrations on the activity of K_{Na}. An inside-out patch of membrane was excised from a trigeminal neurone and the cytoplasmic face of the patch was exposed either to 37.5 mM sodium or 50 mM sodium. Eight consecutive recordings are plotted to illustrate the behaviour of K_{Na} in both situations. The pipette solution was as described in Fig. 3. The superfusion solution contained (mM): KF (75), EGTA (20), HEPES (5) and either NaCl (37.5) and sucrose (70) or NaCl (50) and sucrose (47).

sets of single-channel recordings obtained with two different sodium concentrations in the same patch. It can be seen that the bursts of channel activity are more frequent and tend to last longer in the presence of the higher concentration of sodium.

In guinea-pig heart myocytes, the highest probability of opening measured was 0.49–0.54 at 100 mM sodium but no saturation was actually demonstrated. There seemed to be little dependence of the probability of opening on the voltage (Kameyama *et al.*, 1984). At 0 mV and in the presence of 100 mM intracellular sodium, the K_D was 66 mM and cooperativity was suggested by a Hill coefficient of 2.8. The mean open and closed times were 8.4 and 6.5 ms, respectively. In trigeminal neurones, the probability for the K_{Na} channel to be open tends to saturate for sodium concentrations greater than 50 mM and the Hill coefficient suggests a cooperative binding of 2–3 sodium ions to the channel (Haimann *et al.*, 1990). Also, at a given sodium concentration, the probability for K_{NG} channels to be open increased with depolarization.

6.6 EFFECT OF VOLTAGE AND POTASSIUM CONCENTRATION ON THE CHANNEL CONDUCTANCE

In guinea-pig heart myocytes, the conductance of the K_{Na} channel decreases as the voltage is made more positive than +20 mV (Kameyama *et al.*, 1984). In chick brainstem neurones, the K_{Na} channel shows a very discrete inward rectification over the voltage range studied (Dryer *et al.*, 1989). Our data suggest that the conductance of K_{Na} in chick trigeminal neurones has little voltage dependence over a voltage range between −75 and +50 mV (see Fig. 6.5).

The elementary conductance of K_{Na} depends on the extracellular concentration of potassium on both sides of the membrane. When the extracellular potassium concentration is reduced and the intracellular concentration kept constant, the elementary conductance drops. Similar observations were reported in heart cells (Kameyama *et al.*, 1984) and in brainstem neurones (Dreyer *et al.*, 1989). Studies with another type of channel, which is also modulated by extracellular potassium concentrations (DiFrancesco, 1982), indicated that the elementary conductance could be extremely small under physiological conditions (DiFrancesco, 1986). This prompted us to determine the value of the conductance of K_{Na} with more physiological potassium concentrations (i.e. with a K_o of 5.4 mM and a K_i of 150 mM). We found that with these potassium concentrations the conductance of K_{Na} was 50 pS. Therefore, even under physiological conditions the elementary K_{Na} conductance is rather high. In a reversed condition (K_o 150 mM and K_i 5), Dryer *et al.* (1989) reported a conductance of 50 pS.

6.7 DISCUSSION

Sodium-activated potassium currents have now been reported in four different preparations: guinea-pig heart myocytes (Kameyama *et al.*, 1984), invertebrate neurones (Hartung, 1985) and avian ganglionic (Bader *et al.* 1985; Haimann *et al.*, 1990) and brainstem neurones (Dryer *et al.*, 1989). At the single-channel level and in the presence of high potassium concentrations, the elementary conductance is found to be high in heart and neuronal cells. In heart myocytes the elementary conductance was 207 pS (K_i 49 mM, K_o 150 mM), in brainstem neurones it was 105 pS (K_i 75, K_o 150) and in trigeminal neurones we found 170 pS (K_i 75, K_o 150). It remains to be determined whether these differences are due to channels that are distinct in their protein structure or to post-transcriptional modifications of the channels that depend upon the type of cell studied. Under physiological potassium concentrations the elementary conductance in trigeminal neurones is near 50 pS. Our data suggest that the K_{Na} channel may have several conductance levels.

A difference between K_{Na} channels in neurones may be their ability to be activated by lithium. Dryer *et al.* (1989) report that lithium is as effective as sodium in carrying the inward current during depolarizing steps in brainstem neurones. However, in these cells, lithium did not activate K_{Na} in both whole-cell and single-channel recordings. In contrast, in whole-cell recording from invertebrate neurones, Hartung (1985) reported that K_{Na} was activated by lithium. Our results with single-channel recording from trigeminal neurones indicate that lithium is much less efficient than sodium in activating K_{Na} but there remains some bursting K_{Na} activity (see Fig. 6.3B). Results with lithium in whole-cell recording in the same preparation were unconclusive. First, K_{Na} was not detected in all well-clamped neurones, which complicates the statistical evaluation of the effect of sodium substitution with lithium in a simple screening procedure. It must be emphasized that the failure to detect K_{Na} does not necessarily imply that the current is not present. Fig. 6.7 illustrates this point using a theoretical example. In Fig. 6.7A, the pure sodium current (labelled Na), the pure K_{Na} and the sum of these two currents (labelled sum) are traced. It can be seen that the 'sum' has a time course similar to that illustrated in Fig. 6.1B, i.e. an inward current followed by an outward current. Fig. 6.7B shows that when the amplitude of K_{Na} is reduced the outward component in the 'sum' becomes undetectable. Another reason why we are uncertain about the effect of lithium is that we could record from neurones without K_{Na} in the presence of lithium but we also observed neurones in which K_{Na} was still present even after a long exposure to lithium. It is possible that the effect of lithium may be dose dependent but additional studies are needed to settle this issue.

A question that was already discussed by others (Kameyama *et al.*, 1984; Hartung, 1985; Dryer *et al.*, 1989) concerns the sodium concentration required to activate K_{Na} channels. The high sodium concentration needed to activate K_{Na} in heart myocytes led Kameyama *et al.*. (1984) to conclude that this type of channel might become active only under pathological conditions. In neurones, results of Dryer *et al.* (1989) and our own data suggest that the channel may be active already under physiological conditions. Indeed, if the resting sodium concentration is near 12 mM, K_{Na} channel openings at a low frequency may contribute to the resting potential of neurones.

During synaptic activity, the concentration of sodium may be expected to change in dendritic processes, due to the relatively small volume involved. There are also reasons to believe that the sodium concentration changes at the soma membrane after a single action potential (Bader *et al.*, 1985). The amount of sodium entering the soma will induce little change in the overall sodium concentration. However, it is possible, as proposed by Hartung (1985), that diffusion barriers underneath the membrane permit a local build-up of the sodium concentration. Another factor that would favour activation of K_{Na} is a co-localization of sodium channels with K_{Na} channels in the plasma membrane (Hartung, 1985; Dryer *et al.*, 1989) which would expose K_{Na} channels to a relatively high local sodium concentration. Indeed, at the

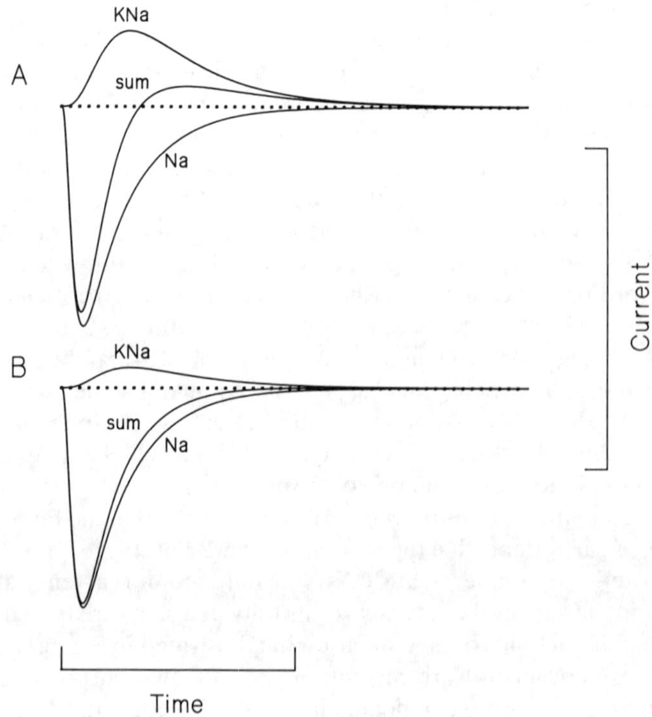

Fig. 6.7 — Simulation showing how K_{Na} might be present but not easily detectable in certain circumstances. The theoretical time-course of the pure sodium current (Na) and of the pure sodium-activated potassium current (K_{Na}) and of the toal current (sum) were computed in two situations. In (A) the peak amplitude of K_{Na} was made four times larger than in (B). It can be seen that while in (A) the trace labelled 'sum' is made up of an inward and then of an outward current, in (B) 'sum' is exclusively made up of an inward current. If a current similar to 'sum' had been obtained experimentally as the current suppressed by tetrodotoxin (see Fig. 6.1B), one would have wrongly concluded that K_{Na} was absent. In such a case, the presence (or absence) of K_{Na} can only be demonstrated by using pharmacological tools (Bader *et al.*, 1985) or special voltage-clamp procedures.

node of Ranvier of myelinated axons, where sodium channels are highly concentrated, the intracellular concentration of sodium changes considerably after only a few action potentials (Bergman, 1970).

What is the density of K_{Na} channels in the neuronal membrane? One way to answer this question is to compare K_{Na} in whole-cell and in single-channel recordings. In intact trigeminal neurones, the K_{Na} conductance activated by 50 mM Na^+in the patch electrode was 2.4±1 nS (n=7; Bader *et al.*, 1985). In this experimental condition the elementary conductance should have been in the vicinity of 50 pS (see section 6.5, above). Therefore, with a probability for the channel to be in the maximally conducting state in the order of 0.2 at 50 mM (Haimann *et al.*, 1990), we can calculate that the number of available channels is low — 200 channels approximately for the entire neuronal soma. Given an estimated cell surface of 1250 μm^2, the channel density of K_{Na} would be 0.16 channel$\times\mu m^{-2}$. It must be emphasized that the value for the probability (0.2) assumes that the sodium concentration at the membrane was indeed 50 mM. During whole-cell recording, the actual sodium concentration may be lower at sites distant from the pipette because of diffusion from the point source and of the continuous functioning of the Na/K pump. Therefore, the number of K_{Na} channels per neurone given above may be a low estimate. Another way of estimating the channel density is to consider the percentage of patches with K_{Na} in outside-out patches. In a series of 40 patches, 13 (32.5%) had at least one K_{Na} channel. In other words, we found at least 0.325 K_{Na} channel per patch. If we assume that the surface of the membrane in the patch is approximately equal to the surface of the opening of our electrodes (0.8 μm^2), the density would be of at least 0.4 K_{Na} channel$\times\mu m^{-2}$. This would correspond to 500 channels per neurone. The presence of more than one K_{Na} channel in several patches may indicate that the density is actually higher and that there may be some clustering of the channels.

An open question is whether K_{Na} channels can be found elsewhere than at the cell soma. The current has not been reported in any axon and in warm-blooded animals even voltage-gated potassium channels do not seem to be present at the nodal region (Chiu *et al.*, 1979) but rather in the paranodal region (Chiu and Ritchie, 1982). For reasons explained in the next paragraph. K_{Na} channels could be useful at the synaptic level, either postsynaptically, in dendrites, or at presynaptic nerve terminals. To date, however, there exists no description of K_{Na} in these structures.

The functional role of K_{Na} is not yet known and it will be difficult to establish until a specific blocking agent is found. So far, no pharmacological studies on this channel are available; the fact that 4AP or a combination of 4AP and TEA block K_{Na} simply indicates that this channel has some similarities with voltage-gated potassium channels. Therefore, for the moment, one can only speculate on the possible role of K_{Na}. Results in whole-cell recording suggest that this current accelerates the repolarizing phase of the action potential. The time-course of K_{Na}, somewhat similar to that of the A current (Connor and Stevens, 1971), may confer to K_{Na} a similar role, i.e. slowing down the initial frequency of discharge during a depolarization. However, unlike the A current, K_{Na} does not inactivate during a steady depolarization. More generally, it can be said that whenever the intracellular sodium increases, K_{Na} promotes a hyperpolarization of the neurone. This may contribute to a reactivation of those currents that were inactivated by depolarization (sodium current, A current and transient calcium current or T current (Carbone and Lux, 1984; Nowycky *et al.*, 1985)).

ACKNOWLEDGEMENTS

We would like to thank Dr S. E. Dryer for kindly sending us a manuscript in press. We thank Professor F. Baumann and Dr A. C. Kato for their comments on the manuscript. This work was supported by grants from the Swiss National Science Foundation no. 3.651.0.84 and no. 3.594.0.87.

REFERENCES

Bader, C. R., Bernheim, L., and Bertrand, D. (1985) Sodium-activated potassium current in cultured avian neurons. *Nature* **317** 540–542.

Bader, C. R., Bertrand, D., and Schlitcher, R. (1987) Calcium-activated chloride current in cultured sensory and parasympathetic quail neurones. *J. Physiol.* **394**, 125–148.

Bergman, C. (1970) Increase of sodium concentration near the inner surface of the nodal membrane. *Pflügers Arch.* **317** 287–302.

Bertrand, D., and Bader, C. R. (1986) DATAC, a multipurpose biological data analysis program based on a mathematical interpreter. *Int. J. Biomed. Comp.* **18** 193–202.

Bertrand, D., Cand, P., Henauer, R., and Bader, C. R. (1983) Fabrication of glass microelectrodes with microprocessor control. *J. Neurosci. Meth.* **7** 171–183.

Brennecke, R., and Lindemann, B. (1974) Theory of a membrane voltage clamp with discontinuous feedback through a pulsed current clamp. *Rev. Sci. Instrum.* **45** 184–188.

Carbone, E., and Lux, H. (1984) A low-voltage-activated, fully inactivating Ca channel in vertebrate sensory neurones. *Nature* **310** 501–502.

Chiu, S. Y., and Ritchie, J. M. (1982) Evidence for the presence of potassium channels in the internode of frog myelinated nerve fibres. *J. Physiol.* **322** 485–501.

Chiu, S. Y., Ritchie, J. M., Rogart, R. B., and Stagg, D. (1979) A quantitative description of membrane currents in rabbit myelinated nerve. *J. Physiol.* **292** 149–166.

Connor, J. A. and Stevens, C. F. (1971) Voltage clamp studies of a transient outward membrane current in gasteropod neural somata. *J. Physiol.* **213** 21–30.

DiFrancesco, D. (1982) Block and activation of the pace-maker channel in calf Purkinje fibres: effects of potassium, caesium and rubidium. *J. Physiol.* **329** 485–507.

DiFrancesco, D. (1986) Characterization of single pacemaker channels in cardiac sino-atrial node cells. *Nature* **324** 470–473.

Dryer, S. E., Fujii, J. T., and Martin, A. R. (1989) A Na^+-activated K^+ current in cultured brainstem neurones from chicks. *J. Physiol.* **410** 283–296.

Haimann, C., Bernheim, L., Bertrand, D., and Bader, C. R. (1990) Potassium current activated by intracellular sodium in quail trigeminal ganglion neurons. *J. Gen. Physiol.* (in press).

Hamill, O. P., Marty, A., Neher, E., Sakmann, B., and Sigworth, F. J. (1981) Improved patch clamp techniques for high resolution current recording from cells and cell-free membrane patches. *Pflügers Arch.* **329** 85–100.

Hartung, K. (1985) Potentiation of a transient outward current by Na^+ influx in crayfish neurones. *Pflügers Arch.* **404** 41–44.

Hodgkin, A. L. and Huxley, A. F. (1952) A quantitative description of membrane current and its application to conduction and excitation in nerve. *J. Physiol.* **117**, 500–544.

Kameyama, M., Kakei, M., Sato, R., Shibasaki, T., Matsuda, H., and Irisawa, H. (1984) Intracellular Na^+ activates a K^+ channel in mammalian cardiac cells. *Nature* **309** 354–356.

Nowycky, M. C., Fox, A. P., and Tsien, R. W. (1985) Three types of neuronal calcium channels with different calcium agonist sensitivity. *Nature* **316** 440–443.

Oxford, G. S., Wu, C. H., and Narahashi, T. (1978) Removal of sodium channel inactivation in squid axons by *N*-bromoacetamide. *J. Gen. Physiol.* **71** 227–247.

Patlak, J., and Horn, R. (1982) Effect of *N*-bromoacetamide on single sodium channel currents in excised membrane patches. *J. Gen. Physiol.* **79** 333–351.

Schwindt, P. C., Spain, W. J., and Crill, W. E. (1989) Long-lasting reduction of excitability by a sodium-dependent potassium current in cat neo-cortical neurons. *J. Neurophysiol.* **61** 233–244.

7

Non-specific cation channels

D. Swandulla and **L. D. Partridge**

7.1 INTRODUCTION

The specific characteristics of ion permeation are a very important element in defining ion channels. Many channels are named for the specific ion to which they are most permeant. While rather selective ion channels serve a multitude of important physiological roles, it is becoming increasingly evident that there is an important, albeit rather loosely defined, category of channels that are rather non-specific. These channels pass at least K^+ and Na^+ and often many other ionic species. We will review, in this chapter, several non-selective cation channels.

One of the earliest non-specific membrane currents to have been described was the acetylcholine (ACh) activated postsynaptic current at the neuromuscular junction (Takeuchi and Takeuchi, 1960). In their classic paper, these authors showed the endplate current to be carried approximately equally well by Na^+ and K^+ but not by Cl^-. The ACh-activated channel and other transmitted-gated ion channels have been extensively investigated since that time. We will concentrate our discussion here on a second group of non-specific channels — those that are activated by intracellular second messengers. In particular we will consider channels that are activated by intracellular Ca^{2+} and by intracellular cyclic nucleotides.

7.2 CALCIUM-ACTIVATED NON-SPECIFIC (CAN) CHANNELS

One very important category of non-specific channels is the Ca^{2+}-activated non-specific or CAN channel. These channels are significant both because of their ubiquitous distribution across tissues and species and because of the fact that they can produce depolarization of the membrane potential while utilizing intracellular Ca^{2+} as a second messenger.

Early descriptions of CAN currents were in cardiac cells (Kass *et al.*, 1978; see also Reuter, 1984; Boyden *et al.*, 1983; Wier and Hess, 1984; Fedida *et al.*, 1987) but CAN currents have since been shown to play a role in the function of numerous other cell types, including skeletal muscle (Sheu and Blaustein, 1983) oocytes (Barish, 1984; Moody, 1985), eggs (Jaffe *et al.*, 1986), neurone somata (Hofmeier and Lux, 1981; Byerly and Moody, 1984, 1986; Lewis, 1984, Kramer and Zucker, 1985; Swandulla and Lux, 1985; Jones *et al.*, 1986; Jones and Lewis, 1988), pituitary cells (Adler *et al.*, 1983), brush border cells (Vaandrager *et al.*, 1986), photoreceptors (Payne *et al.*, 1986) and taste receptors (Kinnamon and Roper, 1987). CAN currents have been found to be associated with two general types of cell function, stimulus–secretion coupling, and the maintenance of a depolarized state of the membrane potential. The hallmark of their action is excitation that is maintained for a length of time that would permit closing of many voltage-dependent channels.

The most unambiguous data about CAN channels comes from experiments in which isolated channels are studied in either the inside-out or outside-out configuration. Under these recording conditions channel conductance, selectivity, kinetics and Ca^{2+} sensitivity can be accurately assessed. The constancy, across tissues and species, observed in such studies leads one to consider CAN channels to be a unique class of channels. Fig. 7.1A illustrates CAN whole-cell and single-channel currents recorded in a *Helix* burster neurone, and Tables 7.1(a) and 7.1(b) summarizes findings from CAN channels in 17 different tissues. We will compare here some of the characteristics of these ubiquitous channels.

7.2.1 Single-channel conductance

Single-channel conductances range from less than 20 pS in aortic endothelial cells (Fichtner *et al.*, 1987) and neutrophils (von Tscharner *et al.*, 1986) to more than 30 pS in macrophages (Lipton, 1986) and loop of Henle cells (Teulon *et al.*, 1987). The majority of values for single-channel conductance of CAN channels are close to 30 pS. However, Lemos *et al.* (1986) reported a maxi CAN channel in peptidergic nerve terminals of a crustacean gland, which has a single-channel conductance of ≈200 pS in symmetrical K^+.

7.2.2 Voltage dependency

Two aspects of the voltage dependence of CAN channels have been investigated: *I–V* linearity and channel-opening kinetics. Current–voltage data have been assessed for the majority of CAN currents. In all but two examples, CAN channels have not been found to rectify when measured with either ramp voltages or with discrete steps. The two exceptions are Schwann cells and neutrophils. In Schwann cells the channel conductance decreases with depolarization (Bevan *et al.*, 1984). On the other hand, in neutrophils, the conductance of CAN channels decreases with hyperpolarization (von Tscharner *et al.*, 1986).

The effect of voltage on channel-opening probability has been measured in only a few preparations. In the majority of cases studied, opening probability was found to be independent of voltage but in a few instances it was found to be voltage dependent. In Schwann cells (Bevan *et al.*, 1984), insulinoma cells (Sturgess *et al.*, 1987) and in neutrophils (von Tscharner *et al.*, 1986) probability of opening was found to increase as the patch was depolarized. In the remaining instance — the aortic endothelial cell (Fichtner *et al.*, 1987) — opening probability increased as the patch was hyperpolarized.

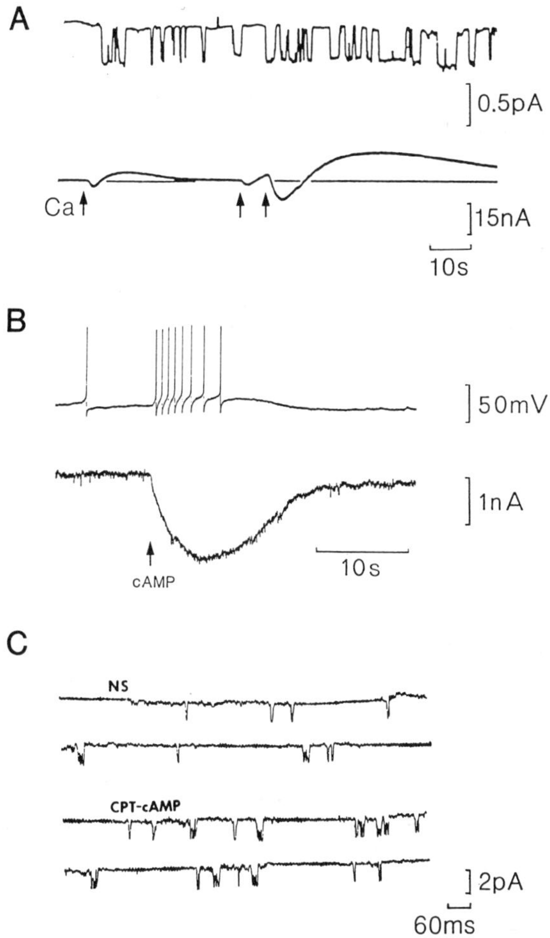

Fig. 7.1 — (A) Simultaneous recordings of single inward channel current (upper trace, cell-attached patch) and whole-cell current (lower trace) activated by intracellular pressure injections of $CaCl_2$ (indicated by arrows) into a bursting neurone of *Helix pomatia*. The cell was voltage-clamped to -50 mV and the patch was clamped to 0 mV. The single-channel current was filtered at 50 Hz. Following the first Ca^{2+} injection the opening probability of the channel was 0.29 and fell to 0.18 at the end of the whole-cell response. During the second double injection, the opening probability rose to 0.48. (B) Pressure injection of cyclic AMP into a neurone of the same type as in (A) was followed by a long-lasting membrane depolarization (upper trace). Under voltage-clamp (holding potential = -50 mV) an inward current was elicited. (C) Effects of membrane permeable CPT–cAMP on single-channel activity in a cell-attached patch. CPT–cAMP increased single-channel activity while the cell was voltage-clamped at -50 mV. Upper two traces are continuous baseline recordings of single-channel activity in normal saline (NS). Lower two traces are continuous recordings 10 min after bath addition of 10^{-4} M CPT-cAMP. (C taken, with permission, from Green and Gillette, 1983).

7.2.3 Channel kinetics

CAN channels have rather slow kinetics. This may reflect an inherently slow gating process of the channel or it may indicate a slow step in the linkage between Ca^{2+}'s

Table 7.1(a) — CAN channels

Preparation	γ(pS)	Ca^{2+} sensitivity	Cation selectivity	Anion selectivity	Ca^{2+} selectivity	Voltage dependence	
						I–V	P_o
Cardiac	30	1 μM	Na, K	No Cl	—	Linear	None
Pancreatic	30	50 nM	Na, K, Rb	No Cl	Low	Linear	None
Neuroblastoma	22	1 μM	Na, K, Cs, Li	No Cl	Very low	Linear	None
Schwann	32	100 μM	Na, K	$P_{Ca}<0.05P_{Na}$	$P_{Cl}<0.2P_{Na}$	↓ g Depol.	$0.76 - E_m$ $0.99 + E_m$
Lacrimal gland	26	10 μM	Na, K	No Cl	—	Linear	—
Lens	25	—	Na, K	None	—	Linear	—
Thyroid	35	1 μM	Na, K	—	—	Linear	None
Insulinoma	25	1 mM	—	—	—	—	↑ Depol.
Mast Cell	30	1 μM	Na, K	—	—	Linear	—
Macrophage	35–45	200 nM	Na, K, Cs, TEA	None	—	Linear	None
Neutrophils	18–25	10 μM	Na, K	No Cl	$P_{Ca} \simeq P_{Na}$	↓ g Hyperpol.	↑ Depol.
Helix burster	30	100 nM	Na, TEA, Tris	—	—	Linear	—
Dorsal root ganglion	20	100 nM	Na, K, Cs, TEA	—	—	Linear	None
Brown Adipose	30	?	Na, K	No Cl	—	Linear	—
Retinal ganglion neurite	30	200 nM	Na, K, Cs	No Cl	Perhaps	Linear	—
Aortic endothelial	12.5	100 nM	Na, K	—	—	—	↑ Hyperpol.
Loop of Henle	22–23	0.5 μM	Na, K, Cs, Li	None	—	Linear	—
Nerve terminal	213	1 μM	Na, K, Cs	—	—	Linear	—

Continued next page

Table 7.1(b) — CAN channels

Preparation	Kinetics (ms)	Modulation (function)	References
Cardiac	τ_o=100–300	(Transient inward current)	Colquhoun *et al.* (1981); Hill *et al.* (1988)
Pancreatic	τ_o=930	Indirect activation by CCK and ACh (acinar fluid secretion)	Gallacher *et al.* (1984); Maruyama and Petersen (1982, 1984); Petersen and Maruyama (1983)
Neuroblastoma	τ_o=50–200	—	Yellen (1982)
Schwann	τ_o0.47 and 14.9	Not blocked by TTX or TEA	Bevan *et al.* (1984)
Lacrimal gland	τ_o=225	(Sustained salt and water secretion)	Marty *et al.* (1984)
Lens	—	—	Rae and Levis (1984a,b)
Thyroid	—	(Stimulant-evoked depol.)	Maruyama *et al.* (1985)
Insulinoma	—	Inhibition by adenine derivatives	Ashford *et al.* (1986); Sturgess *et al.* (1986, 1987)
Mast cell	—	(Osmotic regulation)	Lindau and Fernandez (1986)
Macrophage	—	(Antibody-mediated phagocytosis)	Lipton (1986)
Neutrophils	τ_o ms to s	Indirect activation IP_3 (neutrophil activation)	von Tscharner *et al.* (1986)
Helix burster	τ_o=880	(Depol. drive in burst)	Partridge and Swandulla (1987)
Dorsal root ganglion	—		Simonneau *et al.* (1987)
Brown adipose	τ_o=25–75		Siemen and Reuhl (1987)
Retinal ganglion neurite	—	(Depol. drive in burst)	Lipton (1987)
Aortic endothelial	—	(Na influx)	Fichtner *et al.* (1987)
Loop of Henle	—	(No normal physiol. role)	Teulon *et al.* (1987)
Nerve terminal	Openings for sec.	(Depol. drive in burst)	Lemos and Nordmann (1986); Lemos *et al.* (1986).

interaction with its receptor and the subsequent opening of the channel. With the exception of CAN channels in Schwann cells (Bevan *et al.*, 1984) and those in brown adipose tissue (Siemen and Reuhl, 1987), mean channel open times are in the hundreds of milliseconds. Channel gating kinetics are more complex than would be expected from a kinetic scheme with a single rate constant. Schwann cell CAN channels exhibit a double exponential open probability distribution (Bevan *et al.*, 1984), cardiac cells (Colquhoun *et al.*, 1981) have clusters of openings with long-lived closed states, while bursting neurones are but one example of a cell that exhibits rapid flickerings to a transient closed state (Partridge and Swandulla, 1987; see also Lemos *et al.*, 1986).

7.2.4 Calcium sensitivity

All CAN channels are, by definition, sensitive to Ca^{2+} on their cytoplasmic surface. Little can be said at the present time as to whether there is a Ca^{2+} receptor as part of the channel molecule or whether there is some intermediate step between the Ca^{2+} receptor and the channel gate. However, functioning of most of the channels in inside-out patches indicates that the basic elements for channel gating were present in the isolated membrane patch and that opening of the channels did not require any cytoplasmic substance other than Ca^{2+}. While some investigators have used Ca^{2+} levels as high as 10^{-3} M to ensure channel opening, CAN channels probably require less than 10^{-6} M Ca^{2+} to significantly increase their opening probability. Ca^{2+} sensitivity of 100 nM or less has been found in pancreatic acinar cells (Maruyama and Petersen, 1984), bursting neurones (Partridge and Swandulla, 1987), dorsal root ganglion cells (Simonneau *et al.*, 1987) and aortic endothelial cells (Fichtner *et al.*, 1987). One interesting aspect of the channel's Ca^{2+} sensitivity was found by Maruyama and Petersen (1984) in pancreatic cells. In cell-attached patches, these channels open in reponse to a Ca^{2+} influx from a bath containing only 50 nM Ca^{2+}. When the patch was then excised, however, the sensitivity fell from an initial level of 100 nM immediately following excision to micromolar levels within a few minutes. These observations suggest either a dependence of some cytoplasmic structure in Ca^{2+} sensitivity or a dependence upon the state of the channel when it is excised.

7.2.5 Channel selectivity

As is the case with Ca^{2+} sensitivity, channel selectivity is a defining parameter for CAN channels. While no extensive selectivity studies have been reported, several monovalent cations, chloride and Ca^{2+} have been tested in several different preparations. All channels tested have been found to be approximately equally permeable to Na^+ and K^+. In addition, various channels have been shown to conduct Rb^+, Cs^+, Li^+, TEA and Tris. The only impermeant cation that has been reported is *N*-methyl-D-glucamine (Bevan *et al.*, 1984). A simple molecular sieve model for the permeation filter would suggest a selectivity filter similar to that of the endplate channel with a 4 Å minimum dimension. In all channels tested, Cl^- has been found to be impermeable. The permeability of the channel to Ca^{2+} has been tested in only a few instances. The CAN channels of pancreatic (Petersen and Maruyama, 1983), and neuroblastoma cells (Yellen, 1982) have very low, if any, Ca^{2+} permeability. Schwann cells (Bevan *et al.*, 1984) may have some Ca^{2+} permeability but P_{Ca} is less than one-fifth of P_{Na}. Retinal ganglion neurites (Lipton,

1987) too may have some Ca^{2+} permeability. Only in neutrophils (von Tscharner *et al.*, 1986) do CAN channels appear to have a significant Ca^{2+} permeability. In this case P_{Ca} is about equal to P_{Na}. CAN channels that do have a significant Ca^{2+} permeability, especially since they appear to lack an inactivation gating mechanism, must provide a positive feedback pathway for membrane depolarization.

7.3 CYCLIC NUCLEOTIDE ACTIVATED NON-SPECIFIC (cNAN) CHANNELS

Cyclic nucleotide activated non-specifc (cNAN) channels have been characterized in a number of tissues. Table 7.2 summarizes some of these findings. An important

Table 7.2 — cNAN channels

Preparation	γ(pS)	cN sensitivity	Cation selectivity	Anion selectivity	Voltage dependence		Kinetics
					I–V	P_o	
Rod outer segments	≃ 20	20–60 μM cGMP	Na, K (Ca)	—	—	None	(Perhaps rapid flickering)
Olfactory receptor		cGMP or cAMP	Na, K (Ca)	—	None	—	—
Gastropod neurones	20	10–100 μM cAMP	Na, K (Ca) No H	No Cl	None − 20 to − 100 mV	None	—

early observation regarding cNAN channels was that the intracellular injection of cylic GMP into rod outer segments depolarized the membrane of these cells (Miller and Nicol, 1979; Waloga, 1983). Subsequent voltage-clamp experiments indicated that cyclic nucleotides activated an inward current in these cells (Lipton, 1983; Pinto and Brown, 1984). More recently, single-channel analyses of photoreceptor membranes have shown that cyclic GMP can activate voltage-insensitive cation conductances in inside-out membrane patches (Fesenko *et al.*, 1985; Haynes and Yau, 1985).

A cyclic nucleotide-gated conductance has been recently identified in patches isolated from olfactory receptor cilia (Nakamura and Gold, 1987). This conductance is activated by either cyclic AMP or cyclic GMP. A depolarizing action of cyclic AMP, shown in Fig. 7.1B, has also been observed in gastropod neurones (Liberman *et al.*, 1975; Kononeńko and Mironov, 1981; Aldenhoff *et al.*, 1983; Hockberger and Connor, 1983, 1984). The cNAN current in these gastropod neurones shares many similar characteristics of the currents seen in olfactory and photoreceptors.

A number of different voltage- or transmitter-gated ion channels have been shown to be modified by the presence of cyclic nucleotides, often through the action of protein kinases (e.g. Huganir, 1987). cNAN channels, however, appear to be directly activated from the inner side of the membrane by cyclic nucleotides. In almost every case, there is evidence from measurements with inside-out patches that the activating messengers act directly to gate the ion channels (Hockberger and Swandulla,, 1987). More complex reactions, for instance protein kinase activation, are not involved in these gating processes.

7.3.1 Single-channel conductance

The single-channel conductance of gastropod cNAN channels, measured in cell-attached patches (see Fig. 7.1C) was found to be 20 pS (Green and Gillette, 1983). Measurements of single-channel conductance in receptor cells are more difficult, particularly in the presence of external divalent ions, which drastically reduce single-channel current similar to what was reported for the cyclic AMP-activated conductance in molluscan neurones (Aldenhoff *et al.*, 1983). However, in nominally Ca^{2+} and Mg^{2+}-free bathing solutions single-channel conductances of ≈ 20 pS have been measured (Matthews and Watanabe, 1987).

7.3.2 Voltage dependence

The gastropod cNAN conductance was tested over the voltage range -20 to -100 mV and, over this range, the gating process showed no voltage sensitivity (Aldenhoff *et al.*, 1983; Connor and Hockberger, 1984; see also Green and Gillette, 1983). The olfactory receptor channel exhibits similar voltage insensitivity (Nakamura and Gold, 1987), as did the photoreceptor channel in the absense of divalent cations (Matthews and Watanabe, 1987).

7.3.3 Channel kinetics

Little is known about kinetics of cNAN channels, although in toad retinal rods bursts of openings lasting only milliseconds were recorded (Matthews and Watanabe, 1987). These recordings were observed in $0\,Ca^{2+}$ and $0\,Mg^{2+}$ external solutions; however, it remains unclear whether flickering represents gating transitions intrinsic to the channel or transient blockage of the channel by small amounts of contaminating Ca^{2+} and Mg^{2+} in the nominally divalent free external saline.

7.3.4 Sensitivity to cyclic nucleotides

The concentration of cyclic GMP required to activate the photoreceptor current has been estimated to be in the 100 μM range (Woodruff *et al.*, 1977). Single-channel analyses of isolated patches and recordings from reconstituted channels from rod photoreceptors have confirmed these estimates, showing the most significant change in open probability to occur when the cytoplasmic surface of the patch is exposed to between 20 and 60 μM cyclic GMP (Fesenko *et al.*, 1985; Haynes and Yau, 1985; Cook *et al.*, 1987). The gastropod cNAN channel is activated by cyclic AMP in concentrations between 10 and 100 μM (Connor and Hockberger, 1984; Hockberger and Yamane, 1987).

7.3.5 Channel selectivity

The selectivity of the cNAN channel has not been tested for as many different cations as has the CAN channel. All of the cNAN channels which have been tested have been found to be permable to Na^+ as well as to K^+. In addition, these channels, like the CAN channels, exhibit some permeability to Ca^{2+} ions (see, for example,

Swandulla and Lux, 1984; Yau and Nakatani, 1984). In the gastropod, at least, cNAN channels are impermeant to H^+ and Cl^- ions (Swandulla, 1987).

7.4 OTHER CATION CHANNELS

As we have already mentioned, the ACh-activated channel at the neuromuscular endplate shows very little selectivity among cations, while discriminating well between cations and anions. Extensive permeability measurements have revealed more than 50 organic and inorganic cations that are measurably permeant in endplate channels (Adams *et al.*, 1980; Dwyer, *et al.*, 1980). These permeant ions can be as large as choline or histidine.

Analaysis of the ionic conductances underlying the depolarizations evoked by excitatory amino acids, such as *N*-methyl-D-aspartate (NMDA), has provided evidence for the cationic nature of the channels involved (for an excellent review see Mayer and Westbrook, 1987). While it is still a matter of discussion as to what extent Ca^{2+} ions can pass through these channels, it may be that, as in the case of the ACh channel, the CAN channel and the cNAN channel, a significant Ca^{2+} permeability does exist.

7.5 PHYSIOLOGICAL ROLE OF SECOND MESSENGER-ACTIVATED NON-SPECIFIC CATION CHANNELS

A number of obvious excitatory functions can be attributed to selective channels whose permeant ions have a very positive equilibrium potential, such as Na^+ and Ca^{2+}, and equally obvious inhibitory or stabilizing functions can be attributed to selective channels whose permeant ions have a very negative equilibrium potential, such as K^+ or Cl^-. Non-selective cation channels like the CAN and cNAN channels have generally excitatory actions that are more directly tied to effector functions of the cell than to rapid changes of membrane potential. Furthermore, functions that are dependent on these channels are not as directly related to concentrations of any particular ionic species.

Much of the function of CAN channels can be linked to either stimulus–secretion coupling or to regulation of cellular electrical activity (see Table 7.1). As examples of their role in stimulus–secretion coupling they are responsible for maintained salt and water secretion in lacrimal gland cells (Maruyama and Petersen, 1984), release of mediators of inflammation from macrophages (Lipton, 1986), pancreatic acinar cell fluid secretion and the modulation of salt permeabilities of intestinal brush-border cells (Vaandrager *et al.*, 1986). As generators of a depolarizing drive, CAN currents play an important role in bursting neurones (Swandulla and Lux, 1985), and in thyroid follicular cells (Maruyama *et al.*, 1985). They probably underlie long-lasting plateau potentials in neuropeptide-releasing nerve terminals (Lemos *et al.*, 1986; Lemos and Nordmann, 1986), as well as skeletal muscle depolarizing afterpotentials and action potential plateaus (Salkoff and Wyman, 1983; Suarez-Kurtz, 1979), cardiac pacemaker potentials (Reuter, 1984) and pacemaker depolarization of pituitary cells (Adler *et al.*, 1983). In each case, CAN channels provide a means by which intracellular Ca^{2+}, acting as second messenger, can transduce an effector function of the cell involved.

cNAN channels serve a role similar to that of CAN channels, only using cyclic nucleotides as a second messenger. cNAN currents may play an important role in modulating endogeneous bursting activity (Swandulla, 1987) and in action potential duration (Swandulla and Lux, 1984) in gastropod neurones (Fig. 7.2). In photoreceptors and in olfactory receptors cNAN channels are the basis of sensory transduction.

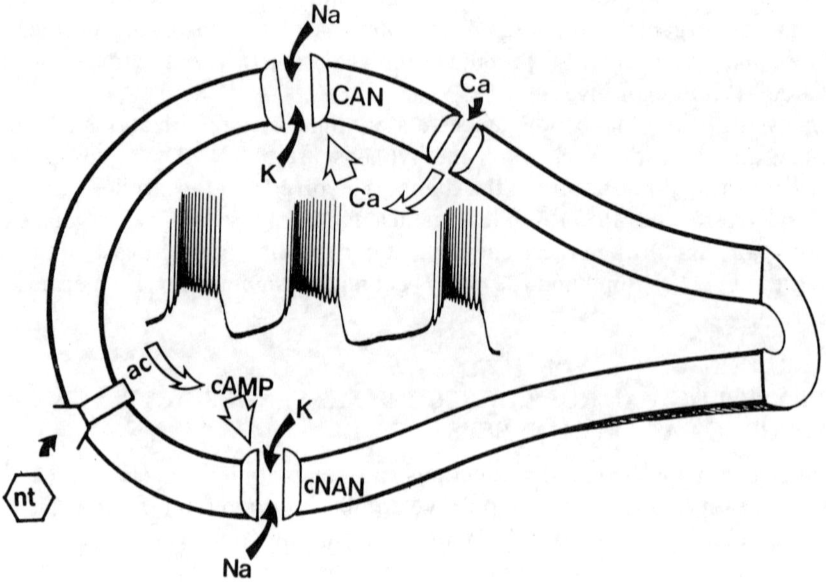

Fig. 7.2 — Schematic representation of the functioning of CAN and cNAN channels in bursting pacemaker neurones of *Helix*. Both channels provide a regulated depolarizing drive that is insensitive to rapid changes in membrane potential. CAN channels produce the depolarizing wave underlying the bursts (for details see Swandulla and Lux, 1985; Partridge and Swandulla, 1987). Bursting amplifies and prolongs the Ca^{2+} signal necessary for the secretion of a peptide from these cells. The interval between bursts can potentially be modulated by cNAN channels that respond to neurotransmitters (nt) such as serotonin or dopamine, which stimulate the adenylate cyclase system (ac).

These two groups of non-specific cation channels may represent important evolutionary adaptations of a general class of ancient non-specific channels. Hille (1984) speculates that an archetypical non-selective cation channel was the evolutionary precursor to more specific ion-selective channels. If so, CAN and cNAN channels may represent a very early specialization for second-messenger ion channel gating.

REFERENCES

Adams, D. J., Dwyer, T. M., and Hille, B. (1980) The permeability of endplate channels to monovalent and divalent metal cations. *J. Gen. Physiol.* **75** 493–510.
Adler, M., Wong, B. S., Sabol, S. L. Busis, N., Jackson, M. B., and Weight, F. F.

(1983) Action potentials and membrane ion channels in clonal anterior pituitary cells. *Proc. Natl Acad. Sci. USA* **80** 2086–2090.

Aldenhoff, J., Hofmeier, G., Lux, H. D., and Swandulla, D. (1983) Stimulation of a sodium influx by cAMP in *Helix* neurons. *Brain Res.* **276** 289–296.

Ashford, M. L. J., Hales, C. N., and Sturgess, N. C. (1986) Adenosine triphosphate inhibits the activity of the calcium-activated non-selective cation channel in a rat insulinoma cell line. *J. Physiol.* **381** 116P.

Barish, M. E. (1984) Calcium-sensitive action potential of long duration in the fertilized egg of the ctenophore *Mnemiopsis leidyi*. *Dev. Biol.* **105** 29–40.

Bevan, S., Gray, P. T. A., and Ritchie, J. M. (1984) A calcium activated cation-selective channel in rat cultured Schwann cells. *Proc. Roy. Soc.* (*London*) **222** 349–355.

Boyden, P. A., Cranefield, P. F., Gadsby, D. C., and Wit, A. L. (1983) The basis for the membrane potential of quiescent cells of the canine coronary sinus. *J. Physiol.* **339** 161–183.

Byerly, L., and Moody, W. J. (1984) Intracellular calcium ions and calcium currents in perfused neurones of the snail, *Lymnaea stagnalis*. *J. Physiol.* **352** 637–652.

Byerly, L., and Moody, W. J. (1986) Membrane currents of internally perfused neurones of the snail, *Lymnaea stagnalis*, at low intracellular pH. *J. Physiol.* **376** 477–491.

Colquhoun, D., Neher, E., Reuter, H., and Stevens, C. F. (1981) Inward current channels activated by intracellular Ca in cultured cardiac cells. *Nature* **294** 752–754.

Connor, J., and Hockberger, P. (1984) A novel membrane sodium current induced by injection of cyclic nucleotides into gastropod neurones. *J. Physiol.* **354** 139–162.

Cook, N., Hanke, W., and Kaupp, U. B. (1987) Identification, purification and functional reconstitution of the cyclic GMP-dependent channel from rod photoreceptors. *Proc. Natl Acad. Sci. USA* **84** 585–589.

Dwyer, T. M., Adams, D. J., and Hille, B. (1980) The permeability of the endplate channel to organic cations in frog muscle. *J. Gen. Physiol.* **75** 469–492.

Fedida, D., Noble, D., Rankin, A. C., and Spindler, A. J. (1987) The arrhythmogenic transient inward current i_{TI} and related contraction in isolated guinea-pig ventricular myocytes. *J. Physiol.* **392** 523–542.

Fesenko, E. E., Kolesnikov, S. S., and Lyubarsky, A. L. (1985) Induction by cyclic GMP of cationic conductance in plasma membrane of retinal rod outer segment. *Nature* **313** 310–313.

Fichtner, H., Fröbe, U., Busse, R., and Kohlhardt, M. (1987) Single nonselective cation channels and Ca^{2+}-activated K^+ channels in aortic endothelial cells. *J. Membr. Biol.* **98** 125–133.

Gallacher, D. V., Maruyama, Y., and Petersen, O. H. (1984) Patch-clamp study of rubidium and potassium conductances in single cation channels from mammalian exocrine acini. *Pflügers Arch.* **401** 361–367.

Green, D., and Gillette, R. (1983) Patch- and voltage-clamp analysis of cyclic AMP-stimulated inward current underlying neurone bursting. *Nature* **306** 784–785.

Haynes, L., and Yau, K. (1985) Cyclic GMP-sensitive conductance in outer segment membrane of catfish cones. *Nature* **317** 61–64.

Hill, J. A., Coronado, R., and Strauss, H. C. (1988) Reconstitution and characterization of a calcium-activated channel from heart. *Circ. Res.* **62** 411–415.
Hille, B. (1984) *Ionic channels of excitable membranes. Sinauer Associates Inc.*, Washington.
Hockberger, P., and Connor, J. (1983) Intracellular calcium measurements with arsenazo III during cyclic AMP injections into molluscan neurons. *Science* **219** 869–871.
Hockberger, P., and Connor, J. (1984) Dose-dependent effects of intracellular cyclic AMP on nerve membrane conductances and internal pH. In: *Primary neural substrates of learning and behavioral change*, pp. 337–358, D. Alkon and J. Farley (eds). Cambridge University Press, Cambridge.
Hockberger, P., and Yamane, T. (1987) Compartmentalization of cyclic AMP elevation in neurons of *Aplysia californica. Cell. Mol. Neurobiol.* **7** 19–33.
Hockberger, P., and Swandulla, D. (1987) Direct ion channel gating: A new function for intracellular messengers. *Cell. Mol. Neurobiol.* **7** 229–236.
Hofmeier, G., and Lux, H. D. (1981) The time courses of intracellular free calcium and related electrical effects after injection of $CaCl_2$ into neurons of the snail, *Helix pomatia. Pflügers Arch.* **391** 242–251.
Huganir, R. L. (1987) Biochemical mechanisms that regulate the properties of ion channels. In: *Neuromodulation*, pp. 64–85. Kaczmarek, L. K. and Levitan, I. B. (eds). Oxford University Press, Oxford.
Jaffe, L. A., Kado, R. T., and Kline, D. (1986). A calcium-activated sodium conductance produces a long-duration action potential in the egg of a nemertean worm. *J. Physiol.* **381** 263–278.
Jones, L. S., and Lewis, D. V. (1988) A calcium-activated, nonselective cationic conductance in *Aplysia* silent neurons. *Brain. Res. Bull.* **20** 607–609.
Jones, L. S., Evans, G. J., Wilson, W. A., and Lewis, D. V. (1986) A novel, calcium-induced triphasic current response in silent, non-bursting *Aplysia* neurons. *Brain Res.* **385** 46–50.
Kass, R. S., Lederer, W. J., Tsien, R. W., and Weingart, R. (1978) Role of calcium ions in transient inward currents and after contractions induced by strophantidin in cardiac purkinje fibres. *J. Physiol.* **281** 187–208.
Kinnamon, S. C., and Roper, S. D. (1987) Passive and active membrane properties of mudpuppy taste receptor cells. *J. Physiol.* **383** 601–614.
Kononenko, N., and Mironov, S. (1981) Effect of intracellular injection of cyclic AMP on electrical characteristics of identified neurons in *Helix pomatia. Neurophysiol.* (*USSR*) **12** 332–338.
Kramer, R. H. and Zucker, R. S. (1985) Calcium-dependent inward currents in *Aplysia* bursting pace-maker neurons. *J. Physiol.* **362** 107–130.
Lemos, J. R., and Nordmann, J. J. (1986) Ionic channels and hormone release from peptidergic nerve terminals. *J. Exp. Biol.* **124** 53–72.
Lemos, J. R., Nordmann, J. J., Cooke, I. M., and Stuenkel, E. L. (1986) Single channels and ionic currents in peptidergic nerve terminals. *Nature* **319** 410–412.
Lewis, D. V. (1984) Spike aftercurrents in R 15 of *Aplysia*: their relationship to slow inward current and calcium influx. *J. Neurophysiol.* **51** 387–403.
Liberman, Y., Minina, S., and Golubtsov, K. (1975) Study of the metabolic synapse. I. Effect of intracellular microinjection of 3′,5′-AMP. *Biophys.* (*USSR*)

20 457–463.
Lindau, M., and Fernandez, J. M. (1986) A patch-clamp study of histamine-secreting cells. *J. Gen. Physiol.* **88** 349–368.
Lipton, S. A. (1983) cGMP and EGTA increase the light-sensitive current of retinal rods. *Brain Res.* **265** 41–48.
Lipton, S. A. (1986) Antibody activates cationic channels via second messenger Ca^{2+}. *Biochim. Biophys. Acta* **856** 59–67.
Lipton, S. A. (1987) Bursting of calcium-activated cation-selective channels is associated with neurite regeneration in a mammalian central neuron. *Neurosci. Lett.* **82** 21–28.
Marty, A., Tan, Y. P., and Trautmann, A. (1984) Three types of calcium-dependent channel in rat lacrimal glands. *J. Physiol*, **357** 293–325.
Maruyama, Y., and Petersen, O. H. (1982) Cholecystokinin activation of single-channel currents is mediated by internal messenger in pancreatic acinar cells. *Nature* **299** 159–161.
Maruyama, Y., and Petersen, O. H. (1984) Single calcium-dependent cation channels in mouse pancreatic acinar cells. *J. Membr. Biol.* **81** 83–87.
Maruyama, Y., Moore, D., and Petersen, O. H. (1985) Calcium-activated cation channel in rat thyroid follicular cells. *Biochim. Biophys. Acta* **821** 229–232.
Matthews, G., and Watanabe, S. (1987) Properties of ion channels closed by light and opened by guanosine 3′,5′-cyclic monophosphate in toad retinal rods. *J. Physiol.* **389** 691–715.
Mayer, M. L., and Westbrook, G. L. (1987) The physiology of excitatory amino acids in the vertebrate central nervous system. *Progr. Neurobiol.* **28** 197–276.
Miller, W., and Nicol, G. (1979) Evidence that cyclic GMP regulates membrane potential in rod photoreceptors. *Nature* **280** 64–66.
Moody, W. J. (1985) The development of calcium and potassium currents during oogenesis in the starfish, *Leptasterias hexactis*. *Dev. Biol.* **112** 405–413.
Nakamura, T., and Gold, G. H. (1987) A cyclic nucleotide-gated conductance in olfactory receptor cilia. *Nature* **325** 442–444.
Partridge, L. D., and Swandulla, D. (1987) Single Ca-activated cation channels in bursting neurons of *Helix*. *Pflügers Arch.* **410** 627–631.
Payne, R., Corson, D. W., and Fein, A. (1986) Pressure injection of calcium both excites and adapts *Limulus* ventral photoreceptors. *J. Gen. Physiol.* **88** 107–126.
Petersen, O. H., and Maruyama, Y. (1983) What is the mechanism of the calcium influx to pancreatic acinar cells evoked by secretagogues? *Pflügers Arch.* **396** 82–84.
Pinto, L., and Brown, J. (1984) Pressure injection of 3′,5′-cyclic GMP into solitary rod photoreceptors of the tiger salamander. *Brain Res.* **304** 197–200.
Rae, J. L., and Lewis, R. A. (1984a) Patch clamp recordings from the epithelium of the lens obtained using glasses selected for low noise and improved sealing properties. *Biophys. J.* **45** 144–146.
Rae, J. L., and Levis, R. A. (1984b) Patch voltage clamp of lens epithelial-cells. Theory and practice. *Mol. Physiol.* **6** 115–162.
Reuter, H. (1984) Ion channels in cardiac membranes. *Ann. Rev. Physiol.* **46** 473–484.
Salkoff, L. B., and Wyman, R. J. (1983) Ion currents in *Drosphila* flight muscles. *J.*

Physiol. **337** 687–709.

Sheu, S. S., and Blaustein, M. P. (1983) Effects of calcium on membrane potential and sodium influx in barnacle muscle fibers. *Am. J. Physiol.* **244** C297.

Siemen, D., and Reuhl, T. (1987) Non-selective cationic channel in primary cultured cells of brown adipose tissue. *Pflügers Arch.* **408** 534–536.

Simonneau, M., Distasi, C., Tauc, L., and Barbin, G. (1987) Potassium channels in mouse neonate dorsal root ganglion cells: a patch-clamp study. *Brain Res.* **412** 224–232.

Sturgess, N. C., Hales, C. N., and Ashford, M. L. J. (1986) Inhibition of a calcium-activated, non-selective cation channel, in a rat insulinoma cell line, by adenine derivatives. *FEBS Lett* **208** 397–400.

Sturgess, N. C., Hales, C. N., and Ashford, M. L. J. (1987) Calcium and ATP regulate the activity of a non-selective cation channel in a rat insulinoma cell line. *Pflügers Arch.* **409** 607–615.

Suarez-Kurtz, G. (1979) The depolarizing afterpotential of crab muscle fibres. A sodium-dependent process mediated by intracellular calcium. *J. Physiol.* **286** 317–329.

Swandulla, D. (1987) Cationic membrane conductances induced by intracellularly elevated cAMP and Ca^{2+}: measurements with ion-selective microelectrodes. *Can. J. Physiol. Pharmacol.* **65** 898–903.

Swandulla, D., and Lux, H. D. (1984) Changes in ionic conductances induced by cAMP in *Helix* neurons. *Brain Res.* **305** 115–122.

Swandulla, D., and Lux, H. D. (1985) Activation of a non-specific cation conductance by intracellular Ca^{2+} elevation in bursting pacemaker neurons of *Helix pomatia*. *Neurophysiol.* **54** 1430–1443.

Takeuchi, A., and Takeuchi, N. (1960) On the permeability of endplate membrane during the action of transmitter. *J. Physiol.* **154** 52–67.

Teulon, J., Paulais, M., and Bouthier, M. (1987) A Ca^{2+}-activated cation-selective channel in the basolateral membrane of the cortical thick ascending limb of Henle's loop of the mouse. *Biochim. Biophys. Acta* **905** 125–132.

Vaandrager, A. B., Ploemacher, M. C., and Dejonge, H. R. (1986) Modulation of salt permeabilities of intestinal brush-border membrane vesicles by micromolar levels of internal calcium. *Biochim. Biophy, Acta* **856** 325–336.

von Tscharner, V., Prod'hom, B., Baggiolini, M., and Reuter, H. (1986) Ion channels in human neutrophils activated by a rise in free cytosolic calcium concentration. *Nature* **324** 369–372.

Waloga, G. (1983) Effects of calcium and guanosine-3′;5′-cyclic-monophosphoric acid on receptor potentials of toad rods. *J. Physiol.* **341** 341–357.

Wier, W. G., and Hess, P. (1984) Excitation–concentration coupling in cardiac purkinje fibres. *J. Gen. Physiol.* **83** 395–415.

Woodruff, M. L., Bownds, D., Green, S. H., Morrisey, J. L., and Shedlovsky, A. (1977) Guanosine 3′,5′-cyclic monophosphate and the *in vitro* physiology of frog photoreceptor membranes. *J. Gen. Physiol.* **69** 667–679.

Yau, K., and Nakatani, K. (1984) Cation selectivity of light-sensitive conductance in retinal rods. *Nature* **309** 352–354.

Yellen, G. (1982) Single Ca^{2+}-activated nonselective cation channels in neuroblastoma. *Nature* **296** 357–359.

8

Potassium channel pharmacology

Nigel S. Cook and **Ulrich Quast**

8.1 INTRODUCTION

Membrane K^+ channels essentially set the resting membrane potential of the cell and influence parameters as different as cell excitability and cell volume. The field of K^+ channel pharmacology is an active and rapidly evolving area of research which is helping to provide an understanding of the role of K^+ channels in these different cellular processes. This review is divided into two parts. First, a fairly extensive overview of the agents known to block K^+ channels is presented. This is dealt with in the context of the selectivities of these agents for different channels (and thereby their usefulness in classifying the K^+ channels present in a particular tissue), rather than concentrating on the molecular or biophysical aspects of their interaction with the channels, which are discussed in other chapters. Actions of these drugs other than on K^+ channels are also described, which might influence the interpretation of results obtained with them experimentally.

The second part of this review deals with agents which increase membrane K^+ permeability. Opening of K^+ channels will bring the membrane potential closer to the Nernst potential for K^+ (usually −80 to −90 mV) and oppose depolarization. Many neurotransmitters, neuropeptides and hormones open membrane K^+ channels, generally via activation of a G-protein (see Chapter 5). In this chapter we rather concentrate on the *in vitro* activity of a new class of drugs which open K^+ channels, preferentially in smooth muscle (but also in other tissues). After a detailed consideration of the pharmacological properties of cromakalim, the prototype of this class of drugs, the other K^+ channel openers (KCOs) are briefly described. Most of these agents were previously referred to as 'direct acting vasodilators', but the recent application of electrophysiological and/or tracer flux techniques has identified compounds as chemically diverse as pinacidil, diazoxide and minoxidil sulphate to be members of this class. Evidence for and against K^+ channel opening being solely responsible for the mechanical effects of these drugs is reviewed. We consider the possible ways in which this mechanism may lead to a modulation of smooth muscle tone and the attempts to identify the nature of the relevant K^+ channel opened by these drugs. The review is concluded by a description of some of the endogenous ligands thought to open K^+ channels in smooth muscle (see also Chapters 5, 9 and 13).

8.2 POTASSIUM CHANNEL BLOCKERS

8.2.1 Classical potassium channel blockers

8.2.1.1 Quaternary ammonium (QA) ions

Tetraethylammonium chloride (TEA) was the first and is still the most frequently used K^+ channel blocker. It blocks most classes of K^+ channels, albeit with varying affinities. The relative potencies for the block of several different K^+ channels by QA ions is shown in Table 8.1. It would appear from this table that no general rule can be applied concerning the activities of these drugs as K^+ channel blockers. For instance, the delayed rectifier (K_V) of squid axon is readily blocked by TEA_i but insensitive to TEA_o, whereas the same channel in other tissues is blocked by TEA_o. Furthermore, whilst the potency of internally applied QA ions at blocking K_V increases as the alkyl chain length is increased, when applied to the outside of the membrane TEA is the most potent blocker of both the K_V and the *Aplysia* (I)K_{Ca} channel (described in detail in Chapter 2). In contrast, inhibition of cromakalim-induced $^{86}Rb^+$ efflux from rat aorta by extracellularly applied QA increases markedly as the alkyl chain length is increased from methyl to pentyl (see section 8.3.1.4).

Effects of TEA on the apamin-sensitive small conductance Ca^{2+}-activated K^+ channels (SK_{Ca}) deserve special mention, since it is claimed that TEA and apamin can be used by their blocking activity to selectively distinguish between BK_{Ca} and SK_{Ca} channels, respectively. In cultured rat skeletal muscle and pheochromocytoma cells, TEA concentrations up to 20 mM were unable to block SK_{Ca} (Romey *et al.*, 1984; Blatz and Magleby, 1986; Schmid-Antomarchi *et al.*, 1986). However, in guinea-pig hepatocytes (where TEA also displaces ^{125}I-apamin binding), neuronal (NG108-15) cells and clonal anterior pituitary (GH3) cells, TEA blocks apamin-sensitive SK_{Ca} channels with IC_{50}s of 3–10 mM (Cook and Haylett, 1985; Higashida and Brown, 1986; Lang and Ritchie, 1988). This is only tenfold higher than the TEA concentration required to block BK_{Ca} channels in GH3 cells (Lang and Ritchie, 1988), illustrating the degree of discrimination which one can expect with TEA between these two channel subtypes in certain cells. Ca^{2+}-activated K^+ channels of intermediate conductance (IK_{Ca}) in molluscan neurones and red blood cells also differ widely in their sensitivities to TEA (Hermann and Gorman, 1981a; Deitmer and Eckert, 1985; Lew and Ferreira, 1978; Stanfield, 1983).

The ATP- dependent K^+ channel (K_{ATP}) in pancreatic β cells and metabolically exhausted skeletal muscle is only blocked about 25% by 10 mM TEA_o (Cook and Hales, 1984; Castle and Haylett, 1987; see also Table 8.1), whilst in ventricular cells a concentration of 1 mM TEA caused an almost complete block of K_{ATP} (Kakei *et al.*, 1985). Of the other K^+ channels where TEA effects have been studied, the serotonin-inactivated current ($I_{K(s)}$) of *Aplysia* neurones is fairly insensitive to TEA_o (IC_{50}>75 mM; Klein and Kandel, 1980; Brezina *et al.*, 1987), Na^+-activated K^+ channels (K_{Na}) are blocked by 20 mM TEA_o (with 4 mM 4-aminopyridine also

Table 8.1 — Structure–activity relationships for block of K^+ channels by quaternary ammonium ions applied to the inner or outer side of membrane

K^+ current	Tissue	Extracellular side[a]	Intracellular side[a]	References
Delayed rectifier (K_V)	Frog node of Ranvier	TEA (0.4)>TPe(≈1)=C9(1.0)> TPrA(60)=TBA(60)>TMA (500)	C9 (≈1.0)>TEA (≈20)	Hille (1967), Armstrong and Hille (1972)
	Squid axon	Insensitive up to 100 mM TEA_o	THA(irrev.)>TBA(0.03 ≈TBA(0.04)>TEA(0.36) ≈TPrA(0.6)>TMA(≈100)	Armstrong (1971), French and Shoukimas (1981)
	Rat alveolar epithelial cells	TBA (est. 0.03)=TPeA=THA		Decoursey *et al.* (1989)
	Mouse neuroblastoma cells	TBA_o (≈0.03) more potent than TBA_i, TPeA equiactive from both sides		Decoursey *et al.* (1989)
M current (I_M)	Mouse spinal cord neurones	Blocked by intracellular TEA injection but only weakly inhibited by 100 mM TEA_o		Nowak and Macdonald (1982)
Maxi K channel	Rabbit sarcoplasmic reticulum		(*Trans* side), TPeA(15)> TBA(26)>TPrA(45) >TEA(85)>TMA(500)	Coronado and Miller (1982)
Ca^{2+}-activated K^+ channel (K_{Ca})	*Aplysia* neurones (probably IK_{Ca})	TEA(0.4)>TPrA(>50)= TBA(>50)>TMA(>500)	Block by TEA_i only at positive membrane potentials	Hermann and Gorman (1981a)
	BK_{Ca} (skeletal and smooth muscle, pancreatic acinar cells, cultured kidney cells)	TEA_o (≈0.3)	Need>10 mM TEA_i for block. TBA(0.5)>TEA(34)>TMA(54)	Vergara *et al.* (1984), Benham *et al.* (1985), Iwatsuki and Petersen (1985), Guggino *et al.* (1987); see Yellen (1987)
ATP-sensitive K^+ channel (K_{ATP})	Frog skeletal muscle	TEA(6.7)	TPA(≈0.1^S)>TEA(1.4^S/ 26^F)≈TMA(3.7^S)	Davies *et al.* (1989a)
CRK-induced $^{86}Rb^+$ efflux	Rat aorta	THA(0.001)<TPeA(0.0001)> TBA(0.004)>TPrA(0.17) >TEA(1.0)>TMA(6.3)		Quast and Webster (1989)

[a]Side of membrane to which drug was applied. CRK, cromakalim; est, estimated; irrev, irreversible; TEA_o/TEA_i, intracellular/extracellular drug application; TMA, tetramethylammonium; TEA, tetraethylammonium; TPrA, tetrapropylammonium; TBA, tetrabutylammonium; TPeA, tetrapentylammonium; THA, tetrahexylammonium; C9, nonyl-triethylammonium. Concentrations causing ≈50% block (in mM) are shown in parentheses — since blockade is in some cases voltage dependent, the original reference should be consulted for the conditions employed. Block of K_{ATP} from the inside is described by two components with ^{S}slow or ^{F}fast kinetics.

present; Bader *et al.*, 1985) and the A current (I_A) is blocked only by high concentrations of TEA_o (50 mM; Thompson and Aldrich, 1980), these K^+ channels being considerably more sensitive to aminopyridines (see later). TEA_i blocks I_A at lower concentrations, comparable with TEA_i block of K_V (Thompson and Aldrich, 1980). The inward rectifying K^+ conductance (I_R) of skeletal muscle, which contributes to the resting K^+ conductance, is blocked by TEA with an IC_{50} of 20 mM (see Stanfield, 1983). However, it is notable that whereas >10 mM concentrations of TEA are known to cause an increase in smooth muscle tone (for references see Cook, 1989), the resting tension of frog skeletal muscle was apparently unaffected by up to 100 mM TEA (Castle and Haylett, 1987). Different K^+ channels would thus appear to be involved in regulation of resting tone (resting K^+ permeability) by these two muscle types. (TEA may indirectly cause skeletal muscle twitching though, via an effect on nerve K^+ channels; see Stanfield, 1983).

TEA shows a number of other ('non-specific') effects at concentrations similar to those used to block K^+ channels. Adams *et al.* (1982) reported that the inhibition of muscarine effects on the M current in bullfrog ganglia was primarily due to a block of muscarinic receptors. TEA also inhibits nicotinic ACh receptors both at autonomic ganglia and at the neuromuscular junction, where it additionally blocks the endplate ion channel (see Stanfield, 1983). Furthermore, being a quaternary ammonium ion, TEA antagonizes choline uptake by nerve terminals, red cells and skeletal muscle. Some of the quaternary ammonium compounds also block Na^+ channels of the squid axon at millimolar concentrations, although with TEA this action is only seen after proteolytic treatment of the nerve (Stanfield, 1983). These various non-specific actions should not be overlooked when evaluating TEA effects on different tissues.

8.2.1.2 Aminopyridines

The aminopyridines, notably 4-aminopyridine (4AP), constitute the next most commonly used group of K^+ channel blockers. They are more selective and usually more potent in their K^+ channel-blocking activity than TEA. 3,4-Diaminopyridine (3,4DAP) is the most potent of this group at blocking K_V (Kirsch and Narahashi, 1978). Delayed rectifier K^+ channels are blocked by aminopyridines with widely differing affinities; K_D at the frog node of Ranvier is 10 μM for 4AP, whilst K_V in certain other tissues is insensitive to 10 mM 4AP (for review see Rudy, 1988). The squid axon K^+ current is incompletely blocked by aminopyridines (Yeh *et al.*, 1976), possibly due to the contribution of other K^+ channels to the overall outward current. Being membrane permeable, the aminopyridines block the squid axon when applied to either side, although their binding site is probably located on the inner side (Yeh *et al.*, 1976).

4AP blocks I_A in many different tissues with an IC_{50} of around 2 mM (Rogawski, 1985), the block increasing as the membrane is hyperpolarized (Thompson and Aldrich, 1980). This effect is similar to squid axon K_V block by 4AP (Yeh *et al.*, 1976). However, 4AP is generally a more potent blocker of I_A than of $I_{K(Ca)}$ (Thompson and Aldrich, 1980; Rudy, 1988) and it has consequently become a useful tool in dissecting out the different K^+ currents contributing to the outward K^+ current in neuronal preparations. The pharmacological subclassification of these neuronal K^+ currents is discussed in section 8.2.2.4 (see also Chapters 2 and 14). Some of them, for example the slowly inactivating outward current, I_{Ks}), are highly sensitive to 4AP (1–30 μM; Stansfield *et al.*, 1986). These authors concluded that block of I_{Ks}, and not the block of I_A, might explain the convulsive actions of 4AP. Of the different voltage-dependent K^+ channels, the inward rectifier in skeletal muscle would seem to be an exception, since it is not blocked by 4AP (Fink and Wettwer, 1978).

With very few exceptions (Hoshi and Aldrich, 1986), aminopyridines have been shown to be without effect on K_{Ca} (BK_{Ca}: Ritchie, 1987; Mallart, 1985; IK_{Ca}: Lew and Ferreira, 1978; Thompson and Aldrich, 1980; Herman and Gorman, 1981b; Apamin-sensitive SK_{Ca}: Cognard *et al.*, 1984; Cook and Haylett, 1985; Ritchie, 1987). Indeed, in pituitary GH3 cells (Rogawski, 1989) and molluscan neurones (Hermann and Gorman, 1981b), 4AP enhanced current through K_{Ca}, an action which was shown to be independent of a Ca^{2+} influx into the cell, but may have been linked to a reduction in intracellular Ca^{2+} buffering. Effects of 4AP on K_{ATP} are also interesting. In guinea-pig ventricular cells (Kakei *et al.*, 1985) and pancreatic β cells (Cook and Hales, 1984), channel openings were essentially abolished by 1 and 10 mM 4AP, respectively. However, the K^+ conductance induced by metabolic exhaustion of frog skeletal muscle, presumably via K_{ATP}, was reported to be insensitive to similar 4AP concentrations (Fink and Wettwer, 1978; Castle and Haylett, 1987). The serotinin-inactivated K^+ channel is blocked only by high (5–10 mM) concentrations of 4AP, making it less sensitive than K_V or I_M in cells possessing all three K^+ currents (Klein *et al.*, 1982; Brezina *et al.*, 1987).

The excitatory effects of aminopyridines on vascular smooth muscle seem to be due to a combination of actions on several different K^+ channels. 4AP (1 mM) causes membrane depolarization in the guinea-pig pulmonary artery and portal vein (Hara *et al.*, 1980). This action is likely to underlie the increase in resting tone and induction of phasic responses seen with aminopyridines in many vessels, including human coronary arteries (where the phasic contractions have been compared to the situation during vasospasm in patients with variant angina pectoris; Uchida *et al.*, 1986). Vasoconstrictor responses are also potentiated by 0.1 mM or higher aminopyridine concentrations (Glover, 1978; Cook, 1989). Cook (1989) showed that 3,4DAP (0.3 and 1 mM) increased the resting tension of the rabbit aorta and augmented contractile responses to angiotensin II, effects which were only partially inhibited by blocking dihydropyridine-sensitive Ca^{2+} channels. However, TEA concentrations causing a comparable augmentation of angiotensin II responses failed to affect the resting tone. This was interpreted as evidence for (at least) two populations of K^+ channels in this tissue, with those responsible for maintenance of the resting K^+ conductance being more sensitive to 3,4DAP than to TEA (Cook, 1989). Okabe *et al.* (1987) have shown that 10 mM 4AP completely blocks outward K^+ currents in rabbit pulmonary artery, whereas they were only partially inhibited by 100 mM TEA.

Aminopyridines also increase neurotransmitter release during sympathetic nerve stimulation, resulting in an augmentation of vascular tone (Glover, 1978; Hara *et al.*, 1980; Kato and Takata, 1987). This action, presumably reflecting a block of K^+ channels in presynaptic nerve terminals, is seen at 4AP concentrations as low as 10 μM (Hara *et al.*, 1980), i.e. below the concentrations having a direct effect on vascular smooth muscle. The Ca^{2+} antagonists verapamil and nicardipine (both 30 μM) fail to inhibit the increase in [^{3}H]noradrenaline outflow induced by 4AP, although this was inhibited by both diltiazem and *cis*(−)-diltiazem to a similar degree, despite the latter being a much less effective Ca^{2+} antagonist (Kato and Takata, 1987). In summary, exactly which of these different aminopyridine actions might contribute to the overall effect on the tone of a blood vessel is likely to depend upon the particular vessel being studied (quiescent/spontaneously active) and whether or not it is under hormonal or nervous stimulation at the time.

8.2.1.3 Inorganic cations

The effects of inorganic mono- and divalent cations on different K^+ channels have been extensively studied. The interaction of these inorganic cations with delayed and inward rectifying K^+ channels are reviewed comprehensively in Chapter 2 and will thus be dealt with here only briefly. Most small organic cations are able to block the delayed rectifier, examples in the squid axon being Na^+ or Li^+ (applied internally — block similar to TEA), Cs^+ or Ba^{2+} (from either side of the membrane) or Hg^{2+} (irreversible) and Zn^{2+} (being two of the more potent divalent ions) (Gilly and Armstrong, 1982). Whilst providing important information on the kinetic properties of different K^+ channels, these ions are probably too non-selective in their blocking action to be of much use in trying to classify the type of K^+ channel on which they act. Ba^{2+} for instance, in addition to blocking the inward and delayed rectifiers, has also been reported to block I_M, I_A, BK_{Ca}, IK_{Ca} (and the volume-regulated K^+ current), SK_{Ca}, K_{ATP}, the atrial ACh_m channel (K_{Ach}) and the serotonin-inactivated K^+ channel, K_S. (Brezina *et al.*, 1987, describe a complete block of K_S in *Aplysia* neurones by 20 mM Ba^{2+}, whereas Klein *et al.*, 1982, found no block of K_S in the same cells by 11 mM Ba^{2+}). The block of the inward rectifier in frog skeletal muscle by Ba_0^{2+} is worth highlighting, since this occurs with a K_D (at −65 mV) of 14 μM (Standen and Stanfield, 1978). Zn^{2+} has been shown to fairly potently inhibit I_A in rat cultured sympathetic neurones (50% reduction in amplitude at 0.1 mM; 0.2 mM Cd^{2+} was inactive) and also block BK_{Ca}, but did not block I_M in these neurones (Constanti and Smart, 1988). In most other cases though, high micromolar or millimolar cation concentrations are required for K^+ channel blockade.

There have been several recent reports of the block by cations of K_{ATP} in different tissues. Horie *et al.* (1987) showed that millimolar concentrations of ${Mg^{2+}}_i$ (and ${Na^+}_i$) block outward current through K_{ATP} in cardiac muscle without affecting inward K^+ current at potentials negative to E_K, thereby making the channel behave like an inward rectifier. Indeed, block by ${Mg^{2+}}_i$ is now thought to be a general physiological mechanism for inward rectifier K^+ channels (see Stanfield, 1988). K_{ATP} of frog skeletal muscle was voltage-dependently blocked by Cs^+ ($K_D = 4.1$ mM at −62 mV; induced channel 'flickering') and by Ba^{2+} ($K_D = 0.1$ mM; decrease in P_o, long channel closures) (Quayle *et al.*, 1988). ${Ba^{2+}}_o$ also blocked K_{ATP} in cardiac (Kakei and Noma, 1984) and smooth muscle ($K_D \approx 0.1$ mM; Standen *et al.*, 1989), $^{86}Rb^+$ efflux induced by metabolic exhaustion of frog skeletal muscle (Castle and Haylett, 1987) and the hyperpolarizing response in arterial smooth muscle to ACh (thought to be due to endothelial cell release of EDHF, which in turn opens K_{ATP}) and vasoactive intestinal peptide (Standen *et al.*, 1989).

Effects of cations on K_{Ca} are complicated by the fact that (a) they may interfere with the inward movement of Ca^{2+} to activate K_{Ca} (e.g. by La^{3+}, Ni^{2+}, Cd^{2+},

Co^{2+}, or Mn^{2+}; Thompson and Aldrich, 1980), and (b), by effects upon the Ca^{2+} activation mechanism of the K^+ channel (as distinct from a direct channel-blocking effect). For example, Connor (1979) has argued that the apparent block of K_{Ca} in *Helix* neurones by Ba^{2+} is probably due to its being a less effective activator of the conduction mechanism than Ca^{2+}. In red blood cells, Simons (1976) has shown that $Ca^{2+}{}_i$ is around ten-times more effective at activating K^+ efflux than Sr^{2+}, with Ba^{2+} being much less active and Mg^{2+} being an antagonist of Ca^{2+}. Intracellular Mg^{2+} has also been reported to block BK_{Ca} in patch-clamp studies of rat skeletal muscle (Ferguson and Magleby, 1989). In contrast, however, Squire and Petersen (1987) have reported a concentration-dependent activation of BK_{Ca} in salivary acinar cells by $Mg^{2+}{}_i$ (1 μM–1 mM) at a constant $Ca^{2+}{}_i$ of 10 nM. Since intracellular Mg^{2+} concentrations are reported to fall within this range, this suggests a possible physiological role for Mg^{2+} in regulating the activity of these channels, although the apparent discrepancy between effects of $Mg^{2+}{}_i$ on BK_{Ca} in different tissues still needs to be resolved.

In neurones of *Aplysia californica*, Deitmer and Eckert (1985) described a Ca^{2+}-dependent slow outward K^+ current which was essentially insensitive to TEA (50 mM), and was also not blocked by apamin (1 μM), quinine (0.3 mM) or 4AP (3 mM), but which could be inhibited by injection of Cs^+, Cd^{2+} or replacement of $Ca^{2+}{}_i$ by $Ba^{2+}{}_i$. Single-channel studies have demonstrated that the BK_{Ca} channels in rat skeletal muscle and cultured rabbit kidney cells are extremely sensitive to $Ba^{2+}{}_i$, a 50% reduction in P_o being seen with 1 or 0.1 μM $Ba^{2+}{}_i$, respectively (Vergara and Latorre, 1983; Guggino *et al.*, 1987). However, applied externally, around 10 000 times higher Ba^{2+} concentrations are required for the same degree of block (Vergara and Latorre, 1983). This voltage-dependent block of BK_{Ca} by Ba^{2+} was proposed to result from its binding to the conduction pathway rather than to the gating mechanism, since the block could be relieved by increasing the K^+ concentration (Vergara and Latorre, 1983). The ability of Ba^{2+} to block BK_{Ca} in arterial smooth muscle membranes (Benham *et al.*, 1985) probably explains, at least in part, its ability to depolarize and contract vascular tissues such as the rabbit aorta (see Karaki *et al.*, 1986).

8.2.2 Naturally occurring toxins which block potassium channels

8.2.2.1 Apamin and Leiurotoxin I

Apamin is an 18 amino acid basic peptide amidated at the carboxy terminus from the venom of the European honey bee, *Apis mellifera* (Habermann, 1984). Two basic (arginine) residues in positions 13 and 14, and two cysteine–cysteine disulphide bridges (1–11 and 3–15) are essential structural requirements for its biological activity (Vincent *et al.*, 1975). Apamin constitutes only a minor component (about 2% dry weight) of the venom and is unlikely to contribute to any inflammatory reaction to a bee sting. The early observation that apamin could potently block ATP-induced relaxations of certain intestinal smooth muscles led to speculation that it might be a purine receptor antagonist (see Habermann, 1984). However, it was subsequently shown that apamin specifically blocked Ca^{2+}-activated K^+ channels whose opening underlies such relaxant responses (Banks *et al.*, 1979; Maas *et al.*, 1980).

In guinea-pig liver cells, Burgess *et al.* (1981) showed that apamin blocked agonist-induced K^+ efflux without affecting the receptor-mediated changes in $Ca^{2+}{}_i$. Apamin blocks this K^+ permeability with an IC_{50} of 1 nM, which is comparable with the affinity of radiolabelled apamin for liver cells under physiological conditions (Cook *et al.*, 1983). Hepatocytes contain about 2000 apamin-binding sites per cell (Cook *et al.*, 1983). Binding studies performed in lower ionic strength media provide higher estimates for the binding affinity of apamin to various membranes (see Romey *et al.*, 1984). Apamin-binding sites appear to be localized on the outside of the cell surface (Seager *et al.*, 1984) and covalent binding of apamin analogues suggests the apamin-sensitive K_{Ca} channel to be a multimer of different polypeptide subunits, although no consensus has so far been reached on the size of the subunit to which apamin binds (Schmid-Antomarchi *et al.*, 1984; Seagar *et al.*, 1986).

Subsequent investigations have shown apamin to be probably the most selective of all K^+ channel blockers. Single-channel electrophysiological studies have shown that apamin selectively blocks Ca^{2+}-activated K^+ channels of low unitary conductance (SK_{Ca}) in cultured rat skeletal muscle (Blatz and Magleby, 1986) and guinea-pig liver cells (Capiod and Ogden, 1989), as previously proposed by Jenkinson *et al.* (1983) and Romey *et al.* (1984). Apamin is without effect on the intermediary K_{Ca} conductance (IK_{Ca}) in red blood cells (Burgess *et al.*, 1981) and moluscan neurones (Hermann and Hartung, 1983), or on the large-conductance K_{Ca} channels (BK_{Ca}) (Romey *et al.*, 1984; Blatz and Magleby, 1986). All other K^+ channels tested to date have likewise proven to be insensitive to apamin. As pointed out above, apamin-sensitive SK_{Ca} channels are relatively insensitive to blockade by TEA (in some tissues) and aminopyridines.

A number of new toxins with apamin-like activity have been identified in the last few years. An endogenous apamin-like peptide has been isolated from pig brain, having a comparable biological activity to apamin and cross-reacting with apamin antibodies (Fosset *et al.*, 1984). Since the firing rate of many excitable cells is dictated by an apamin-sensitive slow after-hyperpolarization (see Chapter 3), it is interesting to speculate that an endogenous apamin-like peptide may serve as one means of regulating the firing frequency of these neurones. Another toxin possessing apamin-like activity was identified by Castle and Strong (1986) in the venom of the Israeli scorpion, *Leiurus quinquestriatus hebraeus* (LQV). Although found in the same venom as charybdotoxin, this toxin is distinctly different, blocking [^{125}I]apamin binding to and K^+ efflux from guinea-pig hepatocytes with an IC_{50} of <20 nM, but showing no activity on CTX-sensitive IK_{Ca} or BK_{Ca} channels (Castle and Strong, 1986; Chicchi *et al.*, 1988). This toxin has now been purified and named leiurotoxin I (LTX; Chicchi *et al.*, 1988). It is a 3.4-kDa peptide with little structural homology to apamin, this probably explaining the lack of cross-reactivity of LTX with antibodies raised against apamin (Castle and Strong, 1986). The affinity of LTX for SK_{Ca} is between 5- and 20-fold weaker than that of apamin in different assays (Castle and Strong, 1986; Chicchi *et al.*, 1988). As with apamin, LTX inhibits the adrenaline-induced relaxation of the guinea-pig taenia caeci (IC_{50}=6.5 nM) but does not modify the rate or force of contraction in guinea-pig atria or rabbit portal vein (Chicchi *et al.*, 1988). What is not clear at this stage is whether LTX is the component of LQV responsible for the block of cromakalim-induced $^{86}Rb^+$ efflux from smooth muscle (Weir and Strong, 1988; Strong *et al.*, 1989; see 8.3.4.2). This would however seem unlikely since, firstly, LTX is itself without effect on the portal vein (Chicchi *et al.*, 1988) and, secondly, apamin is unable to block effects of cromakalim (Weir and Weston, 1986a; Cook and Hof, 1988).

8.2.2.2 Charybdotoxin

Charybdotoxin (CTX) is a recently identified K^+ channel blocker, constituting about 0.1% of the venom of the Israeli scorpion, *Leiurus quinquestriatus hebraeus*. Miller and colleagues (Miller *et al.*, 1985; Smith *et al.*, 1986) first described its isolation and ability to potently block BK_{Ca} (which is insensitive to apamin) from rat skeletal muscle. Subsequently CTX has been shown to block this channel in many other cells (for review see Moczydlowski *et al.*, 1988). Recently, the amino acid sequence of CTX has been elucidated (Gimenez-Gallego *et al.*, 1988; see also Valdivia *et al.*, 1988), showing it to be a highly basic single polypeptide chain of 37 amino acids (4353 Da) with a pyroglutamate blocked N-terminal end. The toxin shows amino acid sequence homology with two other scorpion venom toxins, noxiustoxin and neurotoxin P2 (Gimenez-Gallego *et al.*, 1988; Valdivia *et al.*, 1988), and the proposed tertiary structure of CTX in solution can be modelled to that of α-bungarotoxin with a striking similarity (Gimenez-Gallego *et al.*, 1988).

CTX block of BK_{Ca} is thought to be due to a direct occlusion of the channel by interaction with a site on the extracellular side of the membrane, whereas intracellular CTX is inactive (MacKinnon and Miller, 1988; Guggino *et al.*, 1987). The association rate of CTX was enhanced seven-fold when the channel was open (Anderson *et al.*, 1988) and the dissociation of CTX from the channel could be increased by raising K^+_i (MacKinnon and Miller, 1988).

However, CTX is not a specific blocker of BK_{Ca}. Castle and Strong (1986) showed that it also blocked IK_{Ca} in human red blood cells. Subsequently, Hermann and Erxleben (1987) have shown CTX block of IK_{Ca} in *Aplysia* neurones, the IC_{50} of 30 nM being somewhat lower than its affinity for BK_{Ca} in different cells (generally around 1–10 nM). This block of IK_{Ca} was voltage dependent and readily reversible on washout (Hermann and Erxleben, 1987). In these neurons CTX also suppressed a 'leak' K^+ conductance normally present at the resting membrane potential, which also appeared to be Ca^{2+}-activated. CTX had no effect upon other K^+ currents ($I_{K(V)}$ or I_A) in these neurons, or on inward Na^+ or Ca^{2+} currents (Hermann and Erxleben, 1987). CTX has been reported to have no effect on the K^+ channels from mammalian sarcoplasmic reticulum, rat muscle Na^+ channels or squid axon K_V (Miller *et al.*, 1985). It is also without effect on the apamin-sensitive SK_{Ca} of guinea-pig hepatocytes (Castle and Strong, 1986) and shows only very weak (K_i= 0.8 μM) inhibition of [^{125}I]apamin binding to rat brain synaptosomes, >20 000 times weaker than its affinity for [^{125}I]CTX binding sites (Chicchi *et al.*, 1988). Interestingly, Lewis and Cahalan (1988) have described a complete block by 5 nM CTX of two voltage-dependent K^+ channels in developing mouse thymocytes, which bear a number of similarities to K_V in other cells. This description of CTX blocking non-Ca^{2+}-activated K^+ channels suggests it may be less specific in its action than at first thought. In view of the inability of CTX to block I_A in molluscan neurones (Hermann and Erxleben, 1987) and *Drosophila* muscle (Elkins *et al.*, 1986), it was somewhat surprising that CTX was recently reported to block with a K_D of 3.6 nM the *Drosophila* shaker clone H4 transient I_A K^+ current, expressed in *Xenopus* oocytes

(MacKinnon *et al.*, 1988). This is interesting, not only because I_A is a Ca^{2+}-insensitive K^+ current, but also because the H4 clone encodes a K^+ channel with a single-channel conductance around 1/20 that of the BK_{Ca} channel. This finding indicates that CTX recognition sites must exist on K^+ channels previously thought to be structurally quite different. Subsequently, MacKinnon *et al.* (1989) have found that other shaker transcripts (A, B, C and D; see Chapter 1 for definition) are also blocked by CTX ($K_D \approx 3$ nM) in a manner similar to CTX block of BK_{Ca}. It is also likely that the recent finding that CTX inhibits [^{125}I]DTX binding to synaptic membranes with an IC_{50} of 10–30 nM is explained by a common affinity of these drugs for neuronal I_A channels (Sorensen *et al.*, 1989; Schweitz *et al.*, 1989).

Using [^{125}I]CTX as a probe, Garcia *et al.* (1989) have recently identified a solubilized sarcolemmal (receptor) fraction from smooth muscle which retained CTX binding activity. Subsequent purification (2000-fold) revealed a protein band of 105 kDa (Garcia *et al.*, 1989). However, Lucchesi *et al.* (1989) report that iodination of CTX greatly reduces its affinity for BK_{Ca} channels (>100-fold). These authors also reported the existence of a second peptide in the venom of *Leiurus quinquestriatus*, having 29 of its 37 amino acid residues identical to those of CTX. This second toxin, which they named CTX-Lq2, blocked skeletal muscle BK_{Ca} with a $K_D = 43$ nM (compared to CTX, $K_D = 1.3$ nM), although both peptides were equieffective at blocking K^+ efflux from red blood cells ($K_D = 43$ nM; Lucchesi *et al.*, 1989).

In screening other scorpion venoms for CTX-like activity, *Centruroides sculpturatus* Ewing venom was found to block BK_{Ca} with a similar potency to *Leiurus quinquestriatus* venom (LQV), although with a much longer single-channel mean blocked time (Moczydlowski *et al.*, 1988). The venom of two Old World scorpions, *Buthacus arenicola* and *Buthus occitanus*, also showed weak blockade of BK_{Ca} (Moczydlowski *et al.*, 1988). Alvarez and Garcia-Sancho (1988) found that the venom of the snake *Notechis scutatus* inhibited red cell K_{Ca} with an IC_{50} of 0.1 μg/ml. Unlike CTX, this inhibition was not reversed by washing. There have been several reports of effects of crude LQV on K^+ permeability under different conditions. However, since this venom contains other K^+ channel-blocking toxins in addition to CTX (Castle and Strong, 1986; Chicchi *et al.*, 1988; Lucchesi *et al.*, 1989), results obtained with crude LQV will not be discussed here.

Beech *et al.* (1987) and Talvenheimo *et al.* (1988) have both shown that CTX blocks BK_{Ca} channels in vascular smooth muscle. In intact vascular tissue, CTX causes contraction of the rabbit aorta (Talvenheimo *et al.*, 1988; Cook, 1989). This might be expected to occur as a consequence of K^+ channel blockade leading to membrane depolarization. However, it is unclear if BK_{Ca} channels, which are relatively insensitive to Ca^{2+}_i (see Chapter 3), would be open at resting potentials, thus leaving the question open of whether CTX contraction of smooth muscle is due to a blockade of these channels. An effect of CTX on some resting (leakage) K^+ conductance, similar to that described in *Aplysia* neurones (Hermann and Erxleben, 1987), may be an alternative explanation for these observations. Furthermore, the dihydropyridine Ca^{2+} antagonist isradipine only partially inhibited the CTX-evoked contraction of the rabbit aorta (Cook, 1989), showing that an indirect enhancement of Ca^{2+} influx through dihydropyridine-sensitive Ca^{2+} channels (as would be expected to ensue from membrane depolarization) does not alone explain this contractile activity of CTX. The simplest explanation of this observation is that other Ca^{2+} channels, insensitive to dihydropyridines, are also opened by the depolarizing effect of CTX blocking K^+ channels.

8.2.2.3 Noxiustoxin

As mentioned above, noxiustoxin (NTX; toxin II-11), a 39 amino acid peptide constituting 2% of the venom protein of the scorpion *Centruroides noxius* (Possani *et al.*, 1982), has a primary structure similar to CTX. Applied externally to the voltage-clamped squid axon, NTX concentration-dependently and reversibly blocked K_V with a K_D of 300 nM (Carbone *et al.*, 1982). Venom of the scorpion *Tityus serrulatus* also contains a toxin (fraction II-9) which has a very similar amino acid sequence and blocks squid axon K^+ channels similarly to NTX (Possani *et al.*, 1982). Low concentrations of NTX also decrease the rate of $^{86}Rb^+$ efflux from mouse brain synaptosomes, thereby enhancing [^{3}H]GABA release (EC_{50}=3 nM; Sitges *et al.*, 1986). The type of K^+ channel responsible for these effects was not characterized.

Recently, Valdivia *et al.* (1988) have shown that NTX also blocks BC_{Ca} in skeletal muscle with a K_D of 450 nM (compared to 2 nM for CTX). The block of BK_{Ca} by NTX was characterized by relatively short blocked periods (mean duration=50 ms) compared to CTX (approximately 10 s). It is presently unknown whether this reflects the differing affinities of NTX and CTX for the same binding site, or for different sites. The block of BK_{Ca} by NTX would appear to distinguish it from *Tityus serrulatus* venom, which is reported not to block this channel (Moczydlowski *et al.*, 1988). Applied to the internal side of the membrane, NTX was ineffective at blocking both K_V of squid axon (Carbone *et al.*, 1982) and rat skeletal muscle BK_{Ca} (Valdivia *et al.*, 1988). Effects of NTX on other K^+ channels have not yet been reported, although it is without effect on Na^+ channels (Carbone *et al.*, 1982).

8.2.2.4 Dendrotoxins and MCD peptide

Dendrotoxin from the green mamba snake and toxin from the black mamba belong to a structurally homologous peptide family of 59–60 amino acids both having their N-terminus blocked with pyroglutamic acid (for review see Harvey and Anderson, 1985), as is also the case for CTX. They are potent convulsants when administered centrally and augment acetylcholine release and twitch amplitude at the neuromuscular junction (Anderson and Harvey, 1988), presumably by blocking a component of the nerve terminal K^+ current and thereby prolonging the inward voltage-dependent Ca^{2+} current. They lack, however, the phospholipase A2 activity characteristic of β-bungarotoxin (β-BTX; see 8.2.2.5).

Block of neuronal K^+ currents by dendrotoxin and toxin I have now been shown in several different preparations. At the frog node of Ranvier, toxin I blocks one of the fast components of the outward K^+ current (I_{Kf1}) with an IC_{50} of 0.4 nM (Benoit and Dubois, 1986), this being probably the main contributor to the action potential K^+ current in myelinated nerves. The other K^+ currents (I_{Kf2} and I_{Ks}) were only affected by toxin I at concentrations around 100 times higher. Similar to I_{Kf1}, I_A of hippocampal neurones was also sensitive to DTX (50–350 nM; Halliwell *et al.*, 1986) but K_{Ca} and I_M in these neurones were not blocked. In dorsal root ganglion neurones, Penner *et al.* (1986) subdivided the non-inactivating current I_{Ks} into two components, one being very sensitive to DTX (0.14–1.4 nM), whereas I_{Kf} was not modified by these DTX concentrations. Stansfield *et al.* (1986) reported a similar block by DTX of I_{Ks} in rat visceral sensory neurones. However, after treatment with DAP no effect of DTX on the dorsal root ganglion is visible, suggesting they block the same K^+ channel (Penner *et al.* 1986). In fact, the blocking activity of the dendrotoxins in many ways resemble that of the aminopyridines (albeit at 500- to 2000-fold lower concentrations). Differences do exist though, such as the additional action of 3,4DAP on I_{Kf} in these neurones, and at the frog node of Ranvier where 4AP blocks both fast K^+ currents whilst DTX blocks only I_{Kf1} (Benoit and Dubois, 1986). Aminopyridines also block the delayed rectifier of nerve axons, whereas DTX does not. In guinea-pig hepatocytes, DTX and toxin I failed to block SK_{Ca} or inhibit apamin binding (Cook and Haylett, 1985).

Despite the selectivity of DTX for certain neuronal K^+ currents, there would appear to be some diversity as to which channel is blocked. This is corroborated by binding studies with [^{125}I]DTX which have defined at least two classes of binding sites. Scatchard analyses of the binding to chick brain membranes resolved two components with K_Ds of 0.5 and 15 nM (Black and Dolly, 1986). β-BTX showed selectivity for inhibition of [^{125}I]DTX binding to the high-affinity site. Rehm and Lazdunski (1988) have recently used a β-BTX affinity column to distinguish two populations of [^{125}I]DTX-binding sites in rat brain; one subtype, constituting around 65% of the total [^{125}I]DTX binding, was inhibited by β-BTX with low affinity (around 500 nM), whereas β-BTX inhibited binding to the other site with a K_I of 15 nM (Rehm and Lazdunski, 1988). Another central convulsant, mast cell degranulating (MCD) peptide, also binds with high affinity to sites in rat brain which are non-competitively inhibited by dendrotoxin (K_I=1.7 nM, around twice as potent as MCD peptide itself; Bidard *et al.*, 1987). Once again β-BTX affinity chromatography suggested a small percentage of the MCD peptide-binding sites to have high β-BTX affinity (10 nM), with the remaining 85% inhibited by β-BTX with an IC_{50} of 560 nM. There would thus appear to be a certain homology between the central binding site for dendrotoxins, MCD peptide and β-BTX. Indeed selected (5000-fold) purification of DTX-binding sites retains the binding activity of both β-BTX and MCD peptide, suggesting them to be on the same protein complex (Rehm and Lazdunski, 1988). MCD peptide does not interact with apamin-sensitive K^+ channels at concentrations up to 1 μM (Cook and Haylett, 1985), despite close structural homology of MCD peptide to apamin (both obtained from the venom of the European honey bee).

8.2.2.5 Other snake venom K^+ channel toxins

The irreversible block of transmitter release at the neuromuscular junction by β-BTX, associated with its phospholipase A2 activity, is preceded by an excitatory phase where acetylcholine release is augmented. This augmentation has recently been ascribed to an irreversible block of one component of the presynaptic nerve terminal K^+ current (I_{Ks}), thereby prolonging the inward voltage-dependent Ca^{2+} current (Dreyer and Penner, 1987; Rowan and Harvey, 1988). (This current has been denoted I_{Ks} or $I_{K,s}$, which should not be confused with the serotonin-inactivated K^+ current $I_{K(S)}$ (see Chapter 5)). Block of I_{Ks} by DTX has been shown in guinea-pig dorsal root ganglion neurones (0.45–45 nM: Petersen *et al.*, 1986) and in mouse motor nerve terminals (5–500 nM: Dreyer and Penner, 1987; Rowan and

Harvey, 1988) although not at the frog neuromuscular junction (Rowan and Harvey, 1988). The K^+ current blocked by β-BTX in these preparations is distinct from the DAP-sensitive fast K^+ current (Dreyer and Penner, 1987) and is not carried through Ca^{2+}-activated K^+ channels (Dreyer and Penner, 1987; Rowan and Harvey, 1988). β-BTX (5 μM) is also without effect on the apamin-sensitive SK_{Ca} of guinea-pig hepatocytes (Cook and Haylett, 1985). There is a strong similarity between the prejunctional K^+ currents blocked by β-BTX and DTX, although DTX has further blocking activity in addition to that of β-BTX (Petersen *et al.*, 1986). Similarly, $[^3H]$β-BTX binding is inhibited by toxin I (Othman *et al.*, 1982) whilst studies with $[^{125}I]$DTX suggest two populations of binding sites in rat brain, one having a low affinity for β-BTX (Rehm and Lazdunski, 1988). Two other snake venom toxins with phospholipase A2 activity which facilitate and then block acetylcholine release — crototoxin (>5 nM) and taipoxin (6.5 nM) — caused twitch augmentation and modification of the perineural wave form similar to β-BTX at the mouse motor nerve terminal (Dreyer and Penner, 1987; Rowan and Harvey, 1988). However, the fact that block by these toxins is still seen at low temperatures and in the absence of Ca^{2+} (Rowan and Harvey, 1988) and that it is not mimicked by bee or cobra venom phospholipases (Dreyer and Penner, 1987) suggests their inhibitory effect on K^+ channels is not associated with their phospholipase A2 activity.

The crude venom of the Gaboon viper (*Bitus gabonica*; 100 μg/ml) irreversibly reduced the inward rectifier K^+ current of guinea-pig ventricular myocytes, resulting in a slowing of repolarization of the cardiac action potential (Busso *et al.*, 1988). The venom also inhibited Ca^{2+} inward currents in these cells.

8.2.2.6 Pandinus imperator scorpion venom (Piv)

Pappone and Cahalan (1987) described the ability of Piv to block a component of the voltage-dependent K^+ current in frog myelinated nerve axons (probably the I_{Kf1} component which is sensitive to toxin I; Benoit and Dubois, 1986). Block by Piv was strongly voltage dependent, 50% inhibition occurring with 50 μg/ml at −50 mV and with 100–250 μg/ml at −30 mV. In GH_3 pituitary cells, Piv similarly blocked a K^+ current which resembled that seen in the frog axon (Pappone and Lucero, 1988). In both cases the blockade occurred after just a few seconds exposure to Piv, but failed to reverse significantly during washout for up to 40 min. This irreversibility distinguishes Piv from most other scorpion toxins which block K^+ channels. The blocking action of Piv appeared to be due to an effect on channel gating, causing a shift in the voltage-dependence of channel closing, without any apparent effect on channel opening (Pappone and Cahalan, 1987). This profile resembles the effect of certain scorpion toxins on Na^+ channels. Although the component of Piv responsible for these effects has not yet been purified, it would seem to show a certain selectivity since Piv had no effect on K^+ currents in voltage-clamped frog skeletal muscle fibres (K_V; Pappone and Cahalan, 1987) or on BK_{Ca} from rat skeletal muscle T-tubules (Moczydlowski *et al.*, 1988). Na^+ currents in frog axons were not affected by Piv (Pappone and Cahalan, 1987). However, in GH_3 cells 100 μg/ml Piv reduced both voltage-dependent Na^+ and Ca^{2+} currents by about 20%, although at least in the case of the Na^+ current this effect appeared to be attributable to a different component of the venom from that which inhibited K^+ channels (Pappone and Lucero, 1988).

8.2.2.7 Conotoxins

To date there have been few investigations of the effects of marine snail (*Conus*) venom on K^+ channels. Chestnut *et al.* (1987) reported the presence of three components in *Conus striatus* venom which modified in different ways the delayed rectifier K^+ currents of *Apylsia* neurones. Once the phospholipase activity of the venom had been suppressed, the crude venom (0.9 μg/ml) increased the frequency of action potentials and depolarized neurones, and the individual venom components (around 0.1 μg/kg) appeared (i) to increase peak K^+ current (ii) to slow K_V activation and inactivation, and (iii) to decrease peak K^+ current. The active components remain to be purified. Moczydlowski *et al.* (1988) saw no effect of venom from different *Conus* snails on BK_{Ca} of rat skeletal muscle.

8.2.2.8 Strychnine

Strychnine was shown by Shapiro (1977a) to block voltage-dependent K^+ currents at the frog node of Ranvier. However, this action was not selective in that it also blocked Na^+ conductance in these nerves (Shapiro, 1977b), and it has been shown to block an ACh-activated ion channel in *Aplysia* neurones (Slater and Carpenter, 1982). Cook and Haylett (1985) found strychnine to inhibit apamin binding to, and K^+ efflux from, guinea-pig hepatocytes with an IC_{50} of 180–190 μM. IK_{Ca} in human red blood cells was, however, only minimally inhibited by 1 mM strychnine (Lew and Ferreira, 1978; Cook and Haylett, 1985).

Effects of strychnine on vascular smooth muscle appear to be complex. On the rat portal vein, Louttit *et al.* (1984) showed that 10–300 μM increased the spontaneous electrical and mechanical activity consistent with K^+ channel blockade and comparable to the effects of similar concentrations of TEA or 4AP. At 1 mM strychnine though, the spontaneous activity was inhibited. Similarly, on the rat aorta strychnine (>100 μM) caused an inhibition of the phasic contractions induced by noradrenaline (Louttit *et al.*, 1984). Whilst a membrane-stabilizing effect (Shapiro, 1977b) may in part be responsible for the inhibitory activity of strychnine, Haeusler (1978) has reported that 1 mM strychnine hyperpolarizes (by 3 mV) the rabbit main pulmonary artery, thereby causing a small rightward shift of the noradrenaline concentration–response curve. This action resembles that of the K^+ channel opener cromekalim on large blood vessels (Cook *et al.* 1988c).

8.2.2.9 Phalloidin

Phalloidin, a liver-toxic bicyclic heptapeptide from the death-cap mushroom *Amanita phalloides*, was shown by Cognard *et al.* (1985, 1986) to reversibly inhibit the fast outward K^+ current of frog skeletal muscle (K_V). Blockade was seen both in intact and detubulated muscle fibres. Half-maximal inhibition occurred with 2 pM phalloidin, the degree of inhibition increasing with depolarization (Cognard *et al.*, 1986), although no use dependence was seen. The reversal potential and activation curve of the K^+ current were unaffected by the toxin, but the inactivation curve was shifted towards negative values. Antamanide, a non-toxic monocyclic decapeptide from the same mushroom as phalloidin, shows a similar blocking activity over the same concentration range (10^{-12}–10^{-8} M) in frog skeletal muscle (Raymond *et al.*, 1987). Hence, these toxins would appear to be the most potent K^+ channel blockers so far described, although their efficacy awaits confirmation in other preparations.

8.2.2.10 Capsaicin

Dubois (1982) studied the effect of capsaicin, the active component of hot peppers, upon K^+ currents in voltage-clamped frog myelinated nerve fibres. Of the three components constituting the voltage-dependent K^+ current, capsaicin (10–20 μM) selectively blocked only the latter of the two fast components (I_{Kf2}) without affecting I_{Kf1} (sensitive to toxin I) or the slow component I_{Ks} (sensitive to β-BTX). Block of I_{Kf2} by capsaicin was reversible and was enhanced when the channels were opened by depolarization. In *Helix pomatia* neurones, capsaicin (60–300 μM) concentration-dependently inhibited the early outward current, I_A, and concentrations above 150 μM caused depolarization of the membrane but had minimal effect upon the Ca^{2+} current or IK_{Ca} (Erdelyi and Such, 1984). Petersen *et al.* (1987) have investigated effects of capsaicin on membrane currents of mammalian sensory neurones. In guinea-pig and chick dorsal root ganglion neurones, 30 μM capsaicin applied extracellularly reduced by 30% the amplitude and accelerated inactivation of the fast K^+ current (I_{Kf}). Intracellular capsaicin (30 μM) was without effect on I_{Kf}, but I_{Ks} was equally blocked by capsaicin from either side of the membrane (Petersen *et al.*, 1987). In contrast, 3,4DAP (0.5 mM) inhibited I_{Kf} (without affecting inactivation) but did not affect I_{Ks}. In addition to the effect described on outward K^+ currents, 30 μM capsaicin also inhibited both tetrodotoxin-sensitive and -resistant Na^+ currents in DRG neurones (Petersen *et al.*, 1987). Since the concentrations of capsaicin shown to block K^+ channels are generally higher than the concentrations causing depolarization of sensory nerves (thought to be due to an increased Na^+ and Ca^{2+} conductance; see Marsh *et al.*, 1987), it appears unlikely that K^+ channel blockade is of primary importance in the excitation and subsequent desensitization of nociceptive fibres by capsaicin.

8.2.2.11 Equinatoxins

Equinatoxins I and II (10–100 nM) from the sea anemone *Actina equina* L. reversibly blocked K^+ conductance in voltage-clamped frog mylinated nerve fibres, but also inhibited the Na^+ currents (Suput, 1988).

8.2.2.12 Polymixin B

Polymixin B (0.08–4 mM), a cyclic peptide antibiotic, inhibited K^+ efflux (IK_{Ca}) of human red blood cells, independent of the mode of induction (ATP depletion, Ca^{2+} ionophore A23187, vanadate or propranolol); Varecka *et al.*, 1987). Vanadate-stimulated uptake of $^{45}Ca^{2+}$ by red cells was not inhibited by polymixin B (Varecka *et al.*, 1987), suggesting the inhibition to be at some subsequent step, possibly related to inhibition by polymixin B of protein kinase C (Wise *et al.*, 1982), which is known to activate certain K^+ channels (Walsh and Kass, 1988).

8.2.3 Sulphonylureas and other blockers of ATP-sensitive potassium channels

Henquin and Meissner (1982) were the first to suggest that the stimulation of insulin release from pancreatic β cells by sulphonylureas could be explained by their ability to reduce the K^+ permeability of these cells, this leading to membrane depolarization and Ca^{2+} entry. However, a clearer understanding of this effect has been complicated by the multiplicity of different K^+ channels found in these cells, such that blockade of one channel type often leads to an increased K^+ (or $^{86}Rb^+$) efflux through other K^+ channels (see Cook, 1988). Recently, variations of the patch-clamp technique have been successfully used to identify the primary site of action of tolbutamide as K_{ATP}, which it blocks with an IC_{50} of approximately 7 μM (Sturgess *et al.*, 1985; Trube *et al.*, 1986).

Structure–activity studies have now been performed with a wide variety of hypoglycaemic drugs, showing excellent correlations between their blockade of K_{ATP} (measured with patch clamp), their affinity for a [^{3}H]glibenclamide binding site, inhibition of $^{86}Rb^+$ efflux and stimulation of insulin secretion from insulinoma cell lines (Schmid-Antomarchi *et al.*, 1987; Sturgess *et al.*, 1988; Zunkler *et al.*, 1988a). These papers also draw attention to the rank order of potency and the free drug concentrations in these *in vitro* experiments, which correlate with their relative therapeutic doses and the plasma concentrations used in man. This provides strong supportive evidence that K_{ATP} is indeed the locus for the clinical activity of these hypoglycaemic agents. Of the various agents tested, glibenclamide appeared to be the most potent, causing half-maximal inhibition of K_{ATP} at 4–27 nM (Sturgess *et al.*, 1988; Zunkler *et al.*, 1988a).

Derivatives of glibenclamide and gliquidone which retain the benzoic acid group, but are devoid of the sulphonylurea moiety (meglitinide/HB699 and UL-DF9, respectively), still elicit similar activity to the parent compounds, albeit at reduced (100-fold) potency (Garrino *et al.*, 1985, 1986). Patch-clamp studies have confirmed that meglitinide blocks K_{ATP} as expected (Sturgess *et al.*, 1988; Zunkler *et al.*, 1988a) and binding studies have shown that it competitively interacts with the [^{3}H]glibenclamide binding site (Schmid-Antomarchi *et al.*, 1987). Recently another benzoic acid derivative, (−)-*AZ*-DF265, has been shown to block islet $^{86}Rb^+$ efflux and promote insulin release with a similar potency to glibenclamide (Garrino and Henquin, 1988), showing that the sulphonylurea group is not a prerequisite for these effects.

Blockade of K_{ATP} by tolbutamide and glibenclamide is seen regardless of which side of the cell membrane the drugs are applied (Trube *et al.*, 1986; Sturgess *et al.*, 1988). Furthermore, the intracellular concentration of ATP does not influence greatly the blocking potency of these drugs (Sturgess *et al.*, 1988; Zunkler *et al.*, 1988a). Covalent binding of [^{3}H]glibenclamide to rat β cell tumor membranes resulted in the labelling of two polypeptides of 140 and 33 kDa (Kramer *et al.*, 1988). Labelling of the larger of the two polypeptides was specifically decreased by glibenclamide (IC_{50}=20 nM) and tolbutamide (IC_{50}=20 μM), but not by various Ca^{2+} channel antagonists, suggesting it is likely to constitute the sulphonylurea receptor of β cell membranes. Bernardi *et al.* (1988) have recently isolated a glibenclamide-binding site from pig brain which was purified as a 150-kDa protein. Cloning of the sulphonylurea receptor (K^+ channel?), with the natural extension to whether genetic differences exist between diabetic and normal animals, may now be possible.

In addition to their blockade of K_{ATP} in pancreatic β cells (and related cell lines), the sulplhonylureas are also able to block K_{ATP} in cardiac myocytes (Belles *et al.*, 1987; Fosset *et al.*, 1988), frog skeletal muscle (Spruce *et al.*, 1987), K^+ efflux due to metabolic exhaustion of frog skeletal muscle (3–30 μM glibenclamide/03–2 mM tolbutamide: Castle and Haylett, 1987), and K_{ATP} in arterial smooth muscle (Standen *et al.*, 1989). In heart cells the rank of order of potency for inhibition of

[^{3}H]glibenclamide binding by various sulphonylureas agrees closely with that found in insulinoma cells (Fosset *et al.*, 1988), although patch-clamp studies show K_{ATP} in cardiac cells to be around 50 times less sensitive to tolbutamide than are pancreatic β cells (Belles *et al.*, 1987). The shortening of the action potential seen in guinea-pig ventricular cells when ATP_i falls is reversed by glibenclamide (Fosset *et al.*, 1988) and it has been suggested that blockers of K_{ATP} may be of therapeutic use in suppressing ischaemia-induced dysrhythmias (Cook, 1988; Kantor *et al.*, 1987).

The sulphonylureas appear to be fairly selective inhibitors of K_{ATP}; they have been shown not to block (maximum concentration of drug tested is shown) BK_{Ca} channels in β cells or insulinoma cells (1 mM tolbutamide, 10 μM glibenclamide; Trube *et al.*, 1986; Sturgess *et al.*, 1988), delayed rectifying K current (K_V) (3 μM glibenclamide, 0.2 mM tolbutamide; Rorsmann and Trube, 1986), non-selective cation channel (1 mM tolbutamide/ 10 μM glibenclamide; Sturgess *et al.*, 1988) or voltage-dependent Ca^{2+} currents (50 nM glibenclamide; Schmid-Antomarchi *et al.*, 1987). In cardiac myocytes the Ca^{2+} inward current and the inward rectifying K^+ current were unaffected by 2 mM tolbutamide (Belles *et al.*, 1987). Tolbutamide (0.3 mM) and sparteine (1 mM; see below) inhibited apamin-sensitive SK_{Ca} in guinea-pig hepatocytes by about 50%, but were inactive at 0.1 and 1 mM, respectively, on IK_{Ca} in human red blood cells (Cook and Haylett, 1985). Effects of these drugs on other K^+ channels have not been described, although they have been reported to show fairly high affinity for α- and β-adrenoceptors (Cherskey and Altszuler, 1984). The sulphonylureas also block effects of cromakalim and other K^+ channel openers (see 8.3.1.4 and 8.3.4.3).

It is convenient here to mention some of the other compounds known to modify the K^+ permeability of mouse islets and thereby induce insulin secretion. Sparteine (0.02–1 mM; Paolisso *et al.*, 1985) and adamantanamine (0.01–1 mM; Garrino and Henquin, 1987) inhibited $^{86}Rb^+$ efflux from islets and caused membrane depolarization in a manner resembling the sulphonylureas. Although not yet studied by the patch-clamp technique, it is likely that both of these agents act by blocking K_{ATP} like the sulphonylureas. Among the many pharmacological activities attributed to adamantanamine, it is noteworthy that it is known to inhibit cardiac K^+ channels, resulting in a prolongation of the action potential (for references see Garrino and Henquin, 1987). Sparteine has also been used clinically to treat cardiac arrhythmias (Böhme, 1964). The analogy between these effects and the cardiac effects of the sulphonylureas is readily apparent. Sparteine also inhibits K_V in the squid axon (Ohta and Narahaski, 1973). Amantidine (1 mM) did not block and sparteine only very weakly inhibited SK_{Ca} in guinea-pig hepatocytes (Cook and Haylett, 1985).

8.2.4 Non-specific ion channel blockers

8.2.4.1 Quinine and local anaesthetics

Quinine and its stereoisomer quinidine, which are generally equipotent at blocking K^+ channels (Lew and Ferreira, 1978; Cook and Haylett, 1985) are often used inadvertently as fairly selective blockers of K_{Ca}. They do indeed block the different K_{Ca} channels: SK_{Ca} in hepatocytes (where quinine also displaces apamin binding with a K_I of 0.5 mM; Cook and Haylett, 1985), IK_{Ca} in red blood cells (IC_{50} = 15 μM, Lew and Ferreira, 1978) and molluscan neurones (Herman and Gorman, 1984) and

BK_{Ca} in cultured kidney cells (Guggino *et al.*, 1987). However, they also have equally potent blocking activity at other K^+ channels, such as K_V in neuroblastoma cells (15 μM, Fishman and Spector, 1981), I_A in molluscan neurones (Hermann and Gorman, 1984) and K_{ATP} in pancreatic β cells (Cook and Hales, 1984). Findlay *et al.* (1985) showed that in an insulin-secreting cell line quinine blocked Ca^{2+}-independent inward rectifier K^+ channels (possibly K_{ATP}) in preference to BK_{Ca}, and Smith and Levinson (1989) have recently reported inhibition by quinine of multiple K^+ and Na^+ transport mechanisms in Ehrlich ascites tumour cells. In addition, quinidine is a potent blocker of cardiac Na^+ channels (Brown *et al.*, 1985) and shows affinity for α-adrenoceptors (Mecca *et al.*, 1980). In human red blood cells (Reichstein and Rothstein, 1981) and lymphocytes (Sarkadi *et al.*, 1985) the potency of quinine at inhibiting IK_{Ca} is greatly increased by lowering K^+_o, although block of SK_{Ca} in guinea-pig hepatocytes by quinine was not dependent upon K^+_o (Cook, 1983). In addition to quinine and quinidine ($IC_{50} = 0.42$ mM), the analogues cinchonine (0.37 mM) and cinchonidine (0.42 mM) and the anti-malarial agents quinacrine (73 μM), chloroquin (0.2 mM) and primaquine (0.97 mM) also interacted with SK_{Ca} in guinea-pig hepatocytes (Cook and Haylett, 1985). In red blood cells, however, cinchonine and cinchonidine were less effective blockers of IK_{Ca} than quinine (Lew and Ferreira, 1978; Cook and Haylett, 1985).

In an investigation of the effects of various K^+ channel blockers on the rabbit aorta, quinine was found to act quite differently from any of the other K^+ channel blockers tested: while TEA, 3,4DAP and a CTX-like toxin all augmented either basal or agonist-induced tone in this vessel, quinine instead attentuated contractions elicited by KCl or angiotensin II (Fig. 8.1A). The non-competitive antagonism of KCl responses by quinine (pD'_2=4.0) resembled the profile seen with a Ca^{2+} antagonist. Subsequent $^{45}Ca^{2+}$ uptake studies with depolarized rat aortic smooth muscle cells showed that quinine indeed acts as a voltage-dependent Ca^{2+} channel blocker in vascular smooth muscle (Fig. 8.1B; U. Ruegg, unpublished). The pIC_{50} value for quinine effects on Ca^{2+} channels (=4.0) is in the range of the concentrations used to block K^+ channels in different cells. Similar conclusions have been arrived at for the action of quinine and quinidine on neurones (Hermann and Gorman, 1984) and the heart (Brown *et al.*, 1985). It is clear that by limiting Ca^{2+} entry into cells, quinine may indirectly affect the opening of certain K_{Ca} channels, and this complication needs to be considered when evaluating effects of quinine on K^+ conductance. The non-competitive inhibition by quinine of angiotensin II contractions of the rabbit aorta (Fig. 8.1A) cannot be explained by a blockade of voltage-dependent Ca^{2+} channels, or of K^+ channels (see Cook *et al.*, 1988c., Cook, 1989). This suggests that some additional action of the drug (e.g. on Ca^{2+} entry through receptor-operated channels, or on intracellular Ca^{2+} mobilization) is present at such (>0.1 mM) concentrations. These effects of quinine on the rabbit aorta are comparable to those of the local anaesthetic agent procaine (0.5–10 mM; Ahn and Karaki, 1988), suggestive of a common membrane stabilizing action.

Although the effect of local anaesthetics upon excitable tissues is predominantly due to their block of fast Na^+ channels, most of these drugs also block K^+ channels at higher concentrations. Almers (1976) described a shift by tetracaine (2 mM) of +25 mV in the voltage dependence of K_V in frog skeletal muscle. Although the gating of the delayed rectifier was thus modified, tetracaine did not influence the

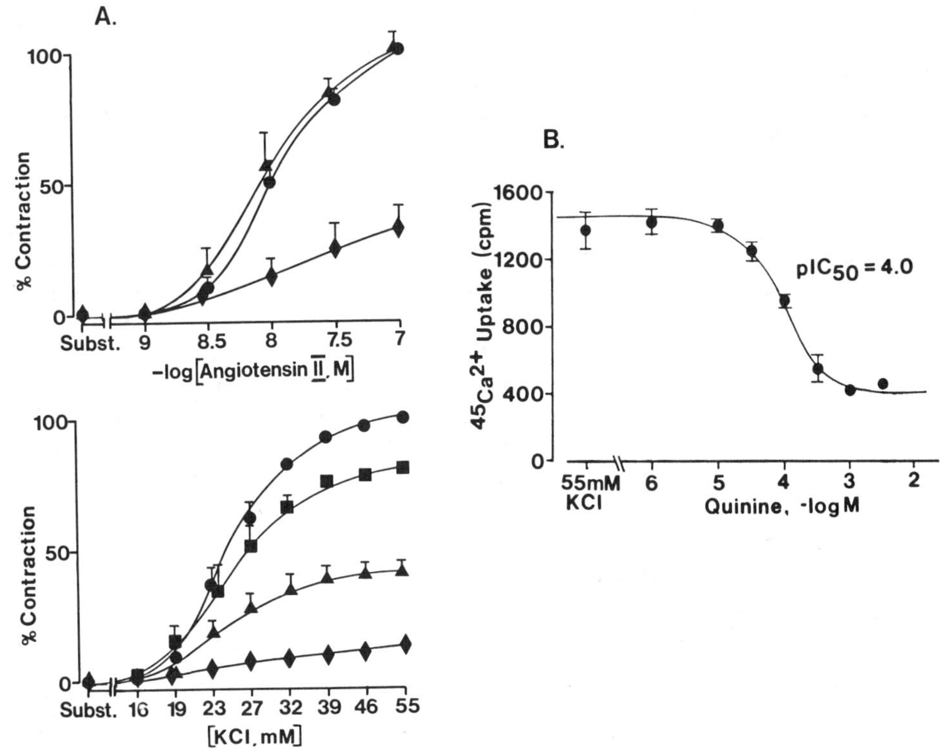

Fig. 8.1. — Effects of quinine on aortic smooth muscle. (A) Quinine (■, 0.03 mM; ▲, 0.1 mM; ⧫, 0.3 mM) caused a concentration dependent non-competitive inhibition of KCl contractions of rabbit aortic rings (lower diagram), the 0.3 mM concentration also inhibiting the angiotensin II concentration–response curve (upper diagram). Control rings (●) received the vehicle for quinine. Means ± s.e.m. of 4 experiments. Removal of the endothelium did not affect the results obtained (not shown). (From Cook *et al.*, 1987; methods as described in Cook (1989)). (B) Quinine concentration-dependently inhibited $^{45}Ca^{2+}$ uptake by rat aortic smooth muscle cells, stimulated by depolarization with 55 mM KCl. Means ± s.e.m. of 4 experiments. (Unpublished work of U. Rüegg; methods as described in Rüegg *et al.*, 1985).

displacement currents in this study, nor the resting K^+ conductance. The K^+ conductance induced by metabolic exhaustion of frog skeletal muscle (thought to be through K_{ATP}) is, however, inhibited by 1 mM tetracaine (62% block) and lignocaine (46% block) (Castle and Haylett, 1987). Local anaesthetics have also been shown to inhibit K_{Ca} in different tissues. Benham *et al.* (1985) showed a dose-dependent inhibition by procaine (1 and 10 mM) of BK_{Ca} current in isolated membrane patches from arterial smooth muscle. This effect was potential- insensitive and the potency of procaine was around 5–10 times greater than that of TEA at these channels. Kitamura *et al.* (1986) also showed depolarization of several rabbit and guinea-pig arterial muscles by procaine (0.5–10 mM) and an inhibition of BK_{Ca} in rabbit portal vein. SK_{Ca} channels were blocked by dibucaine (cinchocaine) with an IC_{50} of 0.8 mM (Cook and Haylett, 1985). In basolateral membranes of epithelial cells, a K^+ channel of intermediate conductance (17 pS) has been identified which is activated during cell swelling. At concentrations >30 μM, lidocaine caused rapid flickering of this channel during bursts of activity, higher concentrations causing the channel to spend progressively more time in the closed state (Richards and Dawson, 1986; Van Driesche, 1986). The resting K^+ conductance in these epithelial cells was not modified by lidocaine.

8.2.4.2 Calcium antagonists

A number of drugs better known for their inhibition of Ca^{2+} channels have been reported at micromolar concentrations to block certain K^+ channels as well (for early references see Gola and Ducreux, 1985). Recently whole-cell patch and voltage-clamp techniques have been applied to try and distinguish such actions from those subsequent to a block of Ca^{2+} channels. Thus Gurney and Nerbonne (1986) reported IC_{50}s of 3–5 μM for inhibition of the K_V and I_A in *Aplysia* bag cells by nifedipine (and nisoldipine), a concentration only two-fold higher than that required to block Ca^{2+} currents in these neurones. Gola and Ducreux (1985) described the inhibition of K_{Ca} (apamin-insensitive) in *Helix* neurones by methoxyverapamil (D600; IC_{50}=130 μM) and Cd^{2+} (22 μM). This block, unlike that of Ca^{2+} channels (IC_{50}s 17 and 22 μM, respectively), was quick to develop and not reversed by washing. Verapamil (100–400 μM) and diltiazem (0.1–0.3 mg/ml) also blocked these K^+ channels, but nifedipine, nicardipine and cinnarizine were inactive at concentrations up to 100 μM. In intestinal smooth muscle, verapamil, diltiazem and nicardipine blocked a voltage-dependent outward K^+ current at concentrations 10–200 times greater than those required to block Ca^{2+} channels (Terada *et al.*, 1987), whilst in cardiac tissue the selectivity of these drugs for Ca^{2+} channels over K^+ channels was even more marked (Hume, 1985). Binding studies with [^{3}H]glibenclamide showed no inhibition by 100 μM verapamil or nifedipine, suggesting they do not interact with the sulphonylurea receptor (K_{ATP}) (Kramer *et al.*, 1988). At a concentration of 100 μM, D600 caused a small (25%) inhibition of [^{125}I]apamin binding to guinea-pig hepatocytes, suggesting a weak affinity for SK_{Ca} (Cook and Haylett, 1985). Hence it is reasonable to assume that effects of these drugs on Ca^{2+} channels will predominate over any effect they might have on K^+ channels in the physiological/clinical situation.

Falipamil (AQA-39) is an analogue of verapamil which has been developed as a specific bradycardic agent because of its ability to selectivity slow the firing frequency of the sinus node. In addition to a strongly voltage-dependent block of myocardial slow Ca^{2+} channels, falipamil (1–100 μM) also increases the duration of the ventricular action potential, presumably by decreasing some K^+ conductance (Trautwein *et al.*, 1981; see also section 8.2.6.2). Osterrieder *et al.* (1981) showed that at low-frequency stimulation of the rabbit sinoatrial node the main effect of falipamil (10 μM) was to inhibit the time-dependent K^+ current (I_K, delayed rectifier), this action presumably prolonging the APD to elicit bradycardia. Only at higher stimulation rates was the Ca^{2+} current affected. Falipamil also blocked the ACh-activated K^+ channel at the sinoatrial node, a combination of both a direct K^+ channel-blocking effect and a competitive antagonism of muscarinic receptors (Osterrieder *et al.*, 1981). Probably all of these different properties together contribute to the overall bradycardic activity of falipamil.

8.2.4.3 Neuromuscular blocking agents

Coronado and Miller (1980) showed blockade of the maxi-K_+ channel of sarcoplasmic reticulum by decamethonium (IC_{50}=120 μM), which induced 'channel flickering' at low concentrations, and hexamethonium (IC_{50}=3.5 mM; open state probability decreased concentration-dependently without any evidence of flickering). In addition to these cholinergic drugs, Cook and Haylett (1983, 1985) described the inhibition of apamin-sensitive K^+ efflux (SK_{Ca}) from, and [^{125}I]apamin binding to, guinea-pig hepatocytes by a series of neuromuscular blocking agents. Most potent among these were tubocurarine, atracurium and pancuronium (K_Is 4–8 μM), with gallamine (14 μM), decamethonium (620 μM) and hexamethonium (760 μM) being less active (Cook and Haylett, 1985). Castle (1987) extended this list to include dequalinium (K_I 1 μM) and vercuronium (4 μM). It was postulated that the ≈11 Å separation of the two positively charged nitrogens in these molecules, thus corresponding to the two arginine residues of apamin, was likely to be of importance for this activity (Cook and Haylett, 1985). A lack of activity in these studies of β-bungarotoxin (0.5 μM), or phencyclidine and amantidine (both 100 μM), suggests that this affinity for apamin-sensitive K^+ channels was neither related to an affinity for endplate nicotinic ACh receptors nor the endplate ion channel (Cook and Haylett, 1985). Subsequent studies have confirmed the block of apamin-sensitive SK_{Ca} by neuromuscular blockers in neurones (10–100 μM, Nohmi and Kuba, 1984; Kawai *et al.*, 1985; 0.1–0.4 μM, Higashida and Brown, 1986) and intestinal smooth muscle (Gater *et al.*, 1985). Gallamine (0.1–10 mM) has also been shown to inhibit voltage-dependent K^+ conductance in frog myelinated nerve fibres, but not in *Myxicola* axons when applied externally or internally (Schauf and Smith, 1981). Decamethonium was found by Castle and Haylett (1987) to be inactive (at 0.3 mM) on K_{ATP} of metabolically exhausted frog skeletal muscle. Effects of these drugs on other K^+ channels have not been reported.

8.2.5 Centrally active drugs which block potassium channels

8.2.5.1 Neuroleptics

The neuroleptics are a diverse group of chemical structures consisting of phenothiazines (chlopromazine, trifluoperazine, fluophenazine), diphenylbutylpiperazines (penfluridol, pimozide, fluspirilene) and butyrophenones (haloperidol), all of which have affinity to calmodulin as well as to a number of neurotransmitter receptors. Their inhibition of dopamine receptors is thought to explain their anti-schizophrenic activity.

Lackinton and Orrego (1981) showed that drugs from each of these classes could inhibit Ca^{2+}-activated K^+ efflux from human red blood cells (through IK_{Ca}). Their relative potencies at inhibiting this K^+ efflux correlated (excepting fluophenazine) with their calmodulin affinity, as measured by their inhibition of red cell Ca^{2+}-ATPase. However, other groups have not always been able to demonstrate such a modulation of red cell IK_{Ca} by calmodulin or phenothiazines (for references see Hoffmann *et al.*, 1984). Subsequently, the regulatory volume decrease of Ehrlich ascites tumour cells and human lymphocytes, where K^+ and Cl^- permeabilities increase in response to hypotonic shock (the K^+ efflux through a channel similar or identical to IK_{Ca} of red cells; see Chapter 3) were also inhibited by these different drugs (Hoffmann *et al.*, 1984; Sarkadi *et al.*, 1985). Most potent among them were pimozide and trifluoperazine, which inhibited the volume decreases of lymphocytes at sub-micromolar concentrations, and another calmodulin antagonist calmidazolium (R24571; IC_{50}=30 nM). McCann and Welsh (1987) have extended these observations by using the patch-clamp technique to demonstrate an inhibition of BK_{Ca} in canine tracheal smooth muscle cells by haloperidol (K_D=1 μM), trifluoperazine (1.4 μM), chlorpromazine (2 μM) and thioridazine (2.4 μM) applied to the cytosolic side of the membrane. Since this rank order of potency differed from their calmodulin-binding affinities, it was concluded that their K^+ channel-blocking effect is probably not mediated through calmodulin. The block by haloperidol was shown to increase with depolarization, suggesting that the binding site was within the channel electrical field (McCann and Welsh, 1987). These authors additionally reported that micromolar concentrations of haloperidol had no effect on a voltage-sensitive K_{Ca} channel from the basolateral membrane of dog and human tracheal epithelium, and Cook and Haylett (1985) found no interaction of chlorpromazine (20 μM) or calmidazolium (10 μM) with apamin binding to guinea-pig hepatocytes.

A recent development is the demonstration that neuroleptics also block K_{Ca} in the CNS. Dinan *et al.* (1987) observed an inhibition of the after-hyperpolarization (AHP) in hippocampal pyramidal cells by 1–10 μM haloperidol, chlorpromazine and trifluoperazine. The specific dopamine-2 receptor antagonist, sulpiride (1 μM), which has minimal anti-calmodulin activity, also markedly reduced the AHP. However, the tricyclic antidepressants imipramine, and desipramine (1 μM) were without effect. Although not addressed in this study, other data suggest the apamin-sensitive SK_{Ca} to be in part responsible for the generation of the slow AHP in neurones (see Chapter 14). Ca^{2+}-activated K^+ efflux from rat brain synaptosomes, which (in view of its sensitivity to TEA, IC_{50}=0.1 μM; Bartschat and Blaustein, 1985) is probably through BK_{Ca} channels, was inhibited by haloperidol (IC_{50}=0.5 μM) and in a biphasic manner by trifluoperazine and fluphenazine (IC_{50}s of 25 nM and 2 μM for the high- and low-affinity components, respectively; Benishin *et al.*, 1988). An effect on voltage-dependent Ca^{2+} influx was seen only at higher drug concentrations. Promethazine, another phenothiazine but having little anti-dopaminergic or anti-psychotic activity, only minimally inhibited synaptosome K_{Ca} at 10 μM (Benishin *et al.*, 1988). However, both sulpiride and the calmodulin antagonists calmidazolium and W7 were relatively weak inhibitors of K_{Ca} in this study, suggesting that neither dopamine-2 receptors nor calmodulin were involved in this action.

As pointed out by others (Dinan, 1987; Benishin *et al.*, 1988) the range of blood concentrations attained during therapy with certain neuroleptics is compatible with the concentrations shown to inhibit K_{Ca} in neuronal tissues. Whether the clinical efficacy of these drugs is in any way related to their blockade of K_{Ca} channels (as recently hypothesized by Dinan, 1987), or whether such an action might underly some of their side effects, now awaits clarification.

8.2.5.2 Phenycyclidine and related analogues

The hallucinogen PCP has multiple pharmacological actions, blocking Na^+ channels, Ca^{2+} channels, neuromuscular endplate ion channels, muscarinic receptors, *N*-methyl D-aspartate-activated excitatory post-synaptic potentials and opiate (σ) receptors, in addition to blocking K^+ channels. What makes the K^+ channel-blocking activity of interest is that it occurs at nanomolar PCP concentrations and, unlike most of the other activities listed above, there appears to be a direct correlation between the ability of different PCP analogues to block K^+ channels and to elicit behavioural effects *in vivo* (Aguayo *et al.*, 1984; Bartschat and Blaustein, 1986).

Albuquerque *et al.* (1981) first showed a block by PCP and related analogues of delayed rectifier K^+ channels in frog skeletal muscle. PCP (100 μM) did not modify the resting membrane potential, but greatly prolonged the action potential duration by blocking the outward K^+ current. Furthermore, depolarization-induced $^{86}Rb^+$ efflux from rat brain synaptosomes was inhibited by PCP with an IC_{50} of 10 nM, with the remainder being inhibited only at micromolar concentrations (Blaustein and Ickowitz, 1983). Behaviourally inactive PCP analogues failed to show the high-affinity inhibition (Bartschat and Blaustein, 1986). Voltage-clamp studies by Aguayo and Albuquerque (1987) in cultured mouse spinal cord neurones showed a block by PCP (0.1 μM) of two K^+ currents: a delayed outward (K_V) and a transient outward current, probably I_A. In mammalian cardiac muscle as in skeletal muscle, PCP (1–100 μM) prolongs the action potential duration without affecting the resting potential, and in voltage-clamped guinea-pig myocytes it was shown to block K^+ channels (in a voltage-dependent manner) and Ca^{2+} channels (Hadley and Hume, 1986).

Adamantane derivatives (ammantadine HCl, bromobenzoylmethyladamantylamine) also appear to block cardiac K^+ channels and elicit a prolongation of the action potential (and hence a positive inotropic effect) similar to phenycylidine (Freeman *et al.*, 1985; Meszaros *et al.*, 1982). As regards other K^+ channels, PCP and amantadine (0.1 mM) failed to block apamin-sensitive SK_{Ca} (Cook and Haylett, 1985) but at very high (>0.1 mM) concentrations PCP has been reported to block BK_{Ca} channels from rat brain incorporated into planar lipid bilayers (Krueger *et al.*, 1982).

Binding studies with [^{3}H]PCP support the evidence from synaptosome $^{86}Rb^+$ efflux studies of a high-affinity site for PCP on presynaptic nerve terminal membranes. [^{3}H]PCP bound to a single site with a K_D of <0.3 μM, the binding being competitively inhibited by various aminopyridine and quaternary ammonium compounds (Sorensen and Blaustein, 1988). The rank order of potency of six PCP analogues for displacing [^{3}H]PCP binding correlated with their relative potencies at inhibiting synaptosome $^{86}Rb^+$ efflux, 1-[1-(2-thienyl)cyclohexyl] piperidine being the most potent (K_D=69 nM, K_I=35 nM; Sorensen and Blaustein, 1988). It will be interesting to see if there is an interaction between PCP and K^+ channel neurotoxins such as DTX, MCD peptide and β-BTX for central binding sites.

8.2.5.3 Ethanol

At a concentration of 58 mM (≈0.33% v/v), ethanol enhanced the responses to a variety of agonists in rabbit aortic strips, and at higher concentrations elicited direct contractile effects (Kalsner, 1970), effects which are comparable with those of known K^+ channel blockers upon this tissue (Cook, 1989). The complex and variable effects of ethanol on membrane conductances include a number of references to blockade of K^+ currents. Takeda *et al.* (1984) showed that 100–500 mM ethanol reversibly prolonged the repolarization phase of the action potential in bullfrog atrial cells, due to a dose-dependent inhibition of an outward K^+ current. At the higher concentrations ethanol also slightly depolarized the membrane, although the inward rectifying K^+ channels (I_{K1} or I_R) were not affected. In *Aplysia* neurones, butanol (40 mM) and hexanol (6 mM) decreased the decay time and amplitude (by about 50%) of I_A, whilst ethanol (up to 400 mM) prolonged only the decay time constant (Treistman and Wilson, 1987).

Ethanol effects on K_{Ca} are complicated by the fact that it is known to increase the Ca^{2+} affinity of Ca^{2+}-binding molecules (see Baker and Schapira, 1980). Indeed, Schwartz (1985), studying K_{Ca} of *Aplysia* bursting pacemaker neurones, and Harris and Coldwell (1985), investigating the Ca^{2+}-dependent K^+ permeability of red cells, both reported an increase in the magnitude of K_{Ca} induced by ethanol (>100 mM) independent of changes in the intracellular Ca^{2+} concentration. Carlen *et al.* (1982) showed a hyperpolarization of rat hippocampal slices and augmentation of the slow AHP by ethanol (5–20 mM≡23–9 mg/100 ml), effects which they attributed to an enhancement of K_{Ca}. It was not possible to discern from these studies whether this was a direct effect of ethanol upon the channels, or on their sensitivity to Ca^{2+}_i. Caldwell and Harris (1985) further demonstrated that the stimulation of $^{86}Rb^+$ efflux from red blood cells by a number of anaesthetic agents (approximate EC_{20}s in mM; ethanol 70, propranolol 40, ether 3, chloroform 1, isoflurane 0.7, enflurane 0.5, halothane 0.03) correlated closely with the *in vivo* plasma concentrations of these drugs required to produce anaesthesia (correlation coefficient=0.9). A recent development in this area is the finding that volatile general anaesthetics reversibly activate a novel voltage-, time- and Ca^{2+}-independent K^+ channel in a molluscan neurone (Franks and Lieb, 1988). The activity of halothane on this channel (IC_{50}=0.0063 atmospheres) is close to the minimum alveolar concentration of this drug during anaesthesia in man.

The opening of apamin-sensitive K_{Ca}, giving rise to slow AHPs, is known to be a key regulator of the firing frequency and duration of repetitive firing of certain central neurones (Fishman and Spector, 1981; see Chapters 3 and 14). Since the stimulation of K_{Ca} by ethanol is generally thought to be due to its Ca^{2+} sensitizing effect, it is therefore not surprising that it also activated apamin-sensitive $^{86}Rb^+$ efflux, as shown in mouse brain synaptosomes by Yamamoto and Harris (1983). Furthermore, the duration of the loss of righting reflex (sleep time) induced by i.p. injection of ethanol in mice was significantly and dose-dependently inhibited by apamin (0.01–0.1 μg/kg, injected intracerebroventricularly, 1 minute before the ethanol; Yamamoto and Harris, 1983). This behavioural evidence therefore points to a role for the apamin-sensitive SK_{Ca} channel in ethanol intoxication and hence to a potential site for therapeutic intervention with selective channel blockers in this situation.

8.2.4.5 Other centrally active drugs which block K^+ channels

Recently there has been much interest in the block of K^+ channels by 9-amino-1,2,3,4-tetrahydroacridine (tacrine, THA), a drug with a similar structure to 4AP, used in the treatment of senile dementia (Alzheimer's disease) (Summers *et al.*, 1986). THA blocks the slow outward K^+ current (K_V) in neurones, thereby prolonging the APD (≥1 μM; Drukarch *et al.*, 1987) and also blocked I_A in hippocampal neurones (IC_{50}=30 μM; Rogawski, 1987), although other effects of

THA at these concentrations (inhibition of cholinesterase activity, block of Na^+ currents) preclude any statement about the relevance of these effects for its clinical efficacy. Osterrieder (1987) has shown that THA (100 μM) also blocks the delayed outward ($I_{K(V)}$) and inward rectifying (I_{K1}) K^+ currents in ventricular cells, leading to a prolongation of APD and the suggestion that this drug might be of benefit in the treatment of ventricular fibrilation (see section 8.2.6.2 and Chapter 13).

The non-tricyclic antidepressant alaprocate is known to potentiate a number of muscarinic cholinergic responses, suggesting a possible use in the treatment of senile dementia, without having any direct action on muscarinic receptors. Hedlund (1987) found that alaprocate (5 μM) prolonged the depolarizing response of neuroblastoma cells elicited by carbachol, without itself modifying the membrane potential of these cells. This response resembled that of other K^+ channel blockers, TEA (5 mM), 4AP (0.5 mM) and quinine (5 mM). Alaprocate did not modify carbachol-induced $^{45}Ca^{2+}$ uptake by these cells, but inhibited the subsequent $^{86}Rb^+$ efflux, suggesting the effect of alaprocate was due to inhibition of a K^+ channel (Hedlund, 1987).

Lackington and Orrego (1985) described an inhibition of red cell K^+ efflux (IK_{Ca}) by methylphenidate (20% reduction at 10 μM, $IC_{50} \approx 190$ μM) but not by the other CNS stimulants nikethamide (1 mM), pentylenetretrazol (1 mM) or nomifensine (0.1 mM). Methylphenidate (1 mM) was also said to inhibit veratridine-stimulated K^+ efflux from rat brain cortex, and it was speculated that this may have been involved in the aggravation of schizophrenia by this drug (Lackington and Orrego, 1985). Caldwell and Harris (1985) reported that two anticonvulsant drugs — phenytoin ($IC_{50} \approx 0.1$ mM) and phenobarbital (0.3 mM) — and also pentobarbital (0.3 mM), inhibited Ca^{2+}-dependent $^{86}Rb^+$ efflux from human red blood cells concentration-dependently. Other anticonvulsants — ethosuximide (highest concentration tested, 3 mM), valproic acid (3 mM), tetrahydrocannabinol and carbamazepine (0.3 mM) — were without effect.

8.2.6 Other potassium channel blockers

8.2.6.1 Forskolin

Forskolin has recently been shown to directly block K_V in *Helix* neurones, pancreatic β cells, chromaffin cells and PC12 cells (Watanabe and Gola, 1987; Zunkler *et al.*, 1988b; Hoshi *et al.*, 1988). Blockade occurred rapidly from the outside of the membrane with IC_{50}s in each cell type of around 15 μM. This K^+ channel-blocking activity was quite independent of the well-known stimulatory effect of forskolin on adenylate cyclase; it could not be mimicked by cyclic AMP or cell-permeable cyclic AMP analogues and was not modified in the presence of phosphodiesterase inhibitors. Furthermore, blockade of K_V by forskolin could be readily shown in the presence of 1 mM (intracellular) cyclic AMP (Zunkler *et al.*, 1988b). Since the concentrations of forskolin blocking this K^+ channel are not dissimilar from those used to activate adenylate cyclase (frequently 10 μM), it is clearly necessary to establish which of these two possible mechanisms is responsible for any effect observed. Evidence for a role of cyclic AMP in a particular cellular event should not therefore be based upon results obtained with forskolin alone.

Block of K_V by forskolin was due to an acceleration of channel inactivation (without affecting single-channel amplitude) leading to a decrease in the mean open time (Zunkler *et al.*, 1988b; Hoshi *et al.*, 1988). In PC12 cells, the block was reversible and neither voltage- nor use-dependent (Hoshi *et al.*, 1988). However, in pancreatic β cells and *Helix* neurones, evidence of both use- and voltage-dependence was seen (Zunkler *et al.*, 1988b; Watanabe and Gola, 1987), the blockade resembling the effects of TEA on K_V in squid axon.

In pancreatic β cells, 100 μM forskolin was without effect on BK_{Ca} or voltage-dependent Ca^{2+} channels and only slightly reduced K_{ATP} (Zunkler *et al.*, 1988b). Forskolin also failed to block IK_{Ca} in *Helix* neurones, current through these channels being if anything increased (Watanabe and Gola, 1987). Forskolin thus shows some selectivity for K_V, although its use as a tool for blocking K^+ channels is likely to be limited by the fact that cyclic AMP itself modulates a number of K^+ channels (see Chapter 5). Effects of forskolin on a variety of K^+ currents in other cells have been reported, but they will not be described here, since it was not established whether this was due to a direct blocking effect or indirect via stimulation of adenylate cyclase. Forskolin also blocks cromakalim-stimulated $^{86}Rb^+$ efflux in smooth muscle independent of its effect on cyclic AMP, with a similar IC_{50} to its block of K_V (Quast, 1988).

8.2.6.2 *Drugs which block K^+ channels in cardiac muscle*

Cardiac electrophysiological events involve the complex interplay of several different K^+ currents, often showing properties which distinguish them from K^+ channels in other tissues (reviewed in Chapter 13). Quinine and other membrane-stabilizing drugs, phencyclidine, sulphonylureas and calcium antagonists also affect cardiac K^+ currents, and these effects are dealt with in other sections of this chapter. Here we wish to concentrate on the class III anti-arrhythmics, which have the characteristic property of prolonging the action potential duration (APD) in cardiac muscle as a consequence of blocking K^+ channels to retard repolarization.

Sotalol, originally tested as a class II anti-arrhythmic because of its β-adrenoceptor blocking activity, has become the prototype of the class III anti-arrhythmic drugs. It is now clear that the APD prolongation is independent of the β-blocking activity of sotalol. Indeed, both D- and L-sotalol prolong the APD *in vitro* (Carmeliet, 1985) and suppress ventricular tachycardia (VT) and fibrillation (VF) induced in infarcted dogs (Lynch *et al.*, 1984), although only the L isomer has relevant affinity for β-adrenoceptors. Carmeliet (1985) localized the electrophysiological effect of sotalol (1–100 μM) to an inhibition of a time-dependent outward K^+ current (I_K; delayed rectifier) activated during the plateau phase of the action potential, and a small reduction of the background (time-dependent) K^+ current (I_{K1}; the inward rectifier).

Other established class III anti-arrhythmics subsequently shown to block cardiac K^+ channels include amiodarone (I_{K1} blocked at 2.5 μM, probably from inside; Sato *et al.*, 1987), the quaternary ammonium compound clofilium (block of I_K but not of I_{K1} at 10–100 μM; Arena and Kass, 1988; Sanguinetti *et al.*, 1989) and the adrenergic neurone blockers bretylium tosylate and bethanidine sulphate (>300 μM block I_K as well as an early outward time-dependent K^+ current, similar to I_A; Bacaner *et al.*, 1986; Bkaily *et al.*, 1988).

Newer class III anti-arrhythmics, in many cases more potent structural analogues of the above prototype compounds, which have been shown to increase APD by blocking K^+ channels are the $5HT_3$ receptor antagonist ICS 205-930 (above 0.1 μM; IC_{50}s for I_{K1} and I_K of 20 and 37 μM, respectively, Scholtysik, 1987; Scholtysik *et al.*, 1988), UK66914 and UK 68798 (both of which, at 2 μM and 50 μM respectively, selectively reduced I_K without affecting background K^+ conductance, I_{K1}; Gwilt *et al.*, 1988, 1989), two tertiary analogues of clofilium — LY 97241 and LY 97119 (both block I_K and I_{K1} at 50 μM; Arena and Kass, 1988) — and E-4031 (claimed by Sawada, 1989, to significantly depress I_K at a concentration of 1 μM, whereas Sanguinetti *et al.* (1989) found no block of I_K at concentrations up to 100 μM, but block of another (4AP-sensitive) K^+ current instead). In this latter study it was claimed that sotalol (100 μM) also blocked this 4AP-sensitive K^+ current and not I_K, whereas clofilium (100 μM) and LY 1901 (10 μM) indeed blocked I_K in these ventricular cells (Sanguinetti *et al.*, 1989).

It is relevant here to reconsider the possible role of K_{ATP} in myocardial dysrhythmias, and the potential therapeutic benefit to be derived from blocking this channel (see Cook, 1988; Quast and Cook, 1989a; biophysics of K_{ATP} in heart reviewed in Chapter 4). Kantor *et al.* (1987) have found that 1 μM glibenclamide reduces the loss of K^+ from rat hearts during the first 5 minutes following global ischaemia and completely prevented the development of VF by these hearts (0/7 compared to 7/7 controls). It is interesting to note that a number of the class III anti-arrhythmics — sotalol, E4031 and CK 1649–C — contain a sulphonanilamide substituent similar to the sulphonylureas (see Steinberg *et al.*, 1986). In this regard, it would be interesting to know if agents such as AZ-DF265, which lack the sulphonylurea moiety of the glibenclamide series but still potently block K_{ATP} (as evidenced by their effects on islet $^{86}Rb^+$ fluxes and insulin secretion; Garrino and Henquin, 1988) are able to block K_{ATP} in the heart and elicit similar anti-arrhythmic effects to glibenclamide. Another new anti-arrhythmic agent currently undergoing clinical evaluation, sodium 5-hydroxydecanoate, has recently been shown (at 100 μM) to inhibit K_{ATP} in guinea-pig heart cells without affecting I_{K1} (Notsu *et al.*, 1989). A compound which (in contrast to the sulphonylurea anti-diabetics) would show a selectivity for K_{ATP} of ventricular cells over the channel in pancreatic β cells would be of particular interest in such indications. It is also noteworthy that K^+ channel-opening drugs such as cromakalim, despite showing a selectivity for smooth muscle, also open ATP-sensitive K^+ channels in cardiac tissue at high concentrations (Escande *et al.*, 1988; see section 8.3.2.1 and Chapter 13). Cromakalim or hypoxia (both causing activation of K_{ATP}; see section 8.3.4.3) have been shown to reverse the prolongation of the APD by sotalol (Siegl *et al.*, 1989) and ICS 205-930 (Scholtysik, 1987), leading to the suggestion that anti-arrhythmic efficacy of such drugs might be compromised under ischaemic or hypoxic situations (Siegl *et al.*, 1989).

8.2.6.3 Dyes

The carbocyanine series of potential-sensitive fluorescent dyes, most notably 3,3-diethylthiadicarbocyanine iodide (diS-C_2-(5)), were shown by Simons (1979) to block IK_{Ca} in red blood cell ghosts when sealed inside them. In a K^+-free medium, diS-C_2-(5) (0.3 μM) caused virtually complete inhibition of K^+ permeability within 1 minute; raising K_o^+ tended to antagonize the block. Since inhibition was most effective when K^+ efflux was maximal, it appears that the dye interacted preferentially with the open channels. Results obtained using the membrane -impermeant anionic dye diS-C_3SO_3-(5) suggested the inhibitory site at which the carbocyanines act is located on the inner surface of the red cell membrane (Simons, 1979), a fact which has probably hindered investigations of these dyes on other K^+ channels. The only other K^+ channel where a blocking effect of diS-C_2-(5) has been reported is the apamin-sensitive SK_{Ca} channel: following a 15-minute pre-incubation of guinea-pig hepatocytes in a low K_o^+ medium containing 50 μM diS-C_2-(5), subsequent phenylephrine-induced K^+ efflux was greatly reduced relative to controls (Cook, 1983). However, this same dye concentration failed to inhibit [^{125}I]apamin binding to these cells, this being the only compound shown to inhibit K^+ efflux without inhibiting apamin binding (Cook and Haylett, 1985). A possible explanation for this would be that the binding sites for apamin and the membrane permeant diS-C_2-(5) were situated on opposite sites of the membrane (extracellular and intracellular, respectively).

Volle (1971) showed an inhibition of the $^{42}K^+$ efflux and depolarization of the membrane of unstimulated frog sartorius muscle fibres by 9-aminoacridine (0.05–1 mM). The type of K^+ channel involved in this action was not defined, although the effect of 9-aminoacridine (0.25 mM) showed certain differences from the block by TEA (10 mM) or Ba^{2+} (0.25 mM). Henquin *et al.* (1979) demonstrated a depolarization of pancreatic β cells by 9-aminoacridine (0.05–1 mM) in the presence of 6 mM glucose and augmentation of insulin release. These effects were attributed to a block of K^+ channels, subsequently confirmed in patch-clamp studies of the effects of 0.1 mM 9-aminoacridine on K_{ATP} in β cells (Cook and Hales, 1984). 9-Aminoacridine also blocks SK_{Ca} in guinea-pig hepatocytes with an IC_{50} of 70 μM (Cook and Haylett, 1985).

Stutzin *et al.* (1983) found that methylene blue (0.1 mM) reversed the blocking effects of guanethidine on the guinea-pig vas deferens in a manner similar to apamin and quinine. They interpreted these effects as a block of K_{Ca} in the presynaptic nerve terminal by methylene blue, thus opposing an opening of the channels by guanethidine to limit noradrenaline release. Subsequently, Castle (1987) showed an inhibition of [^{125}I]apamin binding to guinea-pig hepatocytes by methylene blue and methylene green with K_Is of 22 and 63 μM, respectively. Intracellular recordings from the guinea-pig small intestine myenteric neurones demonstrated a depolarization with an increase in input resistance by methylene blue (10–250 μM), thereby augmenting excitability and increasing transmitter release (Nemeth *et al.*, 1985). Once again this was attributed to an inhibition of K_{Ca}. The possible K^+ channel-blocking action of methylene blue should not be overlooked when using it to antagonize endothelium-dependent effects in vascular smooth muscle. However, it would seem that the K^+ channels opened by EDHF are not sensitive to methylene blue (3 μM; Chen *et al.*, 1988).

8.2.6.4 α-Adrenoceptor antagonists

The alkaloid α_2-adrenoceptor antagonists coryanthine, yohimbine and rauwolscine have recently been shown to directly block BK_{Ca} channels in membranes from the thick ascending limb from rabbit kidney with K_Is of 45, 98 and 310 μM, respectively (Cornejo *et al.*, 1989). The block was manifest as an increased number of rapid closures as the drug concentration was increased, the effect being voltage dependent and reduced as the bath K^+ concentration was raised. Corynanthine applied to the outside of the membrane was >200-fold less active (Cornejo *et al.*, 1989). McPherson and Angus (1989) have recently shown that phentolamine and structurally related analogues (alinidine, ST91) at concentrations >1 μM inhibit the relaxant effects of cromakalim on the dog coronary and rat femoral arteries. The hyperpolarization of the femoral artery by cromakalim was also inhibited by phentolamine. This was suggested to be due to a selective block by phentolamine of the cromakalim-opened K^+ channel, since the relaxant activity of other vasodilators (isoprenaline, sodium nitroprusside) was not inhibited. Prazosin (10 μM), rauwolscine (10 μM) and phenoxybenzamine (1 μM) were devoid of this activity, suggesting that it was not related to an affinity for α-adrenoceptors. Murray *et al.* (1989) have subsequently reported the ability of phentolamine to antagonize cromakalim relaxant effects on the guinea-pig trachea. The block of cromakalim effects by alinidine (which lacks affinity for α-adrenoceptors) is noteworthy, since this drug causes bradycardia (Heinzow *et al.*, 1982), which could conceivably be related to a blockade of cardiac K^+ channels (see section 8.2.4.2).

8.2.6.5 Miscellaneous blockers

Treatment of BK_{Ca} channels with the protein-modifying agent *N*-bromoacetamide (0.3–1 mM) leads to an irreversible abolition (Pallotta, 1985) or reduction (Cornejo, *et al.*, 1987) in the sensitivity of these channels to Ca^{2+}_i. Other characteristics of the channel were not modified appreciably by this treatment.

Treatment of cultured *aplysia* neurones with the non-selective phospholipase inhibitor 4-bromophenacyl bromide was shown by Piomelli *et al.* (1987) to suppress the hyperpolarization response to FMRFamide, an action dependent upon the generation of arachidonate second messengers from membrane phospholipids by phospholipase A_2 (see Chapter 5).

Howland *et al.* (1984) described an inhibition of IK_{Ca} in human red cells by probucol (IC_{50}=0.4 mM), which they considered as possibly being a secondary consequence of an effect of probucol on NADH oxidation in these cells. This K^+ channel blockade might contribute to the ECG changes elicited by probucol clinically.

8.3 POTASSIUM CHANNEL OPENERS

8.3.1 Pharmacological properties of cromakalim in smooth muscle

8.3.1.1 Ion flux studies with cromakalim

The suggestion that cromakalim (see Fig. 8.2 for structure) opens membrane K^+ channels was based on its ability to hyperpolarize various smooth muscle preparations and/or to promote $^{86}Rb^+$ efflux (the latter was a measure of K^+ permeability): rat portal vein (Hamilton *et al.*, 1986); rat aorta (Weir and Weston, 1986b; Hof *et al.*, 1988); rabbit aorta (Kreye and Weston, 1986; Cook *et al.*, 1988c); guinea-pig portal vein (Quast, 1987); guinea-pig trachea (Allen *et al.*, 1986a); guinea-pig taenia caeci (Weir and Weston, 1986a); guinea-pig urinary bladder (hyperpolarization and $^{43}K^+$ but not $^{86}Rb^+$ efflux; Foster and Brading, 1987; Foster *et al.*, 1989). Opening of membrane K^+ channels would be expected to shift the membrane potential towards the Nernst potential for K^+ and therefore lead to a hyperpolarization of the cell. Since stimulation of $^{86}Rb^+$ or $^{42}K^+$ efflux and hyperpolarization are generally

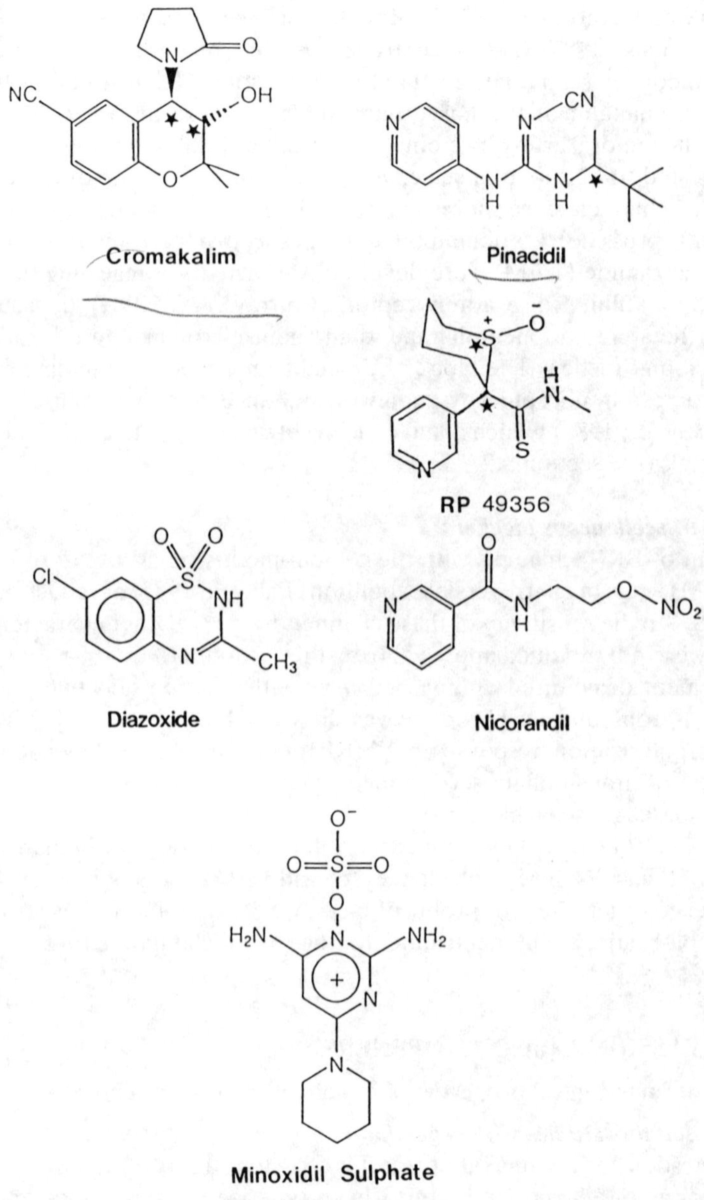

Fig. 8.2—Molecules that open K^+ channels in smooth muscle. Chiral centers are indicated by a star; however, the generic names refer to racemic mixtures.

observed at similar concentrations (>0.1 μM), they would indeed appear to be linked. The hyperpolarization induced by cromakalim is associated with a decrease in membrane resistance (Foster *et al.*, 1989), again consistent with an action of cromakalim on membrane K^+ channels.

In most tracer flux studies with cromakalim, $^{86}Rb^+$ was chosen as a convenient substitute for $^{42}K^+$ or $^{43}K^+$, due to its longer half-life. In the guinea-pig portal vein,

Quast and Baumlin (1988) showed that use of $^{42}K^+$ allowed the detection of an effect of cromakalim on flux at lower concentrations than with $^{86}Rb^+$ (20 nM as opposed to 60 nM). In both the rat aorta and guinea-pig portal vein, the population of K^+ channels opened by cromakalim below 0.1 μM was about three-fold selective for K^+ over Rb^+, whereas at concentrations above 1 μM the selectivity was only 1.5-fold (qualitatively similar findings were reported for pinacidil; see below). This suggested heterogeneity in the K^+ channels opened by cromakalim in these preparations. Although the flux response was quantitatively underestimated by using $^{86}Rb^+$, other characteristics of the flux response (e.g. midpoint of saturation curves, sensitivity to TEA) were similar. Hence, in the case of cromakalim, $^{86}Rb^+$ would seem to be an appropriate substitute for K^+ in these vascular tissues but this appears not to hold in all tissues or for all KCOs; in the guinea-pig trachea, 10 μM cromakalim opens $^{86}Rb^+$-permeable K^+ channels (selectivity for K^+ 1.7-fold) which are blocked by 4.8 mM extracellular Rb^+ (Foster, 1989). In addition, extracellular Rb^+ rendered the sustained relaxation by cromakalim of spontaneous or histamine-induced tone of guinea-pig trachea transient (Morris and Taylor, 1989). In the guinea-pig bladder, cromakalim opens a K^+ channel (causing $^{43}K^+$ efflux) which is essentially impermeable to $^{86}Rb^+$ (Foster and Brading, 1987; Foster *et al.*, 1989), as does minoxidil sulplhate in vascular tissue (see 8.3.3.4).

8.3.1.2 Vasodilator effects of cromakalim

The vasodilator profile of cromakalim shows characteristic features which distinguish it (more or less clearly) from other vasodilators (e.g. Ca^{2+} antagonists, cyclic AMP-elevating agents). Most importantly, cromakalim is able to relax contractions to low (<25 mM) but not high (>30 mM) KCl in various vascular preparations (Hamilton *et al.*, 1986; Weir and Weston, 1986b; Clapham and Wilson, 1987), and the presence of elevated KCl abolishes the vasorelaxant effect of cromakalim against other stimuli (e.g. noradrenaline, angiotensin II; Cook *et al.*, 1988c). This is a behaviour expected for an agent merely acting to open K^+ channels, since in the presence of high KCl concentrations the actual membrane potential approaches the Nernst equilibrium potential for K^+, thus reducing the achievable degree of hyperpolarization (see, for example, Haeusler, 1983, or Nakao *et al.*, 1988). Furthermore, in the presence of elevated K^+ the Nernst potential may be more positive than the threshold potential needed to interfere with Ca^{2+} handling of the cell, thus abolishing the ability of the KCOs to interfere with these processes.

The KCOs have been extensively studied as regards their inhibition of spontaneous activity of the rat or guinea-pig portal vein and inhibition of the tonic noradrenaline contraction of rabbit and rat aortic rings. Cromakalim inhibits the spontaneous activity of portal vein (IC_{50}=20–40 nM) by decreasing the duration of the contractile spikes and the frequency of the contractions, without much effect on their amplitude (Quast 1987; Hof *et al.*, 1988; Quast and Baumlin, 1988). An increase in tracer efflux could, however, be detected only at concentrations where mechanical inhibition was essentially complete (using $^{86}Rb^+$; Quast, 1987) or near the IC_{50} value ($^{42}K^+$; Quast and Baumlin, 1988). A similar situation exists in rat uterus and the guinea-pig bladder (see section 8.3.1.3). The currently favoured theory to explain the relaxation/flux discrepancy is that these drugs preferentially act on K^+ channels in pacemaker cells from where the myogenic contractions originate. Hence the spontaneous activity of the tissue would be inhibited, but (due to the sparsity of these cells) changes in the whole-tissue $^{86}Rb^+$ efflux or membrane potential would be undetectable (Hamilton *et al.*, 1986; Quast, 1987; Hollingsworth *et al.*, 1987, 1988).

A mechanistically interesting effect of cromakalim is seen against the tonic contraction of the rabbit aorta to noradrenaline. In this vessel, noradrenaline does not depolarize the membrane potential (Cauvin *et al.*, 1984; Bray *et al.*, 1988a) and, consequently, dihydropyridine Ca^{2+} antagonists are without effect (see, for example, Cook *et al.*, 1988c). Cromakalim, however, reduces this contraction by about 80% with midpoint at 0.3 μM and this effect was lost when the preparation was partially depolarized by 35 mM KCl (Cook *et al.*, 1988c). In the rat aorta, the tonic noradrenaline contraction can be inhibited about 50% by the dihydropyridine isradipine (midpoint at 1 nM; Quast, unpublished), whereas cromakalim inhibits this contraction by 80% (Buckingham *et al.*, 1989). These data indicate that the cromakalim-induced hyperpolarization affects smooth muscle contraction (and probably Ca^{2+} handling) by as yet unidentified mechanisms in addition to (indirectly) reducing Ca^{2+} influx through dihydropyridine-sensitive Ca^{2+} channels.

Resistance vessels

The potential therapeutic applications of the KCOs in cardiovascular disease (e.g. blood pressure lowering, improved perfusion of ischaemic areas in heart and periphery) depend essentially on the effect of these substances on resistance vessels. Resistance vessels are well known to differ from large vessels in various aspects of Ca^{2+} handling, in particular agonist-induced depolarization (Mulvany and Halpern, 1977; Cauvin *et al.*, 1984, 1989). However, only a few studies have been done with KCOs in such preparations. In rat isolated mesenteric resistance vessels, Weir and Cauvin (1989) found that cromakalim inhibited the phasic and the tonic part of noradrenaline-induced contractions, concomitant with hyperpolarization of the tissue. The concentration–response curves suggested two classes of sites: high-affinity sites ($\approx$0.01 nM, carrying the minor part of total hyperpolarization and relaxation of tonic tension); and low-affinity sites (0.1–0.3 μM). All effects were abolished in the presence of high KCl. The high-affinity site observed in this paper could explain the *in vivo* potency of cromakalim, which is known to lower blood pressure in man at a dose of 0.75 mg/day (Eckl and Greb, 1987). In the rat isolated perfused mesenteric bed (the perfusion pressure of which is governed by resistance vessels) cromakalim inhibited phasic noradrenaline contractions in the presence or absence of Ca^{2+} by about 80% with a monophasic concentration dependence (midpoint at 200 or 60 nM, respectively; Quast, 1989). This effect was stereospecific, inhibitable by glibenclamide and abolished in high KCl. In a similar preparation from spontaneously hypertensive rats, cromakalim inhibited oscillatory responses to prolonged perfusions of noradrenaline but was quite ineffective against the contractions after bolus injections of noradrenaline (Buckingham and Parrott, 1988). Taken together, these studies show that cromakalim is an effective KCO in resistance vessels, as is the case for pinacidil (see 8.3.3.2).

8.3.1.3 Effects of cromakalim in non-vascular smooth muscle

Trachea

In the guinea-pig isolated trachea Allen *et al.* (1986a) showed that cromakalim suppressed spontaneous tone ($IC_{50} \approx 1$ μM) and was also moderately effective against spasmogen-induced tone. At similar concentrations (1–10 μM) cromakalim suppressed spontaneous electrical slow waves, induced TEA- and procaine-sensitive hyperpolarization of the tissue and increased the $^{86}Rb^+$ efflux rate. The authors concluded that the observed relaxation of the trachealis may depend on the opening of $^{86}Rb^+$-permeable K^+ channels. The presence of extracellular Rb^+ inhibited the relaxation of histamine-induced tone of the guinea-pig trachealis by cromakalim (Morris and Taylor, 1989) and blocked the cromakalim-induced $^{42}K^+/^{43}K^+$ and $^{86}Rb^+$ efflux (Foster, 1989). When extracellular K^+ was replaced by Rb^+, the sustained relaxation of spontaneous tone was converted into a transient one (Morris and Taylor, 1989). These experiments were interpreted as cromakalim activating Rb^+-permeable and -impermeable K^+ channels in the guinea-pig trachealis, where the former one is only transiently open. Comparison of cromakalim with nifedipine in various *in vitro* and *in vivo* models of bronchodilation showed that cromakalim had a greater potential as a bronchodilator (Arch *et al.*, 1988).

Bladder

Brading's group (Foster and Brading, 1987; Foster *et al.*, 1989) has shown that cromakalim inhibits quite effectively the spontaneous contractile activity of the guinea-pig urinary bladder ($IC_{50} \approx 0.3$ μM) by reducing frequency and amplitude of the contractions. Again, direct manifestations of K^+ channel opening (hyperpolarization, increase in ^{43}K efflux) were detectable only at higher concentrations (>1 μM), consistent with a preferential action of cromakalim on pacemaker cells. Cromakalim showed minimal efficacy against contractions induced by KCl or transmural nerve stimulation and was ineffective against carbamylcholine. The cromakalim-opened K^+ channel in the guinea-pig bladder appears to be quite impermeable to Rb^+ since the flux response (influx and efflux) was manifest only with $^{43}K^+$ but not with $^{86}Rb^+$. In addition, when K_i^+ was replaced by Rb^+, cromakalim was ineffective in relaxing the tissue or eliciting $^{43}K^+$ efflux. However, in human bladder, pinacidil does appear to open $^{86}Rb^+$-permeable K^+ channels (Fovaeus *et al.*, 1989).

Uterus

Cromakalim relaxes uterine muscle *in vitro* and *in vivo* (Hollingsworth *et al.*, 1987, 1988; Downing *et al.*, 1989). The pharmacological profile *in vitro* resembles that of cromakalim in vascular and tracheal muscle (no effect against spasm to high KCl, inhibition of the relaxant effects by K^+ channel blockers like TEA, 4AP, procaine). Interestingly, cromakalim (10 μM) had only negligible effects on membrane resting potential and failed to increase $^{86}Rb^+$ or $^{42}K^+$ efflux from the tissue (Hollingsworth *et al.*, 1987, 1988). This was explained by a selective action of cromakalim on K^+ channels in pacemaker cells of the uterus (Hollingsworth *et al.*, 1987; cf. portal vein, section 8.2.1.2). Interestingly, the uterine relaxant effect shows tolerance to the drug both in the *in vivo* situation and *ex vivo* (Downing *et al.*, 1989).

Intestinal smooth muscle

Cromakalim inhibits spontaneous tone of the guinea-pig taenia caeci ($pIC_{50} \approx 6$) and increases $^{86}Rb^+$ efflux from this preparation. These effects are not inhibited by apamin, in contrast to similar effects induced by noradrenaline in this tissue (Weir and Weston, 1986a). In addition, cromakalim (1 μM) significantly hyperpolarized the tissue and, surprisingly, increased cyclic AMP 2.5–5-fold at 1.6 and 3.2 μM (McHarg *et al.*, 1989, see section 8.3.4.4). In the isolated guinea-pig ileum, cromakalim inhibited the peristaltic reflex, the action on longitudinal muscle being ten times more potent than on circular muscle (pIC_{50}=5.8 and 4.8, respectively; Buchheit and Bertholet, 1988). These values, together with the high dose (1.2 mg/kg i.p.) required to inhibit small intestinal transit by 50% (Buchheit and Bertholet, 1988) indicated that intestinal muscle may be less sensitive to cromakalim than vascular smooth muscle.

8.3.1.4 Blockers of cromakalim effects on smooth muscle

The effects of cromakalim on K^+ permeability (P_K) and contraction of smooth muscle can be inhibited in parallel by K^+ channel blockers. Quarternary ammonium ions (e.g. TEA), aminopyridines and procaine are effective in the millimolar range (vascular tissue: Coldwell and Howlett, 1987; Quast 1987; Wilson *et al.*, 1988b; Winquist *et al.*, 1988; guinea-pig trachea: Allen *et al.*, 1986a); at the same time they have profound effect on basal tension of smooth muscle. Comparing the potency of (symmetrical) tetraalkylammonium ions as blockers of the cromakalim-induced $^{86}Rb^+$ efflux, Quast and Webster (1989) found that potency increased with increasing chain length up to the *n*-pentyl compound (pIC_{50}=6.9) but was reduced if the chain length was increased further. This indicates that the tetraalkylammonium site of the channel has a strong preference for bulky (but not too bulky) lipophylic cations.

The sulphonylurea glibenclamide, a potent (nM) blocker of the ATP-sensitive K^+ channel in the pancreatic β cell (Schmidt-Antomarchi *et al.*, 1987; Sturgess *et al.*, 1988; Zünkler *et al.*, 1988a) has been shown to inhibit the effects of cromakalim on P_K and/or contraction at 0.1 μM (Quast and Cook, 1988b, 1989a; Wilson *et al.* 1988a; Wilson 1989; Cavero *et al.*, 1989, Eltze, 1989a,b; Winquist *et al.*, 1989) without an effect on basal tension. In a limited concentration range up to 3 μM, the inhibition of glibenclamide of the effects of cromakalim on rat portal vein spontaneous activity and $^{86}Rb^+$ efflux were compatible with competitive antagonism, giving pA_2 values of 7 (Quast and Cook, 1989; see also Buckingham *et al.*, 1989). At high doses glibenclamide also inhibits the blood pressure lowering effect of cromakalim in the rat (Cavero *et al.*, 1988, 1989; Buckingham *et al.*, 1989; Quast and Cook, 1989a). The selectivity of glibenclamide for K_{ATP} suggests that cromakalim may act through ATP-sensitive K^+ channels in smooth muscle (see 8.3.4.3).

8.3.1.5 Stereoselectivity

Cromakalim is a racemate where the substituents in the 3- and 4-positions are in the *trans* configuration (see Fig. 8.2). The effects of cromakalim on vascular P_K and contractility *in vitro* and *in vivo* (blood pressure lowering) are (about 100-fold) stereoselective with the activity residing in the (−)-(3*S*,4*R*) enantiomer (Buckingham *et al.*, 1986; Hof *et al.*, 1988).

8.3.2 Effects of cromakalim in other tissues

8.3.2.1 Heart

Cromakalim, at 10 and 30 μM, reduced action potential duration in guinea-pig papillary muscle with no effect on resting membrane potential and V_{max}, and reduced the slow inward (=Ca^{2+}) current, indicating that at high concentrations cromakalim may increase cardiac P_K (Cain and Metzler, 1985; Grosset and Hicks, 1986). Indeed, cromakalim (300 μM) increased the rate constant of $^{86}Rb^+$ efflux from rat atria by approximately 40% (Quast *et al.*, 1988). The effect on cardiac action potential duration is stereoselectively confined to the (−) enantiomer and reversed by the K^+ channel blockers ICS 205-930 and sotalol (Scholtysik, 1987). Sanguinetti *et al.* (1988) and Mestre *et al.* (1988) showed that the shortening of the cardiac action potential by cromakalim was inhibited by glibenclamide. In experiments on excised cardiac myocyte membranes, 300 μM cromakalim was also shown to increase (single-channel) currents through K_{ATP} (Escande *et al.*, 1988). Cromakalim opens these channels by decreasing their affinity for ATP, thus relieving the blockade of K_{ATP} by physiological levels of ATP in the cell (Thuringer and Escande, 1989).

In isolated guinea-pig ventricular myocytes cromakalim (100 μM) linearized the N-shaped steady-state current–voltage relationship (*I–V* curve) by removing rectification at membrane potentials more positive than the K^+ equilibrium potential and by decreasing P_K at more negative potentials (Osterrieder, 1988). The latter may reflect inhibition of the inwardly rectifying background K^+ current, I_{K1} (Conder and McCullough, 1987; McCullough *et al.*, 1987). In purkinje fibres cromakalim (5 μM) antagonized abnormal pacemaker activity, conferring potentially anti-arrhythmic properties to the drug (Liu *et al.*, 1988).

8.3.2.2 Neurones

High concentrations of cromakalim (≈100 μM) reduced the population spike amplitude in the CA1 region and the bursting rate of pacemaker neurones in the CA3 region of guinea-pig hippocampal slices (Alzheimer and ten Bruggencate, 1988). However, no effect on resting membrane potential or input resistance of the CA3 pacemaker neurones was detected. The compound also reduced hippocampal excitability induced by high extracellular K^+, low Mg^{2+} and low Ca^{2+}. These effects may be due to a weak agonistic action of cromakalim at a voltage- and ligand-sensitive inwardly rectifying K^+ conductance, which is similar to that activated by adenosine (Alzheimer *et al.* 1989).

In mice, cromakalim (≈5 mg/kg i.p. or 0.1 μg i.c.v.) antagonized purposeless chewing induced by the muscarinic agonist pilocarpine (Tricklebank *et al.*, 1988). This effect was not due to an interference with muscarinic agonist binding or the phosphatidylinositol second messenger system but may have been caused by a modulation of neuronal P_K. The low i.v.c. dose (0.1 μg corresponding to 4 μg/kg) makes this result of interest.

Concerning effects in the peripheral nervous system, cromakalim is able to interfere with vagal stimulation in the isolated guinea-pig trachea without significant inhibition of the tone induced by exogenous acetylcholine, indicative of a presynaptic effect of the drug (Hall and Maclagan, 1988; McCaig and De Jonkheere, 1989), possibly due to inhibition of acetylcholine release from the vagus (Hall and Maclagan, 1988; see also section 8.3.5(g)).

8.3.2.3 Secretory systems

Renin

In eight healthy male volunteers, cromakalim (1 mg/day for 5 days) caused a significant increase in plasma renin activity (+122%), angiotensin II (+105%) and noradrenaline (+61%) relative to placebo (Ferrier *et al.*, 1989). This increase in renin activity was considered to be a combination of sympathetic activation together with a direct effect on the release of renin from the juxtaglomerular cells; cromakalim (0.1–10 μM) concentration-dependently increased renin release from cultured rat juxtaglomerular cells (Ferrier *et al.*, 1989). In the pithed rat, where reflex activation of the renin–angiotensin system is prevented, cromakalim (0.1 mg/kg) increased plasma renin activity, indicating a direct effect on juxtaglomerular cells *in vivo*, at blood pressure-lowering doses (Cavero *et al.*, 1988). This effect was prevented by pretreatment with glibenclamide. In this context, it may be noteworthy to recall that the juxtaglomerular cells are derived from vascular smooth muscle cells and that renin secretion is inversely related to the intracellular Ca^{2+} concentration (Kurtz, 1986). Hence membrane hyperpolarization, by reducing the entry of Ca^{2+} into these cells, could account for this stimulation of renin secretion by cromakalim.

Insulin

In the rat, cromakalim at blood pressure-lowering doses, does not increase plasma glucose or decrease plasma insulin levels (Wilson *et al.*, 1988a; Quast and Cook, 1989a). However, at very high concentrations ($\approx$100 μM), cromakalim starts to inhibit insulin secretion *in vitro* (Dunne *et al.*, 1989; Garrino *et al.*, 1989). This selectivity of cromakalim for K^+ channels in smooth muscle is very different from the totally non-selective behaviour of diazoxide in this respect (see below, section 8.3.3.3).

8.3.2.4 Skeletal muscle

Cromakalim in concentrations above 1 μM increased P_K in skeletal muscle fibres from biopsies of patients suffering from various myopathies and having resting membrane potentials from −85 to −60 mV (Spuler *et al.*, 1989). The hyperpolarization evoked by cromakalim (the magnitude of which depended on the difference between the resting potential and the K^+ equlibrium potential) was tolbutamide-sensitive, indicating that the K^+ channel opened in this tissue may be K_{ATP} (Spuler *et al.*, 1989). The enhancement of P_K in skeletal muscle may be a novel therapeutic approach in cases where myopathies are accompanied by membrane depolarization.

8.3.3 Other potassium channel openers

8.3.3.1 Nicorandil

Nicorandil was in fact the first vasodilator compound shown to increase P_K (Furukawa *et al.*, 1981); in addition though, it is a nitro-compound increasing cyclic GMP by stimulation of soluble guanylate cyclase (Holzmann, 1983). The relative contribution of each mechanism appears to depend on the particular smooth muscle being studied and also on the conditions: after development of tachyphylaxis to other nitrovasodilators, nicorandil remains an effective vasodilator both *in vitro* (Kukovetz and Holzmann, 1987) and *in vivo* (Häusler, private communication). Evidence for the involvement of K^+ channel opening in the smooth muscle relaxant action of nicorandil has been obtained in many vascular preparations (for references, see

Weston, 1987), in intestinal smooth muscle (Yamanaka *et al.*, 1985; Weir and Weston 1986a) and in trachea (Inoue *et al.*, 1983; Allen *et al.*, 1986b). However, little or no evidence for K^+ channel opening was found in the rat aorta (Weir and Weston, 1986b; Weston, 1987) and the vasorelaxant effect in the guinea-pig pulmonary artery was insensitive to glibenclamide (Eltze, 1989b). In the rabbit mesenteric artery, the vasorelaxant effect of nicorandil (but not the increase in $^{86}Rb^+$ efflux) was poorly antagonized by procaine, 4-aminopyridine and TEA (Wilson *et al.*, 1989b). Phentolamine has been shown to inhibit the vasorelaxant effects of cromakalim in different blood vessels, by a presumed K^+ channel-blocking effect (see section 8.2.6.4), but it did not block relaxant effects of nicorandil (McPherson and Angus, 1989).

In smooth muscle, nicorandil is generally 30–100 times weaker than cromakalim, i.e. relaxant effects are seen typically in the range 1–10 μM (Weston, 1987). In the heart, nicorandil (30–300 μM) is negatively inotropic, shortens the action potential duration (Yanagisawa and Taira, 1980; Imanishi *et al.*, 1983), and decreases the rate of spontaneous firing in canine purkinje fibres (Yanagisawa and Taira, 1981). These and other cardiac effects (see Taira, 1987, for a review) are linked to an increase in P_K in the heart, independent of an increase in cyclic GMP (Yanagisawa *et al.*, 1988). In guinea-pig isolated ventricular cells, nicorandil (300 μM) activates a hitherto unknown voltage-independent K^+ channel with a unitary conductance of 4.6 pS (Kakei *et al.*, 1986).

In general the dual mechanism of action of nicorandil appears to be reflected in a unique *in vivo* profile (see, for example, Taira, 1987). Much of the extensive knowledge on nicorandil is summarized in a supplement volume of *J. Cardiovasc. Pharmacol.* (**10**, suppl. 8, 1987).

8.3.3.2 Pinacidil

Smooth muscle

Evidence that pinacidil, a 4-pyridyl-cyanoguanidine compound (see Fig. 8.2), exerts its vasorelaxant effects by opening membrane K^+ channels was presented recently from several laboratories (Bray *et al.*, 1987; Cook *et al.*, 1988a; Southerton *et al.*, 1988; Videbaek *et al.*, 1988; Hermsmeyer, 1988; Wilson *et al.*, 1988b). Generally, pinacidil is about three times weaker than cromakalim in its action (Cook *et al.*, 1988a). Again, the K^+ channel opened by pinacidil is Rb^+-permeable but prefers K^+, the selectivity for K^+ being more marked at lower pinacidil concentrations (Videbaek *et al.*, 1988). In this study, which was performed in rat isolated mesenteric resistance vessels, it was also shown that pinacidil relaxed a tonic contraction to noradrenaline with a monophasic concentration dependence (IC_{50}=0.3 μM). The vasorelaxant effect *in vitro* (Eltze, 1989b; Wilson, 1989) and *in vivo* (Cavero *et al.*, 1988) and the increase in $^{86}Rb^+$ efflux (Quast, unpublished) are inhibited by glibenclamide (*in vitro* $pA_2 \approx 7$). The stereospecificity of the effects of pinacidil *in vitro* and *in vivo* is less than that of cromakalim; values ranging from 2 (Arrigoni-Martelli *et al.*, 1980) to approximately 10 (Cook *et al.*, 1989) and 25 (Southerton *et al.*, 1988) have been reported, with the (−) enantiomer of pinacidil being more potent.

Recent *in vitro* studies with the enantiomers of pinacidil on the rabbit aorta have revealed the existence of an additional vasodilator mechanism, seemingly unrelated to K^+ channel opening (Cook *et al.* 1989). Both (−)- and (+)-pinacidil inhibit the tonic noradrenaline contraction concentration-dependently, a sixfold potency difference separating the two enantiomers (Fig. 8.3). However, in vessels depolarized by

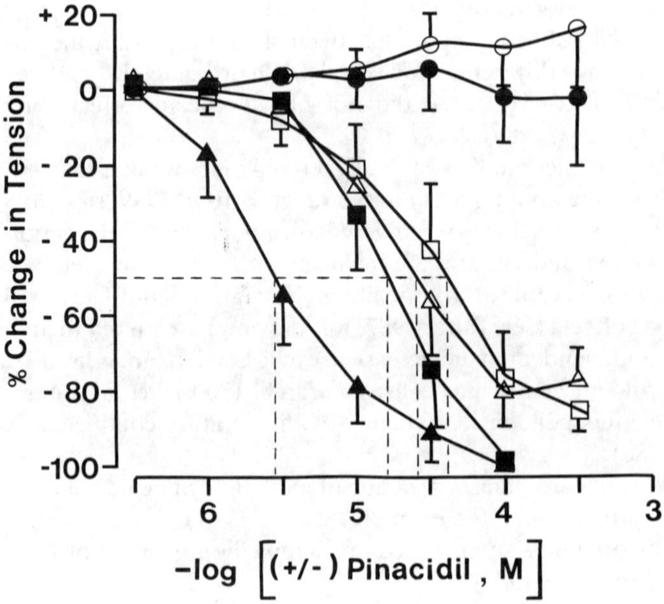

Fig. 8.3 — Inhibition of tonic contractions to noradrenaline in the rabbit aorta by (+)- and (−)-pinacidil. Rabbit aortic rings were contracted by 1 μM noradrenaline as described in Cook *et al.* (1988c). Both (+)- and (−)-pinacidil (■, ▲) inhibited this contraction completely, with IC_{50} values of 16 and 2.5 μM. When the tissue was partially depolarized by 35 mM KCl in the presence of 0.1 μM isradipine, (+)- and (−)-pinacidil still inhibited this contraction (□, △, IC_{50} values=40 and 25 μM). (●, ○) solvent controls in the normal and partly depolarized state. Means ± s.e.m. of 4 experiments. (From Cook *et al.* 1989; methods as described in Cook *et al.*, 1988c).

elevated KCl, where cromakalim loses its relaxant effect against noradrenaline (see section 8.3.1.2), the inhibition curves of the pinacidil enantiomers were shifted to the right but not abolished (Fig. 8.3). No stereoselectivity of this additional mechanism (which begins at 3 μM) was observed. This second mechanism probably explains the greater inhibition of angiotensin II and high KCl contractions of the rabbit aorta by (±)-pinacidil than by cromakalim (Cook *et al.*, 1988a,c). It may also explain why Wilson *et al.* (1988b) found K^+ channel blockers to be less effective blockers of pinacidil (and nicorandil) vasorelaxant effects than of cromakalim. Hence, on the basis of these *in vitro* studies, one would expect there to be significant *in vivo* haemodynamic differences between the two enantiomers of pinacidil and also differences between cromakalim and higher doses of (±)-pinacidil. The mechanism of the additional vasodilator activity of pinacidil has not yet been defined; it is not linked to an inhibition of dihydropyridine-sensitive Ca^{2+} channels, but could conceivably be through an effect on receptor-operated Ca^{2+} channels or on cyclic nucleotides (Cook *et al.*, 1989). Indeed, Kaufmann *et al.* (1986) found that pinacidil, at 100 μM, significantly increased cyclic AMP levels in the rat aorta. At lower concentrations there was no effect on cyclic nucleotide levels (Kauffman *et al.*, 1986; Southerton *et al.*, 1988; Nakajima *et al.*, 1989).

There are relatively few studies with pinacidil in non-vascular smooth muscle. In guinea-pig trachealis, pinacidil relaxes spontaneous tone ($IC_{50} \approx 1$ μM) but is weaker against tone induced by various asthma mediators or KCl (Kudsk *et al.*, 1988; Bray *et al.*, 1987). In guinea-pig taeni caeci, pinacidil inhibited spontaneous activity in a apamin-insensitive manner (Bray *et al.*, 1987) and was inactive in saponin-skinned preparations. (Nakajima *et al.*, 1989). In human bladder preparations, 10–30 μM pinacidil significantly increased $^{86}Rb^{+}$ efflux (Fovaeus *et al.*, 1989; but see the different result obtained with cromakalim in the guinea-pig bladder, section 8.3.1.3). In rats with detrusor instability induced by partial urethra ligation, pinacidil and cromakalim (1 mg/kg p.o.) decreased spontaneous bladder contractions (Malmgren *et al.*, 1989) but voiding ability remained, pointing to interesting therapeutic possibilities for the KCOs in the treatment of incontinence.

Heart

The effects of pinacidil and congeners (i.e. other pyridyl cyanoguadines) have been investigated in detail. As with other KCOs, pinacidil and congeners reduced action potential duration in canine Purkinje fibres and ventricle and decreased the contractile force of cat papillary muscle (Steinberg *et al.*, 1988). The concentrations needed to elicit these effects were only 6–10 times higher than those required to relax phenylephrine-contracted veins and an excellent correlation was obtained when the potencies for these effects were compared. Cromakalim and nicorandil also fitted this correlation (Steinberg *et al.*, 1988) and a similarly good correlation was observed when the *in vivo* correlates of these effects (i.e. fall in blood pressure and decrease in left ventricular d*p*/d*t*) were compared (Smallwood and Steinberg, 1988).

In isolated guniea-pig ventricular cells it was found that pinacidil (10 μM) increased the inward rectifier K^{+} current I_{K1}, reducing its rectification properties without influencing the delayed outward K^{+} current (Iijima and Taira, 1987). The increase in cardiac P_K explains the APD shortening and negative inotropic effect of pinacidil. This change in the current–voltage relationship by 10 μM pinacidil resembles that found with 10 μM cromakalim (Osterrieder, 1988), except at membrane potentials negative to the Nernst potential.

Insulin

Pinacidil (500 μM) increased $^{86}Rb^{+}$ efflux from rat pancreatic islets, but only when these were perfused with glucose, 2-ketioisocaproate or tolbutamide, i.e. under conditions where K_{ATP} in the β cell is inhibited (Lebrun *et al.*, 1988). Concomitantly, $^{45}Ca^{2+}$ efflux and insulin secretion were inhibited by pinacidil. These findings suggest that at these concentrations pinacidil can open K_{ATP}, thereby inhibiting insulin secretion similarly to diazoxide (Trube *et al.*, 1986; see section 8.3.3.3 and Chapter 11).

An update on recent results obtained with pinacidil is presented in a supplement volume of *J. Cardiovasc. Pharmacol.* (**12**, suppl. 12, 1988).

8.3.3.3 Diazoxide

Diazoxide is an antihypertensive whose chronic administration has been precluded by its marked hyperglycaemia, an effect due to reduced insulin secretion through diazoxide-induced opening of ATP-sensitive K^+ channels in the pancreas (see Chapter 11). Evidence that its vasodilator activity may be due to the opening of K^+ channels has recently been obtained (Cook *et al.*, 1988b; Bray *et al.*, 1988c; Trieschmann *et al.*, 1988; Quast and Cook, 1989a; Winquist *et al.*, 1989). The vasorelaxant effects *in vivo* and *in vitro* (including $^{86}Rb^+$ efflux) are detected at micromolar concentrations, i.e. about 200-fold higher than those used for cromakalim (Cook *et al.*, 1988b; Quast and Cook, 1989a; Bray *et al.*, 1988c; Newgreen *et al.*, 1989; Winquist *et al.*, 1989). The vascular effects of diazoxide *in vitro* and *in vivo* are inhibited by glibenclamide (*in vitro* with $pA_2=7$; Quast and Cook, 1989a; Newgreen *et al.*, 1989; Winquist *et al.*, 1989). In rabbit aorta, elevated KCl completely abolished the vasorelaxant effect of diazoxide against noradrenaline (Quast and Cook, 1989a), whereas in rat aorta, diazoxide (100 μM) relaxed by 75% the contraction induced by 80 mM KCl, indicating an additional vasorelaxant mechanism at this concentration (Bray *et al.*, 1988c; Quast, unpublished). K_{ATP} in pancreatic β cells is opened by diazoxide with an EC_{50} of 20 μM in the presence of 0.3 mM ATP (Zünkler *et al.*, 1988a; see Chapter 11), i.e. a concentration 5–10 times higher than that required for vasorelaxation. In the anaesthetized rat, the effects of diazoxide on plasma insulin and glucose levels are detected at the same doses as those that lower blood pressure. This suggests that diazoxide poorly discriminates between the ATP-sensitive K^+ channel in the pancreas and the K^+ channel(s) that these compounds open in smooth muscle (Quast and Cook, 1989a), thus corroborating the clinical experience with diazoxide.

In addition to the clinical use of diazoxide as a vasodilator in the treatment of hypertensive crisis and as a hyperglycaemic agent, a patent has recently been filed for use of diazoxide to promote hair growth (*Chem. Eng. News* **64** 32, 1986; see also minoxidil sulphate, section 8.3.3.4).

8.3.3.4 Minoxidil sulphate

The vasorelaxant profile of minoxidil sulphate, the active metabolite of minoxidil, resembles that of the other KCOs in that it is ineffective against high KCl and that (most of) its relaxant effect against noradrenaline is abolished in depolarizing solutions or in the presence of 10 mM TEA (Meisheri and Cipkus, 1987). Definitive proof of the ability of this compound to increase P_K came from $^{42}K^+$ efflux and membrane potential measurements (Bray *et al.*, 1988c; Meisheri *et al.*, 1988). In some experimental models, the drug is more, and in others it is less potent than cromakalim (Bray *et al.*, 1988c). Most interestingly, minoxidil sulphate is unable to increase the permeability of vascular tissue for $^{86}Rb^+$ (although it does so for $^{42}K^+$), indicating that it acts on K^+ channels different from those opened by cromakalim, pinacidil and diazoxide (Newgreen *et al.*, 1989). In addition, unlike cromakalim and diazoxide, the vasorelaxant effect of minoxidil sulphate in rat blood vessels is inhibited by glibenclamide in a non-competitive manner (Newgreen *et al.*, 1989; Winquist *et al.*, 1989) and it is particularly sensitive to this blocker (Winquist *et al.*, 1989). The parent compound, minoxidil, which is rather inactive *in vitro* (McCall *et al.*, 1983), is launched as an antihypertensive and, as a topical solution, against alopecia areata (male-pattern baldness).

8.3.3.5 RP 49356

RP 49356 appears to be structurally unrelated to cromakalim or pinacidil (Fig. 8.2). Its vasorelaxant profile *in vivo* and *in vitro* corresponds to that of the other KCOs (sensitivity to glibenclamide, no effect against 75 mM KCl) (Mondot *et al.*, 1988; Cavero *et al.*, 1988). The compound is about three times weaker than cromakalim in increasing $^{86}Rb^+$ efflux and inhibiting spontaneous activity of the rat portal vein (Quast, unpublished).

8.3.3.6 Correlation

Fig 8.4 shows that an excellent correlation exists between the potencies of various KCOs to increase $^{86}Rb^+$ efflux from the rat portal vein and to lower blood pressure in anaesthetized rats. This provides strong evidence for the hypothesis that the antihypertensive effect of these compounds (with the exception of (+)-pinacidil and nicorandil, as discussed above) relies on their ability to increase P_K in vascular smooth muscle.

8.3.4 Nature of the potassium channel opened by the KCOs

8.3.4.1 Evidence of heterogeneity

Effects of KCOs have been observed in various smooth muscle preparations (blood vessels, trachea, intestine, baldder) and, generally at higher concentrations, also in heart, pancreas and neuronal tissue. Evidence is emerging for diversity within this group (cf. difference in the effects of cromakalim and minoxidil sulphate on $^{86}Rb^+$ permeability, section 8.3.3.4) and that they may act on different K^+ channels according to the tissue chosen and the concentration used. Examples for the latter include a preferential action on pacemaker cells in portal vein, uterus and bladder at concentrations below those causing a measurable flux response, or the decrease in selectivity for K^+ over Rb^+ with increasing concentrations of cromakalim and pinacidil, as discussed in sections 8.3.1.1 and 8.3.3.2. Electrophysiological evidence for the heterogeneity of K^+ channels opened by cromakalim at high concentrations was obtained by Nakao *et al.* (1988), who found that the hyperpolarizing current evoked by 30 μM cromakalim in the mesenteric vein consisted of two components, only one of which was sensitive to $MnCl_2$. On the other hand, the mesenteric artery displayed only the Mn^{2+}-resistant component (Nakao *et al.*, 1988).

8.3.4.2 Is the channel in vascular smooth muscle Ca^{2+}-dependent?

Electrophysiological studies have clearly shown the ability of the KCOs to open Ca^{2+}-dependent K^+ channels (K_{Ca}) in vascular smooth muscle. In bovine and rabbit aortic cells (Kusano *et al.*, 1987; Gelband *et al.*, 1989) and vascular myocytes from the human mesentery (Trieschmann *et al.*, 1988; Klöckner *et al.*, 1989), submicro-

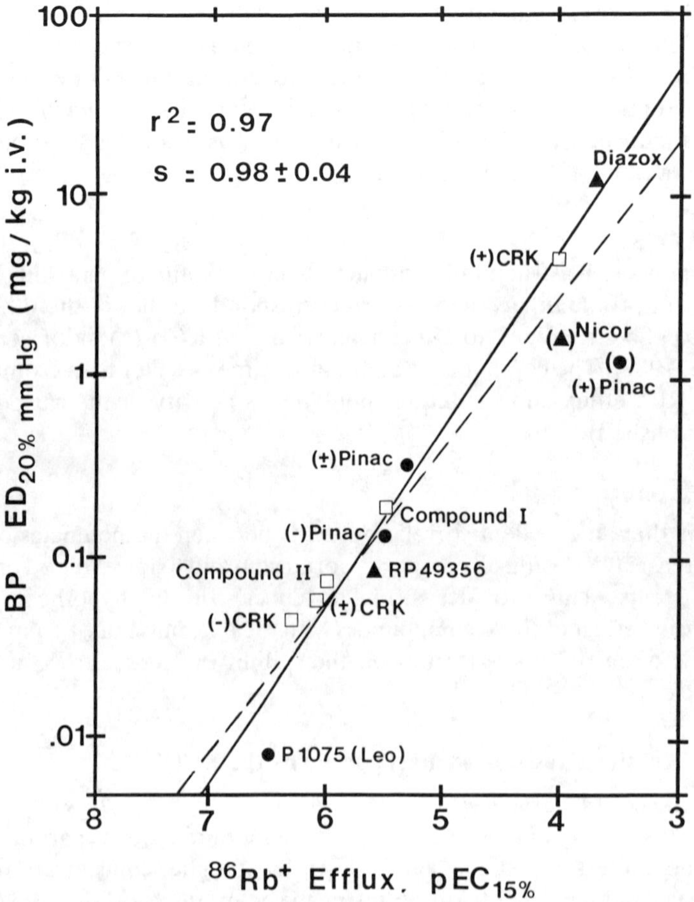

Fig. 8.4 — Correlation between blood pressure-lowering activity of KCOs in anaesthetized rats and their potency at stimulating $^{86}Rb^+$ efflux from the rat portal vein. Blood pressure lowering was determined as described in Cook *et al.* (1988a) and is quantified by the i.v. dose necessary to lower blood pressure by 20%. $^{86}Rb^+$ efflux measurements were performed as described by Quast (1987) and are quantified by the negative logarithm of the concentration (M) which increases efflux by 15%. The broken line is the correlation obtained when all compounds are included (n=13, r^2=0.9, slope=0.84±0.08); the solid line is obtained after exclusion of (+)-pinacidil and nicorandil (n=11, r^2=0.97, s=0.98±0.04). Chemical names of the compounds not appearing in Fig. 8.2: P 1075 (Leo)=*N*-cyano-*N′*-(1,1-dimethylpropyl)-*N″*-3-pyridinylguanidine (prepared according to *J. Med. Chem.* (1978) **21** 773); Compound I=(±)-3,4-dihydro-3-hydroxy-2,2-dimethyl-4-(2-oxo-1-piperazinyl)-2*H*-1-benzopyran-6-carbonitrile (prepared according to European Patent 107 423A); Compound II=(±)-*N*-(6-cyano-3,4-dihydro-3-hydroxy-2,2-dimethyl-2*H*-1-benzopyran-4-yl)-2-furancarboxamide (prepared according to European Patent 126 311A).

molar concentrations of cromakalim have been shown to open BK_{Ca}. Channel conductance was not modified but the duration of the closed periods was reduced in the presence of cromakalim, such that the channel open probability was increased up to 60% (Gelband *et al.*, 1989; Trieschmann *et al.*, 1988). Diazoxide (0.4 mM) and the

Ca^{2+} antagonist (+)-niguldipine (1 μM) acted similarly to cromakalim (1 μM) on BK_{Ca} (Klöckner *et al.*, 1989) and pinacidil also caused an increased opening of large-conductance K^+ channels in single vascular cells, particularly when the cytoplasmic side of the membrane was exposed to >1 μM Ca^{2+} (Hermsmeyer, 1988). The effect of (+)-niguldipine was stereoselective, the (−)-enantiomer acting as an antagonist (Klöckner *et al.*, 1989). Nakao *et al.* (1988) found that the cromakalim-induced hyperpolarization of the guinea-pig mesenteric artery and vein was abolished in Ca^{2+}-free solutions containing 0.5 mM EGTA, which they interpreted as evidence of an action on K_{Ca}. In canine coronary smooth muscle cells, minoxidil sulphate caused a one- to twofold increase of a fast activating K^+ current which was diminished 85% by removing extracellular Ca^{2+} (Wilde and Lee, 1988). On the other hand, there is also electrophysiological evidence against an action of cromakalim on BK_{Ca} channels. Beech and Bolton (1989), studying effects of cromakalim on membrane currents of cells from the rabbit portal vein, found that cromakalim (1–10 μM) caused large membrane hyperpolarizations by increasing a K^+ current which showed little voltage dependence and was insensitive to 0.5 mM TEA, both effects arguing against this channel being a BK_{Ca} channel. The authors suggest that the cromakalim-induced current may be carried by the delayed rectifier (K_V (Beech and Bolton, 1989). Standen *et al.* (1989) also reported that the open probability of single BK_{Ca} channels from rat and rabbit mesenteric arteries was not increased by 10 μM cromakalim.

Work with selective K^+ channel blockers also sheds some light on this question. Following on from the electrophysiological evidence that cromakalim opened BK_{Ca}, we hypothesized that charybdotoxin (CTX, a potent (nM) blocker of BK_{Ca}, see section 8.2.2.2) should therefore block effects of cromakalim. Indeed, an initial investigation showed the ability of crude *Leiurus quinquestriatus* scorpion venom to potently inhibit the cromakalim-evoked $^{86}Rb^+$ efflux from the rat portal vein (Quast and Cook, 1988a). However, isolation of the venom constituent responsible for this effect showed it *not* to be CTX, but instead some other minor component of the venom (Weir and Strong, 1988; Strong *et al.*, 1989). At this stage it is not clear if this cromakalim-blocking toxin is the same as that which blocks apamin-sensitive K^+ channels (Castle and Strong, 1986), recently purified and named leiurotoxin I (Gimenez-Gallego *et al.*, 1988; see section 8.2.2.1). Winquist *et al.* (1988, 1989) have also reported the inability of CTX (0.1 μM) to block the relaxant effects of cromakalim or minoxidil sulphate in the rat portal vein. This, together with the weak potency of TEA as an inhibitor (IC_{50}=10 mM) strongly argues against the K^+ permeability increase due to cromakalim (or its vasorelaxant activity) being mediated through BK_{Ca} channels, at least of the classical type. Apamin, a toxin from bee venom (Habermann, 1984) which selectively blocks SK_{Ca} (see section 8.2.2.1) has been shown not to affect cromakalim activity *in vitro* and *in vivo* (Weir and Weston, 1986a,b; Cook and Hof, 1988).

Many studies show neither the vasorelaxant effects of the KCOs nor the increase in P_K they elicit to be modified by (selective) Ca^{2+} antagonists. The cromakalim-stimulated $^{86}Rb^+$ efflux from the guinea-pig and rat portal vein (Quast, 1987; Quast and Cook, 1988a) and the rabbit aorta (Cook *et al.*, 1988c) was not modified in the presence of isradipine (at concentrations sufficient to block voltage-dependent Ca^{2+} channels). The same is true if $^{42}K^+$ is used instead of $^{86}Rb^+$ (Quast and Baumlin, 1988) or if a different Ca^{2+} antagonist (1 μM nisoldipine) were used instead (Post *et al.*, 1989). This contrasts with results of Kreye and Weston (1986), who found 10 μM nifedipine and gallopamil (D600) to inhibit cromakalim-stimulated $^{86}Rb^+$ efflux from the rabbit aorta. Since both nifedipine and D600 have been shown to directly block K^+ currents in other cells with IC_{50}s values of 4 and 11–17 μM, respectively (Gurney and Nerbonmne, 1986; Gola and Ducreaux, 1985), we feel these effects are most probably related to a non-specific K^+ channel-blocking action, and not associated with their inhibition of voltage-dependent Ca^{2+} channels. Cook *et al.* (1988c) showed that maximally effective (0.1 μM) concentrations of the dihydropyridine Ca^{2+} antagonist isradipine (PN 200–110) did not modify either the anti-vasoconstrictor affinity or efficacy of cromakalim, suggesting that Ca^{2+} entry through dihydropyridine-sensitive Ca^{2+} channels is not required in order for cromakalim to elicit its effects on the aorta. Finally, Coldwell and Howlett (1988) have reported that cromakalim-evoked $^{86}Rb^+$ efflux from the rabbit mesenteric artery is not only well maintained but even increased in a Ca^{2+}-free/EGTA medium (0 Ca^{2+}).

Taken together these data suggest that although there is clear electrophysiological evidence showing that KCOs can open BK_{Ca}, this appears not to be the mechanism through which these drugs exert their pharmacological effects on the vasculature at low concentrations.

8.3.4.3 Is the channel in vascular smooth muscle ATP-sensitive?

This question arose when it was shown that glibenclamide, a potent (nM) inhibitor of K_{ATP} in the pancreas (Schmid-Antomarchi *et al.*, 1987; Zünkler *et al.*, 1988a), potently and apparently competitively (excepting minoxidil sulphate) inhibited the effects of the KCOs on smooth muscle (pA_2=7, Quast and Cook, 1988b, 1989a; Wilson, 1989; Eltze, 1989a,b; Newgreen *et al.*, 1989; Winquist *et al.*, 1989). The similar potency of glibenclamide at inhibiting the different KCOs (with the exception of minoxidil sulphate) suggests they act on the same K^+ channel to elicit their effects. The work with glibenclamide has now been extended to include other sulphonylureas or compounds derived therefrom (Quast and Webster, 1989; Wilson, 1989). This class of blockers of K_{ATP} includes ligands selective for one or both of two putative sites, 'A' (tolbutamide), 'A+B' (glibenclamide) and 'B' (meglitinide/HB699) (see Rufer and Losert, 1979). In a study utilizing four compounds of each type, Quast and Webster (1989) compared their inhibitory potency against cromakalim-induced $^{86}Rb^+$ efflux with their potency of binding to K_{ATP} in a pancreatic cell line (as reported by Schmid-Antomarchi *et al.*, 1987). A highly significant correlation was shown between these two effects, suggesting that the K^+ channel opened by cromakalim in vascular smooth muscle has indeed a binding site which resembles the sulphonylurea site of K_{ATP} in the pancreatic β cell (and in the heart; Fosset *et al.*, 1988).

A second intriguing fact, namely that diazoxide, an opener of K_{ATP} in the pancreas, is also a KCO in smooth muscle has been discussed above. The comparison of cromakalim with diazoxide shows that the K^+ channel in smooth muscle must have a different pharmacology from K_{ATP} in pancreatic β cells, since cromakalim has a minimal effect upon insulin secretion (*in vitro* and *in vivo*) and plasma glucose levels (Wilson *et al.*, 1988a; Cook *et al.*, 1988b; Quast and Cook, 1989a). Cromakalim thus shows selectivity for the K^+ channel in smooth muscle over K_{ATP} in β cells, whereas diazoxide opens both channels with a similar potency. The third intriguing fact is that at least part of the cardiac effects of the KCOs are mediated via the opening of K_{ATP} in the heart, apparently by decreasing the affinity of the channel for ATP (see section 8.3.2.1).

These three lines of evidence argue in favour of the hypothesis that the KCOs interact with an ATP- sensitive K^+ channel in smooth muscle. There is indirect biochemical evidence for the existence of K_{ATP} in smooth muscle since metabolic exhaustion of arterial smooth muscle cells in culture or application of cromakalim (0.1 μM) led to a decrease in intracellular ATP and an increase in $^{86}Rb^+$ efflux (Howlett and Longman, 1989). However, this stimulated $^{86}Rb^+$ efflux was inhibited by TEA (30 mM) but not by glibenclamide (10 μM), casting doubt on the hypothesis that K_{ATP} was involved (Howlett and Longman, 1989).

Very recently, direct evidence at a single-channel level has been obtained for the existence of K_{ATP} in freshly prepared cells from rat or rabbit mesenteric arteries (Standen *et al.*, 1989). This channel, which could be blocked by 'intracellular' ATP (1 mM) and glibenclamide (20 μM), had a single-channel conductance of 135 pS (at 0 mV and 60 mM KCl_o) (Standen *et al.*, 1989), comparable to the conductance of K_{ATP} in cortical neurones (Ashford *et al.*, 1988), but higher than the values reported in other tissues (see Chapter 4). It differed from the BK_{Ca} channels (also having a single-channel conductance of $\approx$150 pS) in that it was not dependent on Ca^{2+}_i or membrane voltage for opening and could be blocked by glibenclamide (Standen *et al.*, 1989). In inside-out patches, where channel activity had been abolished by the presence of 1 mM ATP, cromakalim (1 μM) restored the channel activity, and this could be inhibited by 20 μM glibenclamide. In line with these observations, the relaxation of the noradrenaline-induced tonic contraction of the rabbit mesenteric artery by cromakalim, diazoxide and pinacidil was reversed by glibenclamide (<1 μM), tolbutamide (500 μM) and Ba^{2+} (30–100 μM) but not by TEA (<1 mM) or charybdtoxin (50 nM) (Standen *et al.*, 1989).

The direct electrophysiological demonstration of K_{ATP} in vascular smooth muscle and the effect of cromakalim thereon greatly strengthens the argument that such a channel is indeed the primary target of the KCOs, its opening underlying their mechano-inhibitory effects (but see Beech and Bolton (1989) for a different conclusion). Whether this channel in smooth muscle is also under additional control (e.g. by G-proteins, phosphorylation, other nucleotides, etc.) remains to be seen. The question arises, however, as to what is the physiological or pathophysiological role of this channel, since this may have implications for the therapeutic use of these drugs. It was noted by Quast and Cook (1988b, 1989a) that glibenclamide did not increase vascular tone *in vitro* or *in vivo* at doses blocking effects of cromakalim, showing that the channel was not open under resting conditions, as originally suggested by Hamilton *et al.* (1986). One possibility is that K_{ATP} serves as the target for certain endogenous substances known to cause smooth muscle relaxation. It has been found that the hyperpolarization induced by acetylcholine (via release of endothelium-derived hyperpolarizing factor, EDHF) and vasoactive intestinal peptide (VIP, acting directly on smooth muscle) is seemingly elicited by opening of these K_{ATP} channels (Standen *et al.*, 1989).

In cardiac and skeletal muscle, it has been hypothesized that K_{ATP} might be important for the cell during ischaemia, when changes in the level of intracellular ATP (or other factors, e.g. ATP/ADP ratio) may increase its open probability. This argument is strengthened by the finding that the open probability of these channels in skeletal muscle is increased as the pH is lowered at a constant ATP_i (N. Standen, personal communication). It is presently unclear if K_{ATP} channels in smooth muscle might fulfil a similar role. However, it is noteworthy that in a rat model of occlusive arterial disease, the KCOs cromakalim, pinacidil and nicorandil (in complete contrast to calcium antagonists or hydralazine) selectively enhanced blood cell flux and the pO_2 of hypoxic skeletal muscle (Angersbach and Nicholson, 1988). These authors speculated that the greater efficacy of the KCOs in these (collateral) vessels, as compared to the resistance arterioles supplying normoxic muscle, might be explained if the collaterals were to have a membrane potential further away from the K^+ equilibrium potential. In the light of this new evidence that these drugs open K_{ATP}, we would propose an alternative hypothesis: that during ischaemia or hypoxia a small reduction in the ATP level on the inner side of the membrane (perhaps coupled with changes in pH or other metabolic factors) might place the K_{ATP} channels in a situation where their activation by the KCOs is favoured. This question is certain to receive close attention in the near future as our knowledge of the mechanisms by which these drugs act continues to grow.

8.3.4.4 Cyclic nucleotides, inositol phosphates, G-proteins and protein kinase C

Cyclic nucleotides

Despite the dual ability of nicorandil to increase cyclic GMP levels and open K^+ channels in smooth muscle, there appears to be no obvious link between these two mechanisms. Sodium nitroprusside, for instance, also raises cyclic GMP levels in rat aorta but does not cause membrane hyperpolarization (Taylor *et al.*, 1988). Conversely, no change in tissue cyclic GMP accompanies the cromakalim-evoked $^{86}Rb^+$ efflux in rat aorta (Taylor *et al.*, 1988) or rabbit mesenteric artery (Coldwell and Howlett, 1987), or its vasorelaxant activity in the bovine retractor penis muscle (Gillespie and Sheng, 1988). Cromakalim is also without effect on cAMP levels in various tissues (rat brain slices, Coldwell and Howlett, 1987; rat aorta, Newgreen *et al.*, 1988; bovine retractor penis muscle, Gillespie and Sheng, 1988). Furthermore, cromakalim showed no inhibition of cyclic AMP or cyclic GMP-phosphodiesterase activity (Coldwell and Howlett, 1987), its stimulation of $^{86}Rb^+$ efflux was not modified by cell-permeable derivatives of cyclic AMP or cyclic GMP (Quast, 1987), and its relaxant effects in rat aorta were not modified by either methylene blue or MB 22948 (Newgreen *et al.*, 1988). Hence, cyclic nucleotides appear to play no role in the action of cromakalim on vascular tissue. The situation may, however, be different in the taenia caeci, where McHarg *et al.* (1989) have recently reported increases in cyclic AMP (but not cyclic GMP) levels by 1.6 μM cromakalim. The situation with pinacidil is somewhat less clear cut (cf. section 8.3.3.2).

Phosphoinositides

The accumulation of inositol phosphates by rat cortical slices treated with noradrenaline (in the presence of lithium) was not affected by 10 μM cromakalim (Coldwell and Howlett, 1987). Cromakalim also shows negligible affinity to various receptors coupled to phosphatidylinositol turnover (Coldwell and Howlett, 1987; D. Hoyer, personal communication).

GTP-binding proteins
GTP-binding proteins, at least those sensitive to pertussis toxin (PTX), seem also not to be involved in the actions of cromakalim. Following administration to rats, the ability of cromakalim to elicit $^{86}Rb^+$ efflux (aorta, portal vein, atria) or cause vasorelaxation (aorta, portal vein) was unaffected relative to controls, whereas the effects of carbachol on an atrial K^+ channel (known to be G-protein-coupled) were abolished (Quast *et al.*, 1988).

Protein kinase C
Coldwell (1988, 1989) investigated the interaction between a number of phorbol esters and cromakalim in the rabbit mesenteric artery. Cromakalim inhibited the phorbol 12,13-dibutyrate (PDBu)-induced contraction and, conversely, PDBu inhibited the $^{86}Rb^+$ efflux response to cromakalim concentration-dependently (IC_{50}=70 nM). The latter effect would seem analogous with the K^+ currents in a neuronal cell line, which are inhibited by protein kinase C activation (Higashida and Brown, 1986). Cromakalim did not inhibit [^{3}H]PDBu binding to protein kinase C nor did it inhibit enzyme activity in a phosphorylation assay, suggesting the antagonism between cromakalim and PDBu to be indirect, possibly via a phosphorylation of the cromakalim-opened K^+ channel (Coldwell, 1989). However, phorbol 12-myristate 13-acetate, which is supposedly a more potent kinase C activator, did not inhibit cromakalim-stimulated $^{86}Rb^+$ efflux (Coldwell, 1988).

8.3.4.5 Permeability to Rb^+
In vascular smooth muscle this depends on the agonist. As discussed in sections 8.3.1.1 and 8.3.3.2, the channel population opened by cromakalim and pinacidil is somewhat selective for K^+ over Rb^+, whereas minoxidil sulphate opens Rb^+-impermeable channels. Millimolar concentrations of Rb^+, however, block the vasorelaxant and $^{86}Rb^+$ efflux-stimulating effect of cromakalim in vascular smooth muscle (Quast, unpublished). Similarly, in trachea, cromakalim opens Rb^+-permeable channels which can be blocked by millimolar Rb^+_o (Foster, 1989; Morris and Taylor, 1989; see section 8.3.1.3). In the guinea-pig bladder, the cromakalim-opened channel is essentially impermeable to Rb^+ and blocked by millimolar intra- and extracellular Rb^+ concentrations (Foster *et al.*, 1989). In rat atrium, cromakalim stimulates $^{86}Rb^+$ efflux (Quast *et al.*, 1988).

8.3.5 Consequences of potassium channel opening upon smooth muscle tone
Since mechano-inhibitory effects of cromakalim and other 'pure' KCOs seem invariably to be abolished in depolarized vessels (Cook *et al.*, 1988c), it is reasonable to assume (at least until experimental evidence proves otherwise) that all of its effects are coupled to its ability to open K^+ channels, thereby hyperpolarizing the cell membrane. The potent inhibitory action of K^+ channel openers in spontaneously active vessels such as the portal vein is not unexpected in view of the important role of K^+ permeability in regulating the firing frequency of such tissues (see Cook, 1988). What is more surprising is their ability to inhibit contractions of 'electrically silent' blood vessels such as the rabbit aorta (Cook *et al.*, 1988c; Bray *et al.*, 1988a,b). Noradrenaline contractions of this blood vessel occur without modification of the membrane potential (Cauvin *et al.*, 1984; Bray *et al.*, 1988a), and dihydropyridine-

sensitive Ca^{2+} channels provide only a small fraction of the Ca^{2+} for contraction (Cook *et al.*, 1988c). Cromakalim appears to inhibit with equal efficacy hormonal contractions both with or without a significant contribution from these Ca^{2+} channels (Gillespie and Sheng, 1988). The original postulation, that the relaxant effect of the K^+ channel openers was due to their hyperpolarizing action indirectly closing voltage-sensitive (slow) Ca^{2+} channels (Hamilton *et al.*, 1986), clearly has to be updated to encompass these new findings. We must therefore ask the question, aside from dihydropyridine-sensitive Ca^{2+} channels, which other mechanisms affecting smooth muscle tone are likely to be modified by an increase in P_K and subsequent hyperpolarization of the membrane? Some of these possibilities are considered below (cf. Fig. 8.5).

(a) *Dihydropyridine-insensitive voltage-sensitive calcium channels* have recently been identified in vascular smooth muscle (Bean *et al.*, 1986; Friedman *et al.*, 1986). By hyperpolarizing the membrane, cromakalim might prevent the opening of such channels.

(b) *Receptor-operated cation channels* might be modified by changing the cell membrane potential. At the moment, this possibility is merely speculative since the only such channel that has been characterized electrophysiologically (an ATP-operated cation channel in isolated cells from the rabbit ear artery) is relatively insensitive to changes in membrane voltage (Benham and Tsien, 1987). There may, however, exist other receptor-operated cation channels which show voltage dependence.

(c) *Sodium/calcium exchange*: due to the electrogenic nature of this ion transport system, hyperpolarization of the cell membrane is expected to favour Ca^{2+} extrusion (Mullins, 1979). The relevance of this effect in smooth muscle is difficult to test since specific inhibitors of Na^+/Ca^{2+} exchange are lacking (Kaczorowski *et al.*, 1988). We have found that dichlorobenzamil (10 μM), one of the most potent inhibitors of Na^+/Ca^{2+} exchange, can indeed block the inhibition by cromakalim of angiotensin II contractions of the rabbit aorta (Quast and Cook, 1988b). However, in the same study it was shown that dichlorobenzamil also inhibited the cromakalim-stimulated $^{86}Rb^+$ efflux from the rat portal vein and aorta, suggesting that the former effect may have been due to a direct (K^+) channel-blocking action, rather than to an inhibition of Na^+/Ca^{2+} exchange. The possibility that the anti-vasoconstrictor effects of cromakalim might be indirectly mediated through effects on Na^+/Ca^{2+} exchange therefore requires further investigation.

(d) *Superficially (membrane) bound calcium*: the group of van Breemen has discussed the existence of a 'superficial' (membrane-bound) Ca^{2+} pool in smooth muscle (see, for example, Lodge and van Breemen, 1985; Lukeman and van Breemen, 1985), whose purpose might be to induce Ca^{2+} release from deeper pools within intracellular organelles. By hyperpolarizing the cell membrane cromakalim might interfere with the release of this membrane-bound Ca^{2+}, which would thereby reduce contractions evoked by agonists.

(e) *Refilling of intracellular calcium stores*: evidence has been presented that cromakalim is able to prevent the refilling of intracellular Ca^{2+} stores (Bray *et al.*, 1988b; Cowlrick *et al.*, 1988). The experimental protocol consisted of firstly depleting Ca^{2+} pools by noradrenaline stimulation in Ca^{2+}-free medium (0 Ca^{2+}), followed by a short period of exposure to Ca^{2+} in the absence or presence of cromakalim, before re-examining the contractile effect of noradrenaline in 0 Ca^{2+}.

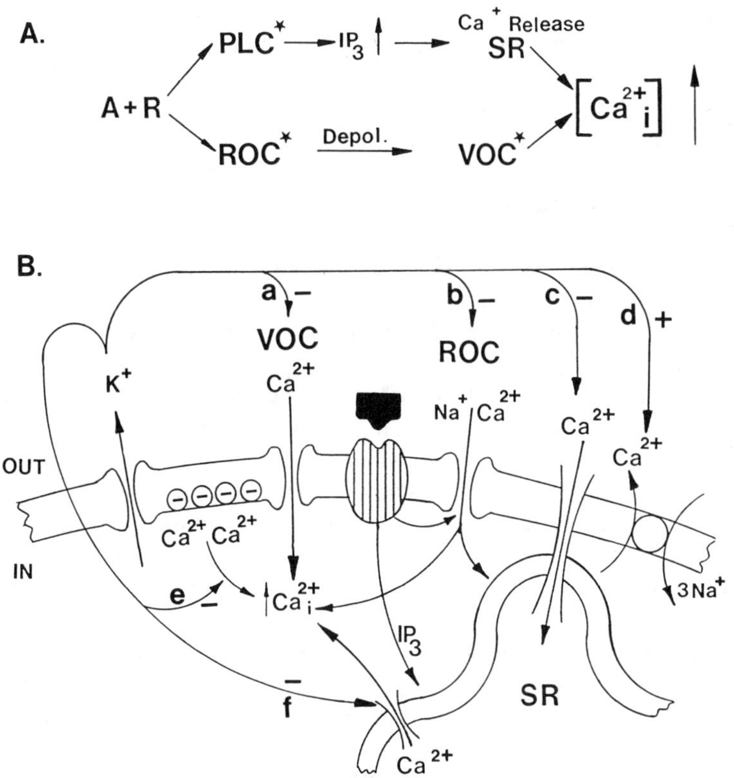

Fig. 8.5 — Possible mechanism by which KCOs may affect Ca^{2+} handling of smooth muscle. (A) Occupation of the receptor R by the agonist A increases intracellular Ca^{2+} by two pathways. Upper: the activated receptor activates a phospholipase C (PLC) via a G-protein (not shown) which leads to an increase in inositol triphosphate (IP_3). IP_3 releases Ca^{2+} from the sarcoplasmic reticulum. Lower: the activated receptor activates a receptor-operated channel (ROC) which depolarizes the membrane and may allow some Ca^{2+} influx. Membrane depolarization then leads to the opening of voltage operated Ca^{2+} channels (VOC). (B) Opening of K^+ channels with subsequent hyperpolarization of the membrane could (hypothetically) oppose: (a) the opening of voltage-sensitive Ca^{2+} channels (VOC) (e.g. L,T type); (b) the opening of receptor-operated channels; (c) the opening of a putative channel which allows Ca^{2+} entry to refill the sarcoplasmic reticulum; (d) Ca^{2+} entry (or facilitate Ca extrusion) via the Na^+–Ca^{2+} exchange, known to be electrogenic; (e) mobilization of Ca^{2+} bound to the negative charges of the inner leaflet of the cell membrane; (f) Ca^{2+} release from the sarcoplasmic reticulum if this process depends on membrane potential. Points (b), (c) and (f) all assume that the respective processes are dependent on the membrane potential, but this has not yet been proven.

The presence of 3–10 μM cromakalim during the period of Ca^{2+} replenishment significantly attenuated the subsequent noradrenaline contractions relative to controls (Bray *et al.*, 1988b; Cowlrick *et al.*, 1988). The mechanism through which cromakalim inhibits this refilling of intracellular Ca^{2+} stores has not yet been elucidated. However, both Chiu *et al.* (1987) and Cox and Ferrone (1988) have reported that cromakalim inhibits $^{45}Ca^{2+}$ influx in response to noradrenaline, which is presumably through dihydropyridine-insensitive Ca^{2+} channels (Cauvin *et al.*, 1984). Inhibition of Ca^{2+} entry through a special 'refilling' channel or through one or other of the pathways discussed above could account for the ability of cromakalim to prevent the refilling of these stores, with the consequence that contractions dependent primarily on the release of intracellular Ca^{2+} would be reduced in the presence of these drugs (Cook *et al.*, 1988c).

(f) *Calcium release from intracellular stores*: caffeine is known to release Ca^{2+} from the sarcoplasmic reticulum. Studies in the rabbit renal artery (Wilson, 1988; Wilson and Hicks, 1989) and aorta (Bray and Weston, 1989) show that cromakalim, pinacidil, RP 49356 and diazoxide inhibit caffeine-induced contractions in a glibenclamide-sensitive manner (Wilson, 1988; Wilson and Hicks, 1989). The data suggest that the KCOs somehow interfere with the ability of caffeine to release Ca^{2+} from the stores. In the perfused mesenteric bed of the rat, contractions to pulses of noradrenaline in 0 Ca^{2+} were inhibited concentration-dependently by cromakalim (Quast, 1989). This apparent inhibition of Ca^{2+} release from intracellular stores by cromakalim was stereoselective and was inhibited by glibenclamide or in depolarized tissues, suggesting that it was associated with K^+ channel opening. However, Wilson (1988) was unable to show such an inhibition of noradrenaline effects by cromakalim in 0 Ca^{2+} in the rabbit renal artery. The noradrenaline-stimulated release of $^{45}Ca^{2+}$ from preloaded rabbit aorta into 0 Ca^{2+} medium was not inhibited by 10 μM cromakalim (Chiu *et al.*, 1987). These apparent discrepancies may be due to cromakalim affecting the release of Ca^{2+} from intracellular stores differently according to the preparation being studied (e.g. large/small vessel).

(g) *Uptake 2 at the synaptic cleft*: in various tissues (smooth muscle, heart, glands but not neurones) there exists a corticoid-sensitive, cocaine-insensitive catecholamine transporter (uptake 2) which translocates catecholamines into the cell, where they are metabolically inactivated. The driving force for this uptake process appears to be the K^+ gradient or the resting membrane potential (Bönisch *et al.*, 1985). Since depolarization inhibits this uptake it is conceivable that hyperpolarization may activate it and, indeed, nicorandil (10 μM) increased catecholamine uptake by rabbit pulomonary artery strips by 17.8±6.8% (Trendelenburg, 1987). The increased clearance of noradrenaline from the synaptic cleft following smooth muscle membrane hyperpolarization could contribute to the relaxant effect of the KCOs.

8.3.6 Endogenous potassium channel openers in smooth muscle

A variety of neurotransmitters and hormones, like acetylcholine, ATP, bradykinin or substance P, can stimulate the endothelium to release relaxing factors (EDRFs). One such factor is nitric oxide, which relaxes smooth muscle by stimulation of soluble guanylate cyclase, leading to an increase in cyclic GMP and vasorelaxation (see Gryglewski *et al.*, 1988 for a recent review). Another as yet unidentified factor, termed EDHF (endothelium-derived hyperpolarizing factor), hyperpolarizes the smooth muscle membrane and increases $^{86}Rb^+$ efflux from the tissue consistent with the opening of Rb^+-permeable K^+ channels (Taylor *et al.*, 1988; Chen *et al.*, 1988), although a contribution to the observed hyperpolarization from an activation of the Na^+/K^+ ATPase cannot be excluded (Feletou and Vanhoutte, 1988). The reversal of the acetylcholine-induced membrane hyperpolarization by glibenclamide suggests that K_{ATP} channels are involved (Standen *et al.*, 1989) and that activation of Na^+/K^+ ATPase by EDHF appears to be of minor importance.

Another factor released from the endothelium is prostacyclin (PGI_2), which acts by stimulating adenylate cyclase (for a review see Gryglewski *et al.*, 1988). In the carotid artery of the dog, PGI_2 and its stable analogue iloprost led to a concentration-dependent hyperpolarization and in increase in $^{42}K^+$ efflux, suggesting that they were opening K^+ channels (Siegel *et al.*, 1987). At the moment, the mechanism of this effect (direct or linked to cyclic AMP) is unclear. In the rat portal vein iloprost behaved like other agents which elevate cyclic AMP; the amplitude of the contractions was reduced, but their frequency increased. A minor increase in $^{86}Rb^+$ efflux was seen only at very high concentrations (>μM) (Quast, unpublished).

Some hormones, like VIP or somatostatin (or the somatostatin analogue SMS 201-995; Lategan *et al.*, 1988), act directly on the smooth muscle cell to produce a hyperpolarization. In the case of VIP, the mechanism of action appears to be an activation of K_{ATP} (Standen *et al.*, 1989). In the heart, somatostatin has actions similar to acetylcholine and adenosine (Webb *et al.*, 1986), both of which open an atrial K^+ channel through a G_i- (or G_k-) protein (see Chapter 5). One might speculate that a similar mechanism may hold for somatostatin, although we are not presently aware of any single-channel data to support this. Another way to open K^+ channels in smooth muscle is by depolarization and/or increasing intracellular Ca^{2+}. The latter pathway is particularly prominent in intestinal muscle, where the receptor-mediated inositol phosphate turnover leads to an increase in Ca^{2+}_i, which in turn opens apamin-sensitive K^+ channels (see Chapter 3). In such cases, neurotransmitters such as ATP and noradrenaline cause myorelaxation instead of contraction (Den Hertog, 1981; see Chapter 3).

8.4 CONCLUDING REMARKS

With the diversity of K^+ channels and their regulation, nature has laid the structural basis for the possibility of intervening with specific K^+ channel-modulating drugs. To date, K^+ channels in the pancreas, smooth muscle and the heart are primary targets of important drugs (sulphonylureas, KCOs, class III anti-arrhythmics) as discussed here and in other parts of this book. The combination of high-resolution electrophysiology, investigations with potent (and hopefully specific) channel toxins, and a wider characterization of individual channels at the molecular (genetic) level will undoubtedly accelerate this development in the years to come.

REFERENCES

Adams, P. R., Brown, D. A., and Constanti, A. (1982) Pharmacological inhibition of the M-current. *J. Physiol.* **332** 223–262.

Aguayo, L. G., and Alburquerque, E. X. (1987) Phenycylidine blocks two potassium currents in spinal neurons in cell culture. *Brain Res.* **436** 9–17.

Aguayo, L. G., Weinstein, H., Maayani, S., Glick, S. D., Warnick, J. E., and Albuquerque, E. X. (1984) Discriminative effects of behaviourally active and inactive analogues of phencylidine on membrane electrical excitability. *J. Pharmac. Exp. Ther.* **228** 80–87.

Ahn, H. Y., and Karaki, H. (1988) Inhibitory effects of procaine on contraction and calcium movement in vascular and intestinal smooth muscles. *Br. J. Pharmacol.* **94** 189–196.

Albuquerque, E. X., Aguayo, L. G., Warnick, J. E., Weinstein, H., Glick, S. D., Maayani, S., Ickowitz, R. K., and Blaustein, M. P. (1981) The behavioural effects of phencyclidines may be due to their blockade of potassium channels. *Proc. Natl. Acad. Sci. USA* **78** 7792–7796.

Allen, S. L., Boyle, J. P., Cortijo, J., Foster, R. W., Morgan, G. P., and Small, R. C. (1986a) Electrical and mechanical effects of BRL 34915 in guinea-pig isolated trachealis. *Br. J. Pharmacol.* **89** 395–405.

Allen, S. L., Foster, R. W., Morgan, G. P., and Small, R. C. (1986b) The relaxant action of nicorandil in guinea-pig isolated trachealis. *Br. J. Pharmac.* **87** 117–127.

Almers, W. (1976) Differential effects of tetracaine on delayed potassium channels and displacement currents in frog skeletal muscle. *J. Physiol.* **262** 613–637.

Alvarez, J., and Garcia-Sancho, J. (1988) Irreversible inhibition of Ca^{2+}-dependent K^+ channels of human red cells by *Notechis scutatus* venom. *J. Physiol.* **407** 97P.

Alzheimer, C., and ten Bruggencate, G. (1988) Actions of BRL 34915 (Cromakalim) upon convulsive discharges in guinea-pig hippocampal slices. *Naunyn–Schmiedeberg's Arch. Pharmacol.* **337** 429–434.

Alzheimer, C., Sutor, B., and ten Bruggencate, G. (1989) Cromakalim (BRL 34915) acts on an inwardly rectifying neuronal K^+ conductance, which is similar to that activated by adenosine. *Pflügers Arch.* **414** (suppl. 1) S121–S122.

Anderson, A. J., and Harvey, A. L. (1988) Effects of the potassium channel blocking dendrotoxins on acetylcholine release and motor nerve terminal activity. *Br. J. Pharmacol.* **93** 215–221.

Anderson, C. S., MacKinnon, R., Smith, C., and Miller, C. (1988) Charybdotoxin block of single Ca^{2+}-activated K^+ channels: Effects of channel gating, voltage, and ionic strength. *J. Gen. Physiol.* **91** 317–333.

Angersbach, D., and Nicholson, C. D. (1988) Enhancement of muscle blood cell flux and pO_2 by cromakalim (BRL 34915) and other compounds enhancing membrane K^+ conductance, but not by Ca^{2+} antagonists or hydralazine, in an animal model of occlusive arterial disease. *Naunyn–Schmiedeberg's Arch. Pharmacol.* **337** 341–346.

Arch. J. R. S., Buckle, D. R., Bumstead, J., Clarke, G. D., Taylor, J. F., and Taylor, S. G. (1988) Evaluation of the potassium channel activator cromakalim (BRL 34915) as a bronchodilator in the guinea-pig: comparison with nifedipine. *Br. J. Pharmacol.* **95** 763–770.

Arena, J. P., and Kass, R. S. (1988) Block of heart potassium channels by clofilium and its tertiary analogs: Relationship between drug structure and type of channel blocked. *Mol. Pharmacol.* **34** 60–66.

Armstrong, C. M. (1971) Interaction of tetraethylammonium ion derivatives with the potassium channels of giant axons. *J. Gen. Physiol.* **58** 413–437.

Armstrong, C. M., and Hille, B. (1972) The inner quaternary ammonium receptor in potassium channels of the node of Ranvier. *J. Gen. Physiol.* **59** 388–400.

Arrigoni-Martelli, E., Nielsen, C. K., Bang Olsen, U. B., and Petersen, H. J. (1980) *N″*-cyano-*N*-4pyridyl-*N′*-1,2,2-trimetylpropylguanidine monohydrate (P1134), a new potent vasodilator. *Experimentia* **36** 445–447.

Ashford, M. L. J., Sturgess, N. C., Trout, N. J., Gardner, N. J., and Hales, C. N.

(1988) Adenosine-5′-triphosphate-sensitive ion channels in neonatal rat cultured central neurones. *Pflügers Arch.* **412** 297–304.

Bacaner, M. B., Clay, J. R., Shrier, A., and Brochu, R. M. (1986) Potassium channel blockade: A mechanism for suppressing ventricular fibrillation. *Proc. Natl Acad. Sci. USA* **83** 2223–2227.

Bader, C. R., Bernheim, D., and Bertrand, D. (1985) Sodium-activated potassium current in cultured avian neurones. *Nature* **317** 540–542.

Baker, P. F., and Schapira, A. H. V. (1980) Anaesthetics increase light emission from aequorin at constant ionised calcium. *Nature* **284** 168–169.

Banks, B. E. C., Brown, C., Burgess, G. M., Burnstock, G., Claret, M., Cocks, T. M., and Jenkinson, D. H. (1979) Apamin blocks certain neurotransmitter-induced increases in potassium permeability. *Nature* **282** 415–417.

Bartschat, D. K., and Blaustein, M. P. (1985) Calcium-activated potassium channels in isolated presynaptic nerve terminals from rat brain. *J. Physiol.* **361** 441–457.

Bartschat, D. K., and Blaustein, M. P. (1986) Phencyclidine in low doses selectively blocks a presynaptic voltage-regulated potassium channel in rat brain *Proc. Natl Acad. Sci. USA* **83** 189–192.

Bean, B. P., Sturek, M., Puga, A., and Hermsmeyer, K. (1986) Calcium channels in muscle cells isolated from rat mesenteric arteries: modulation by dihydropyridine drugs. *Circ. Res.* **59** 229–235.

Beech, D. J., and Bolton, T. B. (1989) Properties of the cromakalim-induced potassium conductance in smooth muscle cells isolated from the rabbit portal vein. *Br. J. Pharmacol.* **98** 851–864.

Beech. D. J., Bolton, T. B., Castle, N. A., and Strong, P. N. (1987) Characterization of a toxin from scorpion (*Leiurus quinquestriatus*) venom that blocks *in vitro* both large (BK) K^+ channels in rabbit vascular smooth muscle and intermediate (IK) conductance Ca^{2+}-activated K^+ channels in human red cells. *J. Physiol.* **387** 32P.

Belles, B., Hescheler, J., and Trube, G. (1987) Changes in membrane currents in cardiac cells induced by long whole-cell recordings and tolbutamide. *Pflügers Arch.* **409** 582–588.

Benham, C. D., and Tsien, R. W. (1987) A novel receptor-operated Ca^{2+}-permeable channel activated by ATP in smooth muscle. *Nature* **328** 275–278.

Benham, C. D., Bolton, T. B., Lang, R. J., and Takewaki, T. (1985) The mechanism of action of Bs^{2+} and TEA on single Ca^{2+}-activated K^+ channels in arterial and intestinal smooth muscle cell membranes. *Pflügers Arch.* **403** 120–127.

Benishin, C. G., Krueger, B. K., and Blaustein, M. P. (1988) Phenothiazines and haloperidol block Ca-activated K channels in rat forebrain synaptosomes. *Am. Soc. Pharmacol. Exp. Ther.* **33** 195–201.

Benoit, E., and Dubois, J.-M. (1986) Toxin I from the snake *Dendroaspis polylepis*: a highly specific blocker of one type of potassium channel in myelinated nerve fiber. *Brain Res.* **377** 374–377.

Bernardi, H., Fosset, M., and Lazdunski, M. (1988) Characterization, purification, and affinity labeling of the brain [^{3}H]glibenclamide-binding protein, a putative neuronal ATP-regulated K^+ channel. *Proc. Natl. Acad. Sci. USA* **85** 9816–9820.

Bidard, J.-N., Mourre, C., and Lazdunski, M. (1987) Two potent convulsant peptides, a bee venom toxin, the MCD peptide and a snake venom toxin, dendrotoxin I, known to block K^+ channels, have interacting receptor sites. *Biochem. Biophys. Res. Commun.* **143** 383–389.

Bkaily, G., Payet, M. D., Benabderrazik, M., Sauve, R., Renaud, J.-F., Bacaner, M., and Sperelakis, N. (1988) Intracellular bretylium blocks Na^+ and K^+ currents in heart cells. *Eur. J. Pharmacol.* **151** 389–397.

Black, A. R., and Dolly, J. O. (1986) Two acceptor sub-types for dendrotoxin in chick synaptic membranes distinguishable by β-bungarotoxin. *Eur. J. Biochem.* **156** 609–617.

Blatz, A. L., and Magleby, K. L. (1986) Single apamin-blocked Ca-activated K^+ channels of small conductance in cultured rat skeletal muscle. *Nature* **323** 718–720.

Blaustein, M. P., and Ickowitz, R. K. (1983) Phencyclidine in nanomolar concentrations binds to synaptosomes and blocks certain potassium channels. *Proc. Natl. Acad. Sci. USA* **80** 3855–3859.

Böhme, H. (1964) Herzrhythmusstörungen mit Sparteinsulfat. *Z. Ges. Inn. Med.* **19** 883–889.

Bönisch, H., Bryan, L. J., Henseling, M., O'Donnell, S. R., Stockmann, P., and Trendelenburg, U. (1985) The effect of various ions on uptake$_2$ of catecholamines. *Naunyn-Schmiederberg's Arch. Pharmacol.* **328** 407–416.

Bray, K. M., Newgreen, D. T., Small, R. C., Southerton, J. S., Taylor, S. G., Weir, S. W., and Weston, A. H. (1987) Evidence that the mechanism of the inhibitory action of pinacidil in rat and guinea-pig smooth muscle differs from that of glyceryl trinitrate. *Br. J. Pharmacol.* **91** 421–429.

Bray, K. M., Weston, A. H., McHarg, A. D., Newgreen, D. T., and Southerton, J. S. (1988a) Analysis of the inhibitory action of BRL 34915 on responses to noradrenaline in rabbit aorta. *Br. J. Pharmacol.* **93** 205P.

Bray, K. M., Weston, A. H., McHarg, A. D., Newgreen, D. T., Southerton, J. S., and Duty, S. (1988b) Further studies on the actions of the K-channel openers cromakalim (BRL34915) and pinacidil, in rabbit aorta. *Pflügers Arch.* **411** R202.

Bray, K. M., Brown, B. S., Duty, S., Kay, P. B., Longmore, J., McHarg, A. D., Newgreen, D. T., Southerton, J. S., Waterfall, J. F., and Weston, A. H. (1988c) Studies on the mode of action of minoxidil sulfate and diazoxide: a comparison with cromakalim. *Br. J. Pharmacol.* **95** 733P.

Bray, K. M., and Weston, A. H. (1989) Effects of K^+ channel openers on the spasmogenic component of caffeine-induced responses in rabbit aorta. *Br. J. Pharmacol.* **96** 220P.

Brezina, V., Eckert, R., and Erxleben, C. (1987) Modulation of potassium conductances by an endogenous neuropeptide in neurones of *Aplysia californica. J. Physiol.* **382** 267–290.

Brown, J., Marshall, R. J., and Winslow, E. (1985) Effects of selective channel blocking agents on contractions and action potentials in K^+-depolarized guinea-pig atria. *Br. J. Pharmacol.* **86** 7–17.

Buchheit, K.-H., and Bertholet, A. (1988) Inhibition of small intestine motility by cromakalim (BRL 34915). *Eur. J. Pharmacol.* **154** 335–337.

Buckingham, R. E., Clapham, J. C., Coldwell, M. C., Hamilton, T. C., and

Howlett, D. R. (1986) Stereospecific mechanism of action of the novel anti-hypertensive agent, BRL 34915. *Br. J. Pharmacol.* **87** 78P.

Buckingham, R. E., and Parrott, L. S. (1988) Cromakalim (BRL 34915) inhibits noradrenaline-evoked vasomotion in the SHR isolated mesenteric vascular bed. *Br. J. Pharmacol.* **94** 355P.

Buckingham, R. E., Hamilton, T. C., Howlett, D. R., Mootoo, S., and Wilson, C. (1989) Inhibition of glibenclamide of the vasorelaxant action of cromakalim in the rat. *Br. J. Pharmacol.* **97** 57–64.

Burgess, G. M., Claret, M., and Jenkinson, D. H. (1981) Effects of quinine and apamin on the calcium dependent potassium permeability of mammalian hepatocytes and red cells. *J. Physiol.* **317** 67–90.

Busso, C., Camino, E., Cedrini, L., and Lovisolo, D. (1988) The effects of gaboon viper (*Bitis gabonica*) venom on voltage-clamped single heart cells. *Toxicon* **26** 559–570.

Cain, C. R., and Metzler, V. (1985) Electrophysiological effects of the anti-hypertensive agent BRL 34915 in guinea-pig papillary muscle. *Naunyn-Schmiedeberg's Arch. Pharmacol.* (Abstr.) **329** R53.

Caldwell, K. K., and Harris, R. A. (1985) Effects of anesthetic and anticonvulsant drugs on calcium-dependent efflux of potassium from human erythrocytes. *Eur. J. Pharmacol.* **107** 119–125.

Capiod, T., and Ogden, D. C. (1989) The properties of calcium-activated potassium channels in guinea-pig isolated hepatocytes. *J. Physiol.* **409** 285–295.

Carbone,, E., Wanke, E., Prestipino, G., Possani, L. D., and Maelicke, A. (1982) Selective blockage of voltage-dependent K^+ channels by a novel scorpion toxin *Nature* **296** 90–91.

Carlen, P. L., Gurevich, N., and Durand, D. (1982) Ethanol in low doses augments calcium-mediated mechanisms measured intracellularly in hippocampal neurons. *Science* **215** 306–308.

Carmeliet, E. (1985) Electriphysiologic and voltage clamp analysis of the effects of sotalol on isolated cardiac muscle and purkinje fibers. *J. Pharmacol. Exp. Ther.* **232** 817–825.

Castle, N. A. (1987) Pharmacological comparison of potassium channels in skeletal muscle, hepatocytes and erythrocytes using arthropod toxins and other ion channel blocking agents. PhD thesis, University of London.

Castle, N. A., and Haylett, D. G. (1987) Effect of channel blockers on potassium efflux from metabolically exhausted frog skeletal muscle. *J. Physiol.* **383** 31–43.

Castle, N. A., and Strong, P. N. (1986) Identification of two toxins from scorpion (*Leiurus quinquestriatus*) venom which block distinct classes of calcium-activated potassium channels. *FEBS Lett.* **209** 117–121.

Cauvin, C., Lukeman, S., Cameron, J., Hwang, O., Meisheri, K., Yamamoto, H., and van Breemen, C. (1984) Theoretical bases for vascular selectivity of calcium antagonists. *J. Cardiovasc. Pharmacol.* **6** S630–S638.

Cauvin, C., Weir, S. W., and Bühler, F. (1989) Differences in Ca^{2+} handling along the arterial tree: an update including studies in human mesenteric resistance vessels. *J. Cardiovasc. Pharmacol.* **12** (suppl. 6) S10–S15.

Cavero, I., Pratz, J., and Mondot, S. (1988) Cardiovascular effects of K-channel openers *in vivo*. In: *Potassium channel modulators: New drugs with novel mechanisms of action*. IBC Symposium, London, December.

Cavero, I., Mondot, S., and Mestre, M. (1989) Vasorelaxant effects of cromakalim in rats are mediated by glibenclamide-sensitive potassium channels. *J. Pharmacol. Exp. Therap.* **248** 1261–1268.

Chen, G., Suzuki, H., and Weston, A. H. (1988) Acetylcholine releases endothelium-derived hyperpolarizing factor and EDRF from rat blood vessels. *Br. J. Pharmacol.* **95** 1165–1174.

Chestnut, T. J., Carpenter, D. O., and Strichartz, G. R. (1987) Effects of venom from *Conus striatus* on the delayed rectifier potassium current of molluscan neurons. *Toxicon* **25** 267–278.

Cherskey, B., and Altszuler, N. (1984) Tolbutamide and glyburide differ in effectiveness to displace α and β-adrenergic radioligands in pancreatic islet cells and membranes. *Diabetes* **33** 499–503.

Chicchi, G. G., Gimenez-Gallego, G., Ber, E., Garcia, M. L., Winquist, R., and Cascieri, M. A. (1988) Purification and characterization of a unique, potent inhibitor of apamin binding from *Leiurus quinquestriatus hebraeus* venom. *J. Biol Chem.* **263** 10192–10197.

Chiu, P. J. S., Tetzloff, G., and Sybertz, E. J. (1987) Effects of BRL 34915 [(+)6-cyano-3,4-dihydro-2,2-dimethyl-*trans*-4-(2-oxo-1-pyrolidyl)-2*H*-benzo (*b*)pyran-3-ol] on transmembrane ^{45}Ca movements in rabbit aortic smooth muscle. *Hypertension* **10** 360.

Clapham, J. C., and Wilson, C. (1987) Anti-spasmogenic and spasmolytic effects of BRL 34915: a comparison with nidedipine and nicorandil. *J. Auton. Pharmac.* **7** 233–242.

Cognard, C., Traoré, F., Potreau, D., and Raymond, G. (1984) Effects of apamin on the outward potassium current of isolated frog skeletal muscle fibres. *Pflügers Arch.* **402** 222–224.

Cognard, C., Ewane-Nyambi, G., Potreau, D., and Raymond, G. (1985) Effects of phalloidin on electrical and mechanical activity of frog skeletal muscle fibers. *Eur. J. Pharmacol.* **110** 89.

Cognard, C., Ewane-Nyambi, G., Potreau, D., and Raymond, G. (1986) The voltage-dependent blocking effect of phalloidin on the delayed potassium current of voltage-clamped frog skeletal muscle fibers. *Eur. J. Pharmacol.* **120** 209–216.

Coldwell, M. C. (1988) Interaction between cromakalim (BRL 34915) and protein kinase C activation in rabbit mesenteric artery. *Br. J. Pharmacol.* **93** 207P.

Coldwell, M. C. (1989) Cromakalim does not interact directly with protein kinase C. *Br. J. Pharmacol.* **96** 254P.

Coldwell, M. C., and Howlett, D. R. (1987) Specificity of action of the novel antihypertensive agent, BRL 34915, as a potassium channel activator. *Biochem. Pharmacol.* **36** 3663–3669.

Coldwell, M. C., and Howlett, D. R. (1988) Potassium efflux enhancement by cromakalim (BRL 34915) in rabbit mesenteric artery: an indirect effect independent of calcium? *Biochem. Pharmacol.* **37** 4105–4110.

Conder, M. L., and McCullough, J. R. (1987) Purported K channel opener, BRL 34915, blocks inwardly rectifying K current in isolated guinea pig ventricular myocytes. *Biophys. J.* **51** 258a.

Connor, J. A. (1979) Calcium current in molluscan neurones: measurement under conditions which maximize its visibility. *J. Physiol.* **286** 41–60.

Constanti, A., and Smart, T. G. (1988) Zinc blocks the A-current rat sympathetic neurones. *J. Physiol.* **396** 159P.

Cook, D. L., and Hales, C. N. (1984) Intracellular ATP directly blocks K^+ channels in pancreatic β-cells. *Nature* **311** 271–273.

Cook, N. S. (1983) Pharmacological characterization of calcium-activated potassium channels in guinea-pig hepatocytes using apamin and other blocking agents. PhD thesis, University of London.

Cook, N. S. (1988) The pharmacology of potassium channels and their therapeutic potential. *Trends Pharm Sci.* **9** 21–28.

Cook, N. S. (1989) The effect of some potassium channel blockers upon contractile responses of the rabbit aorta. *J. Cardiovasc. Pharmacol.* **13** 299–306.

Cook, N. S., and Haylett, D. G. (1983) Effects of tubocurarine and other neurotransmitter blocking agents on Ca-activated K-channels in guinea-pig liver cells. *Br. J. Pharmacol.* **80** 546P.

Cook, N. S., and Haylett, D. G. (1985) Effects of apamin, quinine and neuromuscular blockers on calcium-activated potassium channels in guinea-pig hepatocytes. *J. Physiol.* **358** 373–394.

Cook, N. S., and Hof, R. P. (1988) Cardiovascular effects of apamin and BRL 34915 in rats and rabbits. *Br. J. Pharmacol.* **93** 121–131.

Cook, N. S., Haylett, D. G., and Strong, P. N. (1983) High affinity binding of [^{125}I]monoiodoapamin to isolated guinea-pig hepatocytes. *FEBS Lett.* **152** 265–269.

Cook, N. S., Griffiths, J. L., and Hof, R. P. (1987) The effects of some K^+ channel blockers and the K^+ channel activator BRL 34915 upon the rabbit aorta. *Br. J. Pharmacol.* **90** 130P.

Cook, N. S., Quast, U., Hof, R. P., Baumlin, Y., and Pally, C. (1988a) Similarities in the mechanism of action of two new vasodilator drugs: pinacidil and BRL 34915. *J. Cardiovasc. Pharmacol.* **11** 90–99.

Cook, N. S., Quast, U., and Weir, S. W. (1988b) *In vitro* and *in vivo* comparison of two K^+ channel openers, diazoxide and BRL 34915. *Pflügers Arch.* **411** R46.

Cook, N. S., Weir, S. W., and Danzeisen, M. C. (1988c) Anti-vasoconstrictor effects of the K^+ channel opener cromakalim on the rabbit aorta — comparison with the calcium antagonist isradipine. *Br. J. Pharmacol.* **95** 741–752.

Cook, N. S., Quast, U., and Manley, P. (1989) K^+ channel opening does not alone explain the vasodilator activity of pinacidil and its enantiomers. *Br. J. Pharmacol.* **96** 181P.

Cornejo, M., Guggino, S. E., and Guggino, W. B. (1987) Modification of Ca^{2+}-activated K^+ channels in cultured medullary thick ascending limb cells by *N*-bromoacetamide. *J. Membr. Biol.* **99** 147–155.

Cornejo, M., Guggino, S. E., Sastre, A., and Guggino, W. B. (1989) Isometric yohimbine alkaloids block calcium-activated K^+ channels in medullary thick ascending limb cells of rabbit kidney. *J. Membr. Biol.* **107** 25–33.

Coronado, R., and Miller, C., (1980) Decamethonium and hexamethonium block K^+ channels of sarcoplasmic reticulum. *Nature* **288** 495–497.

Coronado, R., and Miller, C. (1982) Conductance and block by organic cations in a

K^+-selective channel from sarcoplasm reticulum incorporated into planar phospholipid bilayers. *J. Gen. Physiol.* **79** 529–547.

Cowlrick, I. S., Paciorek, P. M., and Waterfall, J. F. (1988) Cromakalim and calcium movements in vascular smooth muscle. *Br. J. Pharmacol.* **95** 640P.

Cox, R. H., and Ferrone, R. A. (1988) Effects of BRL 34915 on ^{86}Rb efflux and active stress in rat vascular smooth muscle. *FASEB J.* **2** A756.

Davies, N. W., Spruce, A. E., Standen, N. B., and Stanfield, P. R. (1989a) Multiple blocking mechanisms of ATP-sensitive potassium channels of frog skeletal muscle by tetraethylammonium ions. *J. Physiol.* **413** 31–48.

Decoursey, T. E., Im, W. B., and Quandt, F. N. (1989) Externally applied tetraalkylammonium ions block open delayed rectifier K channels at both internal and external sites. Biophys J. **55** 540a.

Dietmer, J. W., and Eckert, R. (1985) Two components of Ca-dependent potassium current in identified neurones of *Aplysia californica*. *Pflügers Arch.* **403** 353–359.

Den Hertog, A. (1981) Calcium and the α-action of catecholamines on guinea-pig taenia caeci. *J. Physiol.* **316** 109–125.

Dinan, T. G. (1987) Calcium-activated potassium conductance; an alternative to the dopamine hypothesis of neuroleptic action. *Br. J. Psychiatry* **151** 455–459.

Dinan, T. G., Crunelli, V., and Kelly, J. S. (1987) Neuroleptics decrease calcium-activated potassium conductance in hippocampal pyramidal cells. *Brain Res.* **407** 159–162.

Downing, S. J., Miller, M., and Hollingsworth, M. (1989) Tolerance to cromakalim in the rat uterus *in vivo*. *Br. J. Pharmacol.* **96** 732–738.

Dreyer, F., and Penner, R. (1987) The actions of presynaptic snake toxins on membrane currents of mouse motor nerve terminals. *J. Physiol.* **386** 455–463.

Drukarch, B., Kits, K. S., Van der Meer, E. G., Lodder, J. C., and Stoof, J. C. (1987) 9-Amino-1,2,3,4-tetrahydroacridine (THA), an alleged drug for the treatment of Alzheimer's disease, inhibits acetylcholinesterase activity and slow outward K^+ current. *Eur. J. Pharmacol.* **141** 153–157.

Dubois, J. M. (1982) Capsaicin blocks one class of K^+ channels in the frog node of Ranvier. *Brain Res.* **245** 372–375.

Dunne, M. J., Yule, D. I., Gallacher, D. V., and Petersen, O. H. (1989) Cromakalim (BRL 34915) and diazoxide activate ATP-regulated potassium channels in insulin-secreting cells. *Pflügers Arch.* **414** (suppl. 1) S154–S155.

Eckl, K. M., and Greb, W. H. (1987) Potassium channel activation in the vascular smooth muscle cell-membrane. A new mechanism of action in antihypertensive treatment. Early clinical results with BRL 34915. *Clin. Exp. Hypertens A* **9** 160.

Elkins, T., Ganetsky, B., and Wu, C.-F. (1986) A *Drosophila* mutation that eliminates a Ca^{++}-dependent potassium current. *Proc. Natl Acad. Sci. USA* **83** 8415–8419.

Eltze, M. (1989a) Competitive antagonism by glibenclamide of cromakalim inhibition of twitch contractions in rabbit vs deferens. *Eur. J. Pharmac.* **161** 103–106.

Eltze, M. (1989b) Glibenclamide is a competitive antagonist of cromakalim, pinacidil and 49356 RP in guinea-pig pulmonary artery. *Eur. J. Pharmacol.* **165** 231–239.

Erdelyi, L., and Such, Gy. (1984) The effects of capsaicin on the early outward current in identified snail neurones. *Neuroscience* **48** 349–353.

Escande, D., Thuringer, D., Leguern, S., and Cavero, I. (1988) The potassium channel opener cromakalim (BRL 34915) activates ATP-dependent K^+ channels in isolated cardiac myocytes. *Biochem. Biophys. Res. Commun.* **154** 620–625.

Feletou, M., and Vanhoutte, P. M. (1988) Endothelium-dependent hyperpolarization of canine coronary smooth muscle. *Br. J. Pharmacol.* **93** 515–524.

Ferguson, W. B., and Magleby, K. L. (1989) Magnesium ions reduce the conductance of the large conductance calcium activated potassium channel (BK channel) in cultured rat skeletal muscle. *Biophys. J.* **55** 546a.

Ferrier, C. P., Kertz, A., Lehner, P., Shaw, S. G., Pusterla, C., Saxenhofer, H., and Weidmann, P. (1989) Stimulation of renin secretion by potasium-channel activation with cromakalim. *Eur. J. Pharmacol.* **36** 443–447.

Findlay, I., Dunne, M. J., Ullrich, S., Wollheim, C. B., and Petersen, O. H. (1985) Quinine inhibits Ca^{2+}-independent K^+ channels whereas tetraethylammonium inhibits Ca^{2+}-activated K^+ channels in insulin-secreting cells. *FEBS Lett.* **185** 4–8.

Fink, R., and Wettwer, E. (1978) Modified K-channel gating by exhaustion and the block by internally applied TEA^+ and 4-aminopyridine in muscle. *Pflügers Arch.* **374** 289–292.

Fishman, M. C., and Spector, I. (1981) Potassium current suppression by quinidine reveals additional calcium currents in neuroblastoma cells. *Proc. Natl. Acad. Sci. USA* **78** 5245–5249.

Fosset, M., Schmid-Antomarchi, H., Hugues, M., Romey, G., and Lazdunski, M. (1984) The presence in pig brain of an endogenous equivalent of apamin, the bee venom peptide that specifically blocks Ca^{2+}-dependent K^+ channels. *Proc. Natl. Acad. Sci. USA* **81** 7228–7232.

Fosset, M., DeWille, J. R., Green, R. D., Schmid-Antomarchi, H., and Lazdunski, M. (1988) Antidiabetic sulfonylureas control action potential properties in heart cells via high affinity receptors that are linked to ATP-dependent K^+ channels. *J. Biol. Chem.* **263** 7933–7936.

Foster, C. D., and Brading, A. F. (1987) The effect of potassium channel antagonists on the BRL 34915 activated potassium channel in guinea-pig bladder. *Br. J. Pharmacol.* **92** 751.

Foster, C. D.,, Fujii, K., Kingdom, J., and Brading, A. F. (1989) The effect of cromakalim on the smooth muscle of the guinea-pig urinary bladder. *Br. J. Pharmacol.* **97** 281–291.

Foster, K A. (1989) Cromakalim activation of potassium channels in guinea-pig trachea: effects of extracellular rubidium. *Br. J. Pharmacol.* **96** 233P.

Fovaeus, M., Andersson, K.-C., and Hedlund, H. (1989) The action of pinacidil in the isolated human bladder. *J. Urol.* **141** 637–640.

Franks, N. P., and Lieb, W. R. (1988) Volatile general anaesthetics activate a novel neuronal K^+ current. *Nature* **333** 662–664.

French, R. J., and Shoukimas, J. J. (1981) Blockage of squid axon potassium conductance by internal tetra-*N*-alkylammonium ions of various sizes. *Biophys. J.* **34** 271–291.

Freeman, S. E., Dawson, R. M., Culvenor, A. J., and Keeghan, A. M. (1985)

Interactions of amantadine with the cardiac muscarinic receptor. *J. Mol. Cell. Cardiol.* **17** 9–21.

Friedman, M. E., Suarez-Kurtz, G., Kaczorowski, G. J., Katz, G. M., and Reuben, J. P. (1986) Two calcium currents in a smooth muscle cell line. *Am. J. Physiol.* **250** H699–H703.

Furukawa, K., Itoh, T., Kajiwara, M., Kitamura, K., Suzuki, H., Ito, Y., and Kuriyama, H. (1981) Vasodilating actions of 2-nicotinamidoethyl nitrate on porcine and guinea-pig coronary arteries. *J. Pharm. Exp. Ther.* **218** 248–259.

Garcia, M. L., Vazquez, J., Glavez, A., King, F., Feigenbaum, P., and Kaczorowski, G. J. (1989) Partial purification of the high conductance Ca^{2+}-activated K^+ channel from bovine aortic and tracheal smooth muscle. *Biophys. J.* **55** 541a.

Garrino, M. G., and Henquin, J. C. (1987). Adamantane derivatives: a new class of insulin secretagogues. *Br. J. Pharmacol.* **90** 583–591.

Garrino, M. G., and Henquin, J. C. (1988) Highly potent and stereo-selective effects of the benzoic acid derivative AZ-DF 265 on pancreatic β-cells. *Br. J. Pharmacol.* **93** 61–68.

Garrino, M. G., Schmeer, W., Nenquin, M., Meissner, H. P., and Henquin, J. C. (1985) Mechanism of the stimulation of insulin release *in vitro* by HB 699, a benzoic acid derivative similar to the non-sulphonylurea moiety of glibenclamide. *Diabetologia* **28** 697–703.

Garrino, M. G., Meissner, H. P., and Henquin, J. C. (1986) The non-sulfonylurea moiety of gliquidone mimics the effects of the parent molecule on pancreatic β-cells. *Eur. J. Pharmacol.* **124**, 309–316.

Garrino, M. G., Plant, T. D., and Henquin, J. C. (1989) Effects of putative activators of K^+ channels in mouse pancreatic β-cells. *Br. J. Pharmacol.* **98** 957–965.

Gater, P. R., Haylett, D. G., and Jenkinson, D. H. (1985) Neuromuscular blocking agents inhibit receptor mediated increases in the potassium permeability of intestinal smooth muscle. *Br. J. Pharmacol.* **86** 861–868.

Gelband, G. H., Lodge, N. J., and van Breemen, C. (1989) A Ca^{2+}-activated K^+ channel from rabbit aorta: modulation by cromakalim. *Eur. J. Pharmacol.* **167** 201–210.

Gillespie, J. S., and Sheng, H. (1988) The lack of involvement of cyclic nucleotides in the smooth muscle relaxant action of BRL 34915. *Br. J. Pharmacol.* **94** 1189–1197.

Gilly, W. M. F., and Armstrong, C. M. (1982) Divalent cations and the activation kinetics of potassium channels in squid giant axons. *J. Gen. Physiol.* **79** 965–996.

Gimenez-Gallego, G., Navia, N. A., Reuben, J. P., Katz, G. M., Kaczorowski, G. J., and Garcia, M. L. (1988) Purification, sequence, and model structure of charybdotoxin, a potent selective inhibitor of calcium-activated potassium channels. *Proc. Natl Acad. Sci USA* **85** 3329–3333.

Glover, W. E. (1978) Potentiation of vasoconstrictor responses by 3- and 4-aminopyridine. *Br. J. Pharmacol.* **63** 577–585.

Gola, M., and Ducreux, C. (1985) D600 as a direct blocker of Ca-dependent K currents in *Helix* neurons. *Eur. J. Pharmacol.* **117** 311–322.

Gryglewski, R. J., Botting, R. M., and Vane, J. R. (1988) Mediators produced by the endothelial cell. *Hypertension* **12** 530–548.

Grosset, A., and Hicks, P. E. (1986) Evidence for blood vessel selectivity of BRL

34915. *Br. J. Pharmacol.* **89** 500P.

Guggino, S. E., Guggino, W. B., Green, N., and Sacktor, B. (1987) Blocking agents of Ca^{2+}-activated K^+ channels in cultured medullary thick ascending limb cells. *Am. J. Physiol.* **252** C128–C137.

Gurney, A. M., and Nerbonne, J. M. (1986) Effects of calcium antagonists on neuronal calcium and potassium currents. *Br. J. Pharmacol.* **87** 100P.

Gwilt, M., Dalrymple, H. W., Blackburn, K. J., Burges, R. A., and Higgins, A. J. (1988) UK66914 — A novel class III antiarrythmic agent which blocks potassium channels. *Circulation* (suppl. 2) **78** 150.

Gwilt, M., Dalrymple, H. W., Burges, R. A., Blackburn, K. J., Arrowsmith, J. E., Cross, P. E., and Higgins, A. J. (1989) UK-68,798 is a novel potent and selective class III antiarrhythmic drug. *J. Mol. Cell Cardiol* **21** (suppl. 2) S.11.

Habermann, E. (1984) Apamin. *Pharmac. Ther.* **25** 255–270.

Hadley, R. W., and Hume, J. R. (1986) Actions of phencylidine on the action potential and membrane currents of single guinea-pig myocytes. *J. Pharmacol. Exp. Therap.* **237** 131–136.

Haeusler, G. (1978) Relationship between noradrenaline-induced depolarization and contraction in vascular smooth muscle. *Blood Vessels* **15** 46–54.

Haeusler, G. (1983) Contraction, membrane potential, and calcium fluxes in rabbit pulmonary arterial muscle. *Fed. Proc.* **42** 263–268.

Hall, A. K., and Maclagan, J. (1988) Effect of cromakalim on cholinergic neurotransmission in the guinea-pig trachea. *Br. J. Pharmacol.* **95** 792P.

Halliwell, J. V., Othman, I. B., Pelchen-Mathews, A., and Dolly, J. O. (1986) Central action of dendrotoxin: selective reduction of a transient K conductance in hippocampus and binding to localized acceptors. *Proc. Natl Acad. Sci. USA* **83** 493–497.

Hamilton, T. C., Weir, S. W., and Weston, A. H. (1986) Comparison of the effects of BRL 34915 and verapamil on electrical and mechanical activity in rat portal vein. *Br. J. Pharmacol.* **88** 103–111.

Hara, Y., Kitamura, K., and Kuriyama, H. (1980) Actions of 4-aminopyridine on vascular smooth muscle tissues of the guinea-pig. *Br. J. Pharmacol.* **68** 99–106.

Harris, R. A., and Caldwell, K. K. (1985) Alcohol and the calcium-dependent potassium transport of human erythrocytes. *Alcohol* **2** 149–152.

Harvey, A. L., and Anderson, A. J. (1985) Dendrotoxins: snake toxins that block potassium channels and facilitate neurotransmitter release. *Pharmac. Ther.* **31** 33–55.

Hedlund, B. (1987) Alaprocate inhibits potassium currents and thereby enhances muscarinic stimulation in N1E-115 neuroblastoma cells. *Neuropharmacology* **26** 1535–1540.

Heinzow, B. G. J., Angus, J. A., and Korner, P. L. (1982) Effects of alinidine (ST567) on baroreceptor–heart rate reflexes ands its interactions with clonidine on the baroreflex and on the sympathetic terminals of the isolated atrium. *Eur. J. Pharmacol.* **84** 177–187.

Henquin, J. C., and Meissner, H. P. (1982) Opposite effects of tolbutamide and diazoxide on $^{86}Rb^+$ fluxes and membrane potential in pancreatic β-cells. *Biochem. Pharmacol.* **31** 1407–1415.

Henquin, J. C., Meissner, H. P., and Preissler, M. (1979) 9-Aminoacridine- and tetraethylammonium-induced reduction of the potassium permeability in pancreatic β-cells. *Biochem. Biophys. Acta* **587** 579–592.

Hermann, A., and Erxleben, C. (1987) Charybdotoxin selectively blocks small Ca-activated K channels in aplysia neurons. *J. Gen. Physiol.* **90** 24–47.

Hermann, A., and Gorman, A. L. F. (1981a) Effects of tetraethylammonium on potassium currents in a molluscan neuron. *J. Gen Physiol.* **78** 87–110.

Hermann, A., and Gorman, A. L. F. (1981b) Effects of 4-aminopyridine on potassium currents in a molluscan neuron. *J. Gen Physiol.* **78** 63–86.

Hermann, A., and Gorman, A. L. F. (1984) Action of quinidine on ionic currents of molluscan pacemaker neurons. *J. Gen. Physiol.* **83** 919–940.

Hermann, A., and Hartung, K. (1983) Ca^{2+} activated K^+ conductance in molluscan neurones. *Cell Calcium* **4** 387–405.

Hermsmeyer, R. K. (1988) Pinacidil actions on ion channels in vascular muscle. *J. Cardiovasc. Pharmacol.* **12** S17–S22.

Higashida, H., and Brown, D. A. (1986) Two polyphosphatidylinositide metabolites control two K^+ currents in a neuronal cell. *Nature* **323** 333–335.

Hille, B. (1967) The selective inhibition of delayed potassium currents in nerve by tetraethylammonium ion. *J. Gen Physiol.* **50** 1287–1302.

Hof, R. P., Quast, U., Cook, N. S., and Blarer, S. (1988) Mechanism of action, systemic and regional haemodynamics of the potassium channel activator BRL 34915 and its enantiomers. *Circ. Res.* **62** 679–686.

Hoffmann, E. K., Simonsen, L. O., and Lambert, I. H. (1984) Volume-induced increase of K^+ and Cl^- permeabilities in Ehrlich ascites tumor cells. Role of internal Ca^{2+}. *J. Membr. Biol.* **78** 211–222.

Hollingsworth, M., Amédée, T., Edwards, D., Mirroneau, J., Savineau, J. P., Small, R. C., and Weston, A. H. (1987) The relaxant action of BRL 34915 in rat uterus. *Br. J. Pharmacol.* **91** 803–813.

Hollingsworth, M., Edwards, D., Rankin, J. R., and Weston, A. H. (1988) BRL 34915 and relaxation of rat uterus — role of K^+ channels. *Br. J. Pharmacol.* **93** 199P.

Holzmann, S. (1983) Cyclic GMP as possible mediator of coronary arterial relaxation by nicorandil (SG-75). *J. Cardiovasc. Pharmacol.* **5** 364–370.

Horie, M., Irisawa, H., and Noma, A. (1987) Voltage-dependent magnesium block of adenosine triphosphate-sensitive potassium channel in guinea-pig ventricular cells *J. Physiol.* **387** 251–272.

Hoshi, T., and Aldrich, R. W. (1986) Four distinct classes of voltage-dependent K channels in PC12 cells. *Neurosci. Abstr.* **16** 105.

Hoshi, T., Garber, S. S., and Aldrich, R. W. (1988) Direct effect of forskolin on voltage-gated K^+ channels in PC12 cells. *Biophys. J.* **53** 144A.

Howland, J. L., Daughtey, J. N., Donatelli, M., and Theofrastous, J. P. (1984) Inhibition of the erythrocyte calcium-sensitive potassium channel probucol. *Pharmacol. Res. Commun.* **16** 1057–1064.

Howlett, D. R., and Longman, S. D. (1989) Does cromakalim activate an ATP-sensitive potassium channel in vascular smooth muscle cells? *Br. J. Pharmacol.* **96** 256P.

Hume, J. R. (1985) Comparative interaction of organic Ca^{2+} channel antagonists with myocardial Ca^{2+} and K^+ channels. *J. Pharmacol. Exp. Therap.* **234** 134–140.

Iijima, T., and Taira, N. (1987) Pinacidil increases the background potassium current in single ventricular cells. *Eur. J. Pharmacol.* **141** 139–141.

Imanishi, S., Arita, M., and Kiyosue, T. (1983) Effects of SG-75 (nicorandil) on electrical activity of canine cardiac purkinje fibers: Possible increase in potassium conductance. *J. Pharmacol. Exp. Therap.* **225** 198–205.

Inoue, T., Ito, Y., and Takeda, K. (1983) The effects of 2-nicotinamidoethyl nitrate on smooth muscle cells of the dog mesenteric artery and trachea. *Br. J. Pharmac.* **80** 459–470.

Iwatsuki, N., and Petersen, O. H. (1985) Action of tetraethylammonium on calcium-activated potassium channels in pig pancreatic acinar cells studied by patch-clamp single-channel and whole-cell current recording. *J. Membr. Biol.* **86** 139–144.

Jenkinson, D. H., Haylett, D. G., and Cook, N. S. (1983) Calcium-activated potassium channels in liver cells. *Cell Calcium* **4** 429–437.

Kaczorowski, G. J., Garcia, M. L., King, V. F., and Slaughter, R. S. (1988) Development of inhibitors of sodium, calcium exchange. *Handb. Exp. Pharmacol.* **83** 163–183.

Kakei, M., and Noma, A. (1984) Adenosine-5′-triphosphate-sensitive single potassium channel in the antrioventricular node cell of the rabbit heart. *J. Physiol.* **352** 265–284.

Kakei, M., Noma, A., and Shibasaki, T. (1985) Properties of adenosine-triphosphate-regulated potassium channels in guinea-pig ventricular cells. *J. Physiol.* **363** 441–462.

Kakei, M., Yoshinaga, M., Saito, K., and Tanaka, H. (1986) The potassium current activated by 2-nicotinamidoethyl nitrate (nicorandil) in single ventricular cells of guinea-pigs. *Proc. R. Soc. Lond. B* **229** 331–343.

Kalsner, S. (1970) The potentiating effects of ethanol on responses of aortic strips to stimulant drugs. *J. Pharm. Pharmac.* **22** 877–879.

Kantor, P. F., Coatzee, W. A., Dennis, S. C., and Opie, L. H. (1987) Effects of glibenclamide on ischemic arrhythmias. *Circulation* **76** (suppl. IV) 0065.

Karaki, H., Satake, N., and Shibata, S. (1986) Mechanism of barium-induced contraction in the vascular smooth muscle of rabbit aorta. *Br. J. Pharmacol.* **88** 821–826.

Kato, H., and Takata, Y. (1987) Differential effect of Ca antagonists on the noradrenaline release and contraction evoked by nerve simulation in the presence of 4-aminopyridine. *Br. J. Pharmacol.* **90** 191–201.

Kaufmann, R. F., Schenck, K. W., Conery, B. G., and Cohen, M. L. (1986) Effects of pinacidil on serotonin-induced contractions and cyclic nucleotide levels in isolated rat aortae: comparison with nitroglycerin, minoxidil, and hydralazine. *J. Cardiovasc. Pharmacol.* **8** 1195–1200.

Kawai, T., Oka, J., and Watanabe, M. (1985) Hexamethonium increases the excitability of sympathetic neurons by the blockade of the Ca^{2+}-activated K^+ channels. *Life Sci.* **36** 2339–2346.

Kirsch, G. E., and Narahashi, T. (1978) 3,4-Diaminopyridine, a potent new potassium channel blocker. *Biophys. J.* **22** 507–512.

Kitamura, K., Itoh, T., Ueno, H., Kanmura, Y., Inoue, R., and Kuriyama, H. (1986) The stabilization of vascular smooth muscle by procaine. *Drugs, Exptl.*

Clin. Res. **XII** 773–784.

Klein, M., and Kandel, E. R. (1980) Mechanism of calcium current modulation underlying presynaptic facilitation and behavioural sensitization in *Aplysia. Proc. Natl Acad. Sci. USA* **77** 6912–6916.

Klein, M.,, Camardo, J., and Kandel, E. R. (1982) Serotonin modulates a specific potassium current in the sensory neurons that show presynaptic facilitation in *Aplysia. Proc. Natl Acad. Sci. USA* **79** 5713–5717.

Klöckner, U., Trieschmann, U., and Isenberg, G. (1989) Pharmacological modulation of calcium and potassium channels in isolated vascular smooth muscle cells. *Drug Res.* **39** 120–126.

Kramer, W., Oekonomopulos, R., Punter, J., and Summ, H.-D. (1988) Direct photoaffinity labelling of the putative sulfonylurea receptor in rat β-cell tumor membranes by [^{3}H]glibenclamide. *FEBS Lett.* **229** 355–359.

Kreye, V. A. W., and Weston, A. H. (1986) BRL 34915-induced stimulation of ^{86}Rb efflux in rabbit aorta and its dependence on calcium. *Proc. Physiol. Soc.* **347** 36P.

Krueger, B. K., French, R. J., Blaustein, M. B., and Worley, J. F. (1982) T-AM-Min3 incorporation of Ca^{++}-activated K^+ channels from rat brain, into planar lipid bilayers. *Biophys. J.* **37** 170a.

Kudsk, J. E. N., Mellemkjaer, S., Siggaard, C., and Nielsen, C. B. (1988) Effects of pinacidil on guinea-pig airway smooth muscle contracted by asthma mediators. *Eur. J. Pharmacol.* **157** 221–226.

Kukovetz, W. R., and Holzmann, S. (1987) Cyclic GMP in nicorandil-induced vasodilation and tolerance development. *J. Cardiovasc. Pharmacol.* **10** S25–S30.

Kurtz, A. (1986) Intracellular control of renin-release — an overview. *Klin. Wochenschr.* **64** 838–846.

Kusano, K., Barros, F., Katz, G., Garcia, M., Kaczorowski, G., and Reuben, J. P. (1987) Modulation of K channel activity in aortic smooth muscle by BRL 34915 and a scorpion toxin. *Biophys. J.* **51** 54a.

Lackington, I., and Orrego, F. (1981) Inhibition of calcium-activated potassium conductance of human erythrocytes of calmodulin inhibitory drugs. *Febs Lett.* **133** 103–106.

Lackington, I., and Orrego, F. (1985) Methylphenidate, but not other CNS stimulants, inhibits red blood cell calcium-activated potassium efflux. *Pharmacol. Res Commun.* **17** 63–68.

Lang, D. G., and Ritchie, A. A. (1988) Pharmacological sensitivities of large and small conductance Ca^{2+}-activated K^+ channels in GH3 cells. *Biophys. J.* **53** 144a.

Lategan, T. W., Cauvin, C., Wallnöfer, A., and Ruegg, U. T. (1988) SMS 201-995 may activate the K^+ channels of vascular smooth muscle. *Br. J. Pharmacol.* **94** 418P.

Lebrun, P., Devreux, V., Hermann, M., and Herchuelz, A. (1988) Pinacidil inhibits insulin release by increasing K^+ outflow from pancreatic β-cells. *Eur. J. Pharmacol.* **156** 283–286.

Lew, V. L., and Ferreira, H. G. (1978) Calcium transport and the properties of a calcium activated potassium channel in red blood cell membranes. *Curr. Top.*

Membr. Transp. **10** 217–271.

Lewis, R. S., and Cahalan, M. D. (1988) Subset-specific expression of potassium channels in developing murine T lymphocytes. *Science* **239** 771.775.

Liu, B., Golyan, F., McCullough, J. R., and Vassalle, M. (1988) Electrophysiological and antiarrhythmic effects of the K-channel opener, BRL 34915, in cardiac purkinje fibers. *Drug Development Res.* **14** 123–139.

Lodge, N. J., and van Breemen, C. (1985) Mobilization of extracellularly bound Ca^{2+} during high K^+ and norepinephrine stimulation of the rabbit aorta. *Blood Vessels* **22** 234–243.

Louttit, J. B., Downing, O. A., and Wilson, K. A. (1984) Differential effects of strychnine on two types of vascular muscle. *Eur. J. Pharmacol.* **98** 249–253.

Lucchesi, K., Ravindran, A., Young, H., and Moczydlowski, E. (1989) Analysis of the blocking activity of charybdotoxin homologs and iodinated derivatives against Ca^{2+} activated K^+ channels. *J. Membr. Biol.* **109** 269–281.

Lukeman, S., and van Breemen, C. (1985) Calcium entry pathways and their relationship to calcium antagonist potency in arterial smooth muscle. In: *Recent aspects in calcium antagonism*, pp. 49–63, P. R. Lichtlen (ed.). New York.

Lynch, J. J., Wilber, D. J., Montgomery, D. G., Hsieh, T. M., Patterson, E., and Lucchesi, B. R. (1984) Antiarrhythmic and antifibrillatory actions of the levo- and dextrorotary isomers of sotalol. *J. Cardiovasc. Pharmacol.* **6** 1132–1141.

Maas, A. J. J., den Hertog, A., Raas, R. , and Van den Akker, J. (1980) The effect of apamin on the smooth muscle cells of the guinea-pig taenia caeci. *Eur. J. Pharmacol.* **67** 265–274.

MacKinnon, R., and Miller, C. (1988) Mechanism of charybdotoxin block of the high-conductance Ca^{2+}-activated K^+ channel. *J. Gen Physiol.* **91** 335–349.

MacKinnon, R., Reinhart, P. H., and White, M. M. (1988) Charybdotoxin block of shaker K^+ channels suggests that different types of K^+ channels share common structural features. *Neuron* **1** 997–1001.

MacKinnon, R.,, Miller, C., and Timpe, L. (1989) Charybdotoxin blocks K^+ channels in *Xenopus* oocytes induced by four different shaker cDNA clones. *Biophys. J.* **55** 199a.

Mallart, A. (1985) A calcium-activated potassium current in motor nerve terminals of the mouse. *J. Physiol.* **368** 577–591.

Malmgren, A., Andersson, K.-E., Sjörgen, C., and Andersson, P. O. (1989) Effects of pinacidil and cromakalim on the unstable rat bladder. *Br. J. Pharmacol.* **67** 29P.

Marsh, S. J., Stansfield, C. E., Brown, D. A., Davey, R., and McCarthy, D. (1987) The mechanism of action of capsaicin on sensory c-type neurons and their axons *in vitro*. *Neuroscience* **23** 275–289.

McCaig, D. J., and De Jonckheere, B. (1989) Effect of cromakalim on bronchoconstriction evoked by cholinergic nerve stimulation in guinea-pig isolated trachea. *Br. J. Pharmacol.* **98** 662–668.

McCall, J. M., Aiken, J. W., Chidester, C. G., DuCharme, D. W., and Wendling, M. G. (1983) Pyrimidine triazine 3-oxide sulfates: a new family of vasodilators. *J. Med. Chem.* **26** 1791–1793.

McCann, J. D., and Welsh, M. J. (1987) Neuroleptics antagonize a calcium-activated potassium channel in airway smooth muscle. *J. Gen. Physiol.* **89**

339–352.

McCullough, J. R., Shetty, S. S., Conder, M. L., and Weiss, G. B. (1987) Multiple actions of BRL 34915, a purported K channel opener. *Proceedings 9th Annual Conference of the IEEE Engineering in Medicine and Biology Society*, pp. 134–136.

McHarg, A. D., Southerton, J. S., Weir, S. W., and Weston, A. H. (1989) A comparison of the effects of cromakalim, noradrenaline and isoprenaline in guinea-pig isolated taenia caeci. *Br. J. Pharmacol.* **96** 27P.

McPherson, G. A., and Angus, J. A. (1989) Phentolamine and structurally related compounds selectively antagonize the vascular actions of the K^+ channel opener, cromakalim. *Br. J. Pharmacol.* **97** 941–949.

Mecca, T. E., Elam, J. T., Nash, C. B., and Caldwell, R. W. (1980) α-Adrenergic blocking properties of quinine HCl. *Eur. J. Pharmacol.* **63** 159–166.

Meisheri, K. D., and Cipkus, L. A. (1987) Mionoxidil sulfate acts as a K^+-channel agonist to produce vasodilation. *Fed. Proc.* **46** 553 (Abstract 1383).

Meisheri, K. D., Cipkus, L. A., and Taylor, C. J. (1988) Mechanism of action of minoxidil sulfate-induced vasodilation: a role for increased K^+ permeability. *J. Pharm. Exp. Therap.* **245** 751–760.

Mester, M., Escande, D., and Cavero, I. (1988) Glibenclamide blocks the transmembrane action potential shortening evoked by cromakalim in guinea-pig papillary muscle. *Br. J. Pharmacol.* **95** 571P.

Meszaros, J., Keleman, K., Marko, R., Kecskemeti, V., and Szegi, J. (1982) Inhibition of myocardial K^+ channels by bromobenzoylmethyladamantylamine, an adamantane derivative. *Eur. J. Pharmacol.* **84** 151–160.

Miller, C., Moczydlowski, E., Lattore, R., and Phillips, M. (1985) Charybdotoxin, a protein inhibitor of single Ca^{2+}-activated K^+ channels from mammalian skeletal muscle. *Nature* **313** 316–318.

Moczydlowski, E., Lucchesi, K., and Ravindran, A. (1988) An emerging pharmacology of peptide toxins targeted against potassium channels. *J. Membr. Biol.* **105** 95–111.

Mondot, S., Mestre, M., Caillard, C. G., and Cavero, I. (1988) RP 49356: a vasorelaxant agent with potassium channel activating properties. *Br. J. Pharmacol.* **95** 813P.

Morris, J. E. J., and Taylor, S. G. (1989) Effect of ruibidium on relaxant agents in guinea-pig trachea. *Br. J. Pharmacol.* **96** 232P.

Mullins, L. J. (1979) The generation of electric currents in cardiac fibers by Na–Ca exchange. *Am. J. Physiol.* **236** C103–110.

Mulvany, M. J., and Halpern, W. (1977) Contractile properties of small arterial resistance vessels in spontaneously hypertensive and normotensive rats. *Circ. Res.* **41** 19–26.

Murray, M. A., Boyle, J. P., and Small, R. C. (1989) Cromakalim-induced relaxation of guinea-pig isolated trachealis: antagonism by glibenclamide and by phentolamine. *Br. J. Pharmacol.* **98** 865–874.

Nakajima, S., Kurokawa, K., Imamura, N., and Ueda, M. (1989) A study on the hypotensive mechanism of pinacidil: relationship between its vasodilating effect and intracellular Ca^{2+} levels. *Japan J. Pharmacol.* **49** 205–213.

Nakao, K., Okabe, K., Kitamura, K., Kuriyama, H., and Weston, A. H. (1988)

Characteristics of cromakalim-induced relaxations in the smooth muscle cells of guinea-pig mesenteric artery and vein. *Br. J. Pharmacol.* **95** 795–804.

Nemeth, P. R., Daly, K., Erde, S., and Wood, J. D. (1985) Effects of methylene blue on electrical behaviour of myenteric neurons. *Experientia* **41** 259–261.

Newgreen, D. T., Bray, K. M., Southerton, J. S., and Weston, A. H. (1988) The action of glyceryl trinitrate and sodium nitroprusside on rat aorta: a comparison with nicorandil and BRL 34915. *Br. J. Pharmacol.* **93** 17P.

Newgreen, D. T., Longmore, J., and Weston, A. H. (1989) The effect of glibenclamide on the action of cromakalim, diazoxide and minoxidil sulfate on rat aorta. *Br. J. Pharmacol.* **96** 116P.

Nohmi, M., and Kuba, K. (1984) (+)-Tubocurarine blocks the Ca^{2+}-dependent K^+ channel of the bullfrog sympathetic ganglion cell. *Brain Res.* **301** 146–148.

Notsu, T., Niho, T., Ohhashi, K., Tanaka, I., Yamaguchi, K., Kosuzume, J., and Kanno, M. (1989) Block of the ATP regulated K^+ channels by sodium 5-hydroxydecanoate, a new antiarrhythmic agent, in isolated guinea-pig heart cells. *J. Mol. Cell. Cardiol.* **21** (suppl. 2) S.9.

Nowak, L. M., and MacDonald, R. L. (1982) Substance P: ionic basis for depolarizing responses of mouse spinal cord neurons in cell culture. *J. Neurosci* **2** 1119–1128.

Ohta, M., and Narahaski, T. (1973) Sparteine interaction with nerve membrane potassium conductance. *J. Pharmac. Exp. Therap.* **187** 47–55.

Okabe, K., Kitamura, K., and Kuriyama, H. (1987) Features of 4-aminopyridine sensitive outward current observed in single smooth muscle cells from the rabbit pulmonary artery. *Pflügers Arch.* **409** 561–568.

Osterrieder, W. (1987) 9-Amino-1,2,3,4-tetrahydroacridine (THA) is a potent blocker of cardiac potassium channels. *Br. J. Pharmacol.* **92** 521–525.

Osterrieder, W. (1988) Modification of K^+ conductance of heart cell membrane by BRL 34915. *Naunyn-Schmiedeberg's Arch. Pharmacol.* **337** 93–97.

Osterrieder, W., Pelzer, D., Yang, Q-F., and Trautwein, W. (1981) The electrophysiological basis of the bradycardic action of AQA 39 on the sinoatrial node. *Naunyn-Schmiedeberg's Arch. Pharmacol.* **317** 233–237.

Othman, I. B., Spokes, J. W., and Dolly, J. O. (1982) Preparation of neurotoxic ^3H-β-bungarotoxin: demonstration of saturable binding to brain synapses and its inhibition by toxin I. *Eur. J. Biochem.* **128** 267–276.

Pallotta, B. S. (1985) *N*-Bromoacetamide removes a calcium-dependent component of channel opening from calcium-activated potassium channels in rat skeletal muscle. *J. Gen Physiol* **86** 601–611.

Paolisso, G., Nenquin, M., Schmeer, W., Mathot, F., Meissner, H. P., and Henquin, J. C. (1985) Sparteine increases insulin release by decreasing the K^+ permeability of the β-cell membrane *Biochem. Pharmacol.* **34** 2355–2361.

Pappone, P., and Cahalan, M. D. (1987) *Pandinus imperator* scorpion venom blocks voltage gated potassium channels in nerve fibers. *J. Neurosci.* **7** 3300–3305.

Pappone, P., and Lucero, M. T. (1988) *Pandinus imperator* scorpion venom blocks voltage gated potassium channels in GH3 cells. *J. Gen. Physiol.* **91** 817–833.

Penner, R., Petersen, M., Pierau, F. K., and Dreyer, F. (1986) Dendrotoxin: a selective blocker of a non-inactivating potassium current in guinea-pig dorsal ganglion neurones. *Pflügers Arch.* **407** 365–369.

Petersen, M., Penner, R., Pierau, Fr.-K., and Dreyer, F. (1986) β-Bungarotoxin inhibits a non-inactivating potassium current in guinea-pig dorsal root ganglion neurones. *Neurosci. Lett.* **68** 141–145.

Petersen, M., Pierau, Fr.-K., and Weyrich, M. (1987) The influence of capsaicin on membrane currents in dorsal root ganglion neurones of guinea-pig and chicken. *Pflügers Arch.* **409** 403–410.

Piomelli, D., Volterra, A., Dlae, N., Siegelbaum, S. A., Kandel, E. R., Schwartz, J. H., and Belardetti, F. (1987) Lipoxygenase metabolites of arachidonic acid as second messengers for presynaptic inhibition of aplysia sensory cells. *Nature* **328** 38–43.

Possani, L. D., Martin, B. M., and Svendsen, I. B. (1982) The primary structure of noxiustoxin: A K^+ channel blocking peptide, purified from the venom of the scorpion *Centruroides noxius* Hoffmann. *Carlsberg Res. Commun.* **47** 285–289.

Post, J. M., Smith, J. M., and Jones, A. W. (1988) BRL 34915 (cromakalim) stimulation of $^{42}K^+$ efflux from rabbit arteries is modulated by calcium. *J. Pharmac. Exp. Therap.* **250** 591–597.

Quast, U. (1987) Effect of the K^+ efflux stimulating vasodilator BRL 34915 on $^{86}Rb^+$ efflux and spontaneous activity in guinea-pig portal vein. *Br. J. Pharmacol.* **91** 569–578.

Quast, U. (1988) Inhibition of the effects of the K^+ channel stimulator cromakalim (BRL 34915) in vascular smooth muscle by glibenclamide and forskolin. *Naunyn-Schmiedeberg's Arch. Pharmacol.* **337** (suppl.) R72.

Quast, U. (1989) Cromakalim inhibits phasic contractions of noradrenaline in the rat isolated mesenteric bed even in Ca^{2+}-free medium. *Br. J. Pharmacol.* **96** 114P.

Quast, U., and Baumlin, Y. (1988) Comparison of the effluxes of $^{42}K^+$ and $^{86}Rb^+$ elicited by cromakalim (BRL 34915) in tonic and phasic vascular tissue. *Naunyn-Schmiedeberg's Arch. Pharmacol.* **338** 319–326.

Quast, U., and Cook, N. S. (1988a) *Leiurus quinquestriatus* venom inhibits BRL 34915-induced $^{86}Rb^+$ efflux from the rat portal vein. *Life Sci.* **42** 805–810.

Quast, U., and Cook, N. S. (1988b) Potent inhibitors of the effects of the K^+ channel opener BRL 34915 in vascular smooth muscle. *Br. J. Pharmacol.* **93** 204P.

Quast, U., and Cook, N. S. (1989a) *In vitro* and *in vivo* comparison of two K^+ channel openers, diazoxide and cromakalim, and their inhibition by glibenclamide. *J. Pharmacol. Exp. Therap.* **250** 261–270.

Quast, U., and Cook. N. S. (1989b) Moving together: K^+ channel openers and ATP-sensitive K^+ channels. *Trends Pharm. Sci.* **10** 431–435.

Quast, U., and Webster, C. (1989) Sulfonylureas und tetraalyleammonium ions as inhibitors of cromakalim-stimulated $^{86}Rb^+$ efflux. *Naunyn-Schmiedeberg's Arch. Pharmacol.* **339** (suppl. R64).

Quast, U., Scholtysik, G., Weir, S. W., and Cook, N. S. (1988) Pertussis toxin treatment does not inhibit the effects of potassium channel opener BRL 34915 on rat isolated vascular and cardiac tissues. *Naunyn-Schmiedeberg's Arch. Pharmacol.* **337** 98–104.

Quayle, J. M., Standen, N. B., and Stanfield, P. R. (1988) The voltage-dependent block of ATP-sensitive potassium channels of frog skeletal muscle by caesium and barium ions. *J. Physiol.* **405** 677–697.

Raymond, G., Potreau, D., Cognard, C., Jahn, W., and Wieland, T. (1987)

Antamanide antagonizes the phalloidin-induced negative inotropic effect and blocks voltage dependently the fast outward K^+ current in voltage-clamped frog muscle fibres. Eur. J. Pharmacol. **138** 21–27.

Rehm, H., and Lazdunski, M. (1988) Existence of different populations of the dendrotoxin I binding protein associated with neuronal K^+ channels. *Biochem. Biophys. Res. Commun.* **153** 231–240.

Reichstein, E., and Rothstein, A. (1981) Effects of quinine on Ca^{++}-induced K^+ efflux from human red blood cells. *J. Membr. Biol.* **59** 57–63.

Richards, N., and Dawson, D. C. (1986) Single potassium channels blocked by lidocaine and quinidine in isolated turtle colon epithelial cells. *Am. J. Physiol.* **251** C85–C89.

Ritchie, A. K. (1987) Two distinct calcium-activated potassium currents in a rat anterior pituitary cell line. *J. Physiol.* **385** 591–609.

Rogawski, M. A. (1985) The A-current: how ubiquitous a feature of excitable cells is it? *TINS* **8** 214–219.

Rogawski, M. A. (1987) Tetrahydroamminoacridine blocks voltage-dependent ion channels in hippocampal neurons. *Eur. J. Pharmacol.* **142** 169–172.

Rogawski, M. A. (1989) Aminopyridines enhance opening of calcium-activated potassium channels in GH_3 anterior pituitary cells. *Mol. Pharmacol.* **35** 458–468.

Romey, G., Hugues, M., Schmid-Antomarchi, H., and Lazdunski, M. (1984) Apamin; a specific toxin to study a class of Ca^{2+}-dependent K^+ channels. *J. Physiol.* (*Paris*) **79** 259–264.

Rorsman, P., and Trube, G. (1986) Calcium and delayed potassium currents in mouse pancreatic β-cells under voltage-clamp conditions. *J. Physiol.* **374** 531–550.

Rowan, E. G., and Harvey, A. L. (1988) Potassium channel blocking actions of β-bungarotoxin and related toxins on mouse and frog motor nerve terminals. *Br. J. Pharmacol.* **94** 839–847.

Rudy, B. (1988) Diversity and ubiquity of K channels. *Neuroscience* **25** 729–749.

Rüegg, U. T., Doyle, V. M., Zuber, J.-F., and Hof, R. P. (1985) A smooth muscle cell line suitable for the study of voltage-sensitive calcium channels. *Biochem. Biophys. Res. Commun.* **130** 447–453.

Rufer, C., and Losert, W. (1979) Blood glucose lowering sulfonamides with asymmetric carbon atoms. 3.[1] related *N*-substituted carbamoylbenzoic acids. *J. Med. Chem.* **22** 750–752.

Sanguinetti, M. C., Scott, A. L., Zingaro, G. J., and Siegl, P. K. S. (1988) BRL 34915 (cromakalim) activates ATP-sensitive K^+ current in cardiac muscle. *Proc. Natl Acad. Sci. USA* **85** 8360–8364.

Sanguinetti, M. C., Siegl, P. K. S., and Zingaro, G. J. (1989) The class III antiarrythmic agents, sotalol and E-4031 do not block delayed rectifier K current in guinea pig or ferret ventricular cells. *J. Mol. Cell. Cardiol.* **21** (suppl. 2) S.21.

Sarkadi, B., Cheung, R., Mack, E., Grinstein, S., Gelfand, E. W., and Rothstein, A. (1985) Cation and anion transport pathways in volume regulatory response of human lymphocytes to hyposmotic media. *Am. J. Physiol.* **248** C480–C487.

Sato, R., Hisatome, I., and Singer, D. (1987) Amiodarone blocks an inward-rectifier K^+ channel in guinea-pig ventricular myocytes. *Circulation* **76** (suppl. 4) 150.

Sawada, K. (1989) Depression of the delayed outward K^+ current by a novel class 3 antiarrythmic agent, E-4031, in guinea pig single ventricular cells. *J. Mol. Cell Cardiol.* **21** (suppl. 2) S.20.

Schauf, C. L., and Smith, K. J. (1981) Electrophysiological effects of gallamine triethiodide on mammalian and amphibian myelinated nerve fibres. *J. Physiol.* **312** 40–41P.

Schmid-Antomarchi, H., Hugues, M., Norma, R., Ellory, C., Borsotto, M., and Lazdunski, M. (1984) Molecular properties of the apamin-binding component of the Ca^{2+}-dependent K^+ channel. *Eur. J. Biochem.* **142** 1–6.

Schmid-Antomarchi, H., Hugues, M., and Lazdunski, M. (1986) Properties of the apamin-sensitive Ca^{2+}-activated K^+ channel in PC12 pheochromocytoma cells which hyper-produce the apamin receptor. *J. Biol. Chem.* **261** 8633–8637.

Schmid-Antomarchi, H., DeWille, J., Fosset, M., and Lazdunski, M. (1987) The receptor for antidiabetic sulfonylureas controls the activity of the ATP-modulated K^+ channel in insulin-secreting cells. *J. Biol. Chem.* **262** 15840–15844.

Scholtysik, G. (1987) Evidence for inhibition by ICS 205–930 and stimulation by BRL 34915 of K^+ conductance in cardiac muscle. *Naunyn-Schmiedeberg's Arch. Pharmacol.* **335** 692–696.

Scholtysik, G., Imoto, Y., Yatani, A., and Brown, A. M. (1988) 5-Hydroxytryptamine antagonist ICS 205–930 blocks cardiac potassium, sodium and calcium currents. *J. Pharmacol. Exp. Therap.* **245** 773–778.

Schwartz, M. H. (1985) Effect of ethanol on subthreshold currents of *Aplysia* pacemaker neurons. *Brain Res.* **332** 337–353.

Schweitz, H., Stansfeld, C. E., Bidard, J.-N., Fagni, L., Maes, P., and Lazdunski, M. (1989) Charybdotoxin blocks dendrotoxin-sensitive voltage-activated K^+ channels. *FEBS Lett.* **250** 519–522.

Seagar, M. J., Granier, C., and Couraud, F. (1984) Interactions of the neurotoxin apamin with a Ca^{2+}-activated K^+ channel in primary neuronal cultures. *J. Biol. Chem.* **259** 1491–1495.

Seagar, M. J., Labbe-Juilolie, C., Granier, C., Goll, A., Glossmann, H., Rietschoten, J. V., and Couraud, F. (1986) Molecular structure of rat brain apamin receptor: differential photoaffinity labeling of putative K^+ channel subunits and target size analysis. *Biochemistry* **25** 4051–4057.

Shapiro, B. I. (1977a) Effects of strychnine on the potassium conductance of the frog node of Ranvier. *J. Physiol.* **69** 897–914.

Shapiro, B. I. (1977b) Effects of strychnine on the sodium conductance of the frog node of Ranvier. *J. Gen. Physiol.* **69** 915–926.

Siegel, G., Litza, B., Thiel, M., Carl, A., and Stock, G. (1987) Prostacyclin, a K^+ channel opener in vascular smooth muscle. *Proc. Chin. Acad. Mol. Sci.* **2** 104.

Siegl, P. K. S., Scott, A. L., and Sanguinetti, M. C. (1989) Activation of ATP-sensitive K^+ current ($I_{K(ATP)}$) reverses effects of sotalol on the effective refactory period (ERP) in ferret papillary muscles. *J. Mol. Cell. Cardiol.* **21** (suppl. 2) S.111.

Simons, T. J. B. (1976) Calcium-dependent potassium exchange in human red cell ghosts. *J. Physiol.* **256** 227–244.

Simons, T. J. B. (1979) Actions of a carbocyanine dye on calcium-dependent potassium transport in human red cell ghosts. *J. Physiol.* **288** 481–507.

Sitges, M., Possani, L. D., and Bayon, A. (1986) Noxiustoxin, a short-chain toxin from the Mexican scorpion. *Centuroides noxius*, induces transmitter release by blocking K^+ permeability. *J. Neurosci.* **6** 1570–1574.

Slater, N. T., and Carpenter, D. O. (1982) Blockade of acetylcholine-induced inward currents in *Aplysia* neurones by strychnine and desipramine: Effect of membrane potential. *Cell. Molec. Neurobiol.* **2** 53–58.

Smallwood, J. K., and Steinberg, M. I. (1988) Cardiac electrophysiological effects of pinacidil and related pyridylcyanoguanidines: relationship to antihypertensive activity. *J. Cardiovasc. Pharmacol.* **12** 102–109.

Smith, C., Phillips, M., and Miller, C. (1986) Purification of charybdotoxin, a specific inhibitor of the high-conductance Ca^{2+}-activated K^+ channel. *J. Biol. Chem.* **261** 14607–14613.

Smith, T. C., and Levinson, C. (1989) Quinine inhibits multiple Na^+ and K^+ transport mechanisms in Ehrlich ascites tumor cells. *Biochem. Biophys. Acta* **978** 169–175.

Sorensen, R. G., and Blaustein, M. P. (1988) The rat brain phenycyclidine (PCP) receptor — A putative K^+ channel. *Biochem. Pharmacol.* **37** 511–519.

Sorensen, R. G., Schneider, M. J., and Blaustein, M. P. (1989) *Leiurus quinquestriatus* venom peptides that block brain voltage-gated and Ca-activated K channels also inhibit dendrotoxin binding to synaptic membranes. *Biophys. J.* **55** 550a.

Southerton, J. S., Weston, A. H., Bray, K. M., Newgreen, D. T., and Taylor, S. G. (1988) The potassium channel opening action of pinacidil; studies using biochemical, ion flux and microelectrode techniques. *Naunyn-Schmiedeberg's Arch. Pharmacol.* **338** 310–318.

Spruce, A. E., Standen, N. B., and Stanfield, P. R. (1987) Studies of the unitary properties of adenosine-5′-triphosphate-regulated potassium channels of frog skeletal muscle. *J. Physiol.* **382** 213–236.

Spuler, A., Lehmann-Horn, F., and Grafe, P. (1989) Cromakalim (BRL 34915) restores *in vitro* the membrane potential of depolarized human skeletal muscle fibres. *Naunyn-Schmiedeberg's Arch. Pharmacol.* **339** 327–331.

Squire, L. G., and Petersen, O. H. (1987) Modulation of Ca^{2+}- and voltage-activated K^+ channels by internal Mg^{2+} in salivary acinar cells. *Biochim. Biophys. Acta* **899** 171–175.

Standen, N. B., and Stanfield, P. R. (1978) A potential- and time-dependent blockade of inward rectification in frog skeletal muscle fibers by barium and strontium ions. *J. Physiol.* **280** 169–191.

Standen, N. B., Quayle, J. M., Davies, N. W., Brayden, J. E., Huang, Y., and Nelson, M. T. (1989) Hyperpolarizing vasodilators activate ATP-sensitive K^+ channels in arterial smooth muscle. *Science* **245** 177–180.

Stanfield, P. R. (1983) Tetraethylammonium ions and the potassium permeability of excitable cells. *Rev. Physiol. Biochem. Pharmacol.* **97** 1–67.

Stanfield, P. R. (1988) Intracellular Mg^{2+} may act as a co-factor in ion channel function. *Trends Neurosci.* **11** 475–477.

Stansfield, C. E., Marsh, S. J., Halliwell, J. V., and Brown, D. A. (1986) 4-Aminopyridine and dendrotoxin induce repetitive firing in rat visceral sensory neurones by blocking a slowly inactivating outward current. *Neurosci. Lett.* **64**

299–304.

Steinberg, M. I., Lacefield, W. B., and Robertson, D. W. (1986) Class I and III antiarrhythmic drugs. *Ann. Rep. Med. Chem.* **21** 95–108.

Steinberg, M. I., Ertel, P., Smallwood, J. K., Wyss, V., and Zimmerman, K. (1988) The relation between vascular relaxant and cardiac electrophysiological effects of pinacidil. *J. Cardiovasc. Pharmacol.* **12** S30–S40.

Strong, P. N., Weir, S. W., Beech, D. J., Heistand, P., and Kocher, H. P. (1989) Potassium channel toxins from *Leiurus quinqestriatus hebraeus* venom: purification of charybdotoxin and a second toxin which inhabits cromakalim stimulated ^{86}Rb efflux from aortic smooth muscle. *Br. J. Pharmacol.* **98** 817–826.

Sturgess, N. C., Ashford, M. L. J., Cook, D. L., and Hales, C. N. (1985) The sulphonylurea receptor may be an ATP-sensitive potassium channel. *Lancet* **ii** 474–475.

Sturgess, N. C.,, Kozlowski, R. Z., Carrington, C. A., Hales, C. N., and Ashford, M. L. J. (1988) Effects of sulfonylureas and diazoxide on insulin secretion and nucleotide-sensitive channels in an insulin-secreting cell line. *Br. J. Pharmacol.* **95** 83–94.

Stutzin, A., Paravic, F., Ormeno, G., and Orrego, F. (1983) Guanethidine effects on the guinea-pig vas deferens are antagonized by the blocks of calcium-activated potassium conductance, apamin, methylene blue and quinine. *Mol. Pharmacol.* **23** 409–416.

Summers, W. K., Majovski, L. V., Marsh, G. M., Tachiki, K., and Kling, A. (1986) Oral tetrahydroacminoacridine in long-term treatment of senile dementia, Alzheimer type. *N. Engl. J. Med.* **315** 1241.

Suput, D. (1988) Effects of equinatoxin on single myelinated nerve fibers. *Toxicon* **26** 40–41.

Taira, N. (1987) Similarity and dissimilarity in the mode and mechanism of action between nicorandil and classical nitrates: an overview. *J. Cardiovasc. Pharmacol.* **10** (suppl. 8), S1–S9.

Takeda, R., Momose, Y., and Nakanishi, S. (1984) Effects of ethanol on calcium and potassium currents in single bullfrog atrial cells. *Japan J. Pharmacol.* **36** 422–424.

Talvenheimo, J. A., Lam, G., and Gelband, C. (1988) Charybdotoxin inhibits the 250pS Ca^{2+}-activated K^+ channel in aorta and contracts aorta smooth muscle. *Biophys. J.* **53** 258a.

Taylor, S. G., Southerton, J. S., Weston, A. H., and Baker, J. R. J. (1988) Endothelium-dsependent effects of acetylcholine in rat aorta: a comparison with sodium nitroprusside and cromakalim. *Br. J. Pharmacol.* **94** 853–863.

Terada, K., Kitamura, K., and Kuriyama, H. (1987) Different inhibitions of the voltage-dependent K^+ current by Ca^{2+} antagonists in the smooth muscle cell membrane of rabbit small intestine. *Pflügers Arch.* **408** 558–564.

Thompson, S. H., and Aldrich, R. W. (1980) Membrane potassium channels. In: *The cell surface and neuronal function*, pp. 49–58, C. W. Cotman, G. Poste and G. L. Nicholson, (eds). Elsevier/North Holland Biomedical Press, Amsterdam.

Thuringer, D., and Escande, D. (1989) The potassium channel opener RP 49356 modifies the ATP-sensitivity of K^+-ATP channels in cardiac myocytes. *Pflügers*

Arch. **414** (suppl. 1) S175.
Trautwein, W., Pelzer, D., McDonald, T. F., and Osterrieder, W. (1981) AQA 39, a new bradycardic agent which blocks myocardial calcium (Ca) channels in a frequency- and voltage-dependent manner. *Naunyn-Schmiedeberg's Arch. Pharmacol.* **317** 228–232.
Trendelenburg, U. (1987) The membrane potential of vascular smooth muscle appears to modulate uptake$_2$ of ^{3}H-isoprenaline. *Naunyn-Schmiedeberg's Arch. Pharmacol.* **336** 33–36.
Trieschmann, U., Pichlmaier, M., Klöckner, U., and Isenberg, G. (1988) Vasorelaxation due to K-agonists. Single channel recordings from isolated human vascular myocytes. *Pflügers Arch.* **411** R199.
Treistman, S. N., and Wilson, A. (1987) Effects of long-chain alcohols on I_A in *Aplysia. Biophys. J.* **51** 57a.
Tricklebank, M. D., Flockhart, G., and Freedman, S. B. (1988) The potassium channel activator, BRL 34915, antagonises a behavioural response to the muscarinic receptor agonist, pilocarpine. *Eur. J. Pharmacol.* **151** 349–350.
Trube, G., Rorsman, P., and Ohno-Shosaku, T. (1986) Opposite effects of tolbutamide and diazoxide on the ATP-dependent K^+ channel in mouse pancreatic β-cells. *Pflügers Arch.* **407** 493–499.
Uchida, Y., Nakamura, F., Tomaru, T., Sumino, S., Kato, A., and Sugimoto, T. (1986) Phasic contractions of canine and human coronary arteries induced by potassium channel blockers. *Jpn. Heart J.* **27** 727–740.
Valdivia, H. H., Smith, J. S., Martin, B. M., Coronado, R., and Possani, L. D. (1988) Charybdotoxin and noxiustoxin, two homologous peptide inhibitors of the K^+ (Ca^{2+}) channel. *FEBS Lett.* **226** 280–284.
Van Driessche, W. (1986) Lindocaine blockage of basolateral potassium channels in the amphibian urinary bladder. *J. Physiol.* **381** 575–593.
Varecka, L., Peterajova, E., and Pogady, J. (1987) Polymixin B, a novel inhibitor of red cell Ca^{2+}-activated K^+ channel. *FEBS Lett.* **225** 173–177.
Vergara, C., and Latorre, R. (1983) Kinetics of Ca^{2+}-activated K^+ channels from rabbit muscle incorporated into planar bilayers. Evidence for a Ca^{2+} and Ba^{2+} blockade. *J. Gen Physiol.* **82** 543–568.
Vergara, C., Moczydlowske, E., and Latorre, R. (1984) Conduction, blockade and gating in a Ca^{++}-activated K^+ channel incorporated in planar lipid layers. *Biophys. J.* **45** 73–76.
Videbaek, L. M., Aalkjaer, C., and Mulvany, M. J. (1988) Pinacidil opens K^+-selectivity channels causing hyperpolarization and relaxation of noradrenaline contractions in rat mesenteric resistance vessels. *Br. J. Pharmacol.* **95** 103–108.
Vincent, J., Schweitz, H., and Lazdunski, M. (1975) Structure–function relationships and site of action of apamin, a neurotoxic polypeptide of bee venom with an action on the central nervous system. *Biochemistry* **14** 2521–2525.
Volle, R. L. (1971) Blockade by 9-aminoacridine of potassium fluxes in frog sartorius muscle. *Biochem. Pharmacol.* **20** 315–324.
Walsh, K. B., and Kass, R. S. (1988) Regulation of a heart potassium channel by protein kinase A and C. *Science* **241** 67–69.
Watanabe, K., and Gola, M. (1987) Forskolin interaction with voltage-dependent K

channels in *Helix* is not mediated by cyclic nucleotides. *Neurosci. Lett.* **78** 211–216.

Webb, S. C., Krikler, D. M., Hendry, W. G., Adrian, T. E., and Bloom, S. R. (1986) Electrophysiological actions of somatostatin on the atrioventricular junction in sinus rhythm and reentry tachycardia. *Br. Heart J.* **56** 236–241.

Weir, S. W., and Cauvin, C. (1989) Effects of cromakalim (BRL 34915) on contraction and membrane potential in rat isolated mesenteric resistance vessels. *Naunyn Schmiedeberg's Arch. Pharmacol.* (in press).

Weir, S. W., and Strong, P. N. (1988) Inhibition of BRL 34915-stimulated $^{86}Rb^+$ efflux in rabbit aorta by fractionated *Leirus quinquestriatus hebraeus* scorpion venom. *Br. J. Pharmacol.* **93** 202P.

Weir, S. W., and Weston, A. H. (1986a) Effects of apamin on responses to BRL 34915, nicorandil and other relaxants in the guinea-pig taenia caeci. *Br. J. Pharmacol.* **88** 113–120.

Weir, S. W., and Weston, A. H. (1986b) Effect of apamin on the inhibitory actions of sodium nitroprusside and sodium azide on the guinea-pig taenia caeci. *Br. J. Pharmacol.* **93** 18P.

Weston, A. H. (1987) Some effects of nicorandil on the smooth muscles of the rat and guinea-pig. *J. Cardiovasc. Pharmacol.* **10** (suppl. 8) S56–S61.

Wilde, D. W., and Lee, K. S. (1988) Inward and outward ionic currents of isolated dog coronary smooth muscle cells and effects of ACh and minoxidil sulfate. *Biophys J.* **53** 262a.

Wilson, C. (1988) Comparative effects of cromakalim on contractions to noradrenaline and caffeine in rabbit isolated renal artery, *Br. J. Pharmacol.* **95** 570P.

Wilson, C. (1989) Inhibition by sulphonylureas of vasorelaxation induced by K^+ channel activators *in vitro*. *J. Auton. Pharmac.* **9** 71–78.

Wilson, C., and Hicks, F. (1989) Comparison of K^+ channel activators with other vasorelaxants as inhibitors of caffeine-induced arterial contraction. *Br. J. Pharmacol.* **96** 222P.

Wilson, C., Buckingham, R. E., Mootoo, S., Parrott, L. S., Hamilton, St. C., Pratt, S. C., and Cawthorne, M. A. (1988a) *In vivo* and *in vitro* studies of cromakalim (BRL 34915) and glibenclamide in the rat. *Br. J. Pharmacol.* **93** 126P.

Wilson, C., Coldwell, M. C., Howlett, D. R., Cooper, S. M., and Hamilton, T. C. (1988b) Comparative effects of K^+ channel blockade on the vasorelaxant activity of cromakalim, pinacidil and nicorandil. *Eur. J. Pharmacol.* **152** 331–339.

Winquist, R. J., Heaney, L. A., and Baskin, E. P. (1988) Blockade of the relaxation to BRL 34915 (BRL) and minoxidil by glyburide in isolated rat portal veins. *FASEB J.* **2** A786.

Winquist, R. J., Heaney, L. A., Wallace, A. A., Baskin, E. P., Stein, R. B., Garcia, M. L., and Kaczorowski, G. J. (1989) Glyburide blocks the relaxation responses to BRL 34915 (Cromakalim), minoxidil sulfate and diazoxide in vascular smooth muscle *J. Pharmacol. Exp. Therap.* **248** 149–156.

Wise, R. C., Glass, A. B., Jen-Chou, C.-H., Raynor, R. L., Katoh, M., Schatzmann, R., C., Turner, R. S., Kibler, R. F., and Kuo, J. F. (1982) Phospholipid-sensitive Ca^{2+}-dependent protein kinase from heart II. Substrate specificity and inhibition by various agents. *J. Biol. Chem.* **257** 8489–8495.

Yamamoto, H.-A., and Harris, R. A. (1983) Calcium-dependent $^{86}Rb^+$ efflux and ethanol intoxication: studies of human red blood cells and rodent brain synaptosomes. *Eur. J. Pharmacol.* **88** 357–363.

Yamanaka, K., Furukawa, K., and Kitamura, K. (1985) The different mechanisms of action of nicorandil and adenosine triphosphate on potassium channels of circular smooth muscle of the guinea-pig small intestine. *Naunyn Schmiedeberg's Arch. Pharmacol.* **331** 96–103.

Yanagisawa, T., and Taira, N. (1980) Effect of 2-nicotinamidethyl nitrate (SG75) on the membrane potential of left arterial muscle fibers of the dog. Increase in potassium conductance. *Naunyn-Schmiedeberg's Arch. Pharmacol.* **312** 69–76.

Yanagisawa, T., and Taira, N. (1981) Effect of 2-nicotinamidethyl nitrate SG75 on membrane potentials of canine purkinje fibers. *Jpn. J. Pharmacol.* **31**, 409–417.

Yanagisawa, T., Hashimoto, H., and Taira, N. (1988) The negative inotropic effect of nicorandil is independent of cyclic GMP changes: a comparison with pinacidil and cromakalim in canine atrial muscle. *Br. J. Pharmacol.* **95** 393–398.

Yeh, J. Z., Oxford, G. S., Wu, C. H., and Harahashi, T. (1976) Dynamics of aminopyridine block of potassium channels in squid axon membrane. *J. Gen. Physiol.* **68** 519–535.

Yellen, G. (1987) Permeation in potassium channels: implications for channel structure. *Ann Rev. Biophys. Biophys. Chem.* **16** 227–246.

Zunkler, B. K., Lenzen, S., Manner, K., Panten, U., and Trube, G. (1988a) Concentration-dependent effects of tolbutamide, meglitinide, glipizide, glibenclamide and diazoxide on ATP-regulated K^+ currents in pancreatic β-cells. *Naunyn-Schmiedeberg's Arch. Pharmacol.* **337** 225–230.

Zunkler, B. J., Trube, G., and Ohno-Shosaku, T. (1988b) Forskolin-induced block of delayed rectifying K^+ channels in pancreatic β-cells is not mediated by cAMP. *Pflügers Arch.* **411** 613–619.

Part II

Physiological role of K^+ channels and the therapeutic potential of K^+ channel modulation

9

The role of K^+ channels in the modulation of vascular smooth muscle tone

J. Longmore and **A. H. Weston**

9.1 INTRODUCTION

Vascular smooth muscle tone may be defined as a partially contracted state resulting from any electrical activity generated spontaneously within the muscle together with the influences of circulating and locally released vasoactive substances.

Tone can thus be influenced by one or more of the following factors:

(a) the presence of pacemaker cells which spontaneously generate conducted electrical changes;
(b) the local release of neurotransmitters and the action of circulating hormones;
(c) the release of locally acting vasoactive substances such as endothelium-derived relaxing factor (EDRF);
(d) physical forces which result from changes in blood flow, thus altering the degree of muscle stretch;
(e) spastic disease states like hypertension.

The purpose of this chapter is to discuss the role of K^+ channels in regulating vascular muscle tone, with special reference to the above factors which are summarized in Fig. 9.1.

9.1.1 Role of potassium currents in spike and pacemaker potentials

Vascular smooth muscle tissues can be divided into three groups on the basis of their ability to generate action (spike) potentials: (1) blood vessels typified by the mammalian portal veins which show spontaneous bursts of spike potential discharge; (2) arteries such as rat aorta and guinea-pig mesenteric artery which are normally electrically quiescent, but in which spike potential-like discharge can be evoked by electrical stimulation or by depolarization induced by KCl, by noradrenaline or by

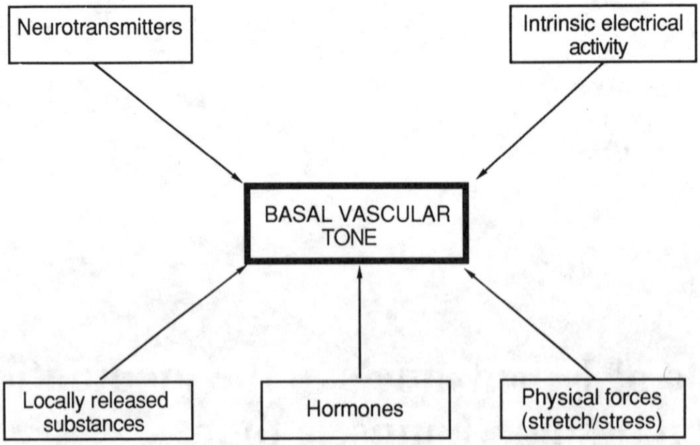

Fig. 9.1 — Factors influencing vascular smooth muscle tone.

K^+ channel-blocking agents; (3) vessels such as rabbit aorta, rabbit main pulmonary artery and rat tail artery in which action potentials are rarely seen.

There is great variability in the time-course and configuration of the spike potentials seen in various vascular smooth muscles and it is possible that the precise ionic mechanisms which underly these events differ from tissue to tissue. However, in most smooth muscles the depolarization phase of the spike consists of an inward current carried largely by Ca^{2+} ions which enter the cell through voltage-operated Ca^{2+} channels (Harder and Sperekalis, 1979; Jones, 1981; Harder, 1983). Depolarization and/or the increase in $[Ca^{2+}]_i$ triggers K^+ outward current(s), of which several types have been identified. In some instances, the characteristics of the K^+ channel associated with a particular current have been identified. However, in many instances the K^+ channel(s) involved in a particular current are not known. Furthermore, the precise function of the current in the gross electrical behaviour of the cell or tissue has not been defined and can only be surmised by analogy with more complete studies in other excitable cells.

The known types of K^+ current (channel) found in smooth muscle are briefly listed below:

(a) A K^+ current carried through Ca^{2+}-dependent K^+ channels, the opening probability of which is low at resting membrane potentials and which are activated as the $[Ca^{2+}]_i$ rises. Both small-conductance and large-conductance Ca^{2+}-dependent K^+ channels have been identified, designated SK_{Ca} (or K_s†) and BK_{Ca} (or K_L, maxi K or BK), respectively. Such channels have been shown to be present in rabbit portal vein (BK_{Ca}, Beech *et al.*, 1987; K_L and K_S, Inoue *et al.*, 1985) and guinea-pig mesenteric artery (BK_{Ca}, Bolton *et al.*, 1985, 1986; Benham *et al.*, 1986).

† Inoue *et al.* (1985) used the abbreviation 'K_S' to denote the small-conductance Ca^{2+}-activated K^+ channels present in the portal vein. This should not be confused with the abbreviation K_S (=S-channel) used elsewhere in this book.

(b) In some smooth muscles (e.g. guinea-pig bladder, Klöckner & Isenberg, 1985) a second K^+ current (the 'delayed rectifier', $I_{K(V)}$) has been identified. This current shows slow activation/inactivation kinetics and voltage but not calcium sensitivity. A current with similar characteristics has been described in rabbit pulmonary artery (Okabe *et al.*, 1987). Furthermore, in guinea-pig mesenteric artery, Bolton *et al.* (1986) have identified small-conductance K^+ channels which are insensitive to changes in $[Ca^{2+}]_i$, but which show voltage sensitivity and are activated only by relatively long periods of depolarization. It is possible that in this tissue these K^+ channels carry an $I_{K(V)}$-type current.

(c) In rabbit pulmonary artery a third type of K^+ current which is rapidly activating and inactivating has also been identified. This current has characteristics similar to those of the neuronal fast transient or I_A current (Okabe *et al.*, 1987). In neuronal cells the I_A current operates in conjunction with the delayed rectifier ($I_{K(V)}$) to modulate the frequency of action potential discharge during a burst (Hille, 1984). Such a function is clearly not subserved by this channel in a non-spiking tissue like pulmonary artery. However, it is possible that this current functions to modulate spike discharge frequency in the multispike complex discharges characteristic of mammalian portal veins.

(d) Electrophysiological recordings from isolated membrane patches of rabbit jejenum smooth muscle cells have demonstrated another K^+ current, the inward or anomalous rectifier, which is activated at potentials substatially more negative than the potassium equilibrium potential. Such a current may serve to modify the slope of pacemaker potentials in this tissue (Benham *et al.*, 1987). However, the physiological role of the inward rectifier is unclear since under physiological conditions such negative membrane potentials are rarely, if at all, observed. More recent electrophysiological studies in guinea-pig submucosal arterioles have detected an inward rectifier current which is activated at normal resting membrane potentials. If such a current is typical of other vascular tissues, it will play an important role in setting the level of resting membrane potential (and tone) of such tissues (Edwards and Hirst, 1988).

(e) Recent patch-clamp studies have demonstrated the existence of a K^+ current in rat and rabbit aorta which is activated by a reduction in 'cytosolic' ATP levels (Standen *et al.*, 1989). This current is carried through K^+ selective channels which possess pharmacological properties similar to the ATP-dependent K^+ channel found in pancreatic β-cells, i.e. they are regulated by changes in intracellular levels of ATP, opened by cromakalim and diazoxide and are blocked by glibenclamide (Standen *et al.*, 1989).

Some cells (i.e. pacemaker cells) from certain vascular tissues show spontaneous bursts of action potentials discharge (or multispike complexes) often associated with an initial slow depolarization or pacemaker potential (Golenhofen *et al.*, 1973; Figure 9.2). In vascular smooth muscle the ionic basis of these pacemaker potentials is not fully understood. However, in other cell types (e.g. molluscan neurones), it is thought to be dependent upon cyclic variations in free $[Ca^{2+}]_i$; during a burst of action potentials Ca^{2+} enters with each action potential faster than the cell can remove it. Consequently $[Ca^{2+}]_i$ rises until the concentration is sufficient to activate Ca^{2+}-dependent K^+ channels, causing membrane hyperpolarization and the closure of voltage-operated Ca^{2+} channels. As the intracellular Ca^{2+} concentration falls, due to internal sequestration and extrusion of Ca^{2+} from the cell, the Ca^{2+}-dependent K^+ channels close and this permits a new cycle of bursting (e.g. Gorman *et al.*, 1981; Hille, 1984). It seems possible that this explanation may also be valid in vascular smooth muscle tissues, although precise evidence is lacking.

A Ca^{2+}-dependent K^+ channel designated K_M on the basis of its ionic conductance (180 pS) in symmetrical high-K^+ conditions has been identified in rabbit portal vein (Inoue *et al.*, 1986). This channel apparently differs from the K_L channel (=BK_{Ca}) in that it is sensitive to $[Ca^{2+}]_o$, but is relatively insensitive to changes in membrane potential and is open at resting membrane potentials (Inoue *et al.*, 1986). It may be speculative to propose that the inactivation of the K_M channel may underlie, at least in part, the origin of the fall in potassium conductance observed during the pacemaker potential. However, this possibility cannot be explored until more specific K^+ channel blockers/openers become available.

Thus, in vascular smooth muscle cells K^+ channels play a critical role in all aspects of electrical activity. Their detailed function in any single cell is still speculative but they are involved in setting the level of the resting potential, in the generation of pacemaker activity, in modulating the frequency of action potential discharge and in the process of repolarization.

9.2 RESTING VASCULAR SMOOTH MUSCLE TONE

9.2.1 Spontaneously active tissues

In tissues which show spontaneous contractile activity, vascular smooth muscle tone is closely related to electrical activity: contractile phases are often accompanied by bursts of action potential discharge and relaxant phases are associated with electrical quiescence. For most smooth muscles an increase in the frequency of bursts of action potential discharge is accompanied by an increase in contractile frequency and duration but not necessarily by an increase in the size of contraction. On the other hand, an increase in the number of action potentials discharged during each burst produces greater tension development (Bolton, 1979). Indeed, in rat portal vein Hermsmeyer (1976b) proposed that the generation of phasic tension waves is dependent upon the frequency and duration of action potential bursts, on the rate of action potential discharge during the burst and also on the number of active pacemaker sites.

The role of K^+ channels in regulating contractile activity in spontaneously active tissues has been investigated using drugs which either block K^+ channels, e.g. TEA, 3,4-diaminopyridine (3,4DAP), or those which open K^+ channels, e.g. cromakalim, diazoxide.

K^+ channel blockers exert a variety of effects on electrical and mechanical activity of portal veins. For example, Southerton *et al.* (1988) reported that in rat portal vein TEA and procaine increase the amplitude and duration of phasic tension waves with little effect on baseline tension. These mechanical effects may reflect the blockade of a group of K^+ channels involved in membrane repolarization, resulting in elevation and prolonged duration of the rise in $[Ca^{2+}]_i$, which in smooth muscle may result in increased amplitude and duration of phasic contractions (Ito *et al.*, 1970). In contrast, Southerton *et al.* (1988) showed that 3,4DAP had no effect on amplitude of contraction but increased baseline tension and contractile frequency. These effects of 3,4DAP may reflect the blockade of different groups of K^+ channel: (1) K^+ channels which are open at resting membrane potential resulting in depolarization accompanied by increased baseline tension; (2) K^+ channels involved in the modulation of the frequency of bursts of action potential discharge resulting in increased frequencies of both multispike complexes and contraction. Indeed, in guinea-pig portal vein, 4-AP, a drug with similar actions to 3,4DAP, caused membrane depolarization and an increase in spike frequency. However, in this tissue 4-AP caused an increase in amplitude of contraction which, unlike the electrical changes, was attenuated by phentolamine and tetrodotoxin, suggesting that these effects were mediated by release of endogenous noradrenaline (Hara *et al.*, 1980).

Recently a number of drugs with K^+ channel-opening properties, e.g. cromakalim, minoxidil sulphate, pinacidil and diazoxide, have become available (see Chapters 8 and 10). Since these agents possess relatively selective actions they have become useful tools in the study of the role of K^+ channels in regulation of vascular tone. In rat portal vein cromakalim exerts a concentration-dependent effect on spontaneous electrical and mechanical activity: low concentrations (0.1 μM) abolish the generation of multispike complexes without any detectable change in membrane potential. Higher concentrations (0.5–5 μM) cause membrane hyperpolarization and abolish mechanical activity (Hamilton *et al.*, 1986; Weir and Weston, 1986). Similar effects of cromakalim and pinacidil in guinea-pig portal vein have also been reported (Quast, 1987; Cook *et al.*, 1988a). These dose-dependent effects suggest that these K^+ channel-opening drugs have an initial action on a group of K^+ channels involved in the generation of spontaneous activity (the pacemaker), whereas the hyperpolarizing effects result from a further action (at higher concentrations) on a second group of K^+ channels which are normally closed at resting membrane potential. These two groups may consist of two different types of K^+ channel which differ in their sensitivity to cromakalim. Alternatively the special electrical properties of pacemaker cells may 'sensitize' them to the effects of this drug (Weir and Weston, 1986; Quast, 1987). The second alternative seems more plausible since it is well known that shifts in pacemaker regions can occur (Ljung, 1970). The involvement of K^+ channels in the generation of pacemaker potentials, which regulate the frequency of action potential discharge, is further demonstrated by recent studies which show that cromakalim and minoxidil sulphate decrease the slope of the pacemaker potential and decrease contractile frequency (Newgreen *et al.*, 1990; Fig. 9.2).

There is little evidence currently available to suggest at which type of K^+ channel pinacidil and cromakalim act. However, in the case of cromakalim, evidence that the channel involved is Ca^{2+}-dependent, yet voltage-insensitive has been described (Beech and Bolton, 1987; Nakao *et al.*, 1988). More recently, the involvement of an ATP-dependent K-channel has been proposed (Standen *et al.*, 1989). Further experiments are clearly required before a definitive statement can be made.

9.2.2 Electrically quiescent tissues

The importance of K^+ channels in regulating the excitability of electrically quiescent tissues is clearly demonstrated by the use of K^+ channel blockers such as TEA^+, 4-AP and procaine. In guinea-pig pulmonary artery 4-AP and procaine cause membrane depolarization accompanied by tension development (Hara *et al.*, 1980).

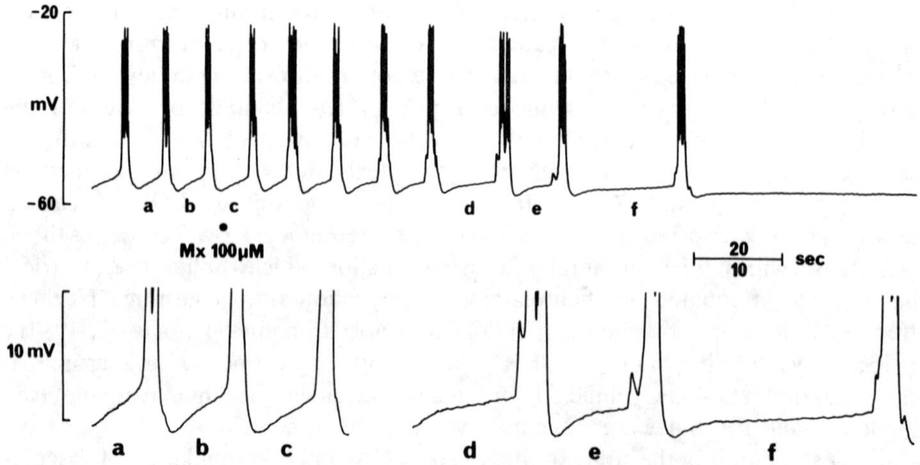

Fig. 9.2 — The effect of minoxidil sulphate (Mx) on pacemaker potentials and multispike complexes in rat portal vein. Upper trace: intracellular electrical changes and their abolition by Mx. Lower panels show the electrical events at (a—f) in the upper panel, but at higher amplification and chart speed. Note that the slope of the pacemaker potentials seen at (a–c) is reduced and finally abolished after exposure to Mx (at dot) (d–f) before membrane hyperpolarisation is observed. From Newgreen *et al.*, 1990.

Furthermore in other tissues these agents may induce action potentials and phasic tension waves, e.g. TEA: guinea-pig basilar artery (Fujiwara and Kuriyama, 1983), rabbit ear artery (Droogmans *et al.*, 1977), rabbit superior mesenteric artery (Harder and Sperekalis, 1979) and rabbit pulmonary artery (Casteels *et al.*, 1977). The ability of TEA to induce action potentials has been attributed to the blockade of K$^+$ channels. This action suppresses outward K$^+$ currents, allowing a net inward Ca^{2+} current which allows the generation of action potentials. In guinea-pig superior mesenteric artery these effects have been attributed to blockade of resting membrane- and voltage-sensitive K$^+$-conductance since in this tissue TEA causes both membrane depolarization and prolongation of action potential duration (Harder and Sperekalis, 1979).

The use of K$^+$ channel blockers makes it possible to identify groups of K$^+$ channels which play different roles in the regulation of smooth muscle tone in spontaneously active and quiescent blood vessels. However, a study of the effects of K$^+$ channel blockers in whole tissues does not allow the type of K$^+$ channel to be identified since in isolated tissue preparations these agents show relatively little selectivity for different types of K$^+$ channels (Cook and Haylett, 1985; Cook, 1988). Furthermore, some of their effects may be mediated by actions on neuronal tissues enhancing the release of neurotransmitters (Hara *et al.*, 1980).

9.3 NEUROTRANSMITTERS AND HORMONES

Vascular smooth muscle tone can be influenced by the release of neurotransmittters such as noradrenaline or acetylcholine and also by the release of blood-borne substances like angiotensin II.

9.3.1 Neurotransmitters

9.3.1.1 Spontaneously active tissues

In spontaneously active tissues such as guinea-pig portal vein, noradrenaline, adrenaline and acetylcholine increase the frequency of bursts of electrical discharge and also the rate of firing of action potentials during each burst. This is accompanied by increased frequency and amplitude of contraction. Higher concentrations of these agents depolarize the membrane to such an extent that action potentials are no longer seen and further membrane depolarization is associated with increased tension development (Bolton, 1979). These effects are accompanied by an elevation of the efflux of radioactively labelled markers for potassium, ^{42}K and ^{86}Rb. This reflects the opening of calcium- and voltage-dependent K^+ channels which opposes the depolarizing action of these neurotransmitters.

The precise identity of the types of K^+ channel involved in these phenomena is not yet clear. However, Den Hertog *et al.* (1984) showed that the ability of adrenaline to depolarize the membrane and to stimulate ^{86}Rb efflux in guinea-pig portal vein is apamin-insensitive, excluding the involvement of the small-conductance Ca^{2+}-dependent K^+ channel (SK_{Ca}) found in some cell types (see Chapter 3). In rat portal vein the involvement of K^+ channels in modulating the effects of noradrenaline has been further demonstrated by the use of cromakalim, which inhibits the mechanical and electrical effects of noradrenaline (Hamilton *et al.*, 1986).

9.3.1.2 Electrically quiescent tissues

In vascular tissues which are normally electrically quiescent, noradrenaline can produce either contractile or relaxant effects. The former are generally mediated via the activation of α_1-adrenoceptors, and may or may not be associated with changes in membrane potential. In tissues in which noradrenaline produces membrane depolarization the exact relationship between changes in membrane potential and tension development is unclear. In rabbit aorta, pulmonary, ear and superior mesenteric arteries (Casteels *et al.*, 1977; Droogmans *et al.*, 1977; Kuriyama and Suzuki, 1978; Bray *et al.*, 1988, 1989), contractile responses to low concentrations of noradrenaline ($\simeq 1\ \mu M$) are not associated with membrane depolarization. However, other reports have shown that in rabbit pulmonary and ear arteries (Haeusler, 1985; Trapani *et al.*, 1981) and most other vascular tissues studied (cat basilar artery, Harder *et al.*, 1981; guinea-pig main pulmonary artery, Suzuki, 1986), noradrenaline-induced contractions are accompanied by membrane depolarization and in some tissues action potential discharge also occurs (guinea-pig pulmonary artery, Suzuki, 1981).

In rabbit pulmonary artery, Haeusler (1985) showed that low concentrations of noradrenaline caused graded membrane depolarization accompanied by corresponding increments in tension development, whereas higher concentrations produced a larger contraction but no additional membrane depolarization. However, in the presence of TEA, high concentrations of noradrenaline did cause further membrane depolarization. This suggests that noradrenaline opens a group of K^+ channels, which are sensitive to TEA, probably as a result of the rise in $[Ca^{2+}]_i$. The opening of these channels thus opposes some of the depolarizing effects of noradrenaline, especially at high concentrations of this agonist.

Changes in electrical and mechanical activity evoked by α_1-adrenoceptor stimulation are associated with an increase in ^{42}K (^{86}Rb) efflux. Indeed Bolton and Clapp (1984) showed that in rabbit aorta noradrenaline-induced tension development is proportional to ^{42}K efflux. Furthermore, the rise in ^{42}K efflux stimulated by noradrenaline is weakly inhibited by TEA, unlike KCl-induced ^{42}K, efflux which is markedly attentuated by this K^+ channel-blocking agent. In the case of KCl, which causes substantial membrane depolarization, the increase in ^{42}K efflux is likely to reflect the opening of potential- and Ca^{2+}-dependent K^+ channels which are subject to blockade by TEA. However, the lack of effect of TEA on noradrenaline-stimulated ^{42}K efflux may reflect the opening of a different group of K^+ channels, insensitive to TEA, which open as a result of the interaction of noradrenaline with its receptor (i.e. a receptor-operated channel). At present it is not clear whether such receptor-operated K^+ channels are Ca^{2+}-dependent and there are conflicting reports in the literature. Bolton and Clapp (1984) reported that noradrenaline-induced ^{42}K efflux was not Ca^{2+}-activated in rabbit aorta, whereas Aaronson and Jones (1985) have shown that in the same tissue this response was Ca-dependent as it was diminished in a Ca^{2+}-free medium. It is possible that different experimental conditions may account for this discrepancy. If Ca^{2+}-dependent K^+ channels are involved in the action of noradrenaline in guinea-pig pulmonary artery the channels are not of the small-conductance type (SK_{Ca}) since in this tissue the depolarization and stimulation of ^{86}Rb efflux by adrenaline are both insensitive to the effects of apamin (Den Hertog *et al.*, 1984), a drug which in other tissue types is a selective blocker of the SK_{Ca} channel (Cook and Haylett, 1985; see Chapter 3).

Thus it may be concluded that in some tissues noradrenaline opens two groups (at least) of K^+ channel (based on their sensitivity to TEA) which serve to oppose the depolarizing actions of noradrenaline.

9.3.2. Hormones

9.3.2.1 Atrial natriuretic peptide

Atriopeptins I, II and III are fragments of atrial natriuretic peptide (ANP, formerly ANF), which is secreted from mammalian atria in response to an increased blood volume. ANP possesses natriuretic, diuretic and vasorelaxant properties. It has been shown to abolish spontaneous mechanical activity in rat portal vein and also to relax rat blood vessels (e.g. renal arteries) and rabbit aorta precontracted with either noradrenaline or angiotensin II (for review see Winquist, 1985). The vasorelaxant effects of ANP are thought to be mediated via a rise in the cytosolic levels of cyclic GMP. Indeed, specific binding sites or 'receptors' for ANP have been identified in a number of vascular tissues (Winquist, 1985) and one of these receptors is coupled to particulate guanylate cyclase (Murad, 1986). Recently, it has been shown that the rise in cyclic GMP levels may promote the opening of membrane K^+ channels. Using patch-clamp recording techniques Williams *et al.* (1988) showed that ANP (and also two other vasodilator agents, nitroprusside and adenosine) increased the duration and frequency of opening of large-conductance Ca^{2+}-dependent K^+ channels in cultured bovine aortic smooth muscle cells. Furthermore, this effect could be mimicked by dibutryl cyclic GMP. These authors proposed that the opening of such K^+ channels would oppose the depolarizing actions of excitatory agents such as noradrenaline, resulting in closure of voltage-operated Ca^{2+} channels and consequently in smooth muscle relaxation.

The involvement of K^+ channels in the vasorelaxant action of ANP in rabbit aorta is supported by the observation that ANP and one of its fragments, atriopeptin II, can inhibit contractile responses induced by low (20 mM) concentrations of KCl but has little effect on contractile responses induced by high (50 mM) concentrations of KCl (Winquist, 1985; Taylor and Meisheri, 1986). Such an action is characteristic of K^+ channel-opening drugs such as cromakalim (Hamilton *et al.*, 1986; Hamilton and Weston, 1989).

The precise role of ANP in modulating vascular smooth muscle tone is at present unclear. However, it is possible that it may play some role in hypertension since plasma levels of ANP are higher in spontaneously hypertensive rats compared to normotensive WKY rats (Gutkowska *et al.*, 1986).

9.3.2.2 Angiotensin II

Angiotensin II is a potent vasoconstrictor agent and recent evidence suggests that changes in K^+ channel opening may be involved in its action. Whole-cell voltage-clamp experiments in cultured smooth muscle cells from rabbit aorta have shown that angiotensin II causes: (a) an increase in a slow inward Ca^{2+} current; and (b) a decrease in an outward K^+ current which resembles I_K, since it is relatively slowly activating and inactivating and can be blocked by TEA and Cs^+ (Bkaily *et al.*, 1988). Furthermore, in rabbit aorta mechanical responses to angiotensin II are inhibited by cromakalim. This effect of cromakalim may be related to its ability to open K^+ channels since its effects are abolished when the tissue is exposed to the depolarizing effects of high KCl concentrations (Cook *et al.*, 1988b). Since it is known that contractile responses of rabbit aorta to angiotensin II are relatively insensitive to the effects of L-type Ca^{2+} channel blockers such as isradipine, it is possible that this relaxant action of cromakalim may not be exerted via the indirect closure of L-type Ca^{2+} channels (Bray *et al.*, 1988, 1989).

9.4 LOCALLY RELEASED HUMORAL SUBSTANCES

9.4.1 Endothelium-derived contracting and relaxing factors

The endothelium plays an important role in the regulation of vascular smooth muscle tone by releasing factors which promote vascular relaxation or contraction.

9.4.1.1 Endogenous contractile factors

At the present time evidence has been obtained for the existence of at least two endothelium-derived contractile factors (EDCFs). These can be released from some blood vessels in response to acetylcholine (rat aorta, canine mesenteric vein), noradrenaline (canine systemic and pulmonary veins), 5-hydroxytryptamine (rat aorta), hypoxia/anoxia (coronary and cerebral arteries) and mechanical stretch (canine cerebral and systemic arteries) (for reviews see Greenberg and Diecke, 1988; Vanhoutte and Katusic, 1988). It is believed that hypoxia and anoxia promote the release of a polypeptide, designated $EDCF_1$, whereas chemical stimuli and mechanical stretch release another substance, possibly an arachidonate derivative or superoxide anions, designated $EDCF_2$ (Vanhoutte and Katusic, 1988). Furthermore, it is known that the endothelium is a source of an extremely potent vasoconstrictor polypeptide, endothelin, which may be distinct from EDCFs, (Vanhoutte and Katusic, 1988; Yanagisawa *et al.*, 1988; Masaki, 1989). Endothelin is thought to act by promoting Ca^{2+} influx into the smooth muscle cells via an action on Ca^{2+} channels which are not dihydropyridine sensitive, a process which can be antagonised by cromakalim (Turner *et al.*, 1989). The mechanisms involved in the action of the EDCFs are unknown and the effects of these substances and of endothelin on K^+ channels have not been described.

9.4.1.2 Endogenous relaxing factors

In 1980 Furchgott and Zadwadski showed that the relaxant effects of acetylcholine in isolated rabbit aorta were dependent on the presence of an intact endothelium. It has since been shown that for most arteries, small resistance vessels and some veins, the relaxant effects of a number of circulating and neuronally released humoral substances, including acetylcholine, serotonin, histamine, braydkinin, Substance P, ATP and calcitonin-gene related peptide (CGRP), are endothelium-dependent (Vanhoutte, 1987; Furchgott, 1984; Peach *et al.*, 1985; Rubanyi and Vanhoutte, 1988; Grace *et al.*, 1987).

Endothelium-dependent relaxation of a number of arterial preparations, at least in the case of acetylcholine, is mediated via activation of guanylate cyclase and a rise in levels of cyclic GMP within the smooth muscle cells, e.g. rat aorta (Taylor *et al.*, 1988), rabbit aorta (Furchgott and Jothiananden, 1983) and bovine coronary artery (Holzmann, 1982). A similar effect of histamine and A23187 (a Ca^{2+} ionophore) has also been demonstrated in rat aorta (Rapaport and Murad, 1983). In some tissues endothelium-dependent relaxation may also be associated with changes in membrane potential. These include the mechano-inhibitory effects of Substance P in pig coronary artery (Beny *et al.*, 1986) and of acetylcholine or carbachol in monkey coronary artery (Mekata, 1986), rat aorta (Taylor *et al.*, 1988; Chen *et al.*, 1988; Chen and Suzuki, 1989) and guinea-pig mesenteric artery (Bolton *et al.*, 1984). In contrast, the relaxant effects of Substance P in guinea-pig mesenteric artery are electrically silent (Bolton and Clapp, 1986).

Taylor *et al.* (1988) showed that in rat aorta with an intact endothelium the peak rise in cyclic GMP levels induced by acetylcholine coincides with peak changes in membrane potential. However, these workers found no direct causal relationship between these two events since: (1) the changes in cyclic GMP levels were longer lasting than the observed electrical changes; (2) endothelium-independent relaxation evoked by sodium nitroprusside was associated with elevated cyclic GMP levels and relatively little change in membrane potential; (3) relaxant effects of cromakalim were accompanied by membrane hyperpolarization without any change in cyclic nucleotide levels. Furthermore, haemoglobin and methylene blue prevent endothelium-dependent rises in cyclic GMP but have no effect on acetylcholine- or histamine-induced transient hyperpolarization or ^{86}Rb efflux (Martin *et al.*, 1985; Taylor *et al.*, 1988; Chen and Suzuki, 1989).

It thus seems that the endothelium releases at least two inhibitory factors, one of which (EDRF) produces an electrically silent inhibitory response associated with the stimulation of guanylate cyclase and a second (now named endothelium-derived hyperpolarizing factor (EDHF) which transiently hyperpolarizes smooth muscle cells, promoting the closure of voltage-operated Ca^{2+} channels and relaxation (Feletou and Vanhoutte, 1986; Komori and Suzuki, 1987; Taylor *et al.*, 1988).

The hyperpolarizing action of acetylcholine or carbachol in vascular smooth muscle cells is closely associated with an increase in membrane permeability to K^+, amongst other ions (Bolton *et al.*, 1984) and a rise in the rate of ^{86}Rb efflux (Taylor *et al.*, 1988), events which are likely to reflect the opening of membrane K^+ channels. Furthermore, these K^+ channels may be Ca^{2+}-dependent since in rabbit aortic strips relaxant effects of acetylcholine are attenuated by drugs such as quinine, TEA and tetramethylammonium (Gebremedhin *et al.*, 1987), which are known to block Ca^{2+}-dependent K^+ channels. However, this interpretation should be viewed with caution since these drugs are known to be non-selective blockers of K^+ channels. In porcine coronary artery there is evidence that the relaxant effects of ATP and bradykinin, but not those of ACh or carbachol, involve the opening of K^+ channels located on the endothelial cells themselves. Bradykinin and ATP, unlike acetylcholine and carbachol, can stimulate ^{86}Rb efflux from endothelial cells taken from this tissue, an action which is likely to reflect the opening of K^+ channels. These K^+ channels may be Ca^{2+}-dependent, since bradykinin- and ATP-stimulated ^{86}Rb efflux from these cells can be blocked by La^{3+}, an inhibitor of transmembrane Ca^{2+} flux (Martin and Gordon, 1983; Gordon and Martin, 1983).

The precise role of EDHF in the control of vascular smooth muscle tone is at the present time unclear. In high-K^+ solutions the relaxant effects of sodium nitroprusside (an endothelium-independent vasodilator) are inhibited by 50%, whereas those of ACh (which releases both EDRF and EDHF) were almost eliminated (Taylor *et al.*, 1988). This suggests that EDHF, which as a K^+ channel opener would not be predicted to relax a high K^+ contraction (Hamilton *et al.*, 1986) contributes to the initiation of mechanical inhibition, since hyperpolarization will result in the closure of voltage-dependent Ca^{2+} channels and consequently in relaxation. This process may be of particular importance in small arterioles which are very dependent on Ca^{2+} influx during contraction.

9.4.2 Calcitonin-gene related peptide (CGRP)

CGRP(s) are a family of peptides found in nerve fibres associated with central and peripheral blood vessels (see Bevan and Brayden, 1986). CGRP exerts potent vasodilatory effects on a variety of vascular tissues, including aorta, coronary and mesenteric arteries in a variety of species (Bevan and Brayden, 1986). In rat aorta the vasorelaxant effects of CGRP are dependent upon the presence of an intact endothelium. However, in contrast to the endothelium-dependent relaxant effects of ACh, this effect of CGRP is not associated with a rise in cyclic GMP levels (Grace *et al.*, 1987). In other tissues such as pig coronary artery the effects of CGRP are endothelium-independent and are accompanied by a rise in the levels of intracellular cyclic AMP which in smooth muscle is known to cause relaxation (Bülbring and Tomita, 1987). CGRP is particularly abundant in coronary blood vessels, which may reflect an important role of this peptide in regulating coronary vascular tone (Shoji *et al.*, 1987).

9.5 PHYSICAL FORCES

There is evidence that vascular smooth muscle tone is dependent upon the local response of the blood vessel wall to the hydrodynamic properties of the blood, i.e. changes in blood pressure and blood flow. The force exerted by blood on a vessel wall consists of two components (Lansman, 1988; Bevan, 1988):

(a) A perpendicular or radial force exerted by blood pressure (transmural pressure), largely on the proteins and elastins of the blood vessel wall. An increase in local blood pressure stimulates vasoconstriction, an effect which is associated with membrane depolarization and Ca^{2+} influx into smooth muscle cells via a pathway distinct from those activated by agonists or by membrane depolarization.
(b) A tangential, frictional force (wall shear stress) produced when a viscous fluid moves across a solid surface. This force falls entirely on endothelial cells.

An increase in local blood flow produces vasodilation, a response which in large arteries (e.g. femoral artery) is entirely endothelium-dependent. However, in smaller arteries a fraction of the flow-induced dilator response persists even after endothelial destruction (Bevan, 1988). There is evidence that a component of flow-induced vasodilation may be associated with activation of a K$^+$ current. Olesen *et al.* (1988) made electrophysiological recordings from cultured bovine aortic endothelial cells (patch-clamp technique; whole-cell configuration) and showed that shear stress caused membrane hyperpolarization and activation of a K$^+$ current (designated *I*K,s)† which showed properties similar to those of inward rectifier currents identified in other cell types. Furthermore, this current was found to be activated at levels of shear stress which were comparable to those found in some arteries *in vivo*. The endothelial-dependent component of flow-induced vasodilation is inhibited by haemoglobin (Bevan, 1988). It is thus possible that this stress-induced hyperpolarization of endothelial cells somehow triggers the release of EDRF(s), which in turn generates vasorelaxation. However, the possibility that relaxation is achieved via local circuit currents which flow from the endothelial cells through gap-junctions to the underlying smooth muscle cells cannot be excluded.

9.6 PATHOPHYSIOLOGY OF POTASSIUM CHANNELS IN VASCULAR SMOOTH MUSCLE

9.6.1 Hypertension

Several animal models, e.g. spontaneously hypertensive rats (SHR), renal hypertensive rats (RH) and deoxycortone acetate (DOCA)-induced hypertensive rats, have been used to investigate changes in the function of K$^+$ channels in hypertensive states. It is well established that vascular smooth muscle from hypertensive rats is characterized by altered ion transport typified by an increase in the steady-state turnover of K$^+$, Na$^+$, Cl$^-$ and Ca^{2+} ions (Jones, 1981). For example, the passive efflux of ^{42}K is greater from aorta, femoral and mesenteric arteries from spontaneously hypertensive rats compared to normotensive controls (Jones, 1974). Furthermore, in rat aorta a correlation exists between the onset of high blood pressure and changes in the transport of Na$^+$ and K$^+$ ions (Jones *et al.*, 1981). Blood vessels from hypertensive rats may also show increased sensitivity to the contractile effects of noradrenaline (e.g. femoral and caudal arteries and mesenteric arterioles; Hermsmeyer, 1976a; Webb and Vanhoutte, 1979; Mulvany *et al.*, 1980) and to 5-hydroxytryptamine (mesenteric artery; Haeusler and Finch, 1972).

† Not to be confused with the slowly activating neuronal K$^+$ current ($I_{K,s}$) described elsewhere in this book.

It has been suggested that changes in membrane permeability in hyptertension closely resemble the difference in membrane permeability between phasic and tonic vascular smooth muscles, and it has been suggested that hypertension represents a transition from a 'tonic' to a 'phasic' state (Jones, 1981). Indeed, Holloway and Bohr (1973) reported that spontaneous contractile activity was more frequently observed in femoral arterial strips from SHR, renal and DOCA hyptertensive rats than normotensive rats. Furthermore, Lamb and colleagues (1985) have shown that tail arteries from a 'stroke prone' strain of SHR (SHRSP) respond to noradrenaline with an initial contractile phase followed by oscillatory contractions (or phasic contractions and relaxation; Myers *et al.*, 1985). This is in contrast to tail arteries from SHR, RH, DOCA and normotensive control rats, which respond to noradrenaline with graded and maintained mechanical responses.

Simultaneous recording of tension and membrane potential from tail arteries in SHRSP rats shows that these oscillatory contractions are associated with pacemaker-like activity (i.e. tension development is often accompanied by bursts of action potential discharge, whereas relaxant phases are electrically quiescent (Lamb and Webb, 1988). The relaxation phases are throught to be mediated by the opening of K^+ channels since in the presence of elevated K^+ levels or some K^+ channel blockers, i.e. Ca^{2+}, sparteine and TEA (30 mM) but not TEA (0.1–10 mM) or 4-AP, the oscillatory contractions are converted to maintained responses (Myers *et al.*, 1985; Lamb *et al.*, 1985).

The excitability of vascular smooth muscles is regulated by K^+ currents which serve to repolarize the membrane following action potentials, to regulate the frequency of action potential discharge and to oppose the depolarizing action of neurotransmitters and other humoral substances (see section 9.3). It appears that although passive ^{42}K efflux and noradrenaline-stimulated ^{42}K efflux are elevated in tissues from hypertensive rats, the outward K^+ currents are insufficient to suppress excitability.

9.6.2 Cerebral vasospasm

Subarachnoid haemorrhage (SAH) may cause and maintain the narrowing of cerebral blood vessels. This abnormal and uncontrolled behaviour of such vessels is generally termed 'cerebral vasospasm', and invariably leads to ischaemia and often cerebral dysfunction. This phenomenon of reduction in cerebral artery calibre is also implicated in migraine and in transient ischaemic attacks. Currently there is a lack of understanding both of the pathophysiology of cerebral vasospasm and a limited knowledge of the mechanisms which underly the condition.

Recent animal studies suggest that smooth muscle K^+ channels are implicated in the development of cerebral vasospasm following SAH. Using an established experimental model of SAH in the dog, in which autologous blood is injected cisternally, Harder *et al.* (1987) reported spasm of basilar arteries *in vivo* (monitored by angiography), and recorded marked membrane depolarization and enhanced electrical spike activity of the arteries when isolated, in comparison to controls. Further analysis supported the concept that the altered muscle cell properties were the result of a reduction in resting K^+ conductance.

Theoretically, a significant reduction of resting K^+ conductance would markedly activate cerebral arterial muscle and also increase its sensitivity to circulating vasoconstrictor agents. In the same study the K^{+} channel opener nicorandil repolarized the muscle cell membrane and abolished the generation of action potentials. In intact animals the K^{+} channel opener nicorandil partially reversed cerebral vasospasm. Further studies are clearly required to establish the role of K^{+} channels in cerebral vasospastic disease.

9.6.3 Vasospastic angina

Uchida (1985) first proposed that coronary vasospasm in man might be generated by a pathological decrease in K-conductance in coronary artery smooth muscle. Such a concept has been extended by Iwaki *et al.* (1987), who showed that the coronary spasm characteristic of some patients could be mimicked using the K^{+} channel blocker TEA. Such a concept is very interesting, but at present there is no direct evidence which links coronary vasospasm to changes in the properties of the K^{+} channels in coronary smooth muscle.

9.7 SUMMARY

K^{+} channels play important roles in regulating vascular smooth muscle tone. They are involved in: (a) regulating tone in vessels which show spontaneous activity; (b) opposing the depolarizing actions of excitatory neurotransmitters (e.g. noradrenaline); (c) the action of locally released vasoactive substances and blood-borne hormones (e.g. CGRP, antiotensin II); (d) autoregulation of blood flow. In addition changes in K^{+} channel functioning have been implicated in certain pathological states involving vascular smooth muscle.

ACKNOWLEDGEMENT

Jenny Longmore was supported by a grant from Rhône-Poulenc.

REFERENCES

Aaronson, P. I., and Jones, A. W. (1985) Calcium regulation of potassium fluxes in rabbit aorta during activation by noradrenaline or high potassium medium. *J. Physiol.* **367** 27–43.

Beech, D. J., and Bolton, T. B. (1987) Effects of BRL 34915 on membrane currents recorded from single smooth muscle cells from the rabbit portal vein. *Br. J. Pharmacol.* **92** 550P.

Beech, D. J., Bolton, T. B., Castle, N. A., and Strong, P. N. (1987) Characterisation of a toxin from scorpion (*Leiurus quinquestriatus*) venom that blocks *in vitro* both large (BK) K^{+}-channels in rabbit vascular smooth muscle and intermediate (IK) conductance Ca^{2+}-activated K^{+} channels in human red cells. *J. Physiol. Proc. Suppl.* **287** 32P.

Benham, C. D., Bolton, T. B., Lang, R. J., and Takewaki, T. (1986) Calcium-activated potassium channels in single muscle cells of rabbit jejunum and guinea-pig mesenteric artery. *J. Physiol.* **371** 45–67.

Benham, D. C., Bolton, T. B., Denbigh, J. S., and Lang, R. J. (1987) Inward rectification in freshly isolated single smooth muscle cells of the rabbit jejunum. *J. Physiol.* **383** 461–476.

Beny, J.-L., Brunet, P. C., and Huggel, H. (1986) Effect of mechanical stimulation, Substance P and vasoactive intestinal polypeptide on the electrical and mechanical activities of circular smooth muscles from pig coronary arteries contracted with acetylcholine: role of endothelium. *Pharmacology* **33** 61–68.

Bevan, J. A. (1988) Basal tone in resistance arteries: role of wall stretch, flow and receptor specialization. In: *Vascular Neuroeffector Mechanisms,* pp. 1–14, J. E. Bevan, H. Majewski, R. A. Maxwell and D. F. Story (eds). IRL Press, Washington.

Bevan, J. A., and Brayden, J. E. (1986) Noradrenergic neural vasodilator mechanisms. *Circ. Res.* **60** 309–326.

Bkaily, G., Peyrow, M., Sculptoreanu, A., Jacques, D., Chahine, M., Regoli, D., and Sperelakis, N. (1988) Angiotensin II increases I_{Si} and blocks I_K in single aortic cell of rabbit. *Pflügers Arch.* **412** 448-450.

Bolton, T. B. (1979) Mechanism of action of transmitters and other substances on smooth muscle. *Physiol. Rev.* **59** 606–718.

Bolton, T. B., and Clapp, L. H. (1984) The diverse effects of noradrenaline and other stimulants on ^{86}Rb and ^{42}K efflux in rabbit and guinea-pig arterial muscle. *J. Physiol.* **355** 43-63.

Bolton, T. B., and Clapp, L. H. (1986) Endothelial-dependent relaxant actions of carbachol and substance P in arterial smooth muscle. *Br. J. Pharmacol.* **87** 713-723.

Bolton, T. B., Lang, R. J., and Takewaki, T. (1984) Mechanisms of action of noradrenaline and carbachol on smooth muscle of guinea-pig arterior mesenteric artery. *J. Physiol.* **351** 549–572.

Bolton, T. B., Lang, R. J., Takewaki, T., and Benham, C. D. (1985) Patch and whole-cell voltage clamp of single mammalian visceral and vascular smooth muscle cells. *Experientia* **41** 887–894.

Bolton, T. B., Lang, R. J., Takewaki, T., and Benham, C. D. (1986) Patch and whole-cell voltage clamp studies on single smooth muscle cells. *J. Cardiovasc. Pharmacol.* **8** (suppl. 8) 520–524.

Bray, K. M., Weston, A. H., McHarg, A. D., Newgreen, D. T., and Southerton, J. S. (1988) Analysis of the inhibitory effects of BRL 34915 on responses to noradrenaline in rabbit aorta. *Br. J. Pharmacol.* **93** 206P.

Bray, K. M., Weston, A. H., Duty, S., Newgreen, D. T., Longmore, J., and Brown, T. (1989) Differences between the effects of cromakalim and nifedipine on agonist-induced responses in rabbit aorta. *Br. J. Pharmacol.* (in press).

Bülbring, E., and Tomita, T. (1987) Catecholamine action on smooth muscle. *Pharmacol. Rev.* **39** 49–96.

Casteels, R., Kitamura, K., Kuriyama, H., and Suzuki, H. (1977) Excitation-coupling in the smooth muscle cells of the rabbit pulmonary artery. *J. Physiol.* **271** 63–79.

Chen, G., and Suzuki, H. (1989) Some electrical properties of the endothelium-dependent hyperpolarisation recorded from rat arterial smooth muscle cells. *J. Physiol.* **410** 91–106.

Chen, G., Suzuki, H., and Weston, A. H. (1988) Acetylcholine releases endothelium-derived hyperpolarising factor and EDRF from rat blood vessels. *Br. J. Pharmacol.* **95** 1165–1174.

Cook, N. S. (1988) The pharmacology of potassium channels and their therapeutic potential. *Tr. Pharmacol. Sci.* **9** 21–28.

Cook, N. S., and Haylett, D. G. (1985) Effect of apamin, quinine and neuromuscular blockers on calcium-activated potassium channels in guinea-pig hepatocytes. *J. Physiol.* **358** 373–394.

Cook, N. S., Quast, U., Hof, R. P., Baumlin, Y., and Pally, C. (1988a) Similarities in the mechanism of action of two new vasodilator drugs, pinacidil and BRL 34915. *J. Cardiovasc. Pharmacol.* **11** 90–99.

Cook, N. S., Wier, S. W., and Danzeisen, M. C. (1988b) Anti-vasoconstrictor effects of the K^+ channel opener cromakalim on the rabbit aorta — comparison with the calcium antagonist isradipine. *Br. J. Pharmacol.* **95** 741–752.

Den Hertog, A., Pielkenrood, J., Ras, R., and van den Akker, J. (1984) The contribution of calcium and potassium to the alpha-action of adrenaline on smooth muscle cells of the portal vein, pulmonary artery and taenia caeci of the guinea-pig. *Eur. J. Pharmacol.* **98** 223–234.

Droogmans, G., Raeymaekers, L., and Casteels, R. (1977) Electro- and pharmacomechanical coupling in the smooth muscle cells of the rabbit ear artery. *J. Gen. Physiol.* **70** 129–148.

Edwards, F. R., and Hirst, G. D. S. (1988) Inward rectification in submucosal arterioles of the guinea-pig ileum. *J. Physiol.* **404** 437-455.

Feletou, M., and Vanhoutte, P. M. (1986) Endothelium-derived factor(s) and membrane potential of vascular smooth muscle. *Blood vessels* **23,** 68.

Fujiwara, S., and Kuriyama, H. (1983) Effects of agents that modulate potassium permeability on smooth muscle cells of the guinea-pig basilar artery. *Br. J. Pharmacol.* **79** 23–25.

Furchgott, R. F. (1984) The role of endothelium in the responses of vascular smooth muscle to drugs. *Ann. Rev. Pharmacol. Toxicol.* **24** 175–197.

Furchgott, R. F., and Jothiananden, D. (1983) Relation of cyclic GMP levels to endothelium-dependent relaxation by acetylcholine in rabbit aorta. *Fed. Proc.* **142** 619.

Furchgott, R. F. and Zawadski, J. V. (1980) The obligatory role of endothelial cells in the relaxation of arterial smooth muscle by acetylcholine. *Nature* **288** 373–376.

Gebremedhin, D., Hadhazy, P., and Magyar, K. (1987) Inhibition by quinine of endothelium-dependent relaxation of rabbit aortic strips. *Br. J. Pharmacol.* **92** 835–841.

Golenhofen, K., Hermstein, N., and Lammel, E. (1973) Membrane potential and contraction of vascular smooth muscle (portal vein) during application of noradrenaline and high potassium and selective inhibitory effects of iproreratil (verapamil). *Microvasc. Res.* **5** 73–80.

Gordon, J. L., and Martin, W. (1983) Endothelium-dependent relaxation of the pig aorta: relationship to stimulation of ^{86}Rb efflux from isolated endothelial cells. *Br. J. Pharmacol.* **79** 531–539.

Gorman, A. L. F., Hermann, A., and Thomas, M. V. (1981) Intracellular calcium and the control of neuronal pacemaker activity. *Fed. Proc.* **40** 2233–2239.

Grace, G. C., Dusting, G. J., Kemp, B. E., and Martin, T. J. (1987) Endothelium and the vasodilator action of rat calcitonin gene-related peptide (CGRP). *Br. J. Pharmacol.* **91** 729–733.

Greenberg, S., and Diecke, F. P. J. (1988) Endothelium-derived relaxing and contracting factors: New concepts and new findings. *Drug Devel. Res.* **12** 131–149.

Gutkowska, J., Horky, K., Lachance, C., Racz, K., Garcia, R., Thibault, G., Kuchel, O., Genest, J., and Cantin, M. (1986) Atrial natriuretic factor in spontaneously hypertensive rats. *Hypertension* **8** (suppl. I) 1–137.

Haeusler, G. (1985) Contraction of vascular muscle as related to membrane potential and calcium fluxes. *J. Cardiovasc. Pharmacol.* **7** (suppl. 6) 53–58.

Haeusler, G., and Finch, L. (1972) Vascular reactivity to 5-hydroxytryptamine and hypertension in the rat. *Naunyn-Schmied. Arch. Pharmacol.* **272** 101–116.

Hamilton, T. C., and Weston, A. H. (1989) Cromakalim, nicrorandil and pinacidil: novel drugs which open potassium channels in smooth muscle. *Gen. Pharmacol.* **20** 1–9.

Hamilton, T. C., Wier, S. W., and Weston, A. H. (1986) Comparison of the effects of BRL 34915 and verapamil on electrical and mechanical activity in rat portal vein. *Br. J. Pharmacol.* **88** 103–111.

Hara, Y., Kitamura, K., and Kuriyama, H. (1980) Actions of 4-aminopyridine on vascular smooth muscle tissues of the guinea-pig. *Br. J. Pharmacol.* **68** 99–106.

Harder, D. R. (1983) Heterogeneity of membrane properties in vascular muscle cells from various vascular beds. *Fed. Proc.* **42** 253–256.

Harder, D. R., and Sperekalis, N. (1979) Action potentials induced in guinea-pig arterial smooth muscle by tetraethylammonium. *Am. J. Physiol.* **237** C75–C80.

Harder, D. R., Abel, P. W., and Hermsmeyer, K. (1981) Membrane electrical mechanism of basilar artery constriction and pial artery dilation by norepinephrine. *Circ. Res.* **49** 1237–1242.

Harder, D. R., Dernbach, P., and Waters, A. (1987) Possible cellular mechanism for cerebral vasospasm after experimental subarchnoid hemorrhage in the dog. *J. Clin. Invest.* **80** 875–880.

Hermsmeyer, K. (1976a) Electrogenesis of increased norepinephrine sensitivity of arterial vascular muscle in hypertension. *Circ. Res.* **38** 362–367.

Hermsmeyer, K. (1976b) Ba^{2+} and K^+ alteration of K^+ conductance in spontaneously active vascular muscle. *Am. J. Physiol.* **230** 1031–1036.

Hille, B. (1984) *Ionic channels of excitable membranes.* Washington, Sinauer.

Holloway, E. T., and Bohr, D. F. (1973) Reactivity of vascular smooth muscle in hypertensive rats. *Circ. Res.* **33** 678–685.

Holzmann, S. (1982) Endothelium-induced relaxation by acetylcholine associated with larger rises in cyclic GMP in coronary arterial strips. *J. Cyclic Nucl. Res.* **8** 409–419.

Inoue, R., Kitamura, K., and Kuriyama, H. (1985) Two Ca-dependent K channels classified by tetraethylammonium distribute on smooth muscle membranes of the rabbit portal vein. *Pflügers Arch.* **405** 173–179.

Inoue, R., Okabe, K., Kitamura, K., and Kuriyama, H. (1986) A newly identified Ca^{2+} dependent K^+ channel in the smooth muscle membrane of single cells dispersed from the rabbit portal vein. *Pflügers Arch.* **406** 138–143.

Ito, Y., Kuriyama, H., and Sakamoto, Y. (1970) Effects of tetraethylammonium chloride on the membrane activity of guinea-pig stomach smooth muscle. *J. Physiol. (Lond.)* **211** 445–460.

Iwaki, M., Mizobuchi, S., Nakaya, Y., Kawano, K., Niki, T., and Mori, H. (1987) Tetraethylammonium induced coronary spasm in isolated perfused rabbit heart: a hypothesis for the mechanism of coronary spasm. *Cardiovasc. Res.* **21** 130–139.

Jones, A. W. (1974) Altered ion transport in large and small arteries from spontaneously hypertensive rats and the influence of calcium. *Circ. Res.* **34** and **35** (suppl. 1) I117–I112.

Jones, A. W. (1981) Vascular smooth muscle and alterations during hypertension. In: *Smooth muscle: An assessment of current knowledge,* pp. 397–430, E. Bülbring, A. F. Brading, A. W. Jones and T. Tomita (eds). Edward Arnold, London.

Jones, A. W., Heidlage, J. F., Meyer, R., Day, B., and Freeland, A. (1981) Nonspecific super-sensitivity of aortic ^{42}K effluxes during DOCA hypertension in the rat and the effects of antihypertensive therapy. In: *New trends in arterial hypertension,* pp. 189–199, M. Worcel *et al.* (eds). Elsevier, Amsterdam.

Klöckner, V., and Isenberg, G. (1985) Action potentials and net membrane currents of isolated smooth muscle cells. *Pflügers Arch.* **405** 329–339.

Komori, K., and Suzuki, H. (1987) Electrical responses of smooth muscle cells during cholinergic vasodilation in the rabbit saphenous artery. *Circ. Res.* **61** 586–593.

Kuriyama, H., and Suzuki, H. (1978) The effects of acetylcholine and norepinephrine on the membrane and contractile properties of smooth muscle cells of rabbit superior mesenteric artery. *Br. J. Pharmacol.* **64** 493–501.

Kuriyama, H., Ito, Y., Suzuki, H., Kitamura, K., and Itoh, T. (1982) Factors modifying contraction–relaxation cycle in vascular smooth muscle. *Am. J. Physiol.* **243** H641–H662.

Lamb, F. S., and Webb, R. C. (1988) Bursting pacemaker activity drives oscillatory contractions to norepinephrine in genetically hypertensive rats. *Hypertension* (in press).

Lamb, F. S., Myers, J. H., Hamlin, M. N., and Webb, R. C. (1985) Oscillatory contractions in tail arteries from genetically hypertensive rats. *Hypertension* **7** (suppl.) I25–I30.

Lansman, J. B. (1988) Endothelial mechanosensors. Going with the flow. *Nature* **331** 481–482.

Ljung, B. (1970) Nervous and myogenic mechanisms in the control of a vascular neuroeffector system. *Acta Physiol. Scand.* (suppl.) **349** 33–69.

Martin, W., and Gordon, J. L. (1983) Differential calcium dependence of contractile responses and ^{86}Rb efflux from the rabbit aorta induced by vasoactive stimuli. *J. Cell. Physiol.* **115** 46–52.

Martin, W., Villani, G. M., Jothianandan, D., and Furchgott, R. F. (1985) Selective blockade of endothelium-dependent and glyceryl trinitrate-induced relaxation by hemoglobin and methylene blue in the rabbit aorta. *J. Pharmacol. Exptl Therap.* **232** 708–716.

Masaki, T. (1989) The discovery, the present state and the future prospects of endothelin. *J. Cardiovasc. Pharmacol.* **13** (Suppl. 5) 51–54.

Mekata, F. (1986) The role of hyperpolarisation in the relaxation of smooth muscle of monkey coronary artery. *J. Physiol.* **371** 257–265.

Mulvany, M. J., Aalkjaer, C., and Christensen, J. (1980) Changes in noradrenaline sensitivity and morphology of arterial resistance vessels during development of high blood pressure in spontaneously hypertensive rats. *Hypertension* **2** 664–671.

Murad, F. (1986) Cyclic guanosine monophosphate as a mediator of vasodilation. *J. Clin. Invest.* **78** 1–5.

Myers, J. H., Lamb, F. S., and Webb, R. C. (1985) Norepinephrine-induced phasic activity in tail arteries from genetically hypertensive rats. *Am. J. Physiol.* **248** H419–H423.

Nakao, K., Okabe, K., Kitamura, K., Kuriyama, H., and Weston, A. H. (1988) Characteristics of cromakalim-induced relaxations in the smooth muscle cells of guinea-pig mesenteric artery and vein. *Br. J. Pharmacol.* **95** 795–804.

Newgreen, D. T., Bray, K. M., McHarg, A. D., Duty, S., Brown, B. S., Kay, P. B., Longmore, J., Southerton, J. S., and Weston, A. H. (1990) The action of diazoxide and minoxidil sulphate on rat blood vessels. *Br. J. Pharmacol.* (in press).

Okabe, K., Kitamura, K., and Kuriyama, H. (1987) Features of 4-aminopyridine sensitive outward current observed in single smooth muscle cells from the rabbit pulmonary artery. *Pflügers Arch.* **409** 561–568.

Olesen, S. P., Clapham, D. E., and Davies, P. F. (1988) Haemodynamic shear stress activates a K^+ current in vascular endothelial cells. *Nature* **331** 168–170.

Peach, M. J., Loeb, A. L., Singer, H. A., and Saye, J. (1985) Endothelium-derived relaxing factor. *Hypertension* **7** (suppl.) I94–I100.

Quast, U. (1987) Effect of the K^+ efflux stimulating vasodilator BRL 34915 on $^{86}Rb^+$ efflux and spontaneous activity in guinea-pig portal vein. *Br. J. Pharmacol.* **91** 569–578.

Rapaport, R. M., and Murad, F. (1983) Agonist-induced endothelium-dependent relaxation in rat thoracic aorta may be mediated through cyclic GMP. *Circ. Res.* **52** 325–357.

Rubanyi, G. M., and Vanhoutte, P. M. (1988) Heterogeneity of endothelium-dependent responses to acetylcholine in canine femoral arteries and veins. *Blood Vessels* **25** 75–81.

Shoji, T., Ishihara, H., Ishikawa, T., Saibo, A., and Goto, K. (1987) Vasodilatory effects of human and rat calcitonin gene-related peptides in isolated porcine coronary arteries. *Naunyn-Schmied. Arch. Pharmacol.* **336** 438–444.

Southerton, J. S., Weston, A. J., Bray, K. M., Newgreen, D. T., and Taylor, S. G.

(1988) The potassium channel opening action of pinacidil, studies using biochemical, ion flux and microelectrode techniques. *Naunyn-Schmied. Arch. Pharmacol.* **338** 310–318.

Standen, N. B., Quayle, J. M., Davies, N. W., Brayden, J. E., Huang, Y., and Nelson, M. T. (1989). Hyperpolarizing vasodilators activate ATP-sensititive K$^+$ channels in arterial smooth muscle. *Science* **245** 177–180.

Suzuki, H. (1981) Effects of endogenous and exogenous noradrenaline on the smooth muscle of guinea-pig mesenteric vein. *J. Physiol.* **321** 495–512.

Suzuki, H. (1986) Increase in membrane resistance during noradrenaline-induced depolarisation in arterial smooth muscle. *Jap. J. Physiol.* **36** 433–440.

Taylor, C. J., and Meisheri, K. D. (1986) Inhibitory effects of a synthetic atrial peptide on contractions and ^{45}Ca fluxes in vascular smooth muscle. *J. Pharmacol. Expt. Therap.* **237** 803–808.

Taylor, S. G., Southerton, J. S., Weston, A. H., and Baker, J. R. J. (1988) Endothelium-dependent effects of acetylcholine in rat aorta: A comparison with sodium nitroprusside and cromakalim. *Br. J. Pharmacol.* **94** 853–863.

Trapani, A., Matsuki, N., Abel, P. W., and Hermsmeyer, K. (1981) Noradrenaline produces tension through electromechanical coupling in the rabbit ear artery. *Eur. J. Pharmacol.* **72** 87–91.

Turner, N. C., Dollery, C. T., and Williams, A. J. (1989) Endothelin-1-induced contractions of vascular and tracheal smooth muscle: effects of nicordipine and BRL 34915. *J. Cardiovasc. Pharmacol.* **13** (Suppl. 5) 5180–5182.

Uchida, Y. (1985) Decreasing potassium conductance — a possible mechanism of phasic coronary vasospasm. *Jap. Circ. J.* **49** 128–139.

Vanhoutte, P. M. (1987) Endothelium and the control of vascular tissue. *NIPS* **2** 18–22.

Vanhoutte, P. M., and Katusic, Z. S. (1988) Endothelium-derived contracting factors: Endothelin and/or superoxide. *Trends Pharmacol. Sci.* **9** 229-230.

Webb, R. C., and Vanhoutte, P. M. (1979) Sensitivity to noradrenaline in isolated tail arteries from spontaneously hyptertensive rats. *Clin. Sci.* **57** 31–33.

Weir, S. W., and Weston, A. H. (1986) The effects of BRL 34915 and nicorandil on electrical and mechanical activity and on ^{86}Rb efflux in rat blood vessels. *Br. J. Pharmacol.* **88** 121–128.

Williams, D. L., Katz, G. M., Roy-Contancin, L., and Reuken, J. P. (1988) Guanosine 5′ monophosphate Ca^{2+}-activated K$^+$ channels in vascular smooth muscle cells. *Proc. Natl. Acad. Sci. USA* **85** 9360–9364.

Winquist, R. J. (1985) The relaxant effects of atrial natriuretic factor on vascular smooth muscle. *Life Sci.* **37** 1081–1087.

Yanagisawa, M., Kurihara, H., Kimura, S., Tomobe, Y., Kobayhashi, M., Mitsui, Y., Yazaki, Y., Goto, K., and Masaki, T. (1988) A novel potent vasoconstrictor peptide produced by vascular endothelial cells. *Nature* **332** 411–415.

10

In vivo studies with drugs which open smooth muscle K^+ channels

R. E. Buckingham

10.1 INTRODUCTION

The discovery of the classical Ca^{2+} antagonists in the 1960s represented a watershed in smooth muscle research. Not only did the introduction of such drugs as verapamil, diltiazem and nifedipine provide clinicians with powerful new drugs to treat such cardiovascular diseases as hypertension, angina, cardiac arrhythmias and hypertrophic obstructive cardiomyopathy, they also provided laboratory scientists with new experimental tools for studying smooth muscle physiology and pharmacology. However, the promise that these 'prototype' Ca^{2+} antagonists might lead to further generations of compounds to treat a more diverse range of disturbances of smooth muscle function (e.g. asthma, irritable bowel syndrone, urinary incontinence, premature labour) remains largely unfulfilled. By contrast, it is becoming increasingly evident that drugs which reduce Ca^{2+} availability indirectly as a consequence of opening or activating certain types of membrane K^+ channels exert this effect in a wide variety of smooth muscle types, suggesting the possibility of a diverse clinical utility. An obvious potential drawback in this scenario would be a high incidence of unwanted side-effects resulting from such a wide diversity of smooth muscle and perhaps other organ involvement.

The ability to open smooth muscle K^+ channels is present in a structurally dissimilar group of drugs; nicorandil, cromakalim (CRK), pinacidil, RP 49356, minoxidil sulphate and diazoxide have all been shown to exhibit this property. For the sake of simplicity, this review will concentrate on the diversity of the *in vivo* pharmacological properties of CRK, pinacidil and nicorandil in animals and, where it is possible to draw analogies, also in man. It is important to stress at the outset, however, that nicorandil's smooth muscle relaxant activity is thought to be due partly to K$^+$ channel activation and partly to a stimulation of guanylate cyclase activity (Holzmann, 1983; Schmidt *et al.*, 1985).

10.2 STUDIES IN ANIMALS

10.2.1 Cardiovascular effects

10.2.1.1 Effects on systemic haemodynamics

CRK is a highly potent blood pressure (BP)-lowering agent in rats, cats, dogs and rabbits (Buckingham *et al.*, 1986a; Cook and Hof, 1988) and its activity both *in vitro* and *in vivo* lies predominantly in the (−)-enantiomer (Buckingham *et al.*, 1986b; Hof *et al.*, 1988). Pinacidil also exhibits BP-lowering activity in a variety of species (Arrigoni-Martelli and Finucane, 1985; Cohen, 1986).

Current data indicate that CRK exhibits greater potency *in vivo* than either pinacidil or nicorandil. By the oral route in conscious renal hypertensive cats, for example, CRK was tenfold more potent than pinacidil in lowering diastolic BP, with pinacidil in turn tenfold more potent than nicorandil (Longman *et al.*, 1988). In conscious female SHR using measurements of area under the control curve of mean arterial BP for 4 hours following oral drug administration, pinacidil was again tenfold more potent than nicorandil, with CRK around threefold more potent than pinacidil (Buckingham, unpublished observations). In the same experiment each compound raised heart rate (HR) dose-dependently, with no obvious difference between them. Although each compound also caused a tachycardia in conscious renal hypertensive cats, low doses of CRK (0.007 and 0.01 mg/kg p.o.) lowered BP without increasing HR (Longman *et al.*, 1988). In conscious normotensive and hypertensive cats it was shown that at equivalent oral hypotensive doses the increase in HR following CRK was less than that observed following pinacidil or nifedipine (Buckingham *et al.*, 1986a; Clapham, 1988; Longman *et al.*, 1988). That the increase in HR caused by CRK is reflexly mediated is suggested by the fact that in conscious hypertensive cats and anaesthetized rats the tachycardia can be prevented by the ß-blockers propranolol or bopindolol, respectively (Buckingham *et al.*, 1986a; Cook and Hof, 1988). Furthermore, Grosset and Hicks (1986) observed that concentrations of CRK up to 30 μM failed to modify the rate of spontaneous beating of guinea-pig isolated right atria, suggesting the absence of a direct cardiac stimulatory effect.

In more detailed studies of the systemic haemodynamic effects of CRK, Clapham and Buckingham (1988) reported that given orally (0.03 and 0.06 mg/kg) to conscious cats and by i.v. infusion (0.002 mg/kg/min for 7–10 min) to anaesthetized open-chest cats, this drug reduced BP and total peripheral resistance whilst increasing cardiac output via both HR and stroke volume. In both models, HR changes were minimal for reductions in BP approaching 20 mmHg. In anaesthetized cats central venous pressure was little influenced by infused CRK.

In anaesthetized open-chest rabbits, Hof *et al.* (1988) examined the haemodynamic effects of CRK in comparison with its enantiomers. Given by i.v. infusion, CRK and its enantiomers reduced BP and increased total peripheral conductance. The racemate was approximately one-half the potency of the (−)-enantiomer and the (+)-enantiomer was more than 100-fold less potent. HR exhibited only a weak tendency to increase following the racemate or (+)-enantiomer but increased more markedly following the (–)-enantiomer (9% following 30 μg/kg). Although myocardial contractile force remained unchanged irrespective of treatment, cardiac output increased dose-dependently. In a comparative study of CRK and the calcium antagonist nitrendipine (1–30 μg/kg i.v.) in anaesthetized normotensive mongrel dogs, Dumez *et al.* (1988) observed that both drugs produced dose-related decreases (10–60%) in BP. However, whereas each dose of nitrendipine caused an acute increase in cardiac output (9–64%), CRK was without effect on this parameter.

The systemic haemodynamic effects of pinacidil were studied by Kawashima and Liang (1985) in conscious normotensive dogs. Cumulative i.v. doses of pinacidil (0.05–0.4 mg/kg) produced dose-dependent falls in BP and total peripheral resistance. Although HR increased more promptly, cardiac output and left ventricular (LV) dP/dt were minimally affected until a total dose of 0.4 mg/kg had been given. Using doses of hydralazine which produced comparable stepwise reductions in total peripheral resistance, the increases in HR, cardiac output and dP/dt were more exaggerated than those seen with pinacidil. Treatment with propranolol significantly reduced the increase in HR and abolished the increases in cardiac output and LV dP/dt due to pinacidil, implying that these responses were attributable to reflex sympathetic stimulation.

Also in conscious dogs, Preuss *et al.* (1985) studied the systemic haemodynamics of nicorandil in comparison with nitroglycerin. Nicorandil (25–300 μg/kg/min) and nitroglycerin (5–60 μg/kg/min) increased HR and reduced aortic and LV pressures. At these infusion rates nicorandil caused greater decreases in aortic and LV pressures than nitroglycerin, but the latter decreased LV end-diastolic pressure to a greater degree. Nicorandil, though not nitroglycerin, increased cardiac output as a result of increasing HR and also increased myocardial contractility (dP/dt_{max}). Pretreatment with propranolol attenuated the nicorandil-related increase in HR, prevented the increase in dP/dt_{max} and reduced the fall in LV end-diastolic pressure.

10.2.1.2 Regional haemodynamic effects

In anaesthetized open-chest rabbits, Cook and Hof (1988) showed that CRK, given by bolus i.v. injection, augmented blood flow to the wall of the stomach and to the small intestine. In support of these findings, Longman *et al.* (1988) demonstrated that when given by i.v. infusion to anaesthetized cats CRK increased superior mesenteric blood flow; pinacidil and nicorandil produced similar effects. Similarly, brain blood flow tended to increase following i.v. CRK injection in the rabbit studies of Cook and Hof (1988) and CRK infusion augmented carotid blood flow in the anaesthetized cat (Buckingham *et al.*, 1986a).

In the intact, unstressed animal, renal blood flow (RBF) is usually maintained within narrow limits by a system of autoregulation. Nonetheless increases in RBF have been observed following i.v. administration of CRK in anaesthetized normotensive cats and dogs (Buckingham *et al.*, 1986a; Longman *et al.*, 1988; Dumez *et al.*, 1988) and following oral administration in conscious normotensive cats (see Fig. 10.1) (Clapham and Buckingham, 1988; Clapham, 1988) though not in anaesthetized open-chest rabbits following i.v. injection (Cook and Hof, 1988). At an equieffective BP-lowering oral dose in conscious normotensive cats, pinacidil failed to increase RBF (Clapham, 1988). Similarly, neither pinacidil nor nicorandil augmented RBF when given by i.v. infusion to anaesthetized cats (Longman *et al.*,

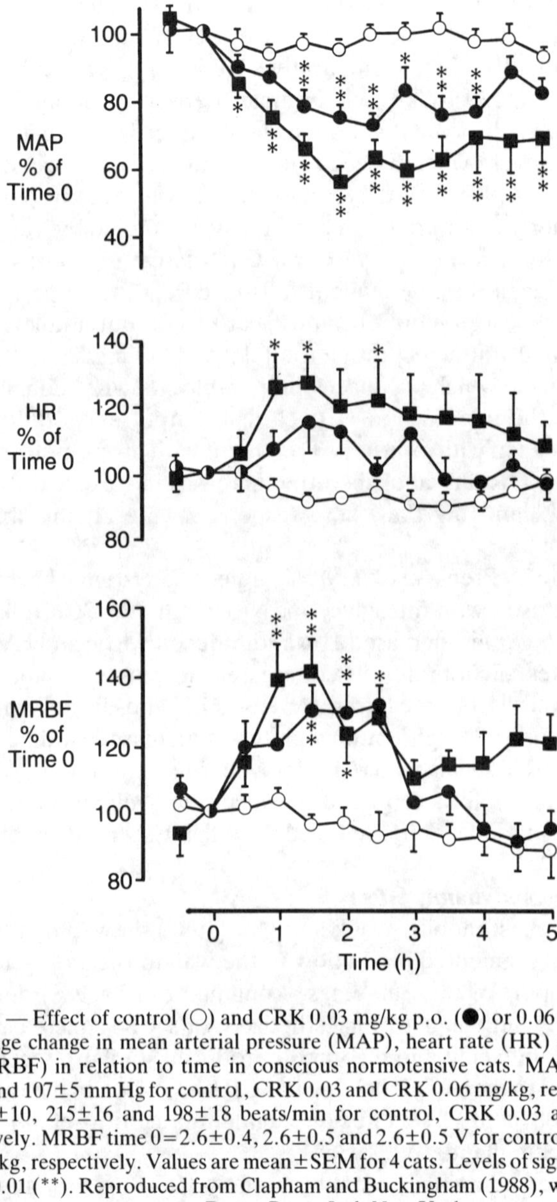

Fig. 10.1 — Effect of control (○) and CRK 0.03 mg/kg p.o. (●) or 0.06 mg/kg p.o. (■) on the percentage change in mean arterial pressure (MAP), heart rate (HR) and mean renal blood flow (MRBF) in relation to time in conscious normotensive cats. MAP at time 0 = 107±7, 104±6 and 107±5 mmHg for control, CRK 0.03 and CRK 0.06 mg/kg, respectively. HR at time 0 = 183±10, 215±16 and 198±18 beats/min for control, CRK 0.03 and CRK 0.06 mg/kg, respectively. MRBF time 0=2.6±0.4, 2.6±0.5 and 2.6±0.5 V for control, CRK 0.03 and CRK 0.06 mg/kg, respectively. Values are mean±SEM for 4 cats. Levels of significance at $p<0.05$ (*) and $p<0.01$ (**). Reproduced from Clapham and Buckingham (1988), with kind permission of Raven Press, Ltd, New York.

1988). Despite these findings in the cat, Olsen and Arrigoni-Martelli (1983) and Morimoto *et al.* (1986) have demonstrated an increase in RBF following the i.v. administration of pinacidil to the anaesthetized dog. These results for pinacidil in the dog would appear to resemble those reported for CRK in this species by Dumez *et al.* (1988).

The relationship between RBF and hypertension in man has been discussed by Hollenberg *et al.* (1978). It follows from their studies that antihypertensive drugs which increase RBF might possess a more desirable haemodynamic profile than those drugs which are ineffective on this parameter.

In both anaesthetized cats and anaesthetized open-chest rabbits CRK was found to have little effect on skeletal muscle blood flow (Buckingham *et al.*, 1986; Cook and Hof, 1988). Olsen and Arrigoni-Martelli (1983) also failed to show an effect of pinacidil on femoral blood flow in anaesthetized dogs. However, in a study in anaesthetized dogs by Shiraki *et al.* (1981), nicorandil caused a dose-related transient increase in femoral blood flow and transient increases also in renal and mesenteric blood flow. Resistances in these vascular beds were reduced for much longer, coinciding with the fall in systemic BP. Sakai *et al.* (1983) noted a biphasic change in femoral blood flow — increase followed by decrease — produced by nicorandil, 0.01–0.1 mg/kg i.v., in anaesthetized mini-pigs. Doses of 0.3 and 1.0 mg/kg, however, produced only a decrease in femoral flow, although again arterial resistance in the recipient vascular bed was reduced at all doses.

The differential effects of calcium antagonists and K^+ channel openers on blood flow to normoxic skeletal muscle is striking. In anaesthetized rats, Angersbach and Nicholson (1988) have demonstrated that at equieffective BP lowering doses nifedipine, verapamil and diltiazem caused greater increases in gastrocnemius muscle blood cell flux than CRK, pinacidil or nicorandil. However, when this skeletal muscle was made ischaemic by unilateral femoral artery ligation 6 weeks earlier, the calcium antagonists no longer increased blood cell flux, and either had no effect upon or reduced (diltiazem) the mean pO_2 of the already hypoxic tissue. By contrast, the K^+ channel activators markedly increased both blood cell flux and tissue pO_2 in the ischaemic muscle. In view of their findings, Angersbach and Nicholson (1988) proposed that the calcium antagonists primarily relax arterial resistance vessels (i.e. those vessels already dilated by the metabolic consequences of hypoxia), whereas the K^+ channel activators may also relax larger arteries, including collateral vessels. These results raise the possibility that drugs such as CRK, pinacidil and nicorandil may be of use in the treatment of peripheral vascular disease.

A coronary vasodilator effect of CRK has been shown in open-chest anaesthetized rabbits by Hof *et al.* (1988), who found that vasodilation was more pronounced in the outer layer of the LV free wall than in the subendocardial layer. Olsen and Arrigoni-Martelli (1983) and Kawashima and Liang (1985) have shown a coronary vasodilator action of pinacidil when given as cumulative i.v. doses to anaesthetized or conscious dogs. In the experiments of Kawashima and Liang (1985), pinacidil caused a dose-dependent increase in coronary blood flow and a corresponding decrease in coronary vascular resistance. Simultaneously, coronary sinus oxygen saturation increased and myocardial oxygen extraction decreased. Myocardial oxygen consumption also increased but LV work did not alter significantly.

The coronary vasodilator properties of nicorandil are now well documented (e.g. Uchida *et al.*, 1978a, b; Taira *et al.*, 1979; Nakagawa *et al.*, 1979; Sakai *et al.*, 1981) and this compound is presently being evaluated primarily as an anti-anginal agent. Increases in coronary blood flow were produced, by i.v. bolus doses or infusions of nicorandil given to anaesthetized mini-pigs (Sakai *et al.*, 1983) or anaesthetized dogs (Taira *et al.*, 1979; Sakai *et al.*, 1981; Preuss *et al.*, 1985), with the greatest increase in perfusion occurring in the subepicardium and midmyocardium (Preuss *et al.*, 1985). In a study by Gross *et al.* (1985) in anaesthetized dogs, transient total occlusion followed by partial occlusion of the left circumflex coronary artery produced marked underperfusion of the subendocardium. Nicorandil, 0.06 mg/kg i.v., augmented blood flow to this ischaemic zone. Further studies in anaesthetized dogs compared the effects of nicorandil and nifedipine on collateral blood flow following acute (Lamping and Gross, 1984) and chronic (Lamping *et al.*, 1984a) occlusion of the left anterior descending (LAD) coronary artery. When BP was held constant by use of an aortic cuff or snare, but not when BP was allowed to fall, nicorandil (25–50 μg/kg/min i.v.) increased collateral blood flow, and in the ischaemic region subendocardial flow was increased to a greater extent than subepicardial flow. Under similar conditions nifedipine (1–3 μg/kg/min i.v.) increased blood flow primarily to the subepicardial region of the ischaemic zone. These results may indicate that nicorandil is a more specific dilator of collateral vessels than of resistance vessels, and Gross *et al.* (1985) also raised the possibility that nicorandil may selectively dilate intramural conductance vessels which distribute blood between subepicardium and subendocardium.

In a further study by Lamping *et al.* (1984b) in anaesthetized dogs, myocardial infarction was produced by a 2-hour occlusion and 30-min reperfusion of the LAD coronary artery. Nicorandil, 100 μg/kg i.v. then 25 μg/kg/min, given shortly after LAD occlusion and throughout the remainder of the experiment, reduced the size of the infarction. This effect, however, was not related to an improvement in myocardial blood flow or to a change in global haemodynamics. In a study by Korb *et al.* (1985) in anaesthetized dogs the myocardium was stressed by repeated 3-min occlusions and reperfusions of the LAD coronary artery. Premedication with nicorandil (0.64 μmol/kg i.v.) led to a significant increase in oxygen debt and free fatty acid metabolism during ischaemia, suggesting that nicroandil may actually exacerbate rather than improve the metabolic and energetic consequences of ischaemia. A further important point to make is that in the study of Gross *et al.* (1985) nicorandil and nitroglycerin produced similar effects on the distribution of coronary blood flow, raising the possibility that this particular profile of activity owes rather less to K^+ channel activation than to the presence of a nitrate group in the molecule.

10.2.1.3 Effects on vasoconstriction evoked by endogenous and exogenous pressor substances

In pithed normotensive rats i.v. bolus injections of pinacidil (0.3–30 mg/kg) or CRK (0.03–0.1 mg/kg) caused dose-dependent rightward displacements of the dose–pressor response curves to α_1- and α_2-adrenoceptor agonists (Thoolen *et al.*, 1983; Donato Di Paola *et al.*, 1987). Pinacidil, 3 mg/kg i.v., also suppressed frequency-dependent pressor responses to electrical stimulation of the spinal sympathetic outflow in this preparation (Thoolen *et al.*, 1983). Other results suggest that the prevailing level of basal BP, which is reduced by pinacidil and CRK, has a substantial influence on the outcome of experiments such as these. Van Meel *et al.* (1983) demonstrated that in pithed normotensive rats, raising the BP with an i.v. infusion of vasopressin reduced the ability of pinacidil, but not of a series of calcium antagonists, to inhibit pressor responses to the α_2-adrenoceptor agonist B-HT 920. Furthermore, Buckingham (1988) reported that in pithed SHR i.v. vasopressin infusion reversed the inhibitory effect of CRK, but not of nifedipine (each at a dose of 0.3 mg/kg i.v.), against frequency-dependent pressor responses evoked by electrical stimulation of the spinal cord sympathetic outflow. Moreover, in conscious SHR at this dose level CRK produced a much greater anti-hypertensive effect than nifedipine. In anaesthetized, ganglion-blocked SHR, both CRK and nifedipine (0.1, 0.3 mg/kg) antagonized pressor responses to incremental infusions of noradrenaline and phenylephrine but only nifedipine also reduced responses to methoxamine, vasopressin and angiotensin II (Buckingham, 1988). In these studies no attempt was made to assess the additional effect of a background infusion of vasopressin on the inhibition by CRK of responses to noradrenaline or phenylephrine. The weak inhibitory effect of K^+ channel activators against pressor responses to angiotensin II *in vivo* has been confirmed by Cook *et al.* (1988). These authors demonstrated that at equivalent BP-lowering doses in anaesthetized normotensive rats, CRK, pinacidil and nicorandil produced less inhibition of pressor responses to bolus injections of angiotensin II than did various calcium antagonists.

Hence, although the K^+ channel activators exhibit potent anti-vasoconstrictor properties *in vitro*, it would appear unlikely that these properties contribute substantially to the falls in peripheral vascular resistance seen with these agents *in vivo*. The fact that both CRK (Buckingham, 1988) and pinacidil (Van Meel *et al.*, 1983) decreased the baseline BP in pithed rats, in which there is no operational sympathetic nervous system, implies that these drugs predominantly relax vascular smooth muscle by a direct action.

10.2.1.4 Effects on aspects of renal function

Hollenberg *et al.* (1978) drew attention to a strong association between the onset of essential hypertension in man and a reduction in RBF. From this association it is possible to make a case for treating hypertensive patients with drugs which augment RBF and thereby reverse a functional abnormality in the renal blood supply. In fact the renal response to drug therapy is likely to be conditioned largely by the fall in BP and the direct and indirect effects of the therapeutic agent(s) in question. Certainly in man vasodilator therapy frequently results in an expansion of plasma volume due to water and salt retention. This effect leads to diminution in anti-hypertensive response and is attributable to several mechanisms: reduced renal perfusion pressure; reflex activation of the renal sympathetic nerves; reflex activation of the renin–angiotensin–aldosterone cascade (Koch-Weser, 1974).

The earlier section of this review, devoted to regional haemodynamics, drew attention to the effects of K^+ channel activators on RBF in conscious and anaesthetized animals. However, comparatively little has been reported thus far on the effects of these drugs on other renal parameters in animals.

As mentioned earlier, in the conscious normotensive cat, oral doses of CRK have been found to augment RBF (Clapham and Buckingham, 1988; Clapham, 1988), whereas at equieffective anti-hypertensive doses pinacidil and nifedipine did not (Clapham, 1988). In this latter study the increases in HR and plasma renin activity (PRA) produced by CRK were less than those seen following pinacidil or nifedipine. In conscious normotensive dogs Arrigoni-Martelli *et al.* (1980) demonstrated that at a dose of pinacidil (0.5 mg/kg i.v.) which caused a fall in systemic BP of approximately 50 mmHg, PRA was increased around threefold. At present, however, there are no reports of comparisons between K^+ channel activators in the context of their liability to increase PRA in conscious dogs.

Of some concern is the report by Ferrier *et al.* (1988) that CRK caused a direct stimulation of renin production by juxtaglomerular cells in culture, though at present there are no reports of appropriate *in vivo* studies to support this observation. As a consequence of renin release *in vivo*, a cascade of events is stimulated in which angiotensin II formation and aldosterone secretion are in turn enhanced. According to Shepherd *et al.* (1988), CRK (10^{-7}, 2×10^{-6} M) marginally increased basal aldosterone secretion but did not affect angiotensin II-stimulated secretion from bovine or rat adrenocortical cells. CRK, 4×10^{-5} M, reduced both basal and angiotensin II-stimulated aldosterone output.

As regards studies of CRK on renal function, only the effect of the drug in rats has been published thus far. Dumez *et al.* (1988) found that in conscious SHR, CRK (0.1, 0.3 mg/kg p.o.) produced a dose-related anti-hypertensive effect and a reduction in urine volume; the higher dose also reduced urinary Na^+ but not K^+ excretion. At 3 mg/kg p.o., CRK produced profound hypotension accompanied by a large diuresis, natriuresis and kaliuresis. These latter results are intriguing and unexplained at present. One possible explanation would be that at sufficiently high dosage CRK inhibits electrolyte transport in the nephron. The fact that diuresis occurred despite a very marked anti-hypertensive effect confirms that the kidneys were still filtering and these observations suggest that the anti-diuretic/anti-natriuretic effect of lower doses of CRK must have been related to activation of counter-regulatory mechanisms.

The effects of pinacidil on renal function have been studied in greater depth. In the study by Arrigoni-Martelli *et al.* (1980) in conscious normotensive dogs, pinacidil, 0.5 mg/kg i.v., increased renal clearances of both *para*-amino hippurate and inulin during the first hour. Higher doses of pinacidil, however, resulted in a decrease of renal clearance. Morimoto *et al.* (1986) investigated the renal haemodynamic and functional effects of pinacidil in anaesthetized dogs. At a dose of 0.25 mg/kg i.v., systemic BP was reduced, RBF was increased but there were no significant changes in glomerular filtration rate (GFR), urine flow rate or urinary electrolyte concentrations. However, at a higher dose (0.5 mg/kg i.v.), BP was further reduced, RBF did not change, but GFR, urine volume and electrolyte excretion were all reduced significantly. Both doses increased PRA. This anti-diuretic effect of pinacidil was not prevented by renal denervation. Administration of pinacidil (2 μg/kg/min) directly into a renal artery produced an increase in RBF, urine volume and electrolyte excretion but no change in PRA, GFR or arterial blood pressure. Overall, the results suggested that the anti-diuretic effect of pinacidil, 0.5 mg/kg i.v., in dogs was due to a fall in renal perfusion pressure rather than to reflex stimulation of renal sympathetic nerve activity. In a further study to investigate the diuretic and natriuretic effect of pinacidil (0.4–10 μg/kg/min) infused into a renal artery of anaesthetized dogs, Morimoto *et al.* (1987) attributed this action partly to an increase in RBF and partly to a decrease in reabsorption of Na^+ and water by the proximal tubule. This latter effect may be due to a direct action on tubular electrolyte handling and/or to an effect on Starling forces in the renal interstitium induced by vasodilation of peritubular capillaries.

10.2.1.5 Effects on cardiac arrhythmias

In vitro studies with K^+ channel activators suggest that they affect cardiac muscle only at much higher concentrations than those active on vascular smooth muscle. Possibly as a result of this relative tissue selectivity, little attention appears to have been given to studying the effects of these drugs on cardiac arrhythmias. However, Kerr *et al.* (1985) have demonstrated that in conscious dogs pinacidil (0.25–1 mg/kg i.v.) produced a dose-related increase in the ratio of sinus to ectopic ventricular beats following two-stage occlusion of the LAD coronary artery 22–24 hours earlier. Although this anti-arrhythmic effect of pinacidil appeared to occur only at doses which lowered BP, doses of hydralazine or sodium nitroprusside which caused similar falls in BP did not show a consistently beneficial effect on the arrthythmia. The anti-arrhythmic effect of pinacidil was not modified by ß-adrenoceptor blockade. In anaesthetized dogs pinacidil had no effect on the arrthythmias induced by i.v. injection of ouabain (Kerr *et al.*, 1985).

10.2.1.6 Effects on baroreceptor sensitivity.

In a study by Clapham and Cooper (1988) in anaesthetized cats, equieffective BP lowering doses of CRK (10 μg/kg i.v.) and pinacidil (60 μg/kg i.v.) were compared for their effects on baroreceptor sensitivity, estimated by relating changes in systolic BP to changes in heart period produced by i.v. doses of phenylephrine or sodium nitroprusside. Using the 'steady state' method of assessment, CRK, but not pinacidil, augmented baroreceptor sensitivity to phenylephrine, whereas pinacidil, but not CRK, increased the sensitivity to sodium nitroprusside. If these results could be extrapolated to hypertensive man, they would suggest that during CRK treatment the baroreceptor control of the circulation might be reset to minimize the increases in BP caused by various presssor stimuli. By contrast, the BP-lowering effects of pinacidil might be diminished by enhanced sympathetic drive to the circulation.

10.2.1.7 **In vivo** *evidence that K^+ channel activation mediates the vascular relaxant effect of CRK*

Wilson *et al.* (1988) and Quast and Cook (1988, 1989) have demonstrated that the anti-diabetic sulphonylurea glibenclamide is an effective inhibitor of CRK's ability to relax vascular smooth muscle and to stimulate ^{86}Rb efflux from vascular tissue *in vitro*. Furthermore, Cavero *et al.* (1988) have now shown that in anaesthetized rats the hypotensive effect of CRK (5 μg/kg/min i.v. for 20 min) was prevented by pretreatment with glibenclamide (20 mg/kg i.v.). In conscious rats Wilson *et al.* (1988) were unable to show an inhibitory effect of glibenclamide, 100 mg/kg p.o., on the hypotensive response to CRK, 0.075 mg/kg i.v. Overall, these results suggest that at sufficiently high dosage *in vivo*, well in excess of that required to stimulate insulin secretion from pancreatic ß-cells, glibenclamide can be shown to block CRK's vascular relaxant effect. These results provide the first corroborative evidence that CRK relaxes vascular smooth muscle *in vitro* and *in vivo* via an activation of membrane K^+ channels. A final point worthy of note is that despite this interaction between CRK and glibencalmide on vascular tissue, Wilson *et al.* (1988) could find no effect of CRK (0.3 mg/kg p.o.) on the plasma insulin concentration in conscious fasted rats during a glucose tolerance test, nor any effect on insulin secretion stimulated by glibenclamide (2.5 mg/kg p.o.). Quast and Cook (1989) reported that a high i.v. dose of CRK (0.3 mg/kg) produced a 25% increase in the resting blood glucose concentation and slightly prolonged the elevation of this parameter evoked by i.v. injection of 0.5 g/kg glucose. It was not established whether this effect, which was minimal in relation to the well-known hyperglycaemic effect of diazoxide, was due to a direct or indirect (reflex) action of CRK.

10.2.2 Bronchial effects

In anaesthetized guinea-pigs, Buckle *et al.* (1987) have shown that CRK (10–400 μg/kg i.v.) caused a dose-related inhibition of the bronchoconstrictor effect of an approximately ED_{50} dose of 5-hydroxytryptamine (5-HT). By contrast, nifedipine (10–300 μg/kg i.v.) produced no inhibition of the 5-HT response. In conscious guinea-pigs, pretreatment with CRK (2.5 and 5 mg/kg p.o.), but not nifedipine (up to 5 mg/kg p.o.), protected against asphyxic collapse caused by inhaled histamine. In a later paper, Arch *et al.* (1988) demonstrated that pinacidil mimicked the anti-bronchoconstrictor effect of CRK in guinea-pigs but was 3-7-fold less potent in its effects against i.v. 5-HT, inhaled histamine or inhaled ovalbumin following earlier sensitization. These results provide justification for the clinical evaluation of the K^+ channel activators as bronchodilator agents.

10.2.3 Effects on bladder and uterus

In a pig model of bladder outflow obstruction with detrusor instability, Foster and Brading (1987) demonstrated that CRK (0.3 mg/kg) abolished the characteristic abnormal spontaneous increases in detrusor pressure without impairing the animals' ability to void. These results provide early support for the notion that K^+ channel activators may find utility in the treatment of urinary incontinence.

In both pregnant and non-pregnant bilaterally ovariectomized rats, equipped with an intrauterine microballoon, CRK (0.1 or 1 mg/kg i.v.) produced an immediate and complete cessation of uterine contractions and the duration of this effect was dose-related (Downing *et al.*, 1988). However, on repeat injection of 1 mg/kg i.v. at 12-hour intervals, rapid tolerance developed to the uterine inhibitory action. This rapid development of tolerance is surprising and the mechanism responsible for it is not understood at present. It would be too soon to suggest that the K^+ channel activators may find a use in the treatment of premature labour.

10.3 CLINICAL EXPERIENCE

10.3.1 Cromakalim

Early cardiovascular studies in man have focused generally on the effects of CRK, 0.5–2 mg orally, given either as a single dose or once daily for several days or weeks. In eight healthy volunteers CRK (2 mg) increased supine and standing HR for 24 hours, indicative of a vasorelaxant effect (Fox *et al.*, 1986). Significant reductions in supine and standing systolic and diastolic BPs were seen only on isolated occasions during this time. In a further 12 normotensive volunteers, Fox *et al.* (1986) demonstrated that following a single oral dose of CRK, 1 or 2 mg, forearm blood flow increased significantly but venous capacitance remained unchanged. In a haemodynamic study conducted by Dixon *et al.* (1988) in eight male patients undergoing routine cardiac catheterization for assessment of ischaemic heart disease, CRK, 0.015 mg/kg infused i.v., caused a short-lived fall in mean BP but a slightly more prolonged fall in systemic vascular resistance. Cardiac output and HR also increased significantly though no changes were observed in mean pulmonary BP and total pulmonary vascular resistance. LV d*P*/d*t* rose slightly but not significantly in all patients, reflecting the trend in HR. In essence these results are similar to the earlier haemodynamic findings in conscious and anaesthetized cats, and support the view that CRK is an arterial vasodilator.

The efficacy of CRK by the oral route in patients with essential hypertension has been established by VandenBurg *et al.* (1978a), who demonstrated that a single dose of 1.5 mg lowered supine and erect mean BP by up to 11 and 13 mmHg, respectively, during 8 hours with only a small increase in HR. The anti-hypertensive effect of CRK had virtually disappeared after 10 hours, despite the fact that the plasma half-life of the active (−)-enantiomer in man is around 18–23 hours (Gill *et al.*, 1988). It needs to be stressed, however, that with a plasma half-life of this order and linear kinetics, the maximum therapeutic effect of CRK would not be expected until at least the fifth day of once-daily dosing. As yet no systematic study has been conducted to establish the relationship between plasma drug concentration and therapeutic response.

In view of the tendency of some vasodilator drugs to become less effective during prolonged treatment because of their stimulatory effects on counter-regulatory mechanisms, it is of importance in the development of a new drug to assess this risk at an early stage in the clinical evaluation programme. There is some evidence that, at least in the short-term, CRK provokes a reflex sympathetic counter-regulatory response. As already stated above, HR is generally elevated and in a study in eight normotensive healthy subjects given a single oral dose of CRK (1 mg), increases were also observed in the plasma noradrenaline concentration (61%), PRA (122%) and the plasma angiotensin II concentration (105%) (Ferrier *et al.*, 1988). In a further study in eight normotensive volunteers, Elliott and Howie (1988) also noted small increases in HR and PRA following a single 1 mg dose or eight daily doses of 1 mg. However, in eight patients with established essential hypertension, Lebel *et al.* (1988) reported that oral CRK, 1.5 mg daily for 3 days, produced small significant falls in means BP and increases in HR but no change in PRA. No change was observed in the plasma aldosterone concentration in this study or in the study by Ferrier *et al.* (1988) in which PRA was increased. A further possible influence on the long-term efficacy of CRK is the drug's effect on renal haemodynamics. In three of eight patients in the study by Lebel *et al.* (1988), effective renal plasma flow was correlated with an increase in the urinary excretion of 6-keto-$PGF_1\alpha$, the stable metabolite of prostacyclin. Overall in this patient group, there was no change in GFR or filtration fraction. It will be recalled that a renal vasodilator action of CRK has also been observed in conscious and anaesthetized cats.

As yet there are no published reports of long-term efficacy trials of CRK in patients and thus it is not known whether tolerance ultimately develops to its anti-hypertensive action. In a multi-centre trial in mild to moderate hypertensive patients reported by VandenBurg *et al.* (1987b), the fall in supine or standing mean BP produced 4 hours after administration of CRK, 1.0 or 1.5 mg once daily (o.d.), was similar on the first and eighth day of dosing. BP was not, however, reduced 24 hours after either the first or final dose of the drug. After the first dose of CRK (0.5, 1.0 or

1.5 mg), supine HR was significantly increased at 4 hours. After the eighth dose, however, there was no increase in supine HR at any dose level, though standing HR was increased following 1.5 mg. An abstract published by Eckl and Greb (1987), mentioning a 6-week trial of oral CRK, 0.75 mg o.d., in patients with primary hypertension stated only that CRK slowly decreased sitting systolic and diastolic BP over 14 days.

The report by VandenBurg *et al.* (1987b) also makes reference to the side-effect profile of CRK during 8 days of treatment. There was no evidence of weight gain at any dose level, though headache was a dose-related event, being reported by 50% of patients taking 2 mg o.d. Other adverse effects of drug treatment were reported as minor and not dose-related.

One further cardiovascular question which has been addressed in man as well as in animals is the possibility that CRK exerts an anti-vasoconstrictor effect. Nguyen *et al.* (1987) reported that CRK 2 mg orally (2 hours earlier) reduced the magnitude of pressor responses evoked by i.v. infusions of noradrenaline (200 ng/kg/min) or angiotensin II (20 ng/kg/min). However, it could be argued that studies such as this may have an intrinsic methodological weakness, making it difficult to draw conclusions. Because i.v. infusion rates of agonists remain fixed throughout the study, and because cardiac output and tissue blood flow may well increase as a consequence of test drug-induced peripheral vasodilatation, the actual arterial plasma concentration of agonist may be reduced during the repeat challenge. This agonist dilution could explain an apparent reduction in vasoconstrictor response.

10.3.2 Pinacidil

Pinacidil, though only recently identified as having K$^+$ channel activator properties (Bray *et al.*, 1987; Cook *et al.*, 1988) has been studied extensively as an antihypertensive agent in man. Early clinical experience with pinacidil has been reviewed by Arrigoni-Martelli and Finucane (1985). In a single oral dose study (37.5 mg) in healthy volunteers, pinacidil, like CRK, increased HR without influencing BP. In the presence of propranolol, the tachycardia was inhibited and supine diastolic and standing systolic BP were reduced (Nicholls *et al.*, 1986a). Given i.v. to normotensive patients undergoing cardiac catheterization, pinacidil, 0.2 mg/kg, caused an immediate fall in BP associated with a fall in systemic vascular resistance and increases in both HR and cardiac index (Nicholls *et al.*, 1986b). Mean pulmonary wedge pressure was persistently reduced and a slight reduction in mean pulmonary artery pressure was observed. Overall this haemodynamic profile is consistent with a selective arterial dilator action of pinacidil.

In chronic efficacy trials in hypertensive patients there seems little doubt that pinacidil is rapidly metabolized. In early trials using a rapidly disintegrating tablet formulation, Ward (1984) considered that in ten patients treated with pinacidil as monotherapy, 40–100 mg daily was required in two or three divided doses to provide satisfactory BP control. Ramsey and Freestone (1983), also using this short-acting formulation, gave pinacidil, 10–30 mg twice daily (b.i.d.), for 3–7 weeks to four patients already receiving a ß-blocker and a diuretic. This study, however, was terminated because of the drug's short duration of action and its severe side-effects. More recent studies, therefore, have utilized a slow-release formulation but doubts remain about its suitability for use as monotherapy. For example, Carlsen *et al.*

(1983) found that in seven of nine patients with essential hypertension given slow-release pinacidil, 20–40 mg b.i.d. for 4 weeks, BP had returned to baseline within 10 hours of each dose. Moreover, although HR was scarcely elevated, oedema was a prominent side-effect in four patients. A high incidence of side-effects, including weight gain, headache, dyspnoea on exertion, insomnia, fatigue, nocturia, dreams, dizziness and palpitations, were reported by Byyny *et al.* (1987) as being a complication of monotherapy in patients receiving up to 75 mg b.i.d. of pinacidil. In this study an anti-hypertensive response was observed in 15 of 17 patients receiving pinacidil alone but the side-effects were best reduced by combination with hydrochlorothiazide and/or propranolol. According to Zachariah *et al.* (1986), oedema was again a prominent side-effect of pinacidil monotherapy in 13 mildly hypertensive patients whose BP was otherwise well-controlled throughout 9 months on a daily intake of 62±18 mg in divided doses. Rather surprisingly, however, HR remained elevated during treatment.

When pinacidil was added to existing inadequate anti-hypertensive drug treatment, D'Arcy *et al.* (1985) reported that a mean of 38 mg b.i.d. resulted in a target supine diastolic BP of <90 mmHg in 19 of 23 patients with only mild and transient side-effects. In moderate to severe hypertension, inadequately controlled by a diuretic or ß-blocker, Koliopoulos *et al.* (1984) observed that pinacidil, 46.7 mg b.i.d. for 4 weeks produced satisfactory BP control for only 5–7 hours after each dose. Again weight gain was a side-effect in some patients. During a 4-month trial by Krusell *et al.* (1986) in ten patients insufficiently controlled by chronic ß-blocker therapy, BP declined to target level within 1 month of initiating pinacidil treatment. However, in seven patients drug dosage needed to be increased to sustain BP control after 2 and 4 months. Once more these results suggest that concomitant diuretic therapy is required to prevent tolerance to pinacidil occurring.

Overall, these results suggest that acute and prolonged treatment with pinacidil are attended by a significant incidence of troublesome side-effects which result from activation of counter-regulatory reflexes. These problems have been highlighted again in Izzo *et al.* (1987), who studied the development of pseudo-tolerance to vasodilator drugs in essential hypertension. These authors reported that following 1 month of treatment at equi-antihypertensive doses pinacidil, but not prazosin or captopril, increased cardiac index. The upright and supine plasma noradrenaline concentrations were increased approximately 50% by pinacidil and prazosin but captopril decreased supine plasma noradrenaline. Pinacidil and prazosin reversibly increased whereas captopril decreased bodyweight; all three compounds caused an expansion of blood volume. The effects of pinacidil on PRA are, however, rather difficult to assess since such studies have often been conducted during concomitant treatment with other anti-hypertensive drugs. For example, Byyny *et al.* (1987) demonstrated that during up to 9 months of treatment with pinacidil PRA was increased by around twofold, though most patients were also receiving either propranolol or hydrochlorothiazide or both drugs. Muiesan *et al.* (1985) reported that when pinacidil treatment (mean daily dose of 36.6 mg) was given to normalize BP in patients already receiving hydrochlorothiazide and amiloride, PRA increased only slightly. HR and the plasma noradrenaline concentration were also only slightly increased by pinacidil in this study.

A rather severe test of pinacidil's effect on PRA was made by Rijk and Thien

(1987), who gave 0.1–0.2 mg/kg i.v. to ten supine untreated hypertensive patients. In this study diastolic BP fell from 108 to 80 mmHg, and at peak effect HR increased by almost 30% and PRA by approximately twofold. Interestingly, the plasma aldosterone concentration exhibited little change. A further i.v. study examined pinacidil's effect on renal function in moderately hypertensive patients inadequately controlled by chronic ß-blocker therapy (Krusell *et al.*, 1986). Pinacidil, 0.1 mg/kg i.v., decreased BP by 30/18 mmHG and raised HR by 21% despite ß-blockade. Renal vascular resistance (RVR) also fell markedly and there was a transient increase in RBF. However, urine volume, Na^+ and K^+ clearance and uric acid excretion all decreased. Retrospectively, both the basal RVR and the fall in RVR produced by acute pinacidil treatment were significantly correlated with the decrease in diastolic BP after 4 months of continued therapy.

10.3.3 Nicorandil

As mentioned earlier, nicorandil is being developed as an anti-anginal drug and there is a strong possibility that its pharmacodynamic effects owe as much to the nitrate, substitutent in the molecule as to its K^+ channel-activating properties. In this regard Belz *et al.* (1985), using non-invasive measuring techniques, compared the haemodynamic changes produced by nicorandil (60 mg sublingually), isosorbide dinitrate (ISDN; 20 mg sublingually) and dihydralazine (10 mg i.v.) in 12 healthy subjects. Dihydralazine produced effects consistent with afterload reduction, whereas ISDN predominantly reduced preload with some additional reduction in afterload. Nicorandil reduced preload to the same extent as ISDN but produced a greater reduction in afterload. Differences between nicorandil and trinitroglycerin (TNG) in their effects on the coronary circulation of anginal patients have been identified by Aizawa *et al.* (1987). These authors showed that nicorandil (4 mg i.v.) increased coronary sinus flow and decreased coronary vascular resistance, whereas TNG (0.3 mg sublingually) exerted opposite effects. In patients with variant angina, TNG was a more effective coronary vasorelaxant than nicorandil. Whereas nicorandil relaxed first-order more than second-order arteries, TNG was more effective in second-order vessels. The fact that nicorandil also reduced coronary vascular resistance suggests that this drug may also relax resistance vessels.

The benefits of nicorandil in stable effort angina have been confirmed by Kato *et al.* (1987). In this multi-centre trial nicorandil, 20 mg orally, was given to 29 patients using a symptom-limited treadmill exercise test. At 1, 3 and 6 hours after administration, nicorandil significantly increased maximal exercise duration and time to 1 mm of ST-segment depression. In a further study (Yokota *et al.*, 1987) in nine supine, exercising patients with old myocardial infarction but without angina pectoris, nicorandil, 15 mg orally, was found to produce a number of beneficial effects: a decrease in pulmonary artery wedge pressure (also seen at rest); improvement in an index of cardiac function obtained from the relationship between pulmonary artery wedge pressure and LV stroke work index; a decrease in the ratio of pressure-rate product to coronary sinus flow (an index of the ratio of myocardial oxygen consumption to myocardial oxygen supply).

In nine anginal patients who reproducibly developed myocardial ischaemia during ventricular pacing at 170 beats/min, Thormann *et al.* (1983) examined the effect of nicorandil, 20 mg sublingually, in the immediate post-pacing period.

Following drug treatment, the anticipated decline in LV function due to ischaemia did not develop and increases were seen in ejection fraction, cardiac index and LV stroke work index. Simultaneously, LV end diastolic pressure decreased and neither angina nor pathological ST segment changes occurred. An improvement in regional wall motion was also observed. Kambara *et al.* (1987) have reported qualitatively similar effects in 11 patients following nicorandil, 5 mg sublingually.

10.3.4 K^+ channel activators in non-cardiovascular indications

The first clinical evidence that K^+ channel activators may have a utility additional to the treatment of cardiovascular disorders has been published by Baird *et al.* (1988). These authors demonstrated that in a single oral dose, in seven healthy volunteers, both cromakalim (2 mg) and salbutamol (4 mg) required the concentration of inhaled histamine to be significantly increased in order to produce a 40% fall in the partial expiratory flow rate. These results suggest that CRK exhibits anti-broncho-constrictor activity in man as well as in animal species. Thus far there are no further published clinical studies which seek to exploit the findings of diverse smooth muscle relaxant properties of K^+ channel activators in animals. In particular, clinical studies in asthma, urinary incontinence and peripheral vascular disease (e.g. intermittent claudication; Raynaud's disease) would seem to be warranted.

10.4 CONCLUSIONS

At present the published clinical experience with CRK, pinacidil and nicorandil is almost exclusively of a vascular nature, though there is an early indication that CRK may be useful in the treatment of bronchial asthma. In view of the diversity of their smooth muscle relaxant effects in animal species *in vitro* and *in vivo*, it will be interesting to observe whether the K^+ channel activators ultimately find an extensive clinical utility.

REFERENCES

Aizawa, T., Ogasawara, K., and Kato, K. (1987) Effects of nicorandil on coronary circulation in patients with ischemic heart disease: comparison with nitroglycerin. *J. Cardiovasc. Pharmacol.* **10** (suppl. 8) S123–S129.

Angersbach, D., and Nicholson, C. D. (1988) Enhancement of muscle blood cell flux and pO_2 by cromakalim (BRL 34915) and other compounds enhancing membrane K^+ conductance, but not by Ca^{2+} antagonists or hydralazine, in an animal model of occlusive arterial disease. *Naunyn-Schmiedeberg's Arch. Pharmacol.* **337** 341–346.

Arch, J. R. S., Buckle, D. R., Bumstead, J., and Taylor, J. F. (1988) Comparison of the effects of cromakalim (BRL 34915) and pinacidil in guinea-pig models of bronchoconstriction. *Proceedings of the British Pharmacological Society*, Nottingham meeting, Abst. P3.

Arrigoni-Martelli, E., and Finucane, J. (1985) Pinacidil. In: *New drugs annual: Cardiovascular drugs*, Vol. 3, A. Scriabine (ed.). Raven Press, New York.

Arrigoni-Martelli, E., Kaergaard Nielsen, Chr., Bang Olsen, U., and Petersen, H.

J. (1980) *N″*-Cyano-*N*-4-pyridyl-*N′*-1,2,2-trimethylpropylguanidine, monohydrate (P1134): a new, potent vasodilator. *Experientia* **36** 445–447.

Baird, A., Hamilton, T. C., Richards, D. H., Tasker, T., and Williams, A. J. (1988) Inhibition of histamine induced bronchoconstriction in normal healthy volunteers by a potassium channel activator, cromakalim, BRL 34915. *Br. J. Clin. Pharmacol.* **25** 114P.

Belz, G. G., Matthews, J. H., Beck, A., Wagner, G., and Schneider, B. (1985) Hemodynamic effects of nicorandil, isosorbide dinitrate and dihydralazine in healthy volunteers. *J. Cardiovasc. Pharmacol.* **7** 1107–1112.

Bray, K. M., Newgreen, D. T., Small, R. C., Southerton, J. S., Taylor, S. G., Weir, S. W., and Weston, A. H. (1987) Evidence that the mechanism of the inhibitory action of pinacidil in rat and guinea-pig smooth muscle differs from that of glyceryl trinitrate. *Br. J. Pharmacol.* **91** 421–429.

Buckingham, R. E. (1988) Studies on the anti-vasoconstrictor activity of BRL 34915 in spontaneously hypertensive rats; a comparison with nifedipine. *Br. J. Pharmacol.* **93** 541–552.

Buckingham, R. E., Clapham, J. C., Coldwell, M. C., Hamilton, T. C., and Howlett, D. R. (1986b) Stereospecific mechanism of action of the novel antihypertensive agent, BRL 34915. *Br. J. Pharmacol.* **87** 78P.

Buckingham, R. E., Clapham, J. C., Hamilton, T. C., Longman, S. D., Norton, J., and Poyser, R. H. (1986a) BRL 34915, a novel antihypertensive agent: comparison of effects on blood pressure and other haemodynamic parameters with those of nifedipine in animal models. *J. Cardiovasc. Pharmacol.* **8** 798–804.

Buckle, D. R., Bumstead, J., Clarke, G. D., Taylor, J. F., and Taylor, S. G. (1987) Reversal of agonist-induced bronchoconstriction in the guinea-pig by the potassium channel activator BRL 34915. *Br. J. Pharmacol.* **92** 744P.

Byyny, R. L., Nies, A. S., Lo Verde, M. E., and Mitchell, W. D. (1987) A double-blind, randomized, controlled trial comparing pinacidil to hydralazine in essential hypertension. *Clin. Pharmacol. Ther.* **42** 50–57.

Carlsen, J. E., Kardel, T., Jensen, H. A. E., Tango, M., and Trap-Jensen, J. (1983) Pinacidil, a new vasodilator: pharmacokinetics and pharmacodynamics of a new retarded release tablet in essential hypertension. *Eur. J. Clin. Pharmacol.* **25** 557–561.

Cavero, I., Mondot, S., Mestre, M., and Escande, D. (1988) Haemodynamic and pharmacological mechanisms of the hypotensive effects of cromakalim in rats: blockade by glibenclamide. *Br. J. Pharmacol.* **95** 643P.

Clapham, J. C. (1988) Renal haemodynamic effects of cromakalim, nifedipine and pinacidil in the conscious cat. *Proceedings of the British Pharmacological Society*, Nottingham meeting. Abst. P2.

Clapham, J. C., and Buckingham, R. E. (1988) The haemodynamic profile of cromakalin in the cat. *J. Cardiovasc. Pharmacol.* **12** 555–561.

Clapham, J. C., and Cooper, S. M. (1988) Effect of cromakalim (BRL 34915) and pinacidil on baroreceptor sensitivity in the anaesthetised cat. *Br. J. Pharmacol.* **93** 200P.

Cohen, M. L. (1986) Current Trends Review. Pinacidil monohydrate — a novel vasodilator: review of preclinical pharmacology and mechanism of action. *Drug Devel. Res.* **9** 249–258.

Cook, N. S., and Hof, R. P. (1988) Cardiovascular effects of apamin and BRL 34915 in rats and rabbits. *Br. J. Pharmacol.* **93** 121–131.

Cook, N. S., Quast, U., Hof, R. P., Baumlin, Y., and Pally, C. (1988) Similarities in the mechanism of action of two new vasodilator drugs: pinacidil and BRL 34915. *J. Cardiovasc. Pharmacol.* **11** 90–99.

D'Arcy, V., Laher, M., McCoy, D., Sullivan, P., Walsh, C. H., and Hickey, M. P. (1985) Pinacidil, a new vasodilator, in the treatment of mild to moderate essential hypertension. *Eur. J. Clin. Pharmacol.* **28** 347–349.

Dixon, M. S., Thomas, P., Winterton, S. J., and Sheridan, D. J. (1988) Acute haemodynamic effects of BRL 34915, a novel vasodilator, in patients with ischaemic heart disease. *Proceedings of the British Pharmacological Society*, Dublin meeting, Abst. P10.

Donato Di Paola, E., Hicks, P. E., and Lorrain, J. (1987) Cardiovascular effects of BRL 34915 in anaesthetised and pithed rats. *Br. J. Pharmacol.* **92** 752P.

Downing, S. J., Hollingsworth, M., and Miller, M. (1988) Cromakalim and the rat uterus *in vivo*. *Proceedings of the British Pharmacological Society*, Nottingham meeting. Abst. C37.

Dumez, D., Zazzi-Sudriez, E., Pautrel, C., Armstrong, J. M., and Hicks, P. E. (1988) Comparison of the cardiovascular and renal effects of BRL 34915 with those of nitrendipine in dogs and SHR. *Br. J. Pharmacol.* **93** 201P.

Eckl, K. M., and Greb, W. H. (1987) Potassium channel activation in the vascular smooth muscle cell-membrane — a new mechanism of action in antihypertensive treatment — early clinical results with BRL 34915. *Clin. Exp. Hypert.* **49** 160.

Elliott, H. L., and Howie, C. A. (1988) Effects of single and multiple doses of BRL 34915 on pressor mechanisms in normotensive males. *Proceedings of International Society of Hypertension*, Kyoto, Japan, Abst. 0484.

Ferrier, C., Kurtz, A., Weidmann, P., Lehner, P., Pusterla, C., Shaw, S., and Saxenhofer, H. (1988) Stimulation of renin production by potassium-channel activation. *Proceedings of International Society of Hypertension*, Kyoto, Japan. Abst. 0154.

Foster, C. D., and Brading, A. F. (1987) An *in vitro* and *in vivo* study of BRL 34915 in guinea-pig and pig bladder smooth muscle. Poster presentation (P1002) to IUPHAR Pharmacology Congress, Sydney.

Fox, J. S., Whitehead, E., and Shanks, R. G. (1986) Cardiovascular effects of the novel anti-hypertensive agent BRL 34915 in normal volunteers. *Br. J. Clin. Pharmacol.* **23** 600P.

Gill, T. S., Davies, B. E., Allen, G. D., and Greb, W. H. (1988) Stereospecific pharmacokinetics of cromakalim enantiomers in healthy male subjects. *Br. J. Clin. Pharmacol.* **25** 669P–670P.

Gross, G. J., Warltier, D. C., Hardman, H. F., and Lamping, K. A. (1985) Enhanced subendocardial perfusion distal to a flow-limiting coronary artery stenosis in dogs: comparative effects of nicorandil, a potential new antianginal agent, and nitroglycerin. *J. Cardiovasc. Pharmacol.* **7** 977–982.

Grosset, A., and Hicks, P. E. (1986) Evidence for blood vessel selectivity of BRL 34915. *Br. J. Pharmacol.* **89** 500P.

Hof, R. P., Quast, U., Cook, N. S., and Blarer, S. (1988) Mechanism of action and systemic and regional hemodynamics of the potassium channel activator

BRL 34915 and its enantiomers. *Circ. Res.* **62** 679–686.

Hollenberg, N. K., Borucki, L. J., and Adams, D. F. (1978) The renal vasculature in early essential hypertension: evidence for a pathogenetic role. *Medicine* **57** 167–178.

Holzmann, S. (1983) Cyclic GMP — a possible mediator of coronary arterial relaxation by nicorandil (SG-75). *J. Cardiovasc. Pharmacol.* **5** 364–370.

Izzo, J. L., Licht, M. R., Smith, R. J., Larrabee, P. S., Radke, K. J., and Kallay, M. C. (1987) Chronic effects of direct vasodilation (pinacidil), alpha-adrenergic blockade (prazosin) and angiotensin-converting enzyme inhibition (captopril) in systemic hypertension. *Am. J. Cardiol.* **60** 303–308.

Kambara, H., Nakamura, Y., Tamaki, S., and Kawai, C. (1987) Beneficial effects of nicorandil on cardiovascular hemodynamics and left ventricular function. *J. Cardiovasc. Pharmacol.* **10** (suppl. 8) S104–S108.

Kato, K., Asanoi, H., Wakabayashi, C., Hosoda, S., Shiina, A., Hosono, K., Kurita, A., Seki, K., Ishida, K., Kuroiwa, A., and Fukumoto, A. (1987) Effect of nicorandil on exercise performance in patients with effort angina: a multicenter trial using a treadmill exercise test, *J. Cardiovasc. Pharmacol.* **10** (suppl. 8) S98–S103.

Kawashima, S., and Liang, C.-S. (1985) Systemic and coronary hemodynamic effects of pinacidil, a new anti-hypertensive agent, in awake dogs: comparison with hydralazine. *J. Pharmacol. Exp. Ther.* **232** 369–375.

Kerr, M. J., Wilson, R., and Shanks, R. G. (1985) Suppression of ventricular arrhythmias after coronary artery ligation by pinacidil, a vasodilator drug. *J. Cardiovasc. Pharmacol.* **7** 875–883.

Koch-Weser, J. (1974) Vasodilator drugs in the treatment of hypertension. *Arch. Intern. Med.* **133** 1017–1027.

Koliopoulos, K., Papadoyannis, D. E., and Karatzas, N. B. (1984) Pinacidil, a new vasodilator, in the treatment of patients with moderate to severe hypertension. *Eur. J. Clin. Pharmacol.* **27** 287–289.

Korb, H., Hoeft, A., Hunneman, D. H., Schraeder, R., Wolpers, H. G., and Hellige, G. (1985) Effectiveness of nicorandil in the preservation of myocardium stressed by transient ischemia and its influence on cardiac metabolism during coronary artery occlusion with subsequent reperfusion: a comparison with isosorbide dinitrate. *Naunyn-Schmiedeberg's Arch. Pharmacol.* **329** 440–446.

Krusell, L. R., Christensen, C. K., and Lederballe Pedersen, O. (1986) Renal effects of pinacidil in hypertensive patients on chronic beta-blocker therapy. *Eur. J. Clin. Pharmacol.* **30** 641–647.

Lamping, K. A., Christensen, C. W., Pelc, L. R., Warltier, D. C., and Gross, G. J. (1984b) Effects of nicorandil and nifedipine on protection of ischemic myocardium. *J. Cardiovasc. Pharmacol.* **6** 536–542.

Lamping, K. A., and Gross, G. J. (1984) Comparative effects of a new nicotinamide nitrate derivative nicorandil SG-75 with nifedipine and nitroglycerin on tissue collateral blood flow following an acute coronary occlusion in dogs. *J. Cardiovasc. Pharmacol.* **6** 601–608.

Lamping, K. A., Warltier, D. C., Hardman, H. F., and Gross, G. J. (1984a) Effects of nicorandil, a new antianginal agent, and nifedipine on collateral blood flow in a chronic coronary occlusion model. *J. Pharmacol. Exp. Ther.* **229** 359–363.

Lebel, M., Grose, J. H., and Lacourcière, Y. (1988) Effect of short-term administration of BRL 34915 on renal hemodynamics and prostanoid excretion in essential hypertension. *Proceedings of International Society of Hypertension*, Kyoto, Japan. Abst. 1022.

Longman, S. D., Clapham, J. C., Wilson, C., and Hamilton, T. C (1988) Cromakalim, a potassium channel activator: a comparison of its cardiovascular haemodynamic profile and tissue specificity with those of pinacidil and nicorandil. *J. Cardiovasc. Pharmacol.* **12** 535–542.

Morimoto, S., Matsumura, Y., Sasaki, Y., Takakuwa, T., and Kageyama, M. (1987) Diuretic effect induced by intrarenal infusion of pinacidil in anaesthetised dogs. *J. Pharmacol. Exp. Ther.* **242** 1050–1055.

Morimoto, S., Sasaki, Y., and Matsumura, Y. (1986) Effects of pinacidil on renal hemodynamics and function in anaesthetized dogs. *J. Pharmacobio-Dyn.* **9** 146–155.

Muiesan, G., Fariello, R., Muiesan, M. L., and Christensen, O. E. (1985) Effect of pinacidil on blood pressure, plasma catecholamines and plasma renin activity in essential hypertension. *Eur. J. Clin. Pharmacol.* **28** 495–499.

Nakagawa, Y., Takeda, K., Katano, Y., Tsukada, T., Kitagawa, T., Otorri, T., and Imai, S. (1979) Effects of 2-nicotinamidoethyl nitrate on the cardiovascular system. *Jpn. Heart J.* **20** 881–895.

Nguyen, P. V., Davis, A., Tasker, T. C. G., and Leenen, F. H. H. (1987) Effects of BRL 34915 on pressor and chronotropic responses to i.v. norepinephrine, angiotensin II, and isoproterenol in normal men. *Cardiovasc. Drugs Ther.* **1** 270.

Nicholls, D. P., McNeill, J., Harron, D. W. G., and Shanks, R. G. (1986a) Cardiovascular effects of pinacidil and propranolol alone and in combination in normal humans. *J. Cardiovasc. Pharmacol.* **8** 51–54.

Nicholls, D. P., Murtagh, J. G., Scott, M. E., Morton, P., and Shanks, R. G. (1986b) Acute haemodynamic effects of pinacidil in man. *Br. J. Clin. Pharmacol.* **22** 287–292.

Olsen, U. B., and Arrigoni-Martelli, E. (1983) Vascular effects in dogs of pinacidil (P1134), a novel vasoactive antihypertensive agent. *Eur. J. Pharmacol.* **88** 389–392.

Preuss, K. C., Gross, G. J., Brooks, H. L., and Warltier, D. C. (1985) Hemodynamic actions of nicorandil, a new antianginal agent, in the conscious dog. *J. Cardiovasc. Pharmacol.* **7** 709–714.

Quast, U., and Cook, N. S. (1988) Potent inhibitors of the effects of the K^+ channel opener BRL 34915 in vascular smooth muscle. *Br. J. Pharmacol.* **93** 204P.

Quast, U., and Cook, N. S. (1989) *In vitro* and *in vivo* comparison of two K^+ channel openers, diazoxide and cromakalim, and their inhibition by glibenclamide. *J. Pharmac. Exp. Therap.* **250** 261–270.

Ramsey, L. E., and Freestone, S. (1983) Preliminary evaluation of pinacidil in hypertension. *Br. J. Clin. Pharmacol.* **16** 336–338.

Rijk, M. C. M., and Thien, Th. (1987) Intravenous pinacidil in the acute treatment of hypertension. *J. Clin. Pharmacol.* **27** 468–474.

Sakai, K., Akima, M., Shiraki, Y., and Hoshino, E. (1983) Effects of a new antianginal agent, nicorandil, on the cardiovascular system of the miniature pig: with special reference to coronary vasospasm. *J. Pharmacol. Exp. Ther.* **227** 220–227.

Sakai, K., Shiraki, Y., and Nabata, H. (1981) Cardiovascular effects of a new coronary vasodilator *N*-(2-hydroxyethyl) nicotinamide nitrate (SG-75): comparison with nitroglycerin and diltiazem. *J. Cardiovasc. Pharmacol.* **3** 139–150.

Schmidt, K., Reich, R., and Kukovetz, W. R. (1985) Stimulation of coronary guanylate cyclase by nicorandil (SG-75) as a mechanism of its vasodilating action. *J. Cyclic Nucleotide Res.* **10** 43–53.

Shepherd, R. M., Williams, B. C., Creedy, A., Fraser, R., and Kenyon, C. J. (1988) The *in vitro* effects of valinomycin and BRL 34915 on angiotensin II-induced aldosterone secretion. *J. Endocrin.* **119** (suppl.) Abst. 106.

Shiraki, Y., Akima, M., Nabata, H., Ohba, Y., Hoshino, E., and Sakai, K. (1981) The hypotensive mechanisms of the new anti-anginal drug, *N*-(2-hydroxyethyl) nicotinamide nitrate (SG-75) in beagle dogs. *Jap. J. Pharmacol.*, **31** 921–929.

Taira, N., Satoh, K., Yanagisawa, T., Imai, Y., and Hiwatari, M. (1979) Pharmacological profile of a new coronary vasodilator drug, 2-nicotinamidoethyl nitrate (SG-75). *Clin. Exp. Pharmacol. Physiol.* **6** 301–316.

Thoolen, M. J. M. C., Van Meel, J. C. A., Wilffert, B., Timmermans, P. B. M. W. M., and Van Zwieten, P. A. (1983) Hemodynamic characterization of pinacidil in rats: comparison with hydralazine. *Pharmacology* **27** 245–254.

Thormann, J., Schlepper, M., Kramer, W., Gottwick, M., and Kindler, M. (1983) Effectiveness of nicorandil (SG-75), a new long-acting drug with nitroglycerin effects, in patients with coronary artery disease: improved left ventricular function and regional wall motion and abolition of pacing-induced angina. *J. Cardiovasc. Pharmacol.* **5** 371–377.

Uchida, Y., Yoshimoto, N., and Murao, S. (1978a) Effect of 2-nicotinamidoethyl nitrate (SG-75) on coronary circulation. *Jpn. Heart J.* **19** 112–124.

Uchida, Y., Yoshimoto, N., and Murao, S. (1978b) Effects of anti-anginal agents on cyclical reductions of coronary blood flow. *Jpn. Heart J.* **19** 904–912.

VandenBerg, M. J., Kakad, J., Wiseman, W. T., Stephens, J. D., Stewart-Long, P. M., and Tasker, T. (1987a) BRL 34915, a novel potassium channel activator: duration of antihypertensive action. *Cardiovasc. Drugs Therapy* **1** 299.

VandenBerg, M. J., Woodward, S. R., Stewart-Long, P., Tasker, T., Pilgrim, A. J., Dews, I. M., and Fairhurst, G. (1987b) Potassium channel activators: antihypertensive activity and adverse effect profile of BRL 34915. *J. Hypert.* **5** (suppl. 5) S193–S195.

Van Meel, J. C., Timmermans, P. B., and Van Zwieten, P. A. (1983) Hypotensive activity of calcium entry blockers in rats. Relationship with depression of alpha$_2$-adrenoceptor-mediated vasopressor responses. *Eur. J. Pharmacol.* **92** 27–34.

Ward, J. W. (1984) Pinacidil monotherapy for hypertension. *Br. J. Clin. Pharmacol.* **18** 223–225.

Wilson, C., Buckingham, R. E., Mootoo, S., Parrott, L. S., Hamilton, T. C., Pratt, S. C., and Cawthorne, M. A. (1988) *In vivo* and *in vitro* studies of cromakalim (BRL 34915) and glibenclamide in the rat. *Br. J. Pharmacol.* **93** 126P.

Yokota, M., Horisawa, T., Iwase, M., Miyahara, T., Yoshida, J., Kamihara, S., Noda, S., Tsunekawa, A., Koide, M., Tsuzuki, M., and Sotobata, I. (1987) Effects of a new vasodilator, nicorandil, on exercise-induced impairment of left ventricular function in patients with old myocardial infarction. *J. Cardiovasc. Pharmacol.* **10** (suppl. 8) S116–S122.

Zachariah, P. K., Sheps, S. G., Schirger, A., Fisher, L. D., Shub, C., Collins, J. B., and Spiekerman, R. E. (1986) Antihypertensive efficacy of pinacidil — automatic ambulatory blood pressure monitoring. *Eur. J. Clin. Pharmacol.* **31** 133–141.

11

Potassium channels and modulation of insulin secretion

M. L. J. Ashford

11.1 INTRODUCTION

The application of the giga-ohm seal patch-clamp recording techniques to endocrine cells has enabled researchers to begin to unequivocally identify the individual channel types that underlie the complex electrical events associated with the detection of stimuli and the consequent release of hormone. This has recently been particularly evident for the pancreatic β-cell and the control of insulin secretion. Therefore my aim is to summarize the recent advances in our understanding of the physiology and pharmacology of insulin release as they pertain to modulation of K^+ channel function in β-cells. Thus much of the following account is concerned with the nutrient and drug modulation of ATP-sensitive K^+ (K_{ATP}) channel activity and the resultant changes in β-cell membrane potential and regenerative electrical activity. Examples of K^+ channels in other endocrine cells are discussed briefly and only where the K^+ channel(s) involved have been reasonably well characterized. More general details of β-cell and general endocrine electrophysiology are given in the reviews by Petersen and Findlay (1987) and Ozawa and Sand (1986), respectively. The biophysical characteristics of K_{ATP} channels are described in detail by Ashcroft (1988) and in Chapter 4 of this book.

11.2 NUTRIENT-INDUCED INSULIN SECRETION

The release of insulin from pancreatic β-cells is controlled by changes in plasma levels of various carbohydrates, of which glucose is the most important physiological stimulus, although fatty acids and amino acids also contribute actions. At low levels of extracellular glucose (~2–3 mM) there is a low or basal level of insulin secretion and the β-cells are mostly electrically silent. As the concentration of glucose is increased the resting membrane potential decreases until a threshold level is reached, at about 4–8 mM glucose (near the normal non-fasting levels), at which point regenerative electrical activity occurs (Dean and Matthews, 1970a; Meissner and Schmeltz, 1974) concomitant with a marked increase in insulin release (Scott *et al.*, 1981). This electrical activity (see Fig. 11.1) takes the form of a slow depolarization (the prepotential) to a threshold potential, followed by a rapid depolarization to a plateau potential upon which are superimposed bursts of action potentials. Repolarization then occurs, to a level more negative than threshold, whereupon the cycle repeats (see reviews by Henquin and Meissner, 1984; Matthews, 1985, for further details). Increasing the extracellular glucose concentration increases the burst frequency, prolongs the burst of action potentials and shortens the interburst duration, until at high concentrations (e.g. 20 mM) there is a sustained depolarization at the plateau level and continuous action potential activity.

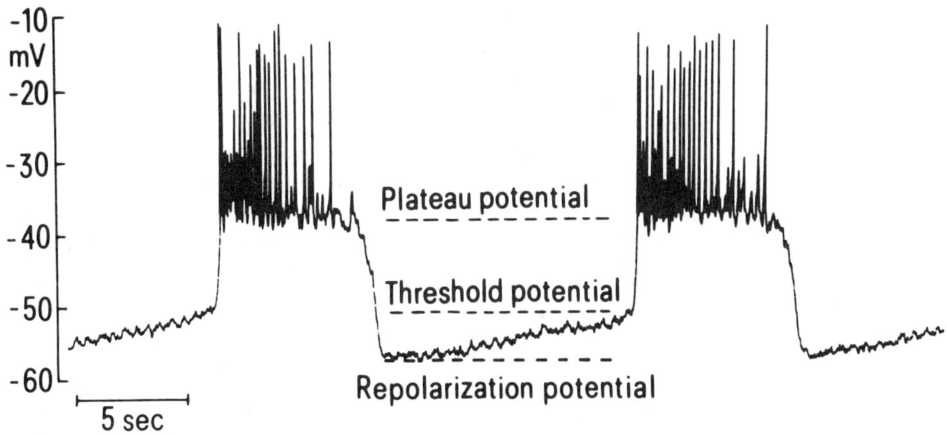

Fig. 11.1 — Electrical activity generated by 10 mM glucose in a single mouse β-cell. Note the slow depolarization to threshold, followed by a rapid depolarization to the plateau potential. (Reproduced, with permission, from Henquin and Meissner, 1984.)

The glucose-induced action potentials are not sodium-dependent (Dean and Matthews, 1970b), are unaffected by TTX (Meissner and Schmeltz, 1974) but are inhibited by various calcium channel blockers and removal of extracellular calcium (Dean and Matthews, 1970b; Matthews and Sakomoto, 1975a). Recent voltage-clamp experiments on insulin-secreting cells have confirmed the existence of voltage-dependent calcium currents which underlie the depolarizing phase of the action potential (Rorsman and Trube, 1986; Satin and Cook, 1988). These authors also demonstrate that a transient sodium current can be elicited from insulin-secreting cells, blockable by TTX. This current is unlikely to contribute significantly to β-cell electrical activity as, at the normal operating range of potentials for β-cells (about −50 to −20 mV) most of the sodium channels would be inactivated (Satin and Cook, 1988). In addition, it has been suggested that the plateau potential is due to the activation of voltage-dependent calcium channels. Calcium entry through these channels contributes to the elevation of internal calcium observed on β-cell stimulation and results ultimately in insulin secretion (Wollheim and Pozzan, 1984; Howell, 1984). Action potential repolarization and termination of the plateau potential, also by repolarization, involve increased membrane K^+ permeability, and it is likely that a voltage-dependent delayed rectifier (Rorsman and Trube, 1986) and a high-conductance K_{Ca} (BK_{Ca}) channel (Atwater *et al.*, 1983) contribute to these actions.

The resting membrane potential of β-cells in the absence of glucose is approximately −70 mV (Rorsman and Trube, 1985; Ashcroft *et al.*, 1988) and in 3–5 mM glucose is closer to −60 mV (Croghan *et al.*, 1986). This suggests that the membrane potential cannot be solely determined by the membrane K^+ permeability (otherwise the value would be closer to E_K; ~−80 mV); a counterbalancing depolarizing current must exist. Although the identity of this current is presently unknown, a persistent background influx of sodium ions could account for it (Matthews and Sakamoto, 1975b; Atwater *et al.*, 1978; Ribalet and Beigelman, 1982). An increase in the extracellular glucose concentration depolarizes the β-cell membrane, an action associated with an increased input resistance (Atwater *et al.*, 1978). Radioactive tracer efflux studies, using either $^{42}K^+$ or $^{86}Rb^+$, indicate that glucose produces a concentration-dependent reduction in membrane K^+ permeability (Sehlin and Täljedal, 1975; Henquin, 1978). Thus the depolarization elicited by glucose, leading to activation of voltage-dependent currents, is due to a K^+ conductance decrease resulting in a shift in the balance towards the background inward current. To initiate this K^+ current inhibition glucose must be metabolized, as inhibitors of the glycolytic pathway such as iodoacetate or mannoheptulose (see below) produce complete block of glucose action (Dean *et al.*, 1975). Certain other carbohydrates (e.g. glyceraldehyde, mannose and inosine) and some amino acids (e.g. leucine and its primary metabolic product, 2-ketoisocaproate) also induce β-cell depolarization and insulin release by a similar mechanism, i.e. a decrease in K^+ permeability (Henquin and Meissner, 1981; see also Chapter 4, section 4.6.2) even though they enter the central metabolic pathway at different points (see below).

Following microelectrode recording studies, a favoured hypothesis to account for the K^+ conductance decrease was that the BK_{Ca} channel contributed to the resting potential and was the target for metabolizable compounds and anti-diabetic drugs such as tolbutamide. The evidence for this stemmed mainly from the observation that quinine, which was suggested to be a specific inhibitor of BK_{Ca} channels (Armando-Hardy *et al.*, 1975), directly depolarizes β-cells and is a good secretagogue. However, this mechanism is now thought to be unlikely, as recent use of patch-clamp methodology has uncovered the identity of another K^+ channel that provides a major contribution to the β-cell resting membrane potential.

Recordings from cell-attached membrane patches of rodent β-cells in short-term tissue culture indicate that the major active channel in glucose-free solution is K^+-selective (Ashcroft *et al.*, 1984; Rorsman and Trube, 1985). Under a symmetrical K^+ gradient (i.e. with 140 mM K^+ in the recording pipette) this K^+ channel is characterized by a conductance of approximately 50 pS. In addition, the open-state probability of the channel exhibits no voltage dependence (Ashcroft *et al.*, 1984; Rorsman and Trube, 1985). The identity of this channel (often termed the G-channel, because of its 'glucose sensitivity') has been determined by recordings from inside-out membrane patches isolated from β-cells. Cook and Hales (1984) described a K^+-selective channel blocked by intracellular ATP, an action not dependent upon membrane potential (Sturgess *et al.*, 1986a) or hydrolysis of the nucleotide. This ATP-sensitive K^+ (K_{ATP}) channel resembles, in appearance, the K^+ channel recorded from cell-attached membrane patches and has also been shown to have a similar slope conductance (56 pS) and single-channel kinetics (Rorsman and Trube, 1985) and have an opening probability independent of membrane potential (Ashcroft *et al.*, 1984). Thus the 'G-channel' and the K_{ATP} channel are considered to be identical. I shall refer to this K^+ channel henceforth as the K_{ATP} channel (for review see Chapter 4).

Superfusion of β-cells, during cell-attached recording, with a solution containing 20 mM glucose resulted in a rapid and reversible inhibition of the channel currents and the appearance of continuous action potential activity. As glucose was present only in the bath solution (and not electrode), and cannot penetrate the pipette–membrane seal (Sakmann and Neher, 1984), channel inhibition must have occurred by an intracellular route. These effects of glucose were reversed on washout of the sugar or if metabolic inhibitors such as mannoheptulose (a hexokinase inhibitor) or sodium azide (an inhibitor of cytochrome A3 in the mitochondrial electron transport pathway) were also present, thus indicating that glucose metabolism is required for block (Ashcroft *et al.*, 1984; Misler *et al.*, 1986). Other cation-selective channels have been observed during recordings from cell-attached membrane patches, such as a high-conductance K channel (Rorsman and Trube, 1985), a 20-pS (possibly K^+-selective) and a 120-pS (activated by depolarization) K^+ channel (Ashcroft *et al.*, 1988). The relatively infrequent occurrence of these channels coupled with a low open state probability at the resting membrane potential suggest that it is unlikely that they contribute significantly to the β-cell resting membrane potential.

The inhibition of the K_{ATP} channel activity by glucose was dose-dependent (Fig. 11.2a), the maximum reduction of channel activity occurring between 0 and 5 mM (approximately 1 mM glucose produced half maximal inhibition) and 10 mM sufficient to induce almost full inhibition (Ashcroft *et al.*, 1988). This corresponds well with the range of glucose concentrations over which pancreatic β-cells exhibit the largest changes in membrane potential and $^{42}K^+$ or $^{86}Rb^+$ efflux (Meissner and Schmeltz, 1974; Sehlin and Täljedal, 1975; Henquin, 1978). The cell-attached recordings clearly show that channel openings occur in discrete bursts and these are separated by long closed periods (interburst intervals) and that within a burst there are many rapid open and closed transitions. Analysis of the underlying channel kinetics indicates that glucose induces channel inhibition by stabilizing the long-lived closed state, resulting in a decreased burst duration and increased interburst interval, but has little apparent effect on the short closures within a burst (Ashcroft *et al.*, 1988).

Accompanying the glucose-induced inhibition of the K_{ATP} channel currents is an increase in the cell input resistance. This is observed in the cell-attached recording configuration by a change in the shape of the single-channel currents (i.e. the current is observed to 'relax' during an opening), and by the appearance of action potentials directly elicited by a single-channel opening (see Fig. 11.2b). The current flowing during a single-channel opening (a few pico-amps) in a cell-attached patch, combined with the increased cellular input resistance (giga-ohm range), is sufficient to depolarize the cell and induce relaxation of the single-channel current (by virtue of a

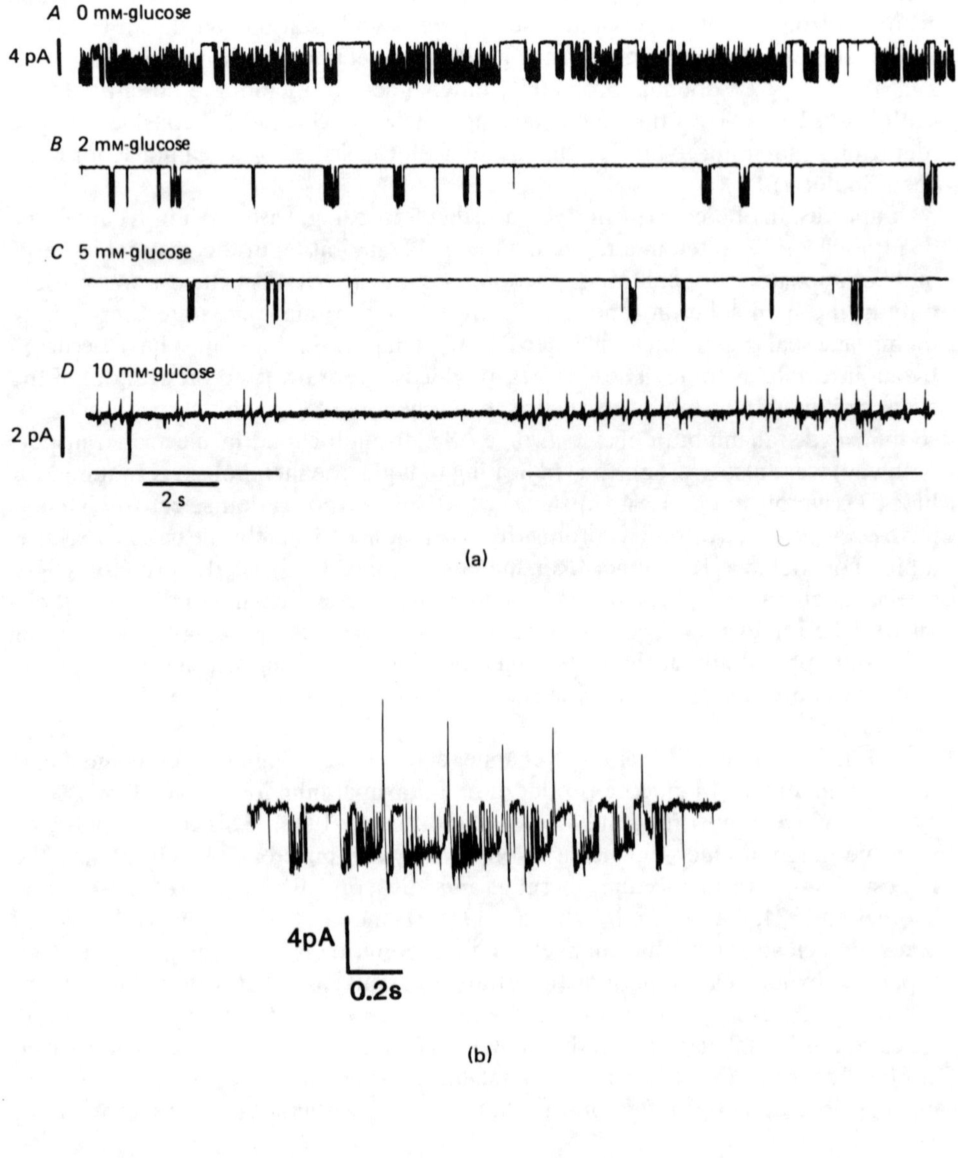

Fig. 11.2 — (a) Dose-dependent actions of glucose on single K_{ATP} channels recorded cell-attached from a rat β-cell. The pipette solution contained 140 mM K^+ and the bath solution 135 mM Na^+; 5 mM K^+. The pipette potential was 0 mV. Note that in the presence of 10 mM glucose, biphasic waveforms representing action potentials are present. (Reproduced, with permission, from Ashcroft *et al.*, 1988.) (b) Cell-attached recording from a CRI-G1 cell, under identical conditions to that described in (a), except that the cell was exposed to 3 mM glucose. Note the 'relaxation' of the single-channel currents and the appearance of action potentials initiated by the current through the open channel. (R. Z. Kozlowski and M. L. J. Ashford, unpublished observations.)

reduced driving force, the cell membrane potential) and generate action potentials (Fenwick *et al.*, 1982). The glucose-induced action potentials recorded from cell-attached patches arise from the unclamped areas of the cell membrane (i.e. outside the patch of membrane in the pipette) undergoing depolarization to threshold. In the absence of extracellular glucose the K_{ATP} channel is not fully activated, in cell-attached recording, as indicated by the increase in average channel activity observed when metabolic inhibitors such as sodium azide, *p*-trifluoromethoxyphenyl hydrazone (FCCP, a mitochondrial uncoupler), KCN (Misler *et al.*, 1986) or rotenone (Ashcroft *et al.*, 1985) are present. These results also suggest a close relationship between cellular metabolism and K_{ATP} channel gating.

Other insulin secretagogues that are utilized by β-cells as metabolic fuel also produce effects identical to glucose. For example, recordings from cell-attached patches of rat primary cultured β-cells show that K_{ATP} channels are closed by metabolizable substrates such as mannose, glyceraldehyde and leucine, whereas non-metabolizable sugars such as D-galactose and 3-*O*-methylglucose are ineffective (Misler *et al.*, 1986). Channels closed by glyceraldehyde cannot be re-opened by mannoheptulose, as this substrate enters glycolysis at the level of glyceraldehyde-3-phosphate, but can be opened by iodoacetate (which is a competitive inhibitor of glyceraldehyde 3-phosphate dehydrogenase). Similarly, leucine-induced inhibition is not reversed by either of these (it enters the metabolic pathway as acetyl CoA) but is by sodium azide. Another mitochondrial substrate, 2-ketoisocaproate, which is a potent insulin-secretagogue (Hutton *et al.*, 1980), also suppresses K_{ATP} channel activity in cell-attached patches and initiates action potential activity (Ashcroft *et al.*, 1987a). In contrast, the amino acid arginine, which is a potentiator of glucose-induced insulin secretion, does not induce channel closure (Ashcroft *et al.*, 1987a). This is in agreement with previous studies, which show that this agent produces only a small depolarization of β-cells (insufficient to generate action potentials) and does not decrease Rb^+ efflux from β-cells (Henquin and Meissner, 1981). It has been argued that arginine-induced depolarization is electrogenic, resulting from intracellular accumulation of the positively charged amino acid.

Essentially identical results have been obtained using insulin-secreting cell lines. For example, 10 mM glyceraldehyde has been reported to inhibit K_{ATP} channels in cell-attached patches from RINm5F cells (Dunne *et al.*, 1986) and 20 mM leucine plus 3 mM glucose to inhibit K_{ATP} channels recorded from CRI-G1 cells (Sturgess *et al.*, 1988; Kozlowski and Ashford, unpublished observations), resulting in the appearance of spontaneous action potential activity (see Fig. 11.2b). Insulin-secreting cell lines such as these, although not responsive (in terms of insulin secretion) to changes in extracellular glucose concentrations over the normal range (3–20 mM), have been observed to respond over the range 0–3 mM (Giroix *et al.*, 1986; C. N. Hales, personal communication) and the presence of glucose does potentiate insulin secretion in response to other secretagogues (e.g. leucine). The altered sensitivity to glucose is considered to be due to the cell lines having a low K_m glucokinase enzyme (resulting in maximum effects at low glucose concentrations) rather than a high K_m glucokinase (as in normal β-cells, giving maximum effects at higher glucose concentrations). Recordings of K_{ATP} channels in cell-attached patches from RINm5F cells have shown that low concentrations of glucose (<3 mM) do produce significant reductions in channel activity. However, it is surprising that the dose–response relationship for glucose-induced inhibition of the K_{ATP} channel in RINm5F cells (Ribalet *et al.*, 1988) is so similar to that for rat pancreatic β-cells (Ashcroft *et al.*, 1988), the concentrations producing half-maximal inhibition of channel activity being 0.5 mM and 1.0 mM, respectively. More detailed studies of the relationship between channel activity, glucose concentration, glucokinase activity and insulin release in these cell lines may help clarify this matter.

It is clear from the evidence described above using cell-attached patches that the K_{ATP} channel is regulated by cellular metabolism. The obvious question that arises is whether ATP is the intracellular mediator that couples nutrient metabolism to channel closure. Overall, the evidence does favour such a proposition. For example, the K_{ATP} channel activity is decreased by the presence of metabolizable substrates known to increase intracellular ATP levels and is increased by metabolic inhibitors which act to lower intracellular ATP concentration. Excision of patches after cell-attached recordings show that the channel is inhibited by ATP and the effects of ATP on the channel kinetics resemble that of glucose, in that long closures appear between shortened bursts of openings (see Chapter 4). For a more detailed discussion of these points see Ashcroft (1988) and Petersen and Findlay (1987).

Although the K_{ATP} channel is the main contributor to the β-cell resting membrane K^+ conductance, it is not the only type of K^+ channel present. Consequently, other channels, activated by depolarization, may have their activity modulated by nutrient metabolism. There is some evidence for such a proposition, as in cell-attached recordings from RINm5F cells a large-conductance (~140 pS) K^+ channel, activated by voltage, has been shown to be inhibited by glucose in a dose-dependent manner (Ribalet *et al.*, 1988). The concentration range for this inhibition is greater than for the K_{ATP} channel, 5–20 mM being required to induce large changes in channel activity. Ribalet *et al.* (1988) have suggested that a K^+ channel with these attributes, i.e. activated by depolarization and modulated (de-activated) by high concentrations of glucose, may contribute to the control of the time-course of the interburst durations as described in Fig. 11.1. Hence, depolarization would activate this channel, resulting in repolarization eventually (end of burst), but in the presence of high glucose concentrations (>5 mM) the activity of the channel would be inhibited, resulting in a shortened interburst duration. This high-conductance channel may be the same as that described by Ashcroft *et al.* (1988) in rat β-cells, although the actions of glucose were not reported. However, in contrast to the observations of Ribalet *et al.* (1988), Rorsman and Trube (1985) reported that a high-conductance K^+ channel (perhaps BK_{Ca}) observed occasionally in cell-attached patches showed no obvious glucose sensitivity.

11.3 HYPOGLYCAEMICS AND K^+ CHANNEL INHIBITION

There are a variety of pharmacological agents which act to alter the β-cell membrane potential and so modify insulin secretion. Amongst those that produce hypoglycaemia (i.e. they act to reduce blood glucose levels), the best known and most commonly prescribed are the sulphonylurea group of compounds. These are used clinically in the treatment of maturity-onset (type 2) diabetes. The sulphonylureas were developed following the finding that certain substituted sulphonamides, used to treat typhoid, also reduced blood glucose levels (see review by Loubatières, 1977).

The structures of some of these compounds are illustrated in Fig. 11.3. Tolbutamide is an example of a first-generation sulphonylurea, and glibenclamide and glipizide are examples of second-generation drugs. Although it is generally accepted that the principle action of the sulphonylureas is to stimulate insulin secretion from the pancreatic β-cells, it should be noted that there is a certain degree of controversy over whether they also act via extrapancreatic mechanisms such as improved insulin action (see, for example, Lebovitz, 1985).

In common with the actions of many nutrients, the sulphonylureas induce a persistent depolarization of β-cells with a concomitant discharge of action potentials (Dean and Matthews, 1970a; Matthews and Sakamoto, 1975a). This depolarization was shown to be associated with a decreased potassium ion permeability (using $^{86}Rb^+$-loaded cells as a marker for potassium movement) of the β-cell (Henquin, 1980; Henquin and Meissner, 1982; Matthews and Shotton, 1984) and an increased input resistance (Ferrer *et al.*, 1984). The result of this depolarization is the consequent activation of voltage-dependent calcium channels (Matthews and Sakamoto, 1975a), calcium influx (Henquin, 1980) and insulin secretion (Howell, 1984). As discussed in the previous section, reduction of an outwardly directed potassium permeability will only result in cell depolarization if acting to counterbalance a steady background depolarizing current.

Unlike the actions of nutrients such as glucose, the sulphonylureas depolarize the β-cell and induce insulin release even in the presence of metabolic poisons, such as iodoacetate, an inhibitor of glycolysis (Dean *et al.*, 1975) or DNP, an uncoupler of oxidative phosphorylation (Ferrer *et al.*, 1984). This indicates that the sulphonylureas do not require to be metabolized to produce their effects and that they act as *initiators* of insulin release. A membrane site of action is also supported by studies indicating that they do not penetrate the β-cell to any significant extent, with the exception of glibenclamide, which does accumulate within the cell (see review by Gylfe *et al.*, 1984), although because tolbutamide is 99% dissociated at physiological pH this may not be easily detectable (Trube *et al.*, 1986). The existence of a specific site on the β-cell plasma membrane was indicated in early studies using ^{35}S-labelled sulphonylyreas (Hellman *et al.*, 1971) and dextran-linked tolbutamide (Bowen and Lazarus, 1974). This has since been confirmed and characterized in more detail using tritiated sulphonylureas (Kaubisch *et al.*, 1982; Geisen *et al.*, 1985; Schmid-Antomarchi *et al.*, 1987). Such studies have shown that there are saturable, high-affinity and reversible binding sites for the sulphonylureas on β-cell plasma membranes. The half-maximal inhibition of [^3H]glibenclamide binding to these sites in rat β-cell tumour membranes by various oral antidiabetic compounds also correlates well with their relative hypoglycaemic activities in vivo (Geisen *et al.*, 1985). These sites therefore fulfil the majority of the criteria for specific receptors.

It follows that if specific sulphonylurea receptors exist in the plasma membrane of β-cells then they must, by virtue of the sulphonylurea-mediated reduction in potassium permeability, be functionally linked to a potassium channel. Initial studies using $^{86}Rb^+$ flux and microelectrode recording methods favoured the BK_{Ca} channel as the target (e.g. Ferrer *et al.*, 1984). However, as described previously, the K_{ATP} channel is the major contributor to the β-cell resting membrane potential and therefore a much more likely target. Indeed, Sturgess *et al.* (1985), using outside-out membrane patches from the rat insulin-secreting cell line CRI-G1, showed that

Tolbutamide

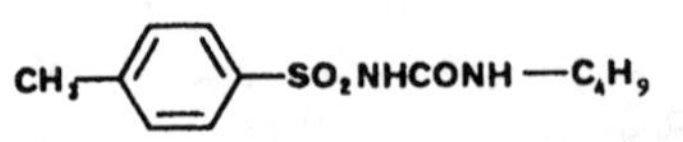

Glibenclamide

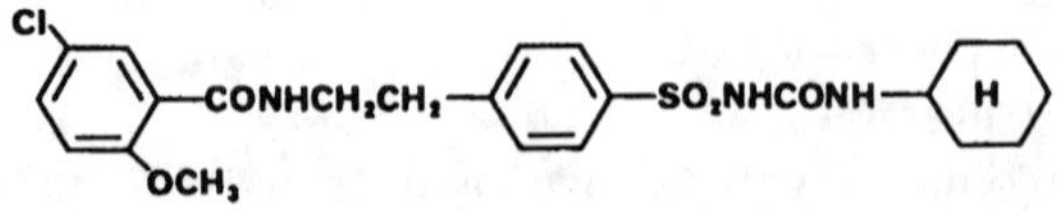

HB 985

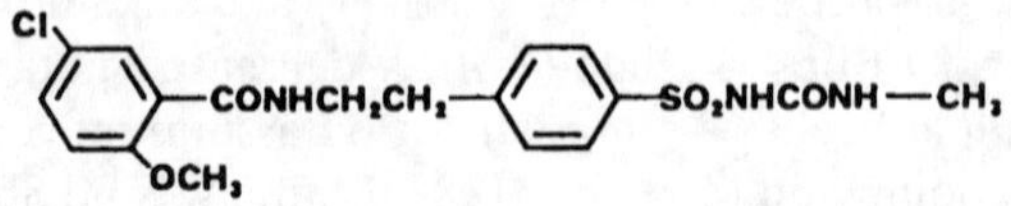

HB 699

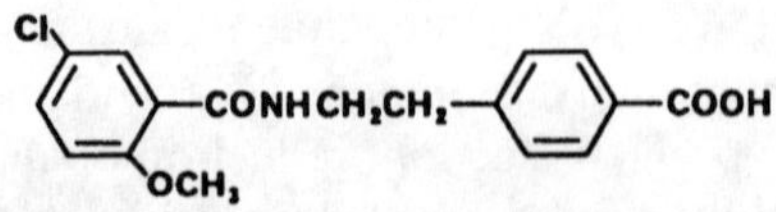

Glipizide

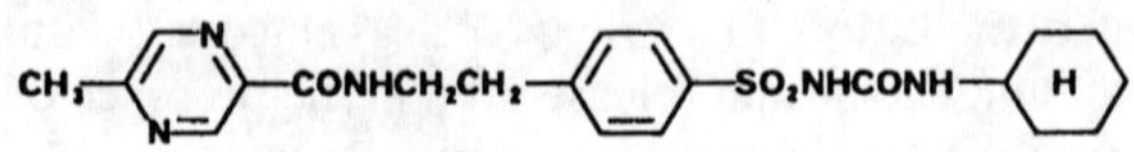

Gliquidone

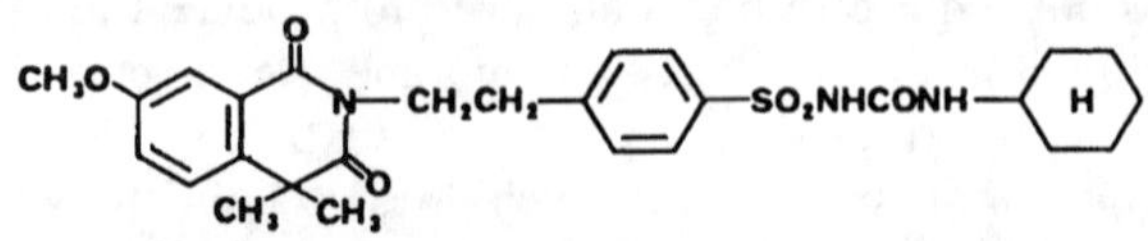

UL-DF 9

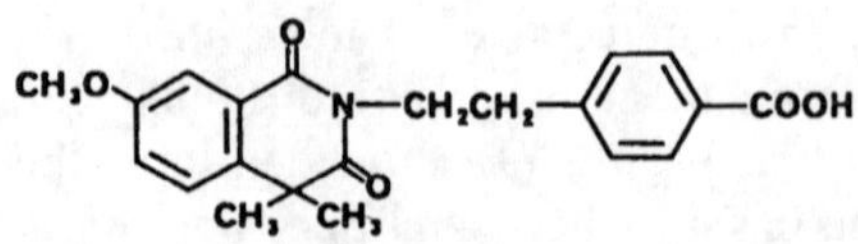

AZ-DF 265

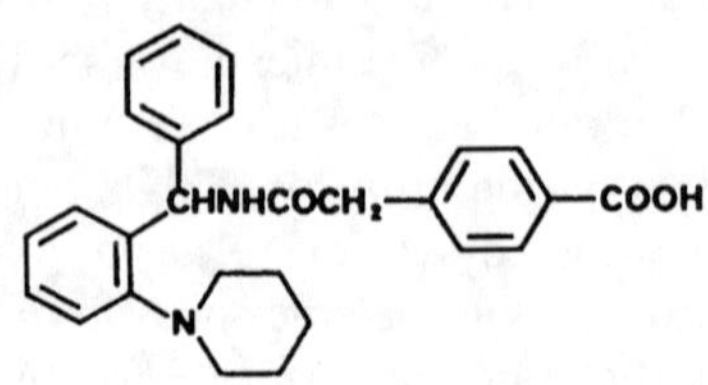

Diazoxide

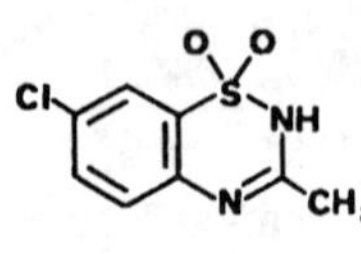

Fig. 11.3 — Structural formulae of the hypoglycaemic drugs and the hyperglycaemic agent diazoxide.

tolbutamide produced a reversible decrease in K_{ATP} channel activity. Further studies using mouse β-cells (Trube *et al.*, 1986; Zunkler *et al.*, 1988a) and this cell line (Ashford *et al.*, 1986a; Sturgess *et al.*, 1988) have confirmed this action and extended it to include other oral hypoglycaemic agents, e.g. glibenclamide and glipizide. In keeping with their relative therapeutic potencies *in vivo* and *in vitro*, the second-generation sulphonylureas inhibit K_{ATP} channels at much lower concentrations than tolbutamide. This has not only been demonstrated on individual K_{ATP} channels using isolated patch recordings, but also using whole-cell voltage-clamped K^+ currents. This latter technique has the advantage that the phenomenon of run-down of the K_{ATP} channel activity can be easily quantified and so allow dose–inhibition curves for the sulphonylureas to be obtained. Typical values for half-maximal inhibition of K_{ATP} currents by tolbutamide are 4–7 μM for mouse pancreatic β-cells (Trube *et al.*, 1986; Zunkler *et al.*, 1988a) and 17 μM for CRI-G1 cells (Sturgess *et al.*, 1988), and for glibenclamide values of 4 nM (Zunkler *et al.*, 1988a) and 27 nM (Sturgess *et al.*, 1988) have been reported. These compare favourably with the concentrations that produce half-maximal stimulation of insulin release from perfused pancreatic islets (Panten *et al.*, 1986; values quoted in Zunkler *et al.*, 1988a) and CRI-G1 cells (Sturgess *et al.*, 1988), and the therapeutic plasma levels of the free drugs (Balant, 1981). However, the half-maximal binding of [^{3}H]glibenclamide to rat β-cell tumour membranes occurs at a concentration of 0.05 nM, a value two orders of magnitude lower than for channel inhibition and therapeutic potency. This discrepancy may simply reflect the conditions under which the binding studies were performed, although Schmid-Antomarchi *et al.* (1987) report a half-maximal inhibition of ATP-sensitive $^{86}Rb^+$ flux from RINm5F cells by glibenclamide of 0.06 nM, a value indistinguishable to that from the binding studies.

The actions of the sulphonylureas on the β-cell appear to be specific for the K_{ATP} channel. Tolbutamide and glibenclamide were shown not to block the voltage-activated K^+ current (delayed rectifier) responsible for repolarization of the action potential or the voltage-dependent calcium current underlying action potential depolarization (Rorsman and Trube, 1986). Similarly the other major potassium channel present in the plasma membrane of insulin-secreting cells, the BK_{Ca}, is unaffected directly by tolbutamide (Trube *et al.*, 1986) or glibenclamide (Sturgess *et al.*, 1988). However, sulphonylurea-induced depolarization of the β-cell will, by virtue of activation of calcium channels, increase internal calcium concentration and so switch on BK_{Ca} channels (an effect potentiated by depolarization) and ultimately lead to cell repolarization. A non-selective cation channel (K/Na_{Ca} or CAN) has also been described in cultured β-cells (Marty and Neher, 1982) and in the cell line CRI-G1 (Ashford *et al.*, 1986b). The latter group report that this channel has a conductance of 25 pS, is activated by intracellular calcium ions and can be inhibited by intracellular adenine nucleotides including ATP (Sturgess *et al.*, 1986b, 1987a). However, this nucleotide-sensitive channel was also shown to be unaffected by both tolbutamide and glibenclamide (Sturgess *et al.*, 1988). These data suggest that the sulphonylureas will be useful specific ligands for the identification and characterization of K_{ATP} channels. In support of this, the K_{ATP} channels of cardiac tissue (Gee and Misler, 1987; Belles *et al.*, 1987) and the K^+ efflux of metabolically exhausted amphibian skeletal muscle (an action attributed to opening of K_{ATP} channels; Castle and Haylett, 1987) are blocked by tolbutamide. The observation that tolbutamide specifically blocks the K_{ATP} channel in mammalian cardiac myocytes (Ca^{2+} currents and the inward rectifying K^+ current are unaffected) has refuelled the controversy of whether diabetics treated with tolbutamide suffer an excess cardiac mortality (Seltzer, 1980). Such an action would depend upon the argument that K_{ATP} channel activation acts as a protective mechanism against the effects of ischaemia (Noma, 1983) and block by tolbutamide overrides this. However, the exact opposite standpoint has also been put forward, that activation of K_{ATP} channels in anoxia results in an increase in extracellular K^+ concentration, which elicits cardiac arrhythmias. Thus tolbutamide block would result in anti-arrhythmic action (Cook, 1988). In addition, Belles *et al.* (1987) argue that the concentration of tolbutmide required to block cardiac K_{ATP} channels is rather high with respect to therapeutic levels of the sulphonylurea. Thus the controversy remains.

At present, the molecular mechanism by which these drugs induce K_{ATP} channel inhibition is unclear. Single-channel recordings have shown that tolbutamide and glibenclamide not only inhibit K_{ATP} channels when applied extracellularly but also when applied intracellularly using inside-out membrane patches (Trube *et al.*, 1986; Sturgess *et al.*, 1988). Using the cell-attached recording configuration, in which there is no direct pathway for the extracellularly applied drug to the channels in the membrane patch because of the pipette–membrane seal (Sakmann and Neher, 1984), tolbutamide and glibenclamide were observed to produce channel closure. Access to the inhibitory site must therefore have occurred via the lipid membrane. This suggests that the effectiveness of the drugs would depend upon their lipophilicity, allowing diffusion through the lipid membrane to their site of action, a contention supported by the correlation between their potency as insulin secretagogues and hydrophobicity (Gylfe *et al.*, 1984). Deleers and Malaisse (1984) investigating the binding of hypoglycaemic sulphonylureas to artificial phospholipid bilayers concluded that insertion into the bilayer was a prerequisite for the initiation of insulin release. It has also recently been suggested that the sulphonylureas are more effective when applied to the intracellular membrane face than the extracellular aspect (Sturgess *et al.*, 1988), indicating that the sulphonylurea receptor may be more easily accessed from the intracellular side. Such an effect may contribute to the often-observed and reported 'slowness' of glibenclamide action. In general, the effects of glibenclamide on membrane potential (Ferrer *et al.*, 1984) and insulin release (Garrino *et al.*, 1985) are slow and have a progressive time-course in comparison to those of tolbutamide, and only when high concentrations (1–100 μM) of glibenclamide are used are their actions comparably rapid (see discussion in Garrino and Henquin, 1988). Recordings of whole-cell voltage-clamp K_{ATP} channel currents from mouse cultured β-cells (Zunkler *et al.*, 1988a) and from CRI-G1 cells (Sturgess *et al.*, 1988) also illustrate this point, both glibenclamide and glipizide having slow onset times for current inhibition. It is possible that these slow, progressive responses to glibenclamide and glipizide at nanomolar concentrations reflect the time dependence of access and binding to the 'sulphonylurea' receptor.

The second-generation sulphonylureas (e.g. glibenclamide) are more potent than tolbutamide and it has been speculated that this is due to their interacting with two distinct sites in the plasma membrane of β-cells (one site recognizing the sulphonylurea moiety, the other a group not present in the first-generation drugs, a benzamide substitutent). Evidence to support this view has derived from studies using the compound HB 699 (a benzoic acid derivative), which is similar to the non-sulphonylurea moiety of glibenclamide (Fig. 11.3). This compound also has hypoglycaemic properties and can elicit insulin release *in vivo* and *in vitro* (e.g. Ribes *et al.*, 1981), an effect ascribable to binding to the non-sulphonylurea site. Further studies have shown that HB 699 can also decrease $^{86}Rb^+$ efflux, depolarize the β-cell membrane, induce electrical activity (similar to that elicited by glucose) and increase insulin release (Garrino *et al.*, 1985). In other words, HB 699 acts in a manner identical to that of the sulphonylureas, although with a potency less than glibenclamide but greater than tolbutamide. Similar data have also been obtained with UL-DF9 (Garrino *et al.*, 1986), the benzoic acid derivative (i.e. corresponding to non-sulphonylurea moiety) of the second-generation sulphonylurea gliquidone. These data have led to speculation that a sulphonylurea group is not a prerequisite for the reduction in K^+ permeability and the subsequent sequence of events giving rise to insulin secretion. Recently, it has also been demonstrated that HB 699 directly inhibits whole-cell recorded K_{ATP} currents (Zunkler *et al.*, 1988a) and single K_{ATP} channel currents recorded from outside-out membrane patches (Sturgess *et al.*, 1988) in a dose-dependent manner indistinguishable from that of the sulphonylureas. Does this mean that there are two active sites, both located on the K_{ATP} channel? Brown and Foubister (1984) have argued that the non-sulphonylurea moiety (the benzamide group) does not play any role in the initiation of insulin release, and is simply a protein-binding site. They suggest that the carboxyl group, having a similar acidity to the sulphonylurea moiety, would occupy the same position on the receptor site and produce an identical action. However, it is unlikely to be this simple as benzoic acid alone is reported to be ineffective (unpublished observations, quoted in Garrino *et al.*, 1986), and the benzoic acid derivative UL DF 9 which has the benzoic acid segment like HB 699 but not the benzamide group is also active (but like HB 699 is 100–200 times less potent than the parent compound). Furthermore AZ-DF 265 (Fig. 11.3), which is a benzoic acid derivative also containing the benzamide group, is reported to be as potent as glibenclamide (Garrino and Henquin, 1988). In contrast, the compound HB 985, which only differs from glibenclamide by the substitution of a cyclohexane ring with a methyl group (but still contains both the benzamide and sulphonylurea moieties) is not nearly as potent as glibenclamide and is also less potent than HB 699, at inducing K_{ATP} channel inhibition (Sturgess *et al.*, 1988). Even more intriguing is that this compound has been reported to have no hypoglycaemic properties (Geisen *et al.*, 1985). In addition, AZ-DF 265 exhibits stereoselectivity, the (+)-enantiomer being about ten times less potent at inhibiting $^{86}Rb^+$ efflux and stimulating insulin release (Garrino and Henquin, 1988). Thus, a sulphonylurea group is *not* a prerequisite for inhibition of the K_{ATP} channel and the resultant sequence of events leading to insulin secretion. Other groups such as the benzamide group (or the carboxyl group of benzoic acid) are likely to be involved in specific interactions with one or more sites associated with the K_{ATP} channel.

It has been reported that the block of K_{ATP} channel currents by tolbutamide is not dependent upon the concentration of ATP at the cytoplasmic membrane surface (Trube *et al.*, 1986; Zunkler *et al.*, 1988a), a concentration of about 1–10 μM tolbutamide being sufficient to induce a 50% inhibition at concentrations of 0, 0.3 mM and 1.0 mM ATP. In support of this, Sturgess *et al.* (1988) show, using voltage-clamped CRI-G1 cells, that tolbutamide produces a 50% block of the K_{ATP} current at a concentration of 17 μM in the absence or presence of 0.1 mM ATP intracellularly. Such evidence supports the suggestion that the ATP and sulphonylurea inhibitory sites associated with the K_{ATP} channel are separate entities. However, contrary to these findings, Zunkler *et al.* (1988c) argue that in the absence of intracellular nucleotides the concentration of tolbutamide required to induce 50% inhibition of K_{ATP} currents is 55 μM and that only when 1 mM ADP (and not AMP, GTP or GDP) is present intracellularly is the half-maximally effective tolbutamide concentration reduced to the previously quoted range of values (~4 μM). They suggest that the K_{ATP} channel, or a closely related structure, contains a receptor accessible to intracellular ADP which controls the sensitivity to tolbutamide. Further experiments are required to elucidate the exact nature of this effect and to determine whether this ADP-modulatory site is different from the ATP binding site.

All studies of sulphonylurea action described so far have used freshly isolated cells, primary cultures or cell lines derived from species other than man (i.e. rat, mouse or hamster). Therefore it is particularly important, with respect to the use of these drugs in the treatment of diabetes mellitus, to determine whether they have a similar mode of action in human β-cells. In two parallel, independent studies using human pancreatic β-cells (Ashcroft *et al.*, 1987b) and human insulin-producing tumour cells (Sturgess *et al.*, 1987b), K_{ATP} channels, indistinguishable in terms of conductance, selectivity, kinetics and ATP sensitivity from previous animal studies, were reported to be present. Both groups also showed that tolbutamide inhibited these K_{ATP} channels reversibly in recordings from isolated patches. Thus it is likely that sulphonylurea-induced inhibition of K_{ATP} channel activity underlies the hypoglycaemic action of these drugs in man.

This also raises the question of whether the primary lesion that results in Type II diabetes mellitus, is some defect in the regulation of this channel. Thus, although capable of being inhibited directly by the sulphonylureas, resulting in insulin secretion, the K_{ATP} channel may be less sensitive to inhibition via glucose metabolism. This could occur, for example, by some failure in a biochemical pathway resulting in insufficient second messenger (ATP?) production linking the metabolic events to channel inhibition or by a defect in the channel itself, or some closely associated protein which results in an inability to recognize and/or transduce the second messenger signal. Recent studies by Rorsman *et al.* (1988) have indicated that the K_{ATP} channel in the β-cell plasma membrane of Type II diabetics shows a normal sensitivity to ATP (as measured by its ability to induce channel closure) and it is suggested that the defect may lie in the glucose metabolism pathways. A similar conclusion was reached by Welsh *et al.* (1987) in order to explain the insensitivity of the K_{ATP} channel activity to glucose, and other nutrient secretagogues, recorded from cell-attached patches of fetal β-cells (which also do not increase insulin secretion in response to glucose). Here too, channel activity recorded from inside-out patches was demonstrated to show a normal sensitivity to ATP, indicating that the lack of responsiveness to glucose was due to immature substrate metabolism.

There are a number of pharmacological agents, other than the sulphonylureas, that have been reported to increase electrical activity and insulin release by blocking K^+ channels. Tetraethylammonium (TEA) does not directly depolarize the β-cell, but in the presence of glucose increases spike amplitude and duration (Atwater *et al.*, 1979a). These actions are consistent with blockade of a voltage-gated delayed rectifier (which can also be blocked by 4-aminopyridine), although it is possible that

BK_{Ca} channels also contribute (Rorsman and Trube, 1986). Forskolin, an adenylate cyclase stimulator, has also been reported to inhibit the delayed outward current in pancreatic β-cells, a direct action specific to this K^+ channel (K_{ATP} and BK_{Ca} are unaffected) and not via a cyclic AMP mechanism (Zunkler *et al.*, 1988b). In contrast, quinine and quinidine directly depolarize the β-cell (in the absence of glucose) and cause electrical activity and insulin release (Atwater *et al.*, 1979b; Henquin *et al.*, 1975) Quinine is supposedly an inhibitor of BK_{ca} channels (Armando-Hardy *et al.*, 1975), but in β-cells it produces a specific block of K_{ATP} channels (the BK_{Ca} channels are instead blocked by TEA; Findlay *et al.*, 1985), thus accounting for its insulin-secreting action. Quinine and quinidine are of interest therapeutically as they are used in the treatment of falciparum malaria and sometimes give rise to severe complications as a result of their stimulation of the β-cells, leading to hyperinsulinaemia and therefore hypoglycaemia (White *et al.*, 1983).

Interestingly, the pancreatic β-cells of the obese ob/ob strain of mice exhibit a high level of electrical activity even in the absence of glucose and show little sensitivity to quinine or glibenclamide, indicating a possible malfunction in the K_{ATP} channels. Further, apamin, a bee venom neurotoxin which is a specific inhibitor of small-conductance K_{Ca} channels (Castle *et al.*, 1989) increases the plateau potential duration in ob/ob mouse β-cells but has little effect on β-cells of normal mice (Rosario, 1986), also suggestive of a major difference in K_{Ca} channels. More detailed studies, including single-channel recordings, are required to elucidate the basis for the modified physiology and pharmacology of the K^+ permeability.

11.4 HYPERGLYCAEMIC AGENTS AND K^+ CHANNEL ACTIVATION

Diazoxide (7-chloro-3-methyl-2-*H*-1,2,4-benzothiadiazine-1, 1-dioxide, see Fig. 11.3), a sulphonamide related to the sulphonylureas produces hyperglycaemia in man (Dollery *et al.*, 1962) and other animals (Talbachnick and Gulbenkian, 1968). It is likely that the principal mechanism by which this increase in blood glucose occurs is via a decrease in insulin secretion from β-cells, and this has been directly demonstrated in man (Seltzer and Allen, 1965; Porte, 1968). However, it should be noted that other mechanisms may also contribute to its hyperglycaemic activity, e.g. stimulation of catecholamine release and hepatic glycogenolysis.

The diazoxide-induced decrease in insulin release is a direct effect on the β-cell, being most pronounced after the cells have been stimulated with nutrients (Milner and Hales, 1969). A number of targets have been suggested for the site of action of diazoxide on the β-cell membrane including adrenergic receptors, but more convincing evidence for a channel site was obtained using $^{86}Rb^+$ flux experiments (Henquin and Meissner, 1982). Diazoxide induced an increase in $^{86}Rb^+$ efflux from β-cells which resulted in a hyperpolarization of the membrane potential, and suppression of the normal electrical activity. Henquin and Meissner (1982) therefore concluded that inhibition of insulin release by diazoxide was due to an increase in K^+ permeability of the β-cell membrane. In addition they also demonstrated that tolbutamide-induced depolarization of the β-cell membrane potential and resultant electrical activity was also suppressed by diazoxide. Thus tolbutamide and diazoxide produce opposing actions on the β-cell plasma membrane K^+ permeability which results in functional antagonism with respect to insulin release.

An obvious corollary of these data, from our knowledge that K_{ATP} channels are the major contributant to the β-cell resting membrane potential and that nutrient metabolism and the sulphonylureas inhibit these channels, is that diazoxide produces the opposite effect (i.e. activation). Whole-cell current clamp recordings from the cell line CRI-G1 clearly illustrate the ability of diazoxide to produce a rapid hyperpolarization of the cell membrane potential, resulting in a cessation of action potential activity (Kozlowski *et al.*, 1989; Fig. 11.4). The hyperpolarization is accompanied by a marked increase in the resting conductance of the cell, a result in keeping with an increase in K^+ permeability. Cell-attached recordings from mouse pancreatic β-cells (Trube *et al.*, 1986) and CRI-G1 cells (Sturgess *et al.*, 1988) show that bath-applied diazoxide reverses nutrient (glucose and leucine) and tolbutamide-induced K_{ATP} inhibition (terminating action potential activity). As diazoxide had no direct access to the channels in the membrane patch, activation must have occurred via the drug entering and moving within the lipid membrane, in a manner similar to the sulphonylureas. The activation of K_{ATP} channel currents and the reversal of tolbutamide inhition were also demonstrated directly using the open-cell attached (Dunne *et al.*, 1987), outside-out membrane patch and whole-cell voltage-clamp configurations (Trube *et al.*, 1986; Zunkler *et al.*, 1988a; Sturgess *et al.*, 1988). However, more recent results (Kozlowski *et al.*, 1989) indicate that diazoxide acts as a more potent activator of K_{ATP} channels when applied to inside-out membrane patches (compared to outside-out patch and whole-cell experiments), which suggests that diazoxide has an internal site of action.

The K_{ATP} channel activation induced by diazoxide has also been shown to be dependent upon the concentration of ATP inhibiting the channel. The dose-dependent enhancement of the K_{ATP} channel current, as measured in whole-cell voltage-clamp experiments, was demonstrated to be greater with a higher concentration of ATP (1 mM compared to 0.3 mM) present internally (Zunkler *et al.*, 1988a). However, an internal concentration of 3 mM ATP was reported (using outside-out membrane patches) to prevent the diazoxide-induced K_{ATP} channel activation (Trube *et al.*, 1986). Furthermore, Sturgess *et al.* (1988) recording from voltage-clamped CRI-G1 cells showed that diazoxide had no significant effect on K_{ATP} currents unless first inhibited by ATP or tolbutamide. Diazoxide did not reverse the inhibition by glibenclamide, probably by virtue of the almost irreversible nature of the glibenclamide block. Therefore it would appear that, in order to elicit an enhancement of K_{ATP} currents, diazoxide must interact with either or both the ATP and sulphonylurea inhibitory sites. In common with the hypoglycaemic sulphonylureas, diazoxide does not appear to effect calcium-activated cation-selective channels (BK_{Ca}; K/Na_{Ca}) present in β-cells and insulin-secreting cell lines (Sturgess, 1988; unpublished observations), and it is not known whether diazoxide is capable of activating K_{ATP} currents in other tissues.

The mechanism by which diazoxide activates K_{ATP} currents is not completely understood. Obviously, from the preceding account, the presence of ATP at the cytoplasmic membrane face is of prime importance. In a detailed study, using the open-cell attached configuration, Dunne *et al.* (1987) investigated the ability of diazoxide to induce K_{ATP} channel activation in the absence and presence of intracellular ATP and Mg^{2+}. They showed that ATP was required for activation to be observed consistently, that high ATP levels (5 mM) suppressed this effect, and

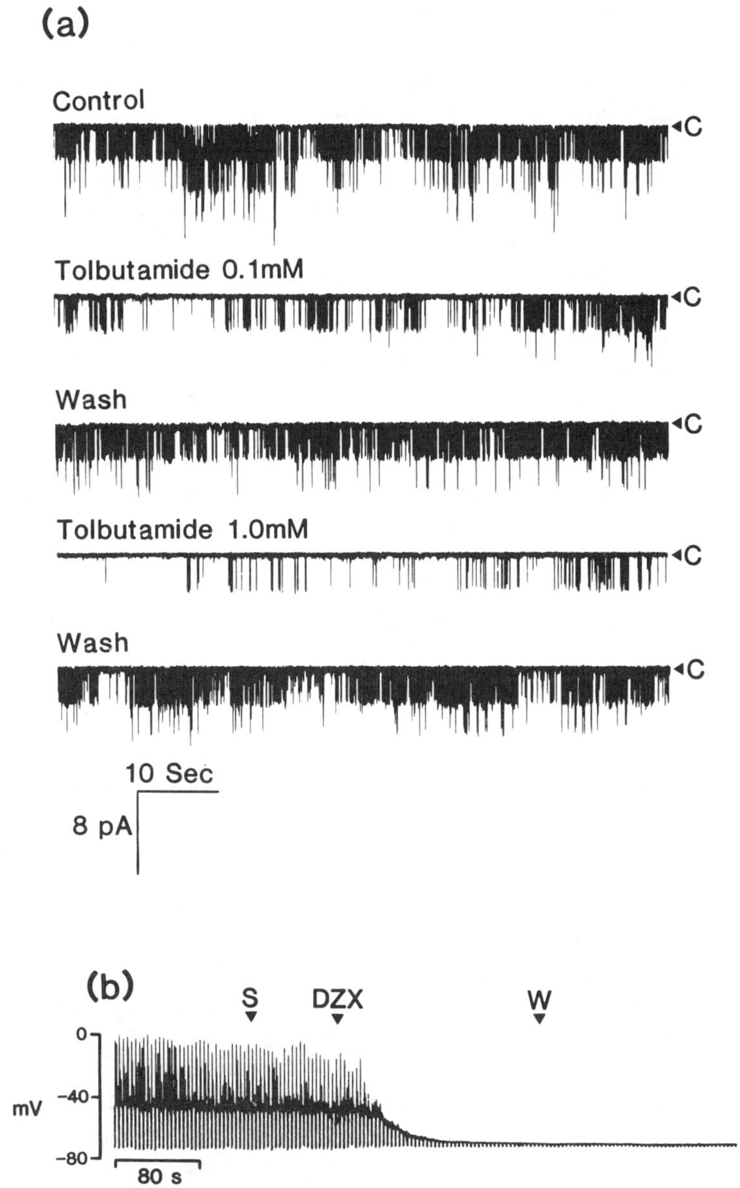

Fig. 11.4 — (a) Inhibition of K_{ATP} channels by tolbutamide. Single-channel currents were recorded from an inside-out patch from a CRI-G1 cell. The patch was exposed to symmetrical 140 mM KCl and held at a membrane potential of −50 mV. Note that the action of tolbutamide is dose-dependent and reversible. (b) Action of diazoxide on a CRI-G1 cell under whole-cell current-clamp conditions. Hyperpolarizing current pulses (10 pA and 1 s duration) were applied every 4 s, often resulting in anodal-break action potentials (although spontaneous action potentials are also evident). Bath application of control saline (S) had no effect, but 0.6 mM diazoxide (DZX) produced an increase in membrane conductance, hyperpolarization and inhibition of firing, effects not reversible on wash (W). (Data in (a) and (b) from R. Z. Kozlowski, and M. L. J. Ashford, M. L. J. unpublished observations.)

that Mg^{2+} was not a necessary co-factor. They concluded that diazoxide interferes directly, or indirectly, with the inhibition of the K_{ATP} channel by the ATP^{4-} species (demonstrated to be a potent inhibitory form, Ashcroft and Kakei, 1987) and that high levels of this ion suppress diazoxide action. However, in direct contrast to these results, Kozlowski *et al.* (1989) show, using isolated patch and whole-cell voltage-clamp recording techniques, that the diazoxide-induced K_{ATP} channel activation was calcium-independent but Mg–ATP dependent. In experiments where the $[ATP^{4-}]$ was maintained relatively constant, activation by diazoxide was observed only when Mg–ATP was also present, otherwise no effect or an inhibition occurred. Diazoxide was also demonstrated to be ineffective when the ATP was substituted with the non-hydrolysable ATP analogue adenylyl imidodiphosphate (AMP–PNP), regardless of the concentration of tetravalent species or whether or not Mg^{2+} was present. Thus the conclusion from this study was that diazoxide enhanced K_{ATP} currents in an Mg–ATP-dependent manner, possibly by means of a phosphorylation. The protein kinase responsible for this phosphorylation must be Ca^{2+}-independent but Mg^{2+}-dependent and be closely associated with the cytoplasmic membrane, as activation is observed in isolated patches. The greater effectiveness observed in inside-out patches also supports this conclusion. These results do not, of course, distinguish between a direct diazoxide-induced stimulation of kinase activity or whether the channel (or some associated protein) becomes susceptible to diazoxide action after phosphorylation. At present it is not clear why a discrepancy exists between the results of the two groups, although the explanation may ultimately lie in the use of differing recording methods.

Diazoxide has a second action on K_{ATP} currents (Kozlowski *et al.*, 1989). In the absence of both Mg^{2+} and ATP from the intracellular environment (particularly the Mg^{2+}), diazoxide inhibits the K_{ATP} current, an action not reversible on wash. Thus, Kozlowski *et al.* (1989) conclude that diazoxide has a dual action on K_{ATP} currents, an activation (which would normally be observed under physiological conditions, i.e. with Mg–ATP present in the cell) and an underlying inhibition (unmasked by the removal of internal Mg^{2+}) which is similar to the effects of the hypoglycaemic sulphonamide derivatives. Is it possible that this latter effect explains why diazoxide reverses tolbutamide inhibition of K_{ATP} currents, simply by competing for a common receptor site?

In man, the free plasma concentration of diazoxide is in the range 6–22 μM (Pruitt *et al.*, 1973). The concentrations used in the aforementioned studies are extremely high in comparison (0.3–0.6 mM, the maximum attainable in solution, for short periods of time). However, Zunkler *et al.* (1988a) showed that the activation of K_{ATP} currents by diazoxide (in the presence of 2 mg/ml albumin, to allow comparison with *in vitro* experiments) was dose-dependent and that the EC_{50} values were 20 μM and 100 μM for pipette (and therefore internal) ATP concentrations of 0.3 mM and 1.0 mM, respectively, although a much greater absolute level of current is observed at the higher ATP concentration. Since intracellular ATP levels in β-cells are considered to be approximately 5 mM (Kakei *et al.*, 1986), these authors considered it likely that an unknown intracellular factor contributes to the action of diazoxide and shifts the concentration–response curve to lower concentrations (they suggest ADP, which acts as a physiological antagonist to ATP; Dunne and Petersen, 1986; Kakei *et al.*, 1986). However, an alternative explanation follows the argument proposed by Cook *et al.* (1988) whereby, under physiological conditions, K_{ATP} currents are approximately 99% inhibited in β-cells and any small change in channel activity will have (by virtue of the very high resting input resistance of the cells) a marked effect on membrane potential. Thus increases in channel activity by much less than 50% of the maximum achievable may well be sufficient to cause a significant hyperpolarization (i.e. the contrary argument with respect to glucose-induced depolarization) at concentrations of diazoxide well within the therapeutic plasma concentration range. The data of Zunkler *et al.* (1988a) support this proposal, in that with higher levels of intracellular ATP diazoxide is actually more effective as a channel activator if one considers a proportional change in current (as opposed to the half-maximal values quoted from absolute maximal changes) such as a 3–5-fold increase from the baseline levels.

Insulin release is also inhibited (thus inducing hyperglycaemia) by hormonal stimuli such as adrenaline, noradrenaline, 5-HT, somatostatin and galanin. It has recently been shown, using the RINm5F cell line, that galanin increases $^{86}Rb^+$ efflux and hyperpolarizes the cell membrane — actions that are reversed by glibenclamide (De Weille *et al.*, 1988). These authors also report that galanin, applied to the external face of outside-out membrane patches, activates K_{ATP} channels formerly inhibited with 2.5 mM ATP at the internal surface. Bullett *et al.* (1989) show that galanin activates K_{ATP} channels in cell-attached patches, but only when present in the electrode, not in the bathing solution, indicating that no soluble second messenger is involved. They conclude that galanin acts via a pertussis toxin-sensitive G-protein because KF, applied to the cytoplasmic membrane face, elicited K_{ATP} channel activation (via production of the AlF_4 complex) and pertussis toxin pretreatment abolished galanin-evoked hyperpolarization, as measured by the membrane potential sensitive fluorescent dye bisoxonol.

Activation of α-adrenergic receptors by adrenaline and noradrenaline has been reported to inhibit insulin secretion, reduce electrical activity and internal calcium levels. These effects are mimicked by clonidine and blocked by yohimbine and phentolamine, indicating they are α_2-mediated (Schuit and Pipeleers, 1986; Nilsson *et al.*, 1988). However, the detailed mechanism(s) whereby α_2-receptor stimulation elicits these actions are not yet clearly defined. Rorsman *et al.* (1987), using whole-cell voltage-clamp recording, show that clonidine also reduces the voltage-dependent calcium current. Recently it has been reported that bath-applied clonidine activates K_{ATP} channels recorded cell-attached from rat β-cells, an action blocked by yohimbine (Ronner *et al.*, 1988), but these observations have yet to be substantiated.

11.5 K$^+$ CHANNELS AND OTHER ENDOCRINE CELLS

The sulphonylureas have been reported to be ineffective depolarizing agents on other endocrine cells, indicating that K_{ATP} channels are unlikely to contribute significantly to their resting membrane potential. Thus the K$^+$ permeability that does underlie the resting membrane potential of such cells likely derives from other types of K$^+$ channels. It is not possible, in the present review, to detail the work pertaining to all the various endocrine cells. However, two interesting examples where single

K^+ channels have been recorded and identified are given below (for several details of other endocrine cells see Ozawa and Sand, 1986).

Parathyroid cells have an unusual K^+-channel linked stimulus–secretion coupling mechanism. As the extracellular calcium concentration is raised, these cells respond by depolarization and a reduction in the rate of release of calcitonin and vice versa for lowered extracellular calcium (i.e. in direct contrast to other endocrine cells, where depolarization elicits hormone release). The depolarization induced by the increased extracellular calcium level is associated with a membrane resistance increase which, in turn, is due to a reduced K^+ permeability. In addition, lowering the extracellular calcium concentration results in increased K^+ permeability and hyperpolarization to a potential close to E_K (Lopez-Barneo and Armstrong, 1983). These actions can be explained by the presence of a K^+ channel regulated by calcium such that it is open in low and closed in high levels of calcium. Preliminary studies by Lopez-Barneo *et al.* (1986) indicated the presence of a BK_{Ca} channel in inside-out patches isolated from bovine dissociated parathyroid gland cells. In a more detailed study Jia *et al.* (1988), also with inside-out membrane patches from bovine parathyroid cells, showed that a large-conductance (150–200 pS in non-symmetrical K^+ solutions) K^+ selective channel is present and that it is sensitive to calcium concentrations but not to voltage. The dependence of the channel open-state probability on calcium concentration is unusual. Although channel opening increases with increasing internal calcium concentration it does so only over a very narrow range (110–160 nM); higher concentrations (160–270 nM) result in a decreased P_{OPEN} and eventual channel closure (>270 nM). The lack of voltage sensitivity and the biphasic concentration dependence of open-state probability to internal calcium exhibited by this BK_{Ca} channel may explain the anomalous calcium dependence of the cell resting membrane potential. At low extracellular calcium levels intracellular calcium will also be low, resulting in active BK_{Ca} channels (membrane potential close to E_K), whereas raised extracellular calcium results in increased intracellular levels which decrease P_{OPEN} of the BK_{Ca} channel and cause depolarization (presumably via some unidentified inward current). However, it is not yet clear how secretion increases as the calcium levels fall and the cells hyperpolarize. Jia *et al.* (1988) suggest that these BK_{Ca} channels may also be present in secretory vesicles and their opening could trigger secretion. Further work is required to validate and extend these observations but they do indicate an unusual coupling between membrane potential and hormone secretion.

Adrenocortical cells are different from other endocrine cells in that secretagogue stimulation increases biosynthesis as well as release of steroids. These actions require external calcium and likely involve the generation of cyclic nucleotides. Adrenocorticotrophin (ACTH) is a potent natural stimulant of steroid secretion. It has been demonstrated to depolarize adrenocortical cells and induce action potential activity (Lymangrover *et al.*, 1982), which leads to increased cytosolic free calcium and release. This depolarization is likely mediated by a decreased K^+ permeability (Kenyon *et al.*, 1985). Payet *et al.* (1987), using the whole-cell voltage-clamp recording technique on rat cultured glomerulosa cells, have shown that ACTH inhibits a transient outward K^+ current in a dose-dependent manner and at concentrations which increase steroid release. This current is blocked by TEA (20 mM), which also acts as a secretagogue (Gallo-Payet and Payet, 1989). Lopez-

Barneo *et al.* (1986) have reported that three types of K^+ channel are present in membrane patches isolated from Y1 cells (a transformed cell line of murine adrenocortical cells which secrete steroids in response to ACTH), the most common being a BK_{Ca} channel with a conductance of 180 pS. Taberes *et al.* (1989) have recently shown, using whole-cell voltage-clamped Y1 cells, that the voltage-dependent K^+ current is mainly due to the activity of BK_{Ca} channels and that it is modulated by a variety of intracellular factors (including $[Mg^{2+}]$, ATP, pH and $[Ca^{2+}]$), suggestive of a link between cell metabolism and membrane potential. Thus the exact nature and mechanism of the ACTH-induced depolarization are not yet completely understood.

11.6 CONCLUSIONS

During the last five years, application of the patch-clamp techniques have allowed a much more detailed analysis of the biophysical and pharmacological characteristics of pancreatic β-cell than had hitherto been possibile. In particular, our knowledge of K_{ATP} channels has advanced significantly by studying these cells (e.g. the discovery of a selective ligand, the hypoglycaemic sulphonylureas). These ligands are currently being used in an attempt to isolate and purify (e.g. Kramer *et al.*, 1988) the K_{ATP} channel (assuming the sulphonylurea binding site and K_{ATP} channel to be closely associated) and recently attempts have been made to clone the K_{ATP} channel. However, our understanding of the role of K^+ channels in stimulus–secretion coupling in other endocrine cells is at present less well advanced. It is to be hoped that the next five years will prove at least as successful in elucidating the mechanism by which various K^+ channels contribute to the control of endocrine cell excitability.

REFERENCES

Armando-Hardy, M., Ellory, J. C., Ferreira, H. G., Fleminger, S., and Lew, V. L. (1975) Inhibition of the calcium-induced increase in the potassium permeability of human red blood cells by quinine. *J. Physiol.* **250** 32–33P.

Ashcroft, F. M. (1988) Adenosine 5′-triphosphate-sensitive potassium channels. *Ann. Rev. Neurosci.* **11** 97–118.

Ashcroft, F. M., and Kakei, M. (1987) Effects of internal Mg^{2+} on ATP-sensitive K^+-channels in isolated rat pancreatic β-cells. *J. Physiol.* **390** 72P.

Ashcroft, F. M., Harrison, D. E., and Ashcroft, S. J. H. (1984) Glucose induces closure of single potassium channels in isolated rat pancreatic β-cells. *Nature* **312** 446–448.

Ashcroft, F. M., Ashcroft, S. J. H., and Harrison, D. E. (1985) The glucose-sensitive potassium channel in rat pancreatic-cells is inhibited by intracellular ATP. *J. Physiol. (Lond.)* **369** 101P.

Ashcroft, F. M., Ashcroft, S. J. H., and Harrison, D. E. (1987a) Effects of 2-ketoisocaproate on insulin release and single potassium channel activity in dispersed rat pancreatic β-cells. *J. Physiol.* **385** 517–529.

Ashcroft, F. M., Kakei, M., Kelly, R. P., and Sutton, R. (1987b) ATP-sensitive K^+ channels in human isolated pancreatic β-cells. *FEBS Lett.* **215** 9–12.

Ashcroft, F. M., Ashcroft, S. J. H., and Harrison, D. E. (1988) Properties of single potassium channels modulated by glucose in rat pancreatic β-cells. *J. Physiol.* **400** 501–527.

Ashford, M. L. J., Sturgess, N. C., Cook, D. L., and Hales, C. N. (1986a) K^+-channels in an insulin-secreting cell line: effects of ATP and sulphonylureas. In: *Biophysics of the pancreatic* β-*cell,* pp. 69–76, I., Atwater, E. Rojas and B. Soria (eds). Plenum Press, New York.

Ashford, M. L. J., Hales, C. N., and Sturgess, N. C. (1986b) A calcium-activated non-selective cation channel in a rat insulinoma cell line. *J. Physiol.* **380** 23P.

Atwater, I., Ribalet, B., and Rojas, E. (1978) Cyclic changes in potential and resistance of the β-cell membrane induced by glucose in the islets of Langerhans from mouse. *J. Physiol.* **278** 117–139.

Atwater, I., Ribalet, B., and Rojas, E. (1979a) Mouse pancreatic β-cells: tetraethylammonium blockage of the potassium permeability induced by depolarization. *J. Physiol.* **288** 561–574.

Atwater, I., Dawson, C. M., Ribalet, B., and Rojas, E. (1979b) Potassium permeability activated by intracellular calcium ion concentration in the pancreatic β-cell. *J. Physiol.* **288** 575–588.

Atwater, I., Rosario, L., and Rojas, E. (1983) Properties of the Ca-activated K^+-channel in pancreatic β-cells. *Cell Calcium* **4** 451–461.

Balant, L. (1981) Clinical pharmacokinetics of sulphonylurea hypoglycaemic drugs. *Clin. Pharmacokin.* **6** 215–241.

Belles, B., Hescheler, J., and Trube, G. (1987) Changes of membrane currents in cardiac cells induced by long whole-cell recordings and tolbutamide. *Pflügers Arch.* **409** 582–588.

Bowen, V., and Lazarus, N. R. (1974) Insulin release from the perfused rat pancreas. Mode of action of tolbutamide. *Biochem. J.* **142** 385–389.

Brown, G. R., and Foubister, A. J. (1984) Receptor binding sites of hypoglycaemic sulphonylureas and related [(Acylamino) alkyl] benzoic acids. *J. Med. Chem.* **27** 79–81.

Bullett, M. J., Dunne, M. J., Li, G., Petersen, O. H., and Wollheim, C. B. (1989) Galanin activates ATP-sensitive K^+ channels via a pertussis toxin-sensitive G-protein in rat insulinoma cells. *J. Physiol.* **410** 19P.

Castle, N. A., and Haylett, D. G. (1987) Effect of channel blockers on potassium efflux from metabolically exhausted frog skeletal muscle. *J. Physiol.* **383** 31–43.

Castle, N. A., Haylett, D. G., and Jenkinson, D. H. (1989) Toxins in the characterization of potassium channels. *Trends Neurosci.* **12** 59–65.

Cook, D. L., and Hales, C. N. (1984) Intracellular ATP directly blocks K^+ channels in pancreatic β-cells. *Nature* **311** 271–273.

Cook, D. L., Satin, L. S., Ashford, M. L. J., and Hales, C. N. (1988) ATP-sensitive K^+ channels in pancreatic β-cells: spare channel hypothesis. *Diabetes* **37** 495–498.

Cook, N. (1988) The pharmacology of potassium channels and their therapeutic potential. *Trends Pharmacol. Sci.* **9** 21–28.

Croghan, P. L., Dawson, C. M., Scott, A. M., and Bangham, J. A. (1986) Contribution of isotope flux studies to understanding the mechanism of the β-cell membrane. In: *Biophysics of the pancreatic* β-cell, pp. 207–223, I. Atwater, E. Rojas and B. Soria (eds). Plenum Press, New York.

Dean, P. M., and Matthews, E. K. (1970a) Glucose-induced electrical activity in pancreatic islet cells. *J. Physiol.* **210** 255–264.

Dean, P. M., and Matthews, E. K. (1970b) Electrical activity in pancreatic islet cells: effects of ions. *J. Physiol.* **210** 265–275.

Dean, P. M., Matthews, E. K., and Sakamoto, Y. (1975) Pancreatic islet cells: effects of monosaccharides, glycolytic intermediates and metabolic inhibitors on membrane potential and electrical activity. *J. Physiol.* **246** 459–478.

Deleers, M., and Malaisse, W. J. (1984) Binding of hypoglycaemic sulphonylureas to an artificial phospholipid bilayer. *Diabetologia* **26** 55–59.

De Weille, J., Schmid-Antomarchi, H., Fosset, M., and Lazdunski, M. (1988) ATP-sensitive K^+ channels that are blocked by hypoglycaemia-inducing sulphonylureas in insulin-secreting cells are activated by galanin, a hyperglycaemia-inducing hormone. *Proc. Natl Acad. Sci. USA* **85** 1312–1316.

Dollery, C. T., Pentecost, B. L., and Samaan, N. A. (1962) Drug induced diabetes. *Lancet* **ii** 735–737.

Dunne, M. J., and Petersen, O. H. (1986) Intracellular ADP activates K^+ channels that are inhibited by ATP in an insulin-secreting cell line. *FEBS Lett.* **208** 59–62.

Dunne, M. J., Findlay, I., Petersen, O. H., and Wollheim, C. B. (1986) ATP-sensitive K^+ channels in an insulin-secreting cell line are inhibited by D-glyceraldehyde and activated by membrane permeabilization. *J. Membr. Biol.* **93** 271–279.

Dunne, M. J., Illot, M. C., and Petersen, O. H. (1987) Interaction of diazoxide, tolbutamide and ATP^{4-} on nucleotide-dependent K^+ channels in an insulin-secreting cell line. *J. Membr. Biol.* **99** 215–224.

Fenwick, E. M., Marty, A., and Neher, E. (1982) A patch-clamp study of bovine chromaffin cells and of their sensitivity to acetylcholine. *J. Physiol.* **331** 577–597.

Ferrer, R., Atwater, I., Omer, E. M., Goncálves, A. A., Croghan, P. C., and Rojas, E. (1984) Electrophysiological evidence for the inhibition of potassium permeability in pancreatic β-cells by glibenclamide. *Q. J. Exp. Physiol.* **69** 831–839.

Findlay, I., Dunne, M. J., Ullrich, S., Wollheim, C. B., and Petersen, O. H. (1985) Quinine inhibits Ca^{2+}-independent K^+ channels whereas tetraethylammonium inhibits Ca^{2+}-activated K^+ channels in insulin secreting cells. *FEBS Lett.* **185** 4–8.

Gallo-Payet, N., and Payet, M. D. (1989) Excitation–secretion coupling: involvement of potassium channels in ACTH-stimulated rat adrenocortical cells. *J. Endocrinol.* (in press).

Garrino, M.-G., and Henquin, J.-C. (1988) Highly potent and stereoselective effects of the benzoic acid derivative AZ-DF 265 on pancreatic β-cells. *Br. J. Pharmacol.* **93** 61–88.

Garrino, M.-G., Schemeer, W., Nenquin, M., Meissner, H.-P., and Henquin, J-C. (1985) Mechanism of the stimulation of insulin release *in vitro* by HB 699, benzoic acid derivative similar to the non-sulphonylurea moiety of glibenclamide. *Diabetologia* **28** 697–703.

Garrino, M.-G., Meissner, H.-P., and Henquin, J.-C. (1986) The non-sulphonylurea moiety of gliquidone mimics the effects of the parent molecule on pancreatic β-cells. *Eur. J. Pharmacol.* **124** 309–316.

Gee, W., and Misler, S. (1987) Tolbutamide inhibits an ATP sensitive K^+ channel in cardiac myocytes. *Biophys. J.* **51** 53a.

Geisen, K., Hitzel, V., Ökomonopoulos, R., Pünter, J., Weyer, R., and Summ, H.-D. (1985) Inhibition of ^{3}H-glibenclamide binding to sulphonylurea receptors by oral antidiabetics. *Arzneimittel-Forschung/Drug Res.* **35** 707–712.

Gillis, K., Gee, W., Falke, L., and Misler, S. (1987) Opposite actions of two structurally similar sulphonamides on an ATP-sensitive K^{+} channel in adult pancreatic β-cells and RINm5F insulinoma cells. *Biophys. J.* **51** 53a.

Giroix, M.-H., Sener, A., Dufrance, J. P., Malaisse-Laqae, F., and Malaisse, W. J. (1986) Glucose metabolism in insulin-producing tumour cells. *Abstracts of 13th International Congress of Biochemistry,* p. 722.

Gylfe, E., Hellman, B., Sehlin, J., and Taljedal, I.-B. (1984) Interaction of sulphonylurea with the pancreatic β-cell. *Experientia* **40** 1126–1134.

Hellman, B., Sehlin, J., and Täljedal, I-B. (1971) The pancreatic β-cell recognition of insulin secretagogues. II. Site of action of tolbutamide. *Biochem. Biophys. Res. Commun.* **45** 1384–1388.

Henquin, J.-C. (1978) D-Glucose inhibits potassium efflux from pancreatic islet cells. *Nature* **271** 271–273.

Henquin, J.-C. (1980) Tolbutamide stimulation and inhibition of insulin release: studies of the under-lying ionic mechanisms in isolated rat islets. *Diabetologia* **18** 151–160.

Henquin, J.-C., and Meissner, H. P. (1981) Effects of amino acids on membrane potential and $^{86}Rb^{+}$ fluxes in pancreatic β-cells. *Am. J. Physiol.* **240** E245–E252.

Henquin, J.-C., and Meissner, H. P. (1982) Opposite effects of tolbutamide and diazoxide on $^{86}Rb^{+}$ fluxes and membrane potential in pancreatic β-cells. *Biochem. Pharmacol.* **31** 1407–1415.

Henquin, J.-C., and Meissner, H. P. (1984) Significance of ionic fluxes and changes in membrane potential for stimulus–secretion coupling in pancreatic β-cells. *Experientia* **40** 1043–1052.

Henquin, J.-C., Horemans, B., Nenquin, M., Verniers, J., and Lambert, A. E. (1975) Quinine -induced modifications of insulin release and glucose metabolism by isolated pancreatic islets. *FEBS Lett.* **57** 280–284.

Howell, S. L. (1984) The mechanism of insulin secretion. *Diabetologia* **26** 319–327.

Hutton, J. C., Sener, A., Herchuelz, A., Atwater, I., Kawazu, S., Boschero, A. C., Somers, G., Devis, G., and Malaisse, W. J. (1980) Similarities in the stimulus–secretion coupling mechanisms of glucose and 2-keto acid-induced insulin release. *Endocrinology* **106** 203–?19.

Jia, M., Ehrenstein, G., and Iwasa, K. (1988) Unusual calcium-activated potassium channels of bovine parathyroid cells. *Proc. Natl Acad. Sci. USA* **85** 7236–7239.

Kakei, M., Kelly, R. P., Ashcroft, S. J. H., and Ashcroft, F. M. (1986) The ATP-sensitivity of K^{+} channels in rat pancreatic β-cells is modulated by ADP. *FEBS Lett.* **208** 63–66.

Kaubisch, N., Hammer, R., Wollheim, C., Rendd, A. E., and Offord, R. E. (1982) Specific receptors for sulphonylureas in brain and in a β-cell tumour of the rat. *Biochem. Pharmacol.* **31** 1171–1174.

Kenyon, C. J., Young, J., and Fraser, R. (1985) Potassium fluxes in bovine adrenal cells during adrenocorticotropin stimulation. *Endocrinology* **116** 109–118.

Kozlowski, R. Z., Hales, C. N., and Ashford, M. L. J. (1989) Dual effects of diazoxide on ATP–K^{+} currents recorded from an insulin-secreting cell line. *Br. J. Pharmacol.* **97** 1039–1050.

Kramer, W., Oekonomopulos, R., Pünter, J., and Summ, H. D. (1988) Direct photo affinity labeling of the putative sulphonylurea receptor in rat β-cell tumor membrane by [^{3}H]glibenclamide. *FEBS Lett.* **229** 355–359.

Lebovitz, H. E. (1985) Oral hypoglycaemic agents. In: *The diabetes annual 1*, pp. 93–110, K. G. M. M. Alberti and L. P. Krall (eds). Elsevier, Amsterdam.

Lopez-Barneo, J., and Armstrong, C. M. (1983) Depolarizing response of parathyroid cells to divalent cations. *J. Gen. Physiol.* **82** 269–294.

Lopez-Barneo, J., Taberes, L., and Castellano, A. (1986) Potassium channels in adrenocortical and parathyroid cells. In: *Biophysics of the pancreatic* B-cell, pp. 125–137, I. Atwater, E. Rojas, and B. Soria (eds). Plenum Press, New York.

Loubatières, A. (1977) Effects of sulphonylureas on the pancreas. In: *The diabetic pancreas,* pp. 489–515, B. W. Volk and K. F. Wellman (eds). Baillière Tindall, London.

Lymangrover, J. R., Matthews, E. K., and Saffran, M. (1982) Membrane potential changes of mouse adrenal zona fasciculata cells in response to adrenocorticotropin and adenosine 3′-5′-monophosphate. *Endocrinology* **110** 462–468.

Marty, A., and Neher, E. (1982) Ionic channels in cultured rat pancreatic islet cells. *J. Physiol.* **326** 36–37P.

Matthews, E. K. (1985) Electrophysiology of pancreatic islet β-cells. In: *The electrophysiology of the secretory cell,* Vol. 2, pp. 93–112, A. M. Poisner and J. M. Trifaró (eds). Elsevier, Amsterdam.

Matthews, E. K., and Sakamoto, Y. (1975a) Electrical characteristics of pancreatic islet cells. *J. Physiol.* **246** 421–437.

Matthews, E. K., and Sakamoto, Y. (1975b) Pancreatic islet calls: electrogenic and electrodiffusional control of membrane potential. *J. Physiol.* **246** 439–457.

Matthews, E. K., and Shotton, P. A. (1984) The control of ^{86}RB efflux from rat isolated pancreatic islets by the sulphonylureas tolbutamide and glibenclamide. *Br. J. Pharmacol.* **82** 689–700.

Meissner, H. P., and Schmeltz, H. (1974) Membrane potentials of beta-cells in pancreatic islets. *Pflügers Arch.* **351** 195–206.

Milner, R. D. G., and Hales, C. N. (1969) The interaction of various inhibitors and stimuli of insulin release studied in rabbit pancreas *in vitro*. *Biochem. J.* **113** 473–479.

Misler, S., Falke, L. C., Gillis, K., and McDaniel, M. L. (1986) A metabolite-regulated potassium channel in rat pancreatic β-cells. *Proc. Natl Acad. Sci. USA* **83** 7119–7123.

Nilsson, T., Arkhammar, P., Rorsman, P., and Berggren, P.-O. (1988) Inhibition of glucose-stimulated insulin release by α_2-adrenoceptor activation is paralleled by both a repolarization and a reduction in cytoplasmic free Ca^{2+} concentration. *J. Biol. Chem.* **263** 1855–1860.

Noma, A. (1983) ATP-regulated K^+ channels in cardiac muscle. *Nature* **305** 147–148.

Ozawa, S., and Sand, O. (1986) Electrophysiology of excitable endocrine cells. *Physiolog. Rev.* **66** 887–952.

Panten, U., Zunkler, B. J., Scheit, S., Kirchhoff, K., and Lenzen, S. (1986) Regulation of energy metabolism in pancreatic islets by glucose and tolbutamide. *Diabetologia* **29** 648–654.

Payet, M. D., Benabderrazik, M., and Gallo-Payet, N. (1987) Excitation–secretion coupling: Ionic currents in glomerulosa cells: Effects of adrenocorticotropin and K^+ channel blockers. *Endocrinology* **121** 875–882.

Petersen, O. H., and Findlay, I. (1987) Electrophysiology of the pancreas. *Physiol. Rev.* **67** 1054–1116.

Porte, D., Jr (1968) Inhibition of insulin release by diazoxide and its relation to catecholamine effects in man. *Ann. NY Acad. Sci.* **150** 281–286.

Pruitt, A. W. Dayton, P. G., and Patterson, J. H. (1973) Disposition of diazoxide in children. *Clin. Pharmacol. Therap.* **14** 73–82.

Ribalet, B., and Beigelman, P. M. (1982) Effects of sodium on β-cell electrical activity. *Am. J. Physiol.* **242** C296–C303.

Ribalet, B., Eddlestone, G. T., and Ciani, S. (1988) Metabolic regulation of the K(ATP) and maxi-K(V) channel in the insulin-secreting RINm5F-cell. *J. Gen. Physiol.* **92** 219–237.

Ribes, G., Trimble, E. R., Blayac, J. P., Wollheim, C. B., Puech, R., and Loubatières-Mariau, M. M. (1981) Effect of a new hypoglycaemic agent (HB 699) on the *in vivo* secretion of pancreatic hormones in the dot. *Diabetologia* **20** 501–505.

Ronner, P., Smith-Maxwell, C., and Kimmich, G. A. (1988) $Alpha_2$-adrenergic agonist clonidine inhibits insulin release by activating ATP-sensitive potassium channels. *Diabetologia* **31** 537A.

Rorsman, P., and Trube, G. (1985) Glucose dependent K^+-channels in pancreatic β-cells are regulated by intracellular ATP. *Pflügers Arch.* **405** 305–309.

Rorsman, P., and Trube, G. (1986) Calcium and delayed potassium currents in mouse pancreatic β-cells under voltage-clamp conditions. *J. Physiol.* **374** 531–550.

Rorsman, P., Arkhammar, P., Berggren, P.-O., and Nilsson, T. (1987) Clonidine inhibition of glucose-stimulated insulin-release involves reduction of the calcium current. *Diabetologia* **30** 575A.

Rorsman, P., Bokvist, K., Arkhammer, P., Berggren, P.-O., Hollerström, C., Nilsson, T., Welsh, M., and Welsh, N. (1988) Defective regulation of glucose and ATP-regulated K^+ channels results in disturbed glucose sensitivity in fetal β cells. *Diabetologia* **31** 537A.

Rosario, L. M. (1986) Electrophysiological measurement show marked differences in the properties of the pancreatic β-cell K-channels from albino mice and a strain of ob/ob (obese) mice. In: *Biophysics of the pancreatic* β-cell, pp. 95–107, I., Atwater, E. Rojas and B. Soria (eds). Plenum Press, New York.

Sakmann, B., and Neher, E. (1984) Patch clamp techniques for studying ionic channels in excitable membranes. *Ann. Rev. Physiol.* **46** 455–472.

Satin, L. S., and Cook, D. L. (1988) Evidence for two calcium currents in insulin-secreting cells. *Pflügers Arch.* **411** 401–409.

Schmid-Antomarchi, H., de Weille, J., Fosset, M., and Lazdunski, M. (1987) The antidiabetic sulphonylurea glibenclamide is a potent blocker of the ATP-modulated K^+ channel in insulin secreting cells. *Biochem. Biophys. Res. Commun.* **146** 21–25.

Schuit, F. C., and Pipeleers, D. G. (1986) Differences in adrenergic recognition by pancreatic A and B cells. *Science* **232** 875–877.

Scott, A. M., Atwater, I., and Rojas, E. (1981) A method for the simultaneous

measurement of insulin release and β-cell membrane potential in single mouse islets. *Diabetologia* **21** 470–475.
Sehlin, J., and Täljedal, I-B. (1975) Glucose-induced decrease in Rb^+ permeability in pancreatic β-cells. *Nature* **253** 635–636.
Seltzer, H. S. (1980) Efficacy and safety of oral hypoglycaemic agents. *Ann. Rev. Med.* **31** 261–272.
Seltzer, H. S., and Allen, E. W. (1965) Inhibition of insulin secretion in 'Diazoxide Diabetes'. *Diabetes* **14** 439.
Sturgess, N. C., Ashford, M. L. J., Cook, D. L., and Hales, C. N. (1985) The sulphonylurea receptor may be an ATP-sensitive K^+ channel. *Lancet* **ii** 474–475.
Sturgess, N. C., Ashford, M. L. J., Carrington, C. A., and Hales, C. N. (1986a) Single channel recordings of potassium currents in an insulin secreting cell line. *J. Endocrinol.* **109** 201–207.
Sturgess, N. C., Hales, C. N., and Ashford, M. L. J. (1986b) Inhibition of calcium-activated non-selective cation channel, in a rat insulinoma cell line, by adenine derivatives. *FEBS Lett.* **208** 397–400.
Sturgess, N. C., Hales, C. N., and Ashford, M. L. J. (1987a) Calcium and ATP regulate the activity of a non-selective cation channel in a rat insulinoma cell line. *Pflügers Arch.* **409** 607–615.
Sturgess, N. C., Carrington, C. A., Hales, C. N., and Ashford, M. L. J. (1987b) Nucleotide-sensitive ion channels in human insulin producing tumour cells. *Pflügers Arch.* **410** 169–172.
Sturgess, N. C., Kozlowski, R. Z., Carrington, C. A., Hales, C. N., and Ashford, M. L. J. (1988) Effects of sulphonylureas and diazoxide on insulin secretion and nucleotide-sensitive channels in an insulin-secreting cell line. *Br. J. Pharmacol.* **95** 83–94.
Taberes, L., Urena, J., and López-Barneo, J. (1989) Properties of calcium and potassium currents of clonal adrenocortical cells. *J. Gen. Physiol.* (in press).
Talbachnik, I. I. A., and Gulbenkian, A. (1968) Mechanism of diazoxide hyperglycaemia in animals. *Ann. NY Acad. Sci.* **150** 204–218.
Trube, G., Rorsman, P., and Ohno-Shosaku, T. (1986) Opposite effects of tolbutamide and diazoxide on the ATP-dependent K^+ channel in mouse pancreatic β-cells. *Pflügers Arch.* **407** 493–499.
Welsh, M., Berggren, P.-O., and Rorsman, P. (1987) Glucose does not block ATP-regulated K^+-channel activity in fetal β cells. *Diabetologia* **30** 596A.
White, N. J., Warrell, D. A., and Chantavanich, P. (1983) Severe hypoglycaemia and hyperinsulinemia in falciparum malariea. *N. Engl. J. Med.* **309** 61–66.
Wollheim, C. B., and Pozzan, T. (1984) Correlation between cytosolic free Ca^{2+} and insulin release in an insulin-secreting cell line. *J. Biol. Chem.* **259** 2262–2267.
Zunkler, B. J., Lenzen, S., Männer, K., Panten, U., and Trube, G. (1988a) Concentration-dependent effects of tolbutamide, meglitinide, glipizide, glibenclamide and diazoxide on ATP-regulated K^+ currents in pancreatic β-cells. *Naunyn Schmiedeberg's Arch. Pharmacol.* **337** 225–230.
Zunkler, B. J., Trube, G., and Ohno-Shosaku, T. (1988b) Forskolin-induced block of delayed rectifying K^+ channels in pancreatic β-cells is not mediated by cAMP. *Pflügers Arch.* **411** 613–619.
Zunkler, B. J., Lins, S., Ohno-Shosaku, T., Trube, G., and Panten, U. (1988c) Cytosolic ADP enhances the sensitivity to tolbutamide of ATP-dependent K^+ channels from pancreatic β-cells. *FEBS Lett.* **239** 241–244.

12

Basolateral membrane potassium channels in secretory epithelia

John D. McCann and **Michael J. Welsh**

In this review we focus on K^+ channels located on the basolateral membrane of secretory epithelia. Rather than attempting to review K^+ channels in all epithelia, we discuss work on several models of Cl^- secreting epithelia, including airway epithelia, intestinal epithelia (T84 cells) and kidney epithelia (Madin Darby canine kidney cells, MDCK). We also discuss fluid secretion in glands, utilizing salivary acinar cells as an example. The role of basolateral membrane K^+ channels in absorptive epithelia (see Schultz, 1981, for a review) and the function of apical membrane K^+ channels in K^+-secreting epithelia (Hunter *et al.*, 1984; Koeppen *et al.*, 1984) will not be discussed, although they share some properties with basolateral membrane K^+ channels.

12.1 A MODEL OF TRANSEPITHELIAL CHLORIDE SECRETION

A discussion of the physiological function of basolateral K^+ channels requires an understanding of the mechanism of Cl^- secretion. Transepithelial Cl^- secretion is made possible by segregation of ion channels, cotransporters and pumps to either the cell's apical or basolateral membrane. Many epithelia that secrete Cl^- from the submucosal (basolateral) surface to the mucosal (apical) surface share a common mechanism of transepithelial Cl^- transport (Welsh, 1987), which is outlined in Fig. 12.1. Although the general mechanism of Cl^- secretion is common to many systems, different epithelia have different transport proteins located in the apical and basolateral membrane. In addition, different model systems have different primary and secondary messenger systems for regulating Cl^- secretion.

At the basolateral membrane Cl^- enters the cell on an electrically neutral Na^+-coupled cotransporter. The identity of the cotransporter is not certain in any Cl^--secreting epithelia and may, in fact, differ from one cell type to another. However, it is likely to be either a Na^+–Cl^- cotransporter or an Na^+–K^+–$2Cl^-$ cotransporter.

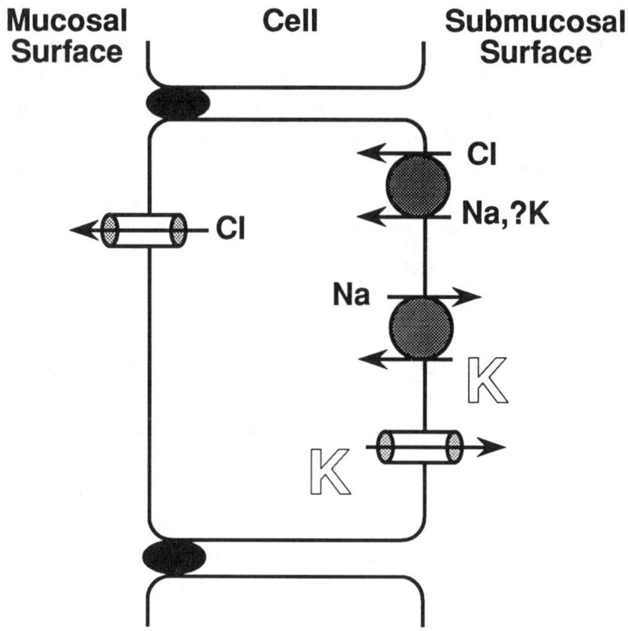

Fig. 12.1 — A model of the mechanism of epithelial Cl^- secretion.

The cotransporter uses the energy in the transmembrane Na^+ gradient to accumulate Cl^- within the cell above electrochemical equilibrium. Cl^- then leaves the cell passively across the apical membrane, moving down a favourable electrochemical gradient through an apical membrane Cl^- channel. The Na^+ gradient that drives Cl^- uptake is maintained by the Na^+–K^+-ATPase located on the basolateral membrane. The Na^+–K^+-ATPase converts the energy in ATP into the electrochemical gradient that supports Cl^- secretion; inhibition of the Na^+–K^+-ATPase with ouabain or a K^+-free solution abolishes Cl^- secretion.

The cell also contains a K^+ channel located in the basolateral membrane. The K^+ channel solves two problems faced by the Cl^--secreting cell. First, when secretion is stimulated K^+ will accumulate in the cell due to K^+ entry on the Na^+–K^+-ATPase and possibly on the Cl^- cotransporter. Activation of basolateral K^+ channels allows K^+ to efflux from the cell, preventing changes in cell volume. Second, stimulation of secretion opens Cl^- channels in the apical membrane, shifting the cell's voltage toward the Cl^- chemical gradient, and as a result diminishing the driving force for Cl^- exit. Activation of basolateral K^+ channels maintains an electrically negative cell membrane voltage, thereby maintaining the driving force for Cl^- exit across the apical membrane.

12.2 EVIDENCE FOR A BASOLATERAL MEMBRANE POTASSIUM CONDUCTANCE IN SECRETORY EPITHELIA

We began by presenting a somewhat teleological explanation of the function of basolateral K^+ channels in Cl^--secreting epithelia; we now review some of the experimental evidence. Much of the early work in this area was aimed at demonstrating that the basolateral membrane of secretory epithelia was preferentially K^+-permeable. The work relied predominantly on three techniques: measurement of tracer ($^{42}K^+$ and $^{86}Rb^+$) efflux; measurement of membrane voltage using conventional microelectrodes; and measurement of intracellular K^+ activity using K^+-selective microelectrodes. The electrophysiological techniques allowed measurement of apical, baslolateral and transepithelial voltages and resistances. In addition, Ba^{2+} has often been used to investigate K^+ channels in epithelia because it blocks nearly all K^+ channels to some extent.

12.2.1 Airway epithelial cells

In airway epithelial cells the intracellular K^+ activity, measured with K^+-selective microelectrodes, is 69 mM (Smith and Frizzel, 1984; Welsh, 1983c). When hormones stimulate transport, the rate of Cl^- secretion increases several-fold, but intracellular K^+ activity remains relatively constant. Because the K^+ activity remains constant, one K^+ must exit the cell for each K^+ that enters if large changes in cell volume are to be avoided. It is reasonable to estimate that for each one Cl^- ion secreted, between 0.66 and 2.0 K^+ ions enter the cell at the basolateral membrane (given some uncertainty about the exact stoichiometry of the Cl^- entry step and the Na^+–K^+-ATPase). However, under these conditions transepithelial K^+ transport accounts for less than 2% of the rate of Cl^- secretion (Smith and Frizzell, 1984; Welsh, 1983c). These observations can only be explained if K^+ is recycled preferentially across the basolateral membrane. However, these observations do not indicate whether K^+ exists via an electroneutral exchanger or an electrogenic channel.

The basolateral membrane voltage (V_b) of airway epithelia bathed in physiological solutions ($[K^+]_o = 5.4$ mM) is approximately −60 mV (Shorofsky *et al.*, 1983; Welsh *et al.*, 1982, 1983). When V_b stimulates Cl^- secretion, the resistance of the basolateral membrane falls, but there is little change in V_b (Shorofsky *et al.*, 1983; Welsh *et al.*, 1982, 1983). Consideration of the absolute value of V_b (much closer to the electromotive force (EMF) of K^+ than to that of Na^+ or Cl^-) suggests that a K^+ permeability dominates the basolateral conductance. This suggestion is supported by the finding that an increase in the submucosal K^+ concentration depolarizes V_b in both secreting and non-secreting epithelia (Smith and Frizzell, 1984; Welsh, 1983b). In contrast, an increase in mucosal K^+ concentration has little effect on the membrane voltage. These results suggest the presence of a basolateral membrane K^+ conductance, but no appreciable apical conductance.

An increase in submucosal $[K^+]$ also inhibits Cl^- secretion (Welsh, 1983b), demonstrating that the basolateral K^+ conductance is responsible for maintaining the negative intracellular voltage and, thereby the driving force for Cl^- exit across the apical membrane. Addition of Ba^{2+} to the submucosal solution increases basolateral membrane resistance, depolarizes V_b, and minimizes the changes in V_b produced by an increase in submucosal K^+ concentration (Smith and Frizzell, 1984; Welsh, 1983a). In contrast, mucosal application of Ba^{2+} has no effect. These data indicate that Ba^{2+} inhibits a basolateral membrane K^+ conductance. Addition of Ba^{2+} to the submucosal solution also inhibits Cl^- secretion (Smith and Frizzell, 1984; Welsh, 1983a), providing further evidence that the basolateral K^+ conductance provides the electrical driving force for Cl^- exit at the apical membrane.

12.2.2 T-84 cells

T-84 cells were derived from a human colon cancer and have been the most extensively studied model of an intestinal secretory cell (see Barrett and Dharmsathaphorn, 1989, for an excellent review). In T-84 cells evidence for a basolateral membrane K^+ permeability has been primarily obtained from measurements of $^{42}K^+$ and $^{86}Rb^+$ efflux rates. In unstimulated monolayers the rate of $^{86}Rb^+$ efflux across the basolateral membrane is ten times the rate of efflux across the apical membrane (Mandel *et al.*, 1986). Hormones that increase the rate of transepithelial Cl^- secretion also increase the rate of $^{86}Rb^+$ efflux across the basolateral membrane (Cartwright *et al.*, 1985; Mandel *et al.*, 1986; Weymer *et al.*, 1985). Serosal Ba^{2+} inhibited hormone-stimulated increases in $^{86}Rb^+$ efflux (Mandel *et al.*, 1986; Weymer *et al.*, 1985). Thus, the predominant K^+ permeability in T-84 cells is in the basolateral membrane.

12.2.3 MDCK cells

In the MDCK renal epithelial cell line, K^+ is taken up across the basolateral membrane via pathways inhibited by ouabain and loop diuretics, suggesting uptake by the Na^+–K^+-ATPase and the Na^+–K^+–2Cl-cotransporter (Aiton *et al.*, 1982). K^+ is not secreted; instead it is recycled across the basolateral membrane. Studies similar to those in T-84 cells showed that $^{86}Rb^+$ efflux across the basolateral membrane increased when the rate of Cl^- secretion was stimulated by hormones (Brown and Simmons, 1982). In addition, the K^+ channel antagonist quinine (1 mM) inhibited hormone-stimulated $^{86}Rb^+$ efflux. Further evidence for a K^+ conductance was obtained by measurements of membrane potential in non-confluent MDCK cells grown on impermeable supports: addition of Ba^{2+} depolarized the resting membrane potential and diminished the depolarizing effect of step elevations in extracellular K^+ concentration (Paulmichl *et al.*, 1986).

12.2.4 Salivary gland cells

Geometric and technical considerations have necessitated the use of some different experimental approaches to demonstrate the existence of basolateral membrane K^+ conductance in salivary glands. (For recent reviews on fluid secretion by salivary acini, see Gallacher *et al.*, 1985, and Petersen and Maruyama, 1984). Salivary acini secrete a NaCl-rich fluid and the rate of secretion is controlled by neurotransmitters. Stimulation of secretion produces K^+ efflux, and an increase in extracellular $[K^+]$ on the basolateral side of the epithelium (Burgen, 1956; Petersen, 1970; Poulsen and Bledsoe, 1978). The net loss of cellular K^+ is accompanied by a large increase in membrane conductance (Ginsborg *et al.*, 1974; Nishiyama and Petersen, 1974). Taken together these results suggest a basolateral conductive pathway for K^+ exit.

12.3 REGULATION OF BASOLATERAL MEMBRANE POTASSIUM CHANNELS

As indicated above, effective transepithelial Cl^- secretion requires coordinated regulation of apical Cl^- channels and basolateral membrane channels. What are the messages that allow cross-talk between the apical and basolateral membrane? We review some of the data on basolateral K^+ channel regulation and, whenever possible, we refer to patch-clamp studies of individual basolateral K^+ channels.

12.3.1 Airway epithelial cells

When neurotransmitters and hormones stimulate Cl^- and fluid secretion, Cl^- channels open in the apical membrane and K^+ channels open in the basolateral membrane. In airway epithelia stimulation of Cl^- secretion induces a fourfold increase in basolateral K^+ conductance (Welsh, 1983a). What is the factor that regulates basolateral K^+ conductance? In many excitable cells K^+ channels are regulated by membrane voltage. However, when adrenaline stimulates Cl^- secretion in airway cells, the basolateral K^+ conductance increases even when basolateral membrane voltage is held constant (Welsh, 1985). These findings indicate that the channel is not primarily regulated by voltage. Other work shows that the K^+ channel is not tightly linked to the activity of the basolateral Na^+–K^+-ATPase: studies using varying basolateral K^+ concentrations and ouabain, to inhibit the Na^+ pump, dissociated the basolateral K^+ conductance and the activity of the Na^+–K^+-ATPase (Welsh, 1985).

Thus, it seemed most likely that the basolateral K^+ channel was regulated by an intracellular second messenger. Two second messengers are reasonable candidates to regulate K^+ channels; cyclic AMP, and the intracellular Ca^{2+} concentration, $[Ca^{2+}]_i$. The suggestion that these two second messengers might be involved could be appreciated from measurement of transepithelial Cl^- secretion alone: stimulation of Cl^- secretion, which requires an increase in basolateral K^+ permeability, is caused by many secretagogues that increase cellular levels of cyclic AMP, by membrane-permeant cyclic AMP analogues, and by the Ca^{2+} ionophore A23187 (Welsh, 1987). At present it is not certain whether hormones that induce an increase in $[Ca^{2+}]_i$ without an increase in cyclic AMP can maintain Cl^- secretion. The simplest mechanism of coordinating the apical Cl^- channel and the basolateral K^+ channel would be to have both channels regulated by the same intracellular second messenger, either cyclic AMP or $[Ca^{2+}]_i$, or both. Precedence for such channels exist: in the land snail *Helix*, a low-conductance K^+ channel can be activated by either Ca^{2+} or by cyclic AMP-dependent phosphorylation (Ewald *et al.*, 1985). To date, only one K^+ channel has been described in airway epithelial cells (Welsh and McCann, 1985). In excised, inside-out patches, the channel had a maximum slope conductance of approximately 25 pS and, in the presence of symmetrical K^+ concentrations, showed inward-rectification. The channel was highly selective for K^+ over Na^+ and did not conduct anions.

In cell-attached patches from non-stimulated cells, the channel was quiescent; no channel activity was observed. Addition of either epinephrine or the Ca^{2+} ionophore A23187 activated the channel (Welsh and McCann, 1985). The effect of Ca^{2+} ionophore was dependent upon the $[Ca^{2+}]$ in the bathing solution. In excised, inside-out patches the channel was regulated by the $[Ca^{2+}]$ bathing the cytosolic surface of the membrane patch. At a $[Ca^{2+}]_i$ of 0.1 μM the channel was rarely open and at 1.0 mM the channel spent most of the time in the open state. Attempts to activate the channel with the cyclic AMP-dependent protein kinase under conditions that activate the apical membrane Cl^- channel produce no change in open state probability. Consistent with results in the intact epithelium, the channel showed no appreciable regulation by voltage. Regulation by Ca^{2+}, but not by voltage is an unusual property for K^+ channels in general (see Chapter 3, sections 3.3.2.1 and 3.3.3). A channel similar to that in airway cells has been reported in HeLa cancer cells (Sauve *et al.*, 1986), which were derived from an epithelia. A different type of Ca^{2+} but not voltage gated K^+ channel has been described in MDCK cells (see below). Thus, the gating of K^+ channels by Ca^{2+} but not voltage would seem to be a property common to different epithelia.

If the basolateral membrane K^+ channel is regulated by Ca^{2+} and the apical membrane Cl^- channel is regulated cyclic AMP, how do secretagogues that utilize cyclic AMP as a second messenger activate basolateral K^+ channels? One hypothesis that we have recently tested is that an increase in cellular levels of cyclic AMP may increase $[Ca^{2+}]_i$. In electrically excitable tissues, there is a precedent; cyclic AMP activation of L-type Ca^{2+} channels causes an influx of Ca^{2+} (Tsein *et al.*, 1986). However, in airway epithelia, cell membrane Ca^{2+} channels have not been demonstrated. To address this issue, $[Ca^{2+}]_i$ was measured in single airway epithelia cells using the flourescent indicator, Fura-2 (McCann *et al.*, 1989). Addition of the β-adrenergic agonist isoprenaline or membrane-permanent cyclic AMP analogues increased $[Ca^{2+}]_i$ by releasing Ca^{2+} from intracellular stores. This observation provides a potential mechanism for coupling cyclic AMP- and Ca^{2+}-regulated processes.

However, the cyclic AMP-dependent increase in $[Ca^{2+}]_i$ was transient, lasting approximately 1 minute. In contrast cyclic, AMP-stimulated Cl^- secretion may continue for hours. How can one explain this discrepancy in time-course? There are several possible alternatives. First, there may be a second class of K^+ channels on the basolateral membrane of airway epithelia that can be activated by phosphorylation with cyclic AMP-dependent protein kinase. It is possible that previous patch-clamp experiments have not detected such a channel because it may only be expressed in polarized epithelia. (In polarized epithelial cells the basolateral membrane is inaccessible to the patch pipette). A second alternative is that Ca^{2+} may be increased in a localized area of the cell that is not reflected by measurement of total $[Ca^{2+}]_i$. For example, a local influx of Ca^{2+} across the basolateral membrane might be immediately recycled; such local fluxes of Ca^{2+} might be sufficient to activate Ca^{2+}-gated K^+ channels, without producing large changes in the average value of $[Ca^{2+}]_i$, which is measured by Fura-2. Other potential mechanisms for regulating basolateral K^+ channels include: changes in cell volume (cell swelling has been proposed to regulate basolateral membrane K^+ channels in *Necturus* small intestine, Lau *et al.*, 1984) and changes in cellular ATP concentrations (an ATP-inhibited K^+ channel might be activated if the ATYP concentration falls when Na^+–K^+-ATPase activity increases, Cook and Hales, 1984 and Weiss and Lamp, 1987). Clearly, further work is needed to completely understand the regulation of basolateral K^+ channels in airway epithelia.

12.3.2 T-84 cells

Most of our understanding of the regulation of basolateral membrane K^+ permeability in T-84 cells has come from studies measuring tracer efflux. T-84 cells secrete Cl^- in response to agonists that increase cellular levels of cyclic AMP (VIP and PGE_1), $[Ca^{2+}]_i$ (carbachol, histamine, A23187), or cyclic GMP (*E. coli* heat-stable enterotoxin) (Barrett and Dharmsathaphorn, 1989). Here we discuss the regulation of K^+ permeability by cyclic AMP and $[Ca^{2+}]_i$ because that has been the focus of much of the work in T-84 cells. However, cyclic AMP and cyclic GMP may share a common mechanism because cyclic AMP-dependent protein kinase and cyclic GMP-dependent protein kinase may phosphorylate the same target proteins (Shenolikar, 1988). An increase in intracellular cyclic AMP or $[Ca^{2+}]_i$ leads to a prolonged stimulation of Ba^{2+}-sensitive Rb^+ efflux across the basolateral membrane (Mandel *et al.*, 1986; McRoberts *et al.*, 1985; Weymer *et al.*, 1985).

However, several observations suggest that cyclic AMP and $[Ca^{2+}]_i$ stimulate different K^+ permeabilities. First, A23187-stimulated Rb^+ efflux was less sensitive to Ba^{2+} than cyclic AMP-stimulated efflux (Mandel *et al.*, 1986). Second, A23187-stimulated Rb^+ uptake was refactory to ATP depletion, whereas cyclic AMP-stimulated efflux was not (McRoberts *et al.*, 1985). Third, A23187-stimulated Rb^+ efflux required extracellular Ca^{2+}, whereas cyclic AMP stimulated efflux did not (McRoberts *et al.*, 1985). This finding suggests that cyclic AMP does not elevate $[Ca^{2+}]_i$ by opening cell membrane Ca^{2+} channels. Moreover measurements of $[Ca^{2+}]_i$ have shown that cyclic AMP does not increase $[Ca^{2+}]_i$ in T-84 cells (Dharmsathaphorn and Pandol, 1986). Fourth, agonists that increase either cyclic AMP or $[Ca^{2+}]_i$ do not produce additive effects on the second messengers, but do produce additive effects in stimulation of Rb^+ efflux (Dharmsathaphorn and Pandol, 1986). Confirmation that T-84 cells contain both Ca^{2+} and cyclic AMP-regulated K^+ channels awaits description of the electrical properties and regulation of the K^+ channels at the single-channel level.

12.3.3 MDCK cells

MDCK cells secrete Cl^- in response to the β-adrenergic agonist isoprenaline and the combined α- and β-adrenergic agonist adrenaline (Brown and Simmons, 1981). The data suggest that both cyclic AMP and $[Ca^{2+}]_i$ play a role in regulating secretion. However, the basolateral membrane K^+ permeability appears to be solely regulated by $[Ca^{2+}]_i$. The Ca^{2+} ionophore A23187, and adrenaline, which raises $[Ca^{2+}]_i$ (Breuer *et al.*, 1988), stimulate $^{86}Rb^+$ efflux across the basolateral membrane of MDCK cells (Brown and Simmons, 1982). Electrophysiological measurements support the notion that secretagogues activate a basolateral K^+ channel: adrenaline-caused a sustained hyperpolarization of membrane voltage in subconfluent MDCK cells (Paulmichl *et al.*, 1986). Interestingly, verapamil (a Ca^{2+} channel blocker) partially antagonized the hyperpolarizing effects of adrenaline with a time-course suggesting that the initial K^+ channel activation was mediated by release of Ca^{2+} from intracellular stores and that a second phase of K^+ channel activation was mediated by Ca^{2+} influx (Paulmichl *et al.*, 1986). Bathing cells in Ca^{2+}-free solution mimicked the effect of verapamil on adrenaline-stimulated hyperpolarization (Paulmichl *et al.*, 1986).

Studies of single K^+ channels with the patch-clamp suggest that two Ca^{2+}-activated K^+ channels might account for the basolateral membrane K^+ permeability (Kolb *et al.*, 1987). One channel has an inwardly rectifying current–voltage relation and one had an outwardly rectifying current–voltage relation. Both channels had larger unitary conductances than the K^+ channel reported in airway epithelial cells.

Both channels were activated in cell-attached patches by adrenaline and both were activated by internal $[Ca^{2+}]_i$ in inside-out patches. The probability of the channels being in the open state was only slightly affected by voltage.

12.3.4 Salivery gland cells

Fluid secretion by mammalian salivary acini is controlled by hormones and neurotransmitters that increase $[Ca^{2+}]_i$ (acetylcholine, noradrenaline, substance P, A23187). In contrast to the other epithelia discussed so far, cyclic AMP primarily mediates protein secretion rather than fluid secretion (Gallacher, 1985). The basolateral membrane of salivary acini is accessible to the patch pipette, which has facilitated description of the channels responsible for the basolateral membrane K^+ permeability. As expected, those channels were Ca^{2+}-activated, but unexpectedly they were also voltage-gated (Maruyama *et al.*, 1983). Thus, they are almost identical to large-conductance K^+ channels (BK_{Ca}) found in electrically excitable tissues (Latorre and Miller, 1983). The channels described in salivary glands have a linear current–voltage relation when bathed in symmetrical K^+-containing solutions and a single-channel conductance of 250 pS. There is a sigmoid relationship between membrane voltage and open-channel probability; the probability of the channels being in the open state increases with membrane depolarization. The relationship is shifted to the left by an increase in $[Ca^{2+}]_i$.

12.4 SUMMARY

The conductance of the basolateral membrane of secretory epithelial cells is dominated by K^+ channels. Activation of basolateral K^+ channels is coupled to activation of apical Cl^- channels, so that large changes in cell volume are prevented and electrical and chemical driving forces are maintained. A common theme in all of these epithelia is that basolateral K^+ channels are regulated by $[Ca^{2+}]_i$; this is supported by evidence obtained in the intact cell and in excised membrane patches. There are suggestions, particularly in T-84 cells, that basolateral K^+ channels may also be regulated by cyclic AMP. However, the responsible channel has not been identified at the single-channel level. With the exception of the salivary gland, it appears that the basolateral membrane K^+ channels that have been described at the single-channel level may be somewhat unique to secretory epithelia in that they are regulated by Ca^{2+} but not voltage. This observation raises the possibility that specific channel agonists and antagonists could be developed for use in clinical settings.

ACKNOWLEDGEMENTS

Portions of the work from the authors' laboratory were supported by grants from the National Institutes of Health (HL29851 and HL42385). John D. McCann was supported by the March of Dimes Birth Defects Foundation.

REFERENCES

Aiton, J. F., Brown, C. D. A., Ogden, P., and Simmons, N. L. (1982). K^+ transport in 'tight' epithelial monolayers of MDCK cells. *J. Membr. Biol.* **65** 99–109.

Barrett, K. E., and Dharmsathaphorn, K. (1990) Mechanism of chloride secretion in the colonic epithelial cell line, T_{84}. In: *Secretory diarrhoea*, E. Lebenthal and M. E. Duffey, (eds.). Raven Press, New York.

Breuer, W. V., Mack, E., and Rothstein, A. (1988) Activation of K^+ and Cl^- channels by Ca^{2+} and cyclic AMP in dissociated kidney epithelial (MDCK) cells. *Pflügers Arch.* **411** 450–455.

Brown, C. D. A., and Simmons, N. L. (1981) Catecholamine-stimulation of Cl^-secretion in MDCK cell epithelium. *Biochem. Biophys. Acta* **649** 427–435.

Brown, C. D. A., and Simmons, N. L. (1982) K^+ transport in 'tight' epithelial monolayers of MDCK cells: Evidence for a calcium-activated K^+ channel. *Biochim. Biophys. Acta* **690** 95–105.

Burgen, A. S. V. (1956) The secretion of potassium in saliva. *J. Physiol. (Lond.)* **132** 20–39.

Cartwright, C. A., McRoberts, J. A., Mandel, K. G., and Dharmsathaphorn, K. (1985) Synergistic action of cyclic adenosine monophosphate- and calcium-mediated chloride secretion in a colonic epithelial cell line. *J. Clin. Invest.* **76** 1837–1842.

Cook, D. L., and Hales, C. N. (1984) Intracellular ATP directly blocks K^+ channels in pancreatic B cells. *Nature* **311** 271–273.

Dharmsathaphorn, K., and Pandol, S. J. (1986) Mechanism of chloride secretion induced by carbachol in a colonic epithelial cell line. *J. Clin. Invest.* **77** 348–354.

Ewald, D. A., Williams, A., and Levitan, I. B. (1985) Modulation of single Ca^{2+}-dependent K^+-channel activity by protein phosphorylation. *Nature* **315** 503–506.

Gallacher, D. V. (1985) Electrophysiology of salivery exocrine cells. In: *The electrophysiology of the secretory cell* , pp. 39–60, A. M. Poisner, and J. M. Trifaro, (eds). Elsevier, Amsterdam.

Ginsborg, B. L., House, C. R., and Silinsky, E. M. (1974) Conductance changes associated with the secretory potential in the cockroach salivary gland. *J. Physiol. (Lond.)* **236** 723–731.

Hunter, M., Lopes, A. G., Boulpaep, E. L., and Giebisch, G. H. (1984) Single channel recordings of calcium-activated potassium channels in the apical membrane of rabbit cortical collecting tubules. *Proc. Natl Acad. Sci. USA* **81** 4237–4239.

Koeppen, B. M., Beyenbach, K. W., Helman, S. I. (1984) Single-channel currents in renal tubules. *Am. J. Physiol.* **247** F380–F384.

Kolb, H. A., Paulmichl, M., and Lang, F. (1987) Epinephrine activities outward-rectifying K channel in Madin–Darby canine kidney cells. *Pflügers Arch.* **408** 584–591.

Latorre, R., and Miller, C. (1983) Conduction and selectivity in potassium channels. *J. Membr. Biol.* **71** 11–30.

Lau, K. R., Hudson, R. L., and Schultz, S. G. (1984) Cell swelling increases a barium-inhibitable potassium conductance in the basolateral membrane of *Necturus* small intestine. *Proc. Natl Acad. Sci. USA* **81** 3591–3594.

Mandel, K. G., McRoberts, J. A., Beuerlein, G., Foster, E. S., and Dharmsathaphorn, K. (1986) Ba^{2+} inhibition of VIP- and A23187-stimulated Cl^- secretion by T_{84} cell monolayers. *Am. J. Physiol.* **250** C486–C494.

Maruyama, Y., Gallacher, D. V., and Petersen, O. H. (1983) Voltage and Ca^{2+}-activated K^+ channel in baso-lateral acinar cell membranes of mammalian salivary glands. *Nature* **302** 827–829.

McCann, J. D., Bhalla, R. C., and Welsh, M. J. (1989) Bradykinin and isoproterenol release intracellular calcium via different second messengers in airway epithelial cells. *Am. J. Physiol.* **257** L116–L124.

McRoberts, J. A., Beuerlein, G., and Dharmsathaphorn, K. (1985) Cyclic AMP and Ca^{2+}-activated K^+ transport in a human colonic epithelial cell line. *J. Biol. Chem.* **260** 14163–14172.

Nishiyama, A., and Petersen, O. H. (1974) Membrane potential and resistance measurement in acinar cells from salivary glands *in vitro*: Effect of acetylcholine. *J. Physiol. (Lond.)* **242** 173–188.

Paulmichl, M., Defregger, M., and Lang, F. (1986) Effects of epinephrine on electrical properties of Madin–Darby canine kidney cells. *Pflügers Arch.* **406** 367–371.

Petersen, O. H. (1970) Some factors influencing stimulation-induced release of potassium from the cat submandibular gland to fluid perfused through the gland. *J. Physiol. (Lond.)* **208** 431–447.

Petersen, O. H., and Maruyama, Y. (1984) Calcium-activated potassium channels and their role in secretion. *Nature* **307** 693–696.

Poulsen, J. H., and Bledsoe, S. W. (1978) Salivary gland K^+ transport: *in vivo* studies with K^+-specific microelectrodes. *Am. J. Physiol.* **234** E79-E83.

Sauve, R., Simoneau, C., Monette, and R. Roy, G. (1986) Single-channel analysis of the potassium permeability in HeLa cancer cells: Evidence for a calcium-activated potassium channel of small unitary conductance. *J. Membr. Biol.* **92** 269–282.

Schultz, S. G. (1981) Homocellular regulatory mechanisms in sodium-transporting epithelia: Avoidance of extinction by 'flush-through'. *Am. J. Physiol.* **241** F579–F590.

Schultz, S. G., and Hudson, R. L. (1986) How do sodium-absorbing cells do their job to survive? *News Physiol. Sci.* **1** 185–189.

Shenolikar, S. (1988) Protein phosphorylation: Hormones, drugs, and bioregulation. *FASEB. J.* **2** 2753–2764.

Shorofsky, S. R., Field, M., and Fozzard, H. A. (1983) Electrophysiology of Cl secretion in canine trachea. *J. Membr. Biol.* **72** 105–115.

Smith, P. L., and Frizzell, R. A. (1984) Chloride secretion by canine tracheal epithelium: IV. Basolateral membrane K permeability parallels secretion rate. *J. Membr. Biol.* **77** 187–199.

Tsein, R. W., Bean, B. P. Hess, P., Lansman, J. B., Nilius, B., and Nowycky, M. C. (1986) Mechanisms of calcium channel modulation by β-adrenergic agents and dihydropyridine calcium agonists. *J. Mol. Cell Cardiol.* **18** 691–710.

Weiss, J. N., and Lamp, S. T. (1987) Glycolysis preferentially inhibits ATP-sensitive K^+ channels in isolated guinea pig cardiac myocytes *Science* **238** 67–69.

Welsh, M. J. (1983a) Barium inhibition of basolateral membrane potassium conductance in tracheal epithelium. *Am. J. Physiol.* **244** F639–F645.

Welsh, M. J. (1983b) Evidence for basolateral membrane potassium conductance in canine tracheal epithelium. *Am. J. Physiol.* **244** C337–C384.

Welsh, M. J. (1983c) Intracellular potassium activities in canine tracheal epithelium. *Am. J. Physiol.* **245** C248–C254.

Welsh, M. J. (1985) Basolateral membrane potassium conductance is independent of sodium pump activity and membrane voltage in canine tracheal epithelium. *J. Membr. Biol.* **84** 25–33.

Welsh, M. J. (1987) Electrolyte transport by airway epithelia. *Physiol. Rev.* **67** 1143–1184.

Welsh, M. J., and McCann, J. D. (1985) Intracellular calcium regulates basolateral potassium channels in a chloride-secreting epithelium. *Proc. Natl Acad. Sci. USA* **82** 8823–8826.

Welsh, M. J., Smith, P. L., and Frizzell, R. A. (1982) Chloride secretion by canine tracheal epithelium: II. The cellular electrical potential profile. *J. Membr. Biol.* **70** 227–238.

Welsh, M. J., Smith, P. L., and Frizzell, R. A. (1983) Chloride secretion by canine tracheal epithelium: III. Membrane resistances and electromotive forces. *J. Membr. Biol.* **71** 209–218.

Weymer, A., Huott, W., Liu, W., McRoberts, J. A., and Dharmsathaphorn, K. (1985) Chloride secretory mechanism induced by prostaglandin E1 in a colonic epithelial cell line. *J. Clin. Invest.* **76** 1828–1836.

13

Therapeutic potential of K^+ channel modulation in heart

W. Osterrieder and **J. F. Waterfall**

13.1 INTRODUCTION

K^+ currents in heart are responsible for the maintenance of a high, stable resting membrane potential in the working myocardium during diastole, as well as for the termination of excitation in both myocardium and pacemaker tissue. Although a large body of knowledge on the variety of K^+ currents and their properties has been accumulated over many years (see, for example, the review by Trautwein, 1973), there has always remained some scepticism on the final relevance of these data because of the limitations of the 'voltage-clamp' technique used to study K^+ currents in multicellular tissue. Shortcomings, such as the accumulation or depletion of K^+ in the interstitial space during voltage steps and voltage inhomogeneity along the preparation, have now been overcome with the availability of isolated, viable cardiac cells and the introduction of the patch-clamp technique (Hamill *et al.*, 1981), allowing the study of K^+ currents even at the level of a single K^+ channel. As a consequence, it has been possible to precisely characterize K^+ currents already known and, in addition, to find some hitherto unknown.

It is one of the purposes of the present chapter to provide a review on the different K^+ currents existing in the heart. Since this area has been extensively reviewed (Carmeliet and Veerecke, 1979; Cohen *et al.*, 1986; Carmeliet *et al.*, 1987; Trube, 1988), we intend to give here a more general overview. Where possible, we will also refer to evidence for the presence of these K^+ channels in human tissue. It is not intended to elucidate the subtleties of the nomenclature of K^+ channels in various parts of the heart and different species.

In the second part of this chapter, the therapeutic potential of K^+ channel modulation in the heart will be addressed. In some aspects, the rationale will be on firm ground, whereas in others it will be highly speculative.

13.2 DIVERSITY OF K^+ CURRENTS IN MAMMALIAN HEART

13.2.1 The cardiac action potential

Before turning to the individual K^+ currents found in the heart, we will briefly describe the configuration of the transmembrane action potential. In the myocardium, the resting membrane potential of around −80 mV in physiological saline solution is predominantly a K^+ potential. The rapid depolarization phase is caused by the fast Na^+ inward current. The following maintained depolarization (plateau phase) is due to a delicate balance between inward (depolarizing) and outward (repolarizing) currents, inward currents being mainly carried by Ca^{2+} ions and outward currents by K^+ ions. The action potential is terminated by a relatively rapid repolarization to the resting potential. The shape of the ventricular action potential is typically seen in many mammalian multicellular preparations, including those from man (Trautwein *et al.*, 1962). It should also be mentioned that the basic mechanisms underlying the action potential, derived from investigations in non-human tissue, are applicable to human heart (Trautwein *et al.*, 1962).

The initial depolarization phase of the Purkinje fibre action potential in various species is followed by a short period of rapid repolarization, which in turn is followed by the plateau phase. The plateau phase can be preceded by a small depolarization so that this sequence of events appears as a 'notch' in the early course of the action potential. Pronounced notches have been shown to exist in isolated adult human atria (Escande *et al.*, 1987). This pattern is caused by a short-lasting efflux of K^+ ions, but since it has not been observed in human ventricular muscle (Trautwein *et al.*, 1962) it is likely that the early K^+ efflux plays no role there.

According to the present state of knowledge, the decay of a time-dependent K^+ current is involved in the generation of slow diastolic depolarization in the pacemaking tissues of the heart (see Noble, 1984).

13.2.2 The inwardly rectifying current (I_{K1})

The steady-state current–voltage relationship in various cardiac preparations from different species displays a typical N-shape (for review, see Trautwein, 1973). The outward current amplitude is relatively small at 0 mV, as compared with −50 mV. Negative of the K^+ reversal potential (E_K), the current–voltage relationship is steep. The typical curvature is mainly caused by the inwardly rectifying K^+ current or inward rectifier, which in heart is now usually termed I_{K1}.

The inward rectifier, as found in heart, has great importance in the animal kingdom. Its role is to permit long depolarizing responses (see Hille, 1984) such as the plateau phase of the cardiac action potential. This is accomplished by only a very small conductance at plateau potentials. On repolarization, membrane conductance then increases, resulting in the rapid terminal phase of the action potential. Small deviations from the zero-current potential (near E_K) in either direction immediately result in a large counteracting current due to the steep slope of the current–voltage relationship, thus guaranteeing a stable resting potential. From this, it is also understandable that the pacemaker cells in heart possess only few I_{K1} channels compared to quiescent cardiac cells (Noma *et al.*, 1984). The long plateau phase renders the myocardium inexcitable until the accompanying contraction has ceased (refractory period).

Increases of the extracellular K^+ shift the zero-current potential in the current–voltage relationship into the positive direction, as expected from the Nernst equation, and increase the total membrane conductance (see Trautwein, 1973). Studies on single I_{K1} channels in isolated cardiac myocytes using the patch-clamp technique (Sakmann and Trube, 1984) have directly demonstrated inward rectification and that the greater the concentration of extracellular K^+ the more K^+ ions can flow through an individual channel.

13.2.3 Delayed outward current (I_K)

Voltage-clamp studies in Purkinje fibres, sinoatrial node, atrioventricular node, atria and ventricles have demonstrated the existence of a K^+ current system with characteristic features. When the membrane potential is, for example, voltage-clamped from −50 mV to more positive potentials, the outward current during the test depolarization rises very slowly, with time constants between about 100 and 500 ms. Because of this kinetic behaviour, the current has been labelled the delayed outward current (I_K). It cannot be activated at potentials negative of about −50 mV, is fully activated positive of 0 to +20 mV and does not inactivate during sustained test depolarizations (for review, see Cohen *et al.*, 1986; Trube, 1988). Thus, this current is operative mainly in the late plateau phase to initiate (but not to participate in) the rapid final repolarization phase of the action potential.

Surprisingly, the knowledge of the delayed outward current is less detailed than that of the inward rectifier. In multicellular tissue, relatively large delayed outward currents during test depolarizations and, correspondingly, large tail currents on repolarization have been described. Mainly based on the presence of two exponentials in the K^+ tail current, the existence of two separate types of delayed outward currents has been proposed (for review, see Cohen *et al.*, 1986). Such a conclusion should be treated cautiously since it is now evident from single-channel recordings that an individual channel can occupy multiple kinetically different states, and that the averaging of consecutive channel openings therefore can result in a complex pattern of the mean current. Another as yet unexplained finding is that the delayed outward current observed in isolated myocytes is generally smaller than expected from studies based on multicellular tissue. In this context no single-channel data on this current component in myocardium have been reported. (One detailed single-channel study on I_K in nodal tissue exists (Shibasaki, 1987)). In a recent report no delayed outward current was seen in voltage-clamped isolated human myocytes (Fig. 4 in Mitchell *et al.*, 1986), although it may be argued that the duration of the test pulses (100 ms) was short. Also the delayed outward current was not observed in human atrial myocytes (mentioned in Carmeliet *et al.*, 1987).

Physiological modulators of the delayed outward current include the catecholamines released from the sympathetic fibres. The increase in the delayed outward current by β-adrenergic agonists is presumably due to phosphorylation of the K^+ channels, because in isolated myocytes injection of (Brum *et al.*, 1983), or intracellular perfusion with (see Fig. 6 in Kameyama *et al.*, 1985), the catalytic subunit of the cyclic AMP-dependent protein kinase had the same effect. An indirect effect by the concomitant increase in inward Ca^{2+} current can be excluded since the voltage dependencies clearly differ (Kameyama *et al.*, 1985). The increase in delayed outward current by catecholamines serves to limit the degree of Ca^{2+} inward current-induced prolongation of action potential during increased sympathetic tone.

13.2.4 Transient (early) outward currents (I_{to})

As mentioned above, a fast early repolarization phase (the 'notch') in the action potential in some cardiac tissues such as Purkinje fibres or atria is caused by an early K^+ efflux. The corresponding current system has been named the transient (or early) outward current (I_{to}). Transient outward currents have recently been reviewed (Carmeliet *et al.*, 1987). We restrict ourselves to the description of I_{to} in human atrium.

Escande *et al.* (1987) carried out a detailed voltage-clamp study in isolated atrial myocytes from man. As in sheep Purkinje fibres (Coraboeuf and Carmeliet, 1982), I_{to} was not carried by a homogeneous population of K^+ channels but could be separated into two types. Both components of current activated positive of -10 mV and inactivated thereafter. One component, I_{lo} (l for long-lasting) had a slow decay and was inhibited by 4-aminopyridine and Ba^{2+} but not by the Ca^{2+} channel blocker Co^{2+}. The other component, I_{to} (b for brief), had a short rise time and a more rapid decay. It could be suppressed by either Co^{2+} or by caffeine, an inhibitor of Ca^{2+} release from the sarcoplasmic reticulum, suggesting that I_{to} is a Ca^{2+}-activated K^+ current (see also Chapter 3).

13.2.5 K^+ current activated by acetylcholine and adenosine (K_{ACh})

Acetylcholine released from the vagal nerves slows spontaneous rate in the sinoatrial node, prolongs conduction in the atrioventricular node and shortens the action potential in the atria (for review, see Trautwein, 1982). Single-channel measurements in sinoatrial (Sakmann *et al.*, 1983) and atrial cells (Soejima and Noma, 1984) showed that acetylcholine specifically increased the open probability of a K^+ channel which differed from other K^+ channels present in these cells. Brief openings with a time constant of about 1 ms are characteristic for this channel.

Adenosine, a physiological metabolite, hyperpolarizes and shortens the atrial action potential like acetylcholine (Johnson and McKinnon, 1956). It has been suggested that adenosine and acetylcholine increase K^+ conductance via a common mechanism (Belardinelli and Isenberg, 1983). It was indeed demonstrated that both activate the same K^+ channel (Kurachi *et al.*, 1986), although both agonists bind to different receptors. The likely explanation is that the muscarinic acetylcholine and the adenosine receptors are linked via GTP-binding proteins to the same population of K^+ channels (Kurachi *et al.*, 1986). (For modulation of K^+ channels by G proteins, see Chapter 5). Adenosine may be of special relevance in hypoxic atria where large quantities are released and the action potential is markedly abbreviated (Szentmiklosi *et al.*, 1979).

13.2.6 The ATP-sensitive K^+ current

ATP-sensitive K^+ channels (see also Chapter 4) have raised considerable interest during the past few years since their original discovery in cardiac muscle (Noma,

1983; Trube and Hescheler, 1983, 1984). In inside-out patches channel activity is depressed when the ATP concentration exceeds 1 mM, the physiological concentration in the cell being 3–4 mM. Half-maximal inhibition of this current occurs at about 0.1 mM ATP. Analysis of the current–voltage relationship indicates that the ATP-dependent K^+ channels have inwardly rectifying properties, but to a much lesser extent than I_{K1}. In other words, when cytosolic ATP falls below a critical value, a considerable amount of outward current flows during the plateau phase of the action potential. As a consequence of this 'ATP-sensing' mechanism, action potential duration is decreased. In this way the ATP-dependent K^+ current forms a link between hypoxia (Trautwein *et al.*, 1954) or metabolic poisoning (Carmeliet, 1978) and electrical membrane effects.

13.2.7 K^+ current activated by intracellular Na^+

One single report describes the existence of a K^+ channel activated by intracellular Na^+ in mammalian cardiac cells (Kameyama *et al.*, 1984; see also Chapter 6). This channel is gated by increases in Na_i^+ over 20 mM and is probably of no importance under physiological conditions. It has been speculated that it is activated when Na_i^+ increases because of prolonged failure of Na^+ pumping. No clinical correlate of such failure is, however, known to us.

13.2.8 Ca^{2+}-activated, non-specific cation current

The Ca^{2+}-activated, non-specific cation current is included in this listing, although it is not a pure K^+ current since K^+ and Na^+ are equally permeable (see also Chapter 7). This current (recently reviewed by Reuter, 1984) is inwardly directed at the resting potential and is responsible for the oscillatory depolarizing after-potentials (for review, see Cranfield, 1977) when free intracellular Ca^{2+} is increased above the normal diastolic level. When after-potentials reach the activation threshold of the action potential, extrasystoles are triggered. This may, in extreme cases, lead to life-threatening ventricular fibrillation.

13.3 EXISTING AND POTENTIAL THERAPEUTIC APPLICATIONS OF K^+ CHANNEL MODULATORS IN HEART

13.3.1 Antimuscarinic agents

Probably the longest-known K^+ channel modulator in therapeutic use is atropine. By displacing acetylcholine from muscarinic receptors, it attenuates the bradycardic effects of vagal stimulation by decreasing K^+ conductance in the sinoatrial and atrioventricular nodes. Atropine therefore is used to treat arrhythmic disturbances such as excessive bradycardia after digitalis intoxication (Hoffman and Bigger, 1985), to improve the clinical condition of patients with early myocardial infarction by relieving sinus or nodal bradycardia or AV block (Weiner, 1985) and to alleviate AV nodal brady-arrhythmias after overdoses of Ca^{2+} antagonists (Kozlowski *et al.*, 1988).

13.3.2 Anti-arrhythmic agents of the class III type

Anti-arrhythmics are nowadays classified into four main categories according to their mode of action (Vaughan Williams, 1975). Drugs of the class III type prolong the myocardial action potential and, consequently, the effective refractory period. They are valuable drugs for the treatment of ventricular tachycardia and ventricular fibrillation. The class III anti-arrhythmics available for therapeutic use have recently been reviewed in detail (Bexton and Camm, 1982; Hartenstein and Wagner, 1986). The compounds which prolong the refractory period by a blockade of K^+ channels are in part dealt with in a preceding section in this book (Chapter 8). They include bretylium, bethanidine, sotalol, melperone and clofilium.

Interestingly, most of these compounds were not developed originally for the indication of arrhythmias. The prototype bretylium, for example, was first introduced as an anti-hypertensive, sotalol as a β-blocker and melperone as a neuroleptic. This indicates that these compounds are pharmacologically non-specific.

Among the K^+ channel blockers with class III properties, clofilium appears to be of particular interest (Steinberg *et al.*, 1984). Most currently used compounds claimed to block cardiac K^+ channels have to be administered in high concentrations (1–100 μM) and show minimal selectivity for individual channel types. Clofilium, in contrast, is a highly potent compound with an EC_{50} value of about 20 nM, for increasing action potential duration in isolated Purkinje fibres (Steinberg *et al.*, 1984). In voltage-clamp studies in isolated guinea-pig cells (Arena and Kass, 1988), clofilium was shown to be highly selective for the delayed rectifier as compared with the inward rectifier. Since clofilium has a quaternary nitrogen, and clofilium derivatives with a tertiary nitrogen were unselective, the pathway of access to the K^+ channels is important for selectivity. The functional significance of selectivity for the delayed outward current is that the action potential is prolonged without effects on the terminal repolarization phase. A homogeneous repolarization within the myocardium is a prerequisite for a class III anti-arrhythmic to be devoid of proarrhythmicity. It should be mentioned that a number of drugs (certain α-adrenoceptor antagonists and adrenergic amines) interfere with the action of clofilium (Steinberg *et al.*, 1984), suggesting that it may have properties other than class III activity. Recently a new class III anti-arrhythmic, UK 66914 has been reported which is presumed to act via K^+ channel blockade (Hogg *et al.*, 1989).

13.3.3 K^+ channel openers

The three K^+ channel openers nicorandil, pinacidil and cromakalim (see also Chapter 8) have been studied for their effects on K^+ currents in single ventricular cells of the guinea-pig heart. The main effect on the steady-state current–voltage relationship of nicorandil (Kakei *et al.*, 1986), pinacidil (Jijima and Taira, 1987) and cromakalim (Osterrieder, 1988) was to partly remove or (for high concentrations of cromakalim) to fully abolish inward rectification. These studies, however, could not determine the precise nature of the K^+ channel affected. Nevertheless, the effects result in a shortening of the action potential.

It is difficult to judge the therapeutic potential of K^+ channel openers in the heart. Because this group of compounds shows a 10–100-fold selectivity for vascular smooth muscle over heart, the reflex response to arterial blood pressure decrease will indirectly influence the heart. We will therefore hypothetically assume that cardioselective K^+ agonists were available.

At present, two potential applications where K^+ channel openers could be of benefit are foreseen. Investigation of the effects of the anticonvulsant diphenylhydantoin has led to speculation that shortening of the action potential duration relative to the effective refractory period may be a mechanism to abolish arrhythmias of the re-entry type (see Fig. 7 in Wit *et al.*, 1975). In the example they discuss, a premature impulse arises at the end of the effective refractory period during the repolarization phase of a Purkinje fibre. As result, the premature action potential has a very slow upstroke velocity and conducts only slowly into a loop of Purkinje fibre bundles. After shortening of the action potential, the premature impulse arises from a higher level of membrane potential with an increased upstroke velocity and therefore conducts more rapidly, activating the Purkinje fibre loop in a normal manner. It appears, however, extremely difficult to prove such a beneficial effect of a K^+ channel opener in this clinical setting. In addition, a general shortening of the action potential is believed to predispose to arrhythmias of the re-entry type.

It is well known that 'diseased' human atrial cells are characterized by low levels of resting membrane potential accompanied by low amplitude and upstroke velocity of the action potential. A decrease of K^+ conductance is involved in this depolarization, and perfusion with the K^+ channel activator acetylcholine (see above) leads to marked increases in resting potential (Imanishi and Arita, 1987). Depolarization can also be induced by stretch (Deck, 1964), as occurs in the atria in heart failure patients. It has long since been known (Trautwein *et al.*, 1962) that non-triggered, abnormal automaticity may develop in such partially depolarized human myocardial tissue (as shown *in vitro*). Furthermore, islets of depolarized tissue in a normally polarized surrounding have been proven to be ectopic foci. K^+ channel openers may prevent arrhythmias of such origin by repolarizing the cells and stabilizing the resting potential, although experimental evidence is still lacking.

13.3.4 Blockers of the ATP-dependent K^+ channel

During ischaemia, the myocardium releases K^+ into the extracellular space and, as a consequence, K_o^+ rises. The time-course of rise is biphasic. In isolated swine and rat heart, an initial increase of K_o^+ to about 10 mM occurs after 5–18 min, followed by a further rise to 20–30 mM K_o^+ after 1 hour (Hill and Gettes, 1980; Knopf *et al.*, 1988). The resulting depolarization frequently causes ventricular fibrillation. (It is not possible to reverse this depolarization by K^+ channel openers, since they can only repolarize to the Nernst potential.)

A novel approach for the prevention of ischaemia-induced ventricular arrhythmias was attempted by Kantor *et al.* (1987). They argued that if ischaemic arrhythmias are caused by K^+ loss, and if K^+ loss occurs through the ATP-dependent channels (see section 13.2.6 above), blockers of these channels could prevent these arrhythmias. The blocker used was the sulphonylurea glibenclamide (see also Chapters 4, 8 and 11). The model used was the globally ischaemic isolated rat heart stimulated at 5 Hz, where ischaemic arrhythmias are consistently observed. Arrhythmias developed in all seven control hearts, whereas none were observed in heart perfused with 1 μM glibenclamide. K^+ efflux during the first 5 min of ischaemia was significantly reduced. The results were confirmed in a more detailed study (Kantor *et al.*, 1988). It was concluded in these studies that the reduction of ischaemic K^+ loss by glibenclamide is the mechanism of its antiarrhythmic effect and that glibenclamide may be the prototype of a new class of antiarrhythmic agents.

13.3.5 Blockers of the Ca^{2+}-activated, non-specific cation channel

There exists as yet no pharmacological tool to directly block the Ca^{2+}-activated, non-specific cation current. Such compounds would be a therapeutic novelty for the prevention of the delayed and/or early depolarizations.

13.4 CONCLUSIONS

At the moment, nine different K^+ currents (or eight different K^+ channels, since the channels activated by acetylcholine and adenosine are identical) have been described in mammalian heart. The main difficulties in the identification of therapeutic targets for K^+ channel modulators are threefold. (1) There is a lack of highly selective K^+ channel modulators of the individual channel types. With the exception of acetylcholine, adenosine and ATP, which act at a receptor level, the pharmacological modulators act at a channel level and do not differentiate grossly between the channel types. (2) There is no evidence of tissue specificity within the heart. For example, if a K^+ channel opener was used as a 'bradycardiac agent' because it moves the diastolic potential away from the firing threshold in the sinoatrial node, it would undoubtedly also increase AV conduction time and shorten the myocardial action potential. (3) The presently available K^+ channel modulators have multiple pharmacological activities. For example, treatment of ischaemic arrhythmias with the ATP-dependent K^+ channel blocker glibenclamide would be associated with hypoglycaemia. Thus the evaluation of the therapeutic potential of K^+ channel modulators in heart has not been and will not be an easy task.

In summary, we can state that K^+ channel modulation at the level of the acetylcholine receptor is a well-established clinical tool. We predict that the search for highly selective, potent blockers of the delayed rectifier as class III anti-arrhythmics will continue and we expect to see more preclinical data on the use of blockers of the ATP-dependent channel. Only time will tell whether blockers of the non-specific cation channel will be found.

Acknowledgements

We wish to thank Dr G. Trube for reading the manuscript.

REFERENCES

Arena, J. P., and Kass, R. S. (1988) Block of heart potassium channels by clofilium and its tertiary analogs: relationship between drug state and type of channel blocked. *Mol. Pharmacol.* **34** 60–66.

Belardinelli, L., and Isenberg, G. (1983) Isolated atrial myocytes: adenosine and acetylcholine increase potassium conductance. *Am. J. Physiol.* **244** (*Heart Circ. Physiol.* **13**) H734–H737.

Bexton, R. S., and Camm, A. J. (1982) Drugs with a class III antiarrhythmic action. *Pharmac. Ther.* **17** 315–355.

Brum, G., Flockerzi, V., Hofmann, F., Osterrieder, W., and Trautwein, W. (1983) Injection of catalytic subunit of CAMP-dependent protein kinase into isolated cardiac myocytes. *Pflügers Arch.* **398** 147–154.

Carmeliet, E. (1978) Cardiac transmembrane potentials and metabolism. *Circ. Res.* **42** 577–587.

Carmeliet, E., and Veerecke, E. (1979) Electrogenesis of the action potential and automaticity. In: *Handbook of physiology, section 2: The cardiovascular system*, pp. 269–334, Vol. I, R. M. Berne, N. Sperelakis, and S. R. Geiger (eds). American Physiological Society, Bethesda.

Carmeliet, E., Biermans, G., Callewaert, G., and Veerecke, J. (1987) Potassium currents in cardiac cells. *Experientia* **43** 1175–1184.

Cohen, I. S., Datyner, N. B., Gintant, G. A., and Kline, R. P. (1986) Time-dependent outward currents in heart. In: *The heart and cardiovascular system*, pp. 637–669, H. A. Fozzard, E. Haber, J. B. Jennings, A. M. Katz and H. E. Morgan (eds). Raven Press, New York.

Coraboeuf, E., and Carmeliet, E. (1982) Existence of two transient outward currents in sheep cardiac Purkinje fibres. *Pflügers Arch.* **392** 352–359.

Cranefield, P. F. (1977) Action potentials, afterpotentials and arrhythmias. *Circ. Res.* **41** 415–423.

Deck, K. A. (1964) Aenderungen des Ruhepotentials und der Kabeleigenschaften von Purkinje-Fäden bei der Dehnung. *Pflügers Arch.* **280** 131–140.

Escande, D., Coulombe, A., Faivre, J.-F., Deroubaix, E., and Coraboeuf, E. (1987) Two types of transient outward currents in adult human atrial cells. *Am. J. Physiol.* **252** (*Heart Circ. Physiol.* **18**), H142–H148.

Hamill, O. P., Marty, A., Neher, E., Sakmann, B., and Sigworth, F. J. (1981) Improved patch-clamp techniques for high-resolution current recording from cells and cell-free membrane patches. *Pflügers Arch.* **391** 85–100.

Hartenstein, J., and Wagner, B. (1986) Antiarrhythmic drugs. In: *Cardiovascular Drugs*, pp. 305–415, J. A. Bristol (ed.). Wiley, New York.

Hill, J. L., and Gettes, L. S. (1980) Effect of acute coronary artery occlusion on local myocardial extracellular K^+ activity in swine. *Circulation* **61** 768–778.

Hille, B. (1984) *Ionic channels of excitable membranes*, pp. 109–112. Sinauer, Sunderland, MA.

Hoffman, B. F., and Bigger, J. T. (1985) Digitalis and allied cardiac glycosides. In: *The pharmacological basis of therapeutics*, 7th edn, pp. 716–747, A. Goodman Gilman, L. S. Goodman, T. W. Rall, and F. Murad (eds). Macmillan, New York.

Hogg, K. J., Thomson, A., Elliott, H. L., Cobbe, S. M., and Reid, J. L. (1989) The pharmacokinetics and pharmacodynamics of a new class III antiarrhythmic UK 66, 914 — Results of a phase I study. *Br. J. Clin. Pharmacol.* (in press).

Imanishi, S., and Arita, M. (1987) Factors related to the low resting membrane potentials of diseased human atrial muscles. *Jpn. J. Physiol.* **37** 393–410.

Jijima, T., and Taira, N. (1987) Pinacidil increases the background potassium current in single ventricular cells. *Eur. J. Pharmacol.* **141** 139–141.

Johnson, E. A., and McKinnon, M. G. (1956) Effect of acetylcholine and adenosine on cardiac cellular potentials. *Nature (Lond.)* **178** 1174–1175.

Kakei, M., Yoshinaga, M., Saito, K., and Tanaka, H. (1986) The potassium current activated by 2-nicotinamidoethyl nitrate (nicorandil) in single ventricular cells of guinea pigs. *Proc. Roy. Soc. Lond. B* **229** 331–343.

Kameyama, M., Hofmann, F., and Trautwein, W. (1985) On the mechanism of β-adrenergic regulation of the Ca channel in the guinea-pig heart. *Pflügers Arch.* **405** 285–293.

Kameyama, M., Kakei, M., Sato, R., Shibasaki, T., Matsuda, H., and Irisawa, H. (1984) Intracellular Na^+ activates a K^+ channel in mammalian cardiac cells. *Nature* **309** 354–356.

Kantor, P. F., Coetzee, W. A., Dennis, S. C., and Opie, L. H. (1987) Effects of glibenclamide on ischaemic arrhythmias. *Circulation* **76** (suppl.) IV–17.

Kantor, P. F., Coetzee, W. A., Carmeliet, E. E., Dennis, S. C., and Opie, L. H. (1989) Reduction in ischaemic K^+ loss and arrhythmias: the effect of the sulphonylurea, glibenclamide. *Circ. Res.* (submitted).

Knopf, H., Theising, R., and Hirche, H. (1988) The effect of desipramine on ischaemia-induced changes in extracellular K^+, Na^+, and H^+ concentrations and noradrenaline release in the isolated rat heart during global ischaemia. *J. Cardiovasc. Pharmacol.* **12** (suppl. 1) S8–S14.

Kozlowski, J. H., Kozlowski, J. A., and Schuller, D. (1988) Poisoning with sustained release verapamil. *Am. J. Med.* **85** 127.

Kurachi, Y., Nakajima, T., and Sugimoto, T. (1986) On the mechanism of activation of muscarinic K^+ channels by adenosine in isolated atrial cells: involvement of GTP-binding proteins. *Pflügers Arch.* **407** 264–274.

Mitchell, M. R., Powell, T., Sturridge, M. F., Terrar, D. A., and Twist, V. W. (1986) Electrical properties and response to noradrenaline of individual heart cells isolated from human ventricular tissue. *Cardiovasc. Res.* **20** 869–876.

Noble, D. (1984) The surprising heart: a review of recent progress in cardiac electrophysiology. *J. Physiol.* **353** 1–50.

Noma, A. (1983) ATP-regulated K^+ channels in cardiac muscle. *Nature* **385** 147–148.

Noma, A., Nakayama, T., Kurachi, Y., and Sugimoto, T. (1984) Resting K conductances in pacemaker and non-pacemaker heart cells of the rabbit. *Jpn. J. Physiol.* **34** 245–254.

Osterrieder, W. (1988) Modification of K^+ conductance of heart cell membrane by BRL 34915. *Naunyn-Schmiedeberg's Arch. Pharmacol.* **337** 93–97.

Reuter, H. (1984) Ionic channels in cardiac membranes. *Ann. Rev. Physiol.* **46** 473–484.

Sakmann, B., and Trube, G. (1984) Conductance properties of single inwardly rectifying potassium channels in ventricular cells from guinea-pig heart. *J. Physiol.* **347** 641–657.

Sakmann, B., Noma, A., and Trautwein, W. (1983) Acetylcholine activation of single muscarinic K^+ channels in isolated pacemaker cells of mammalian heart. *Nature* **303** 250–253.

Shibasaki, T. (1987) Conductance and kinetics of delayed rectifier potassium channels in nodal cells of the rabbit heart. *J. Physiol.* **387** 227–250.

Soejima, M., and Noma, A. (1984) Mode of regulation of the ACh-sensitive K-channel by the muscarinic receptor in rabbit atrial cells. *Pflügers Arch.* **400** 424–431.

Steinberg, M. I., Lindstrom, T. D., and Fasola, A. F. (1984) Clofilium. In: *New drugs annual: Cardiovascular drugs* pp. 103–121, A. Scriabine (ed.). Raven Press, New York.

Szentmiklosi, A. J., Nemeth, M., Szegi, J., Papp, J. G., and Szekeres, L. (1979) On the possible role of adenosine in the hypoxia-induced alterations of the electrical and mechanical activity of the atrial myocardium. *Arch. Int. Pharmacodyn.* **238** 283–295.

Trautwein, W. (1973) Membrane currents in cardiac muscle fibres. *Physiol. Rev.* **53** 793–835.

Trautwein, W. (1982) Effect of acetylcholine on the S-A node. In: *Cellular pacemakers*, Vol. 1, pp. 127–160, D. Carpenter (ed.). Wiley, New York.

Trautwein, W., Gottstein, U., and Dudel, J. (1954) Der Aktionsstrom der Myokardfaser im Sauerstoffmangel. *Pflügers Arch.* **260** 40–60.

Trautwein, W., Kassebaum, D. G., Nelson, R. M., and Hecht, H. (1962) Electrophysiological study of human heart muscle. *Circ. Res.* **10** 306–312.

Trube, G. (1988) Potassium currents in isolated adult cardiac myocytes. In: *Isolated cardiac myocytes,* Vol. II: *Electrophysiology and contractile function,* pp. 75–94. H. M. Piper, and G. Isenberg (eds). CRC Press, Boca Raton, Fla.

Trube, G., and Hescheler, J. (1983) Potassium channels in isolated patches of cardiac cell membrane. *Naunyn-Schmiedeberg's Arch. Pharmacol.* **322** R64.

Trube, G., and Hescheler, J. (1984) Inward-rectifying channels in isolated patches of the heart cell membrane: ATP-dependence and comparison with cell-attached patches. *Pflügers Arch.* **401** 178–184.

Vaughan Williams, E. M. (1975) Classification of antidysrhythmic drugs. *Pharm. Ther. B* **1** 115–138.

Weiner, N. (1985) Atropine, scopolamine and related antimuscarinic drugs: In: *The pharmacological basis of therapeutics*, 7th edn, pp. 130–144, A. Goodman Gilman, L. S. Goodman, T. W. Rall and F. Murad, (eds). Macmillan, New York.

Wit, A. L., Rosen, M. R., and Hoffman, B. F. (1975) Electrophysiology and pharmacology of cardiac arrhythmias. VIII. Cardiac effects of diphenylhydantoin. *Am. Heart J.* **90** 397–404.

14

K^+ channels in the central nervous system

J. V. Halliwell

14.1 INTRODUCTION

In the last 15 years improvements in electrophysiological techniques and the development of specific biochemical probes have led to our rapidly expanding knowledge about the K^+ conductances of the mammalian central nervous system. In this chapter I will review the K^+ conductances that have been identified in a variety of mammalian species, including man, discuss the properties they confer on the particular neurones that express them and describe the modulation that results from neurotransmitter action and exposure to pharmacological agents. Implicit in these considerations is an assessment of the possible physiological roles of central K^+ conductances and their involvement in pathological states.

14.2 METHODS OF INVESTIGATING CENTRAL NEURONAL K^+ CHANNELS

Although the title of this chapter refers specifically to the term 'channel' relatively little of the material that will be reviewed here is concerned with a study of K^+ channel properties directly. It is assumed that channels that are K^+ conducting pathways across the plasma membrane underlie the electrical and chemical phenomena discussed below. Since much of the evidence for a diversity of K^+ channels is indirect, it is worth considering briefly the general strategies that underlie the reported research findings. Broadly speaking, there are two approaches: biochemical, which is concerned with the distribution, chemical structure and expression of channels; and electrophysiological, which addresses the question of channel type and function based on kinetic and permeation data and the use of pharmacological tools. Of course the two approaches cross-fertilize, so that, for instance, electrophysiological screening is used to assess the compounds that are used to label membrane components for biochemical analysis; this strategy has identified potential channel structures which have been cloned, expressed in a biological system and, in turn, subjected to electrophysiological analysis for comparison with the native target channels in situ.

Biochemical approaches to the study of K^+ conductance originated with the study of K^+ movements (or fluxes of markers for K^+) in brain slices, suspensions of cells or subcellular components, and the effects thereon of pharmacological agents whose action had previously been demonstrated electrophysiologically. These studies are useful in indicating the presence of transmembrane pathways for K^+ and currently enjoy somewhat greater time resolution than previously (resolving to about 1 s) so that (relatively speaking) components of transient or sustained flux may be resolved (Bartschat and Blaustein, 1985). Similar experimental procedures employ the measurement of transmitter release combined with pharmacological manipulation to investigate roles of K^+ conductance in presynaptic function. Such methods are of necessity indirect, but are often the only way that presynaptic K^+ channel function can be probed in the CNS. The time resolution of these methods (~1 s) is limited, however, compared with electrophysiology (<1 ms); the methods also rely heavily on K^+ channel blockers that are not absolutely specific for one kind of channel. It is consequently difficult to reconcile exactly data from these kinds of studies with conductances defined electrophysiologically. Therefore, further consideration will not be given to this kind of experiment.

Another biochemical approach is the study of the binding and localization of high-affinity probes in the CNS that, again, have demonstrated K^+ channel activity; often these are plant or animal toxins (see recent review by Castle *et al.*, 1989) evolved to have specific defensive or offensive roles to the advantage of their orginators. An interference with K^+ channel activity is particularly effective in producing lethal or incapacitating states of neuronal dysfunction and indicates the important stabilizing role played by the various K^+ channels in the nervous system. Use of such toxins has allowed affinity purification of membrane proteins which could be channel structures or component parts thereof (Dolly, 1988). Sequencing of potential channel proteins and screening of cDNA libraries allows the possibility of expression of mRNA by systems such as *Xenopus* oocyte to produce additional functional channels. A complementary approach has recently been employed whereby message from cloned mutant genes known to be deficient in certain classes of K^+ channel types is similarly expressed (see Miller, 1988). Messenger RNA derived from rat cerebral cortex has also led to expression of functional channels, the kinetic properties of which are currently being studied (Stuhmer *et al.*, 1988).

Electrophysiological recording has demonstrated the diversity of central K^+ conductances. As central neurones are relatively inaccessible for intracellular recording in situ, intracellular recording *in vitro* has been employed more often. Much description of the various K^+ currents that follows derives from intrasomatic recording from single neurones in brain slice preparations. Direct recordings of slow membrane currents from intact cells is possible thanks to the employment of single-electrode voltage-clamp techniques (e.g. Wilson and Goldner, 1975); in addition, the presence of other (usually kinetically quicker) currents has been inferred from the changes in neuronal behaviour observed after manipulation of cells' resting potential by injected current: so-called current clamp. There are examples of heroic experiments *in vivo* where the same kinds of recording technique have been used (see Crill and Schwindt, 1983). These studies are important in the validation of the more precise experimental approaches undertaken *in vitro*, where the nervous system is subject to more drastic disruption in the interest of experimental protocol.

Macroscopic currents and single-channel currents may be studied more precisely in single-cell preparations dissociated acutely, or grown in short-term or long-term tissue culture, all of which are amenable to the patch-clamp technique (Hammill *et al.*, 1981) as well as conventional microelectrode methods. One should be aware, however, in considering the data derived from such investigations, that the nature of observed channel behaviour may be aberrant as a result of proteolytic digestion used to isolate cells, the frequent (though necessary) use of immature (i.e. embryonic) cells as a starting point for preparing cultures and possible deficiencies in cellular environment during culture. This is especially pertinent when attention is drawn to the question of K$^+$ channel modulation. As molecular biological methods become increasingly applied to investigating the structure of K$^+$ channels, expression of channels in oocytes will be the standard screening for clones; it is almost a truism that post-translational factors and simultaneous expression of mRNA for few of possibly many channel subunits could lead to the production of channel types that are indeed novel and bear little functional relevance to K$^+$ channels *in situ*.

14.3 K$^+$ CHANNEL TYPES IN THE MAMMALIAN CENTRAL NERVOUS SYSTEM

14.3.1 Voltage-activated K$^+$ currents

14.3.1.1 Delayed rectifier

The original description of the action potential mechanism in squid giant axon by Hodgkin and Huxley in 1952 featured just two conductances (see Hodgkin, 1964); in addition to the Na$^+$ current which conferred excitability, a more slowly developing K$^+$ current was activated by depolarization, the role of which appeared to be an aid to axonal repolarization beyond that resulting from Na$^+$ channel inactivation. Voltage-clamp studies in the mammalian CNS reveal that upon depolarization a delayed outward current is produced. However, as we shall see, outward current slow to activate following neuronal depolarization can arise from voltage-dependent membrane behaviour or from the results of Ca^{2+} or Na$^+$ influx. Moreover, when the involvement of these cations is eliminated with specific blockers, several separable K$^+$ conductances may be identified that produce a maintained or slowly declining outward current. These conductances, which are described below, all confer to the current–voltage characteristics of central neurones the property of delayed or outward rectification. That is, the current activated by depolarization of the cell sums with the current that elicits its depolarization, giving the impression that the cell passes outward current more easily than inward current. To embrace the spirit of the Hodgkin–Huxley delayed rectifier (K_V) a central K$^+$ conductance should display the following properties: sigmoidal activation described by first-order kinetics raised to a power, even slower inactivation [>10 s] at depolarized potentials, activation potential threshold more positive than −20 mV, resistance to Ca^{2+} and Na$^+$ channel blockers. Moreover, as demonstrated for other neuronal types (Hille, 1967; Thompson, 1977), K_V in the CNS should be sensitive to blockade by extracellular TEA [~10 mM] and be relatively insensitive to 4-AP. Hippocampal pyramidal cells of rat and guinea-pig display K$^+$ conductances with similar properties to those enumerated above, both in cell culture (Segal and Barker, 1984; Rogawski, 1986) and when freshly dissociated (Numann *et al.*, 1987; Sah *et al.*, 1988). One departure from this list for the forebrain K_V is (i) a lower threshold for activation ~ -50 mV and a steady state inactivation such that only half the channels are available for activation from rest but appreciably more [50–100%] current may be produced by a standard depolarization from around -85 mV (Numann *et al.*, 1987: Segal and Barker, 1984; Sah *et al.*, 1988). Large depolarizing stimuli will, therefore, trigger greater repolarizing action through the K_V conductance mechanism when the starting membrane potential is more negative; this could lead to a sharper cell response when hyperpolarizing inhibitory tone is present. K_V-like conductance has also been described in cultured rat cerebellar granule cells (Cull-Candy *et al.*, 1986) and neocortical cells (Zona *et al.*, 1988) *in vitro* and cat motoneurones *in vivo* (Barrett *et al.*, 1980; Schwindt and Crill, 1981), where it is sensitive to TEA.

It is difficult to ascribe a definitive role for K_V in the CNS. One problem is that while, in all probability, it is ubiquitous, there is no compound that is selective in blocking this conductance. The strategy of observing the behaviour of neurones or groups of neurones following specific pharmacological treatment cannot be adopted. A spike repolarizing action of K_V could be invoked as its major role *pace* squid axon but there is a fast Ca-activated transient K^+ conductance demonstrable in the CNS (see below) that subserves this function more effectively (Zbicz and Weight, 1985; Storm, 1987). The suggestion put forward by Numann *et al.* (1987) that outward current carried by K_V channels is set by the resting potential has implications for control of the duration of (depolarizing) excitatory input, that depends on the neurone's past history. Alternatively K_V could be the K^+ conductance of last resort for stabilizing cell excitability, being, by and large, the final voltage-gated conductance to be activated by depolarization.

Direct modulation of K_V has not been demonstrated in the CNS but any change in membrane potential resulting from other cellular actions could *indirectly* change the activity of K_V channels. This is not to say that direct modulation of K_V does not occur: bag cells of *Aplysia* possess delayed TEA-sensitive K^+ current that is inhibited by raised intracellular cyclic AMP (cAMP) (Strong and Kaczmarek, 1985).

14.3.1.2 Transient K^+ current (I_A)

A K^+ current that activates much more rapidly upon depolarization (activation time constant (T) > 5 ms) than the delayed rectifier (K_V) and which also inactivates in less than $\sim$100 ms was originally discovered in invertebrate neurones (Connor and Stevens, 1971; Neher, 1971). Inactivation of this conductance is complete at membrane potentials positive to -55 mV; hence for a neurone with a resting potential between -60 and -50 mV to be influenced at all by this conductance it must experience a prior hyperpolarization to remove inactivation and then be depolarized again positive to the threshold voltage (~ -50 mV) for activation. This transient current was termed the A current (I_A) and its counterpart has been reported in mammalian central neurones *in situ* (Barrett *et al.*, 1980), in brain slices (Gustafsson *et al.*, 1982; Williams *et al.*, 1984; Aghajanian, 1985; Zbicz and Weight, 1985; Sah *et al.*, 1985; Halliwell *et al.*, 1986; Schwindt *et al.*, 1988b), cell culture (Segal *et al.*, 1984; Segal and Barker, 1984; Cull-Candy *et al.*, 1986; Nakajima *et al.*,

1986; Zona *et al.*, 1988) or when acutely dispersed (Numann *et al.*, 1987). Its presence may be detected from the inspection of voltage trajectories in some neurones (with a near-threshold resting potential) that have been hyperpolarized with steady current and then allowed to relax following termination of the current (see Llinas, 1984). The A current delays the return of voltage to rest and, if maximally activated, causes a characteristic 'notch' or transient hyperpolarization. Similarly, neurones with very negative resting potentials, when rapidly depolarized, display a similar 'A notch' that can delay or cancel the appearance of an action potential (Gustafsson *et al.*, 1982; see Fig. 14.1 a_1). Functionally speaking, the

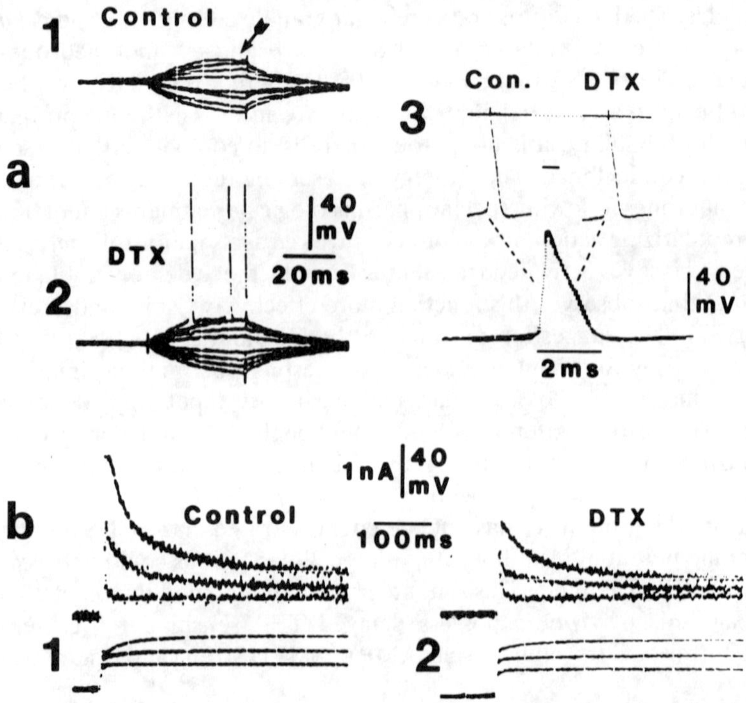

Fig. 14.1 — The A-current and its blockade by DTX in the hippocampal slice. (a) Current clamp records: 1 and 2 show superimposed voltage responses from a CA1 neurone at a resting potential of −75 mV elicited by 30 ms duration current pulses in steps of 0.1 nA from −0.4 nA to +0.4 nA under control conditions (1) and 30 min after 50 nM DTX was included in the medium perfusing the slice. Note evidence of the 'A notch' (arrowed) and the supervention of action potentials in DTX, signifying a lowering of the current threshold for spike initiation by the toxin. Part 3 shows that spikes triggered in DTX-containing medium exhibit no broadening compared to control repsonses. (b) Voltage-clamp records: transient outward currents (upper trace) activated in another CA1 neurone by positive voltage steps (lower trace) from a holding potential of −75 mV and in the presence of 0.5 μM TTX under control conditions (1) and after 30 min exposure to DTX (2). (Part a: J. V. Halliwell and J. O. Dolly, unpublished observations; Part b: from Halliwell *et al.* (1986), Fig. 3).

presence of I_A can influence the threshold of a neurone which has a negative resting potential. This becomes apparent when selective blockers of I_A are applied; dendrotoxin (DTX), a component of the venom of the Eastern green mamba,

Dendroaspis angusticeps, blocks I_A in the somata of hippocampal pyramidal neurones (Dolly *et al.*, 1984; Halliwell *et al.*, 1986; see Fig. 14.1) and reduces the size of a short depolarizing current pulse required to fire such a cell (Fig. 14.1 a_2). The effects of DTX on I_A in hippocampal somata occur at concentrations lower than 50 nM (Halliwell *et al.*, 1986); toxin I, a polypeptide structurally related to DTX and derived from the venom of the black mamba (*Dendraspis polylepis*) has similar actions but is about ten times more potent than DTX (Harvey and Anderson, 1985). 4-AP also blocks I_A with a similar effect on current threshold for firing action potentials and concomitant loss of 'A notch' but with at least 1000-fold less effectiveness (Gustafsson *et al.*, 1982; Segal *et al.*, 1984; Segal and Barker, 1984; Numann *et al.*, 1987).

None of the blockers mentioned so far have actions confined to I_A (at least if one adopts the characteristics listed above as defining the current). Dendrotoxins (Harvey and Anderson, 1985), the aminopyridines (4-AP and 3,4DAP) and tetrahydroaminoacridine (THA: structurally related to 4AP) (Rogawski, 1987) reduce other central K^+ conductances (see below) in addition to I_A. These conductances are more persistent than I_A so that functional changes of neuronal behaviour having durations longer than the time constant of decay for I_A cannot be ascribed to loss of I_A alone. Functional changes, such as the threshold effect described above, which do lie within the time domain of I_A, can reasonably indicate a role for this conductance in the CNS. Another role for I_A originally suggested by Connor and Stevens (1971) was its participation in setting the frequency of neuronal discharge by influencing the interspike voltage trajectory: interplay between the level of hyperpolarization resulting from the spike afterpotential (thereby setting the availability of I_A) and depolarizing current is believed to set the rate at which membrane potential approaches spike threshold. I_A could also participate in generating the hyperpolarizing spike afterpotential in concert with other voltage- and Ca-activated K^+ conductances. In the CNS it is not clear whether spike repolarization is dependent on A channel current, since spike broadening is observed after 4-AP (Storm, 1987) but occurs much less after DTX (Fig. 14.1 a_3); the 4-AP-induced spike broadening could be reduction of I_A or more recently discovered K^+ current I_D (Storm, 1988; see below) that is more sensitive to 4-AP than I_A.

Since it appears likely that both firing patterns and threshold for firing are influenced by the presence of I_A, control over this conductance via neurotransmitter modulation could change the behaviour of central neurones. Noradrenaline acting at β-adrenoceptors (Sah *et al.*, 1985) and acetylcholine (ACh) acting at muscarinic receptors (Nakajima *et al.*, 1986) have been reported to depress I_A in hippocampal pyramidal neurones, whereas noradrenaline acting at α_1-adrenoceptors reduces I_A in dorsal raphe neurones (Aghajanian, 1985); activity in pathways utilizing these transmitters could consequently enhance the excitability of these cell types by this mechanism.

14.3.1.3 Slowly inactivating outward current (I_D)

Storm (1988) has recently reported that in addition to I_A, hippocampal neurones possess another K^+ conductance that activates rapidly (like I_A, within milliseconds) but inactivates over about 1 s. This current is distinguishable from the delayed rectifier on several counts: it is insensitive to TEA, very sensitive to 4-AP (blocked at

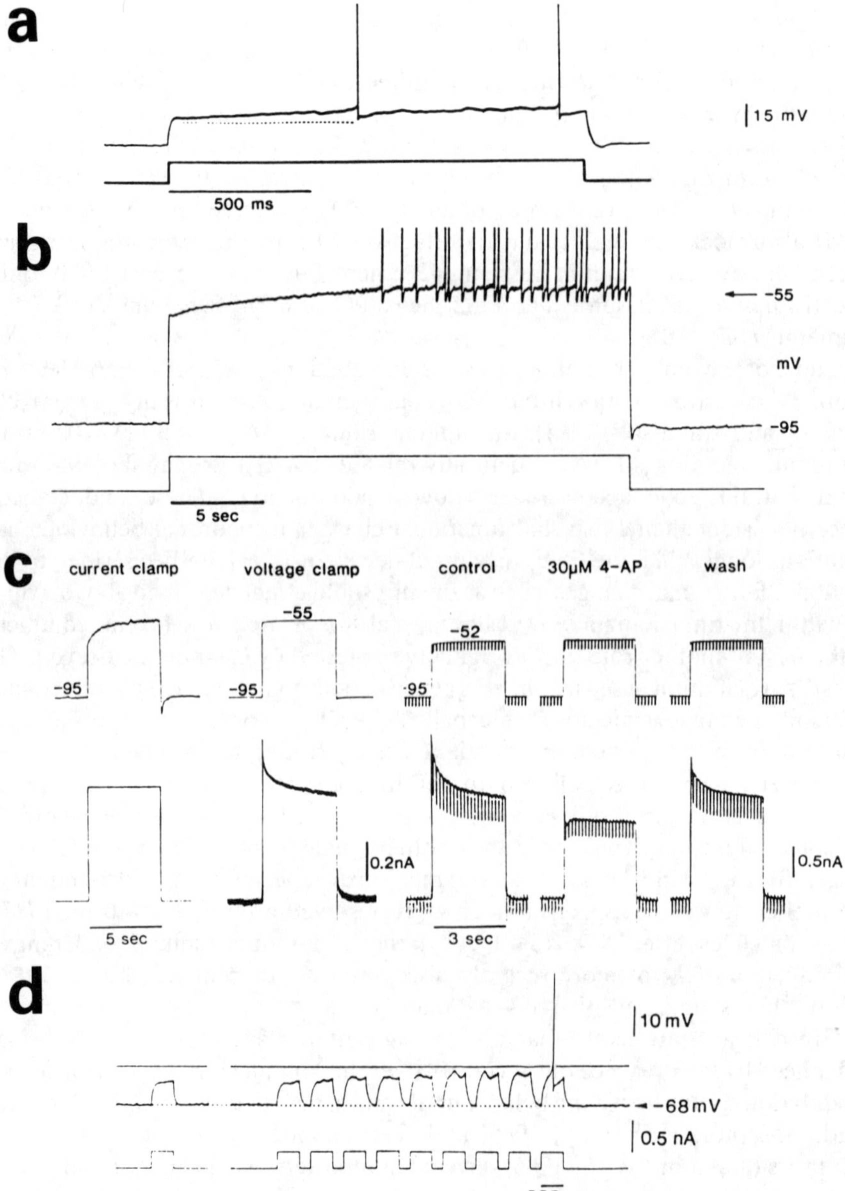

Fig. 14.2 — I_D in hippocampus. Firing is delayed in CA1 hippocampal neurones to a degree that depends on prior hyperopolarization of the cell; the action potential is delayed for a relatively short time when triggered from −72 mV (a) but the delay exceeds 10 s when the cell is hyperpolarized to −95 mV before being excited (b). (c) Comparison of current and voltage-clamp measurements shows that a quickly activating, slowly inactivating outward current (I_D) underlies the voltage trajectory from negative potentials (left-hand side) and that this outward current is blocked by low doses of 4-AP (right-hand side). (d) A single depolarizing pulse fails to fire the cell from −68 mV, but when repeatedly presented a similar stimulus eventually elicits an action potential as inactivation of I_D accumulates with successive depolarizations. This allows temporal integration of excitatory stimuli by the neurone. (From Storm, 1988, Figs 1 and 3.)

30 μM) and is activated at membrane potentials negative to −60 mV (Fig. 14.2c). It contrasts also with I_A in having a 10–20-mV more negative activation and contributing a smaller (one-sixth) overall maximal conductance. It also recovers from inactivation with a prolonged time-course (τ~5 s). The functional role of this current was demonstrated most elegantly by Storm (1988), who showed it to be responsible for the extended delay to the first action potential in a cell injected with maintained, just supra-rheobasic current (Fig. 14.2a,b). He named the conductance I_D (for delay). Another consequence for the control of neuronal firing derives from the kinetic behaviour of I_D: recovery from inactivation is also slow, so that depolarizing pulses of around 200 ms duration which are insufficient for spike generation, yet still initiate inactivation of I_D, can cause this inactivation to accumulate if they are presented repetitively. Inactivation of I_D may summate to the point at which action potentials are generated by stimuli which were initially subthreshold because of the delaying effects of available D channels (Fig. 14.2d). This property allows temporal summation of inputs by the cell. I_D also makes spike discharge frequency sensitive to the prevailing membrane potential, thereby changing the input–output relation for the neurone. It remains to be seen whether this membrane current is modulated by any known neurotransmitter.

14.3.1.4 Inward (anomalous) rectifier ($I_{f.i.r.}$ and I_Q)

The term 'anomalous rectifier' was first coined by Katz (1949) to describe the passage of inward but not outward K^+ currents in skeletal muscle membrane under conditions of symmetrical $[K^+]$; the anomaly referred to the fact that the rectification ran in a direction counter to that predicted by the constant field equation (Hodgkin and Katz, 1949): a mild outward rectification. 'Classical anomalous rectification' has been studied in muscle and egg-cell membrane and is characterized by the following properties: activation by hyperpolarization negative to the reversal potential for K^+ (E_{K^+}); activation potential shifting with E_{K^+}; blockade by Cs^+ and Ba^{2+}; inactivation or blockade by extracellular cations at very negative potentials (> − 120 mV); inward conductance impermeable to Rb^+ (see Stanfield, 1988). Inward K^+ currents activated by hyperpolarization in olfactory cortex neurones (Constanti and Galvan, 1983a), forebrain cholinergic neurones (Stanfield *et al.*, 1985), cells of the locus coeruleus (Inoue *et al.*, 1988) and the raphe nucleus (Williams *et al.*, 1988) and a proportion of human neocortical neurones (Halliwell, 1989) possess all or most of the properties listed above and could legitimately be termed 'anaomalous rectifiers' (I_{K1}). Thus, these currents are typified by the olfactory cortex current of Constanti and Galvan (1983a) and named by them $I_{f.i.r.}$ (for fast inward rectifier): a rapidly activating (τ: 50 ms at − 100 mV) inward current carried by K^+ with increased conductance and positively shifted activation in high $[K^+]_o$, blocked by both Cs^+ and Ba^{2+} (see Fig. 14.3a).

Non-linear current–voltage characteristics determined from intracellular recording in central neurones frequently display a decreased slope resistance negative to rest when compared with that positive to rest. Such non-linearities are often based not only on the existence of K^+ channels activated by hyperpolarization but also *apparent* increases in cell input resistance in the potential range between rest and spike threshold. Nevertheless, the latter phenomenon has also been referred to as 'anomalous rectification' when encountered in hippocampal cells and has been

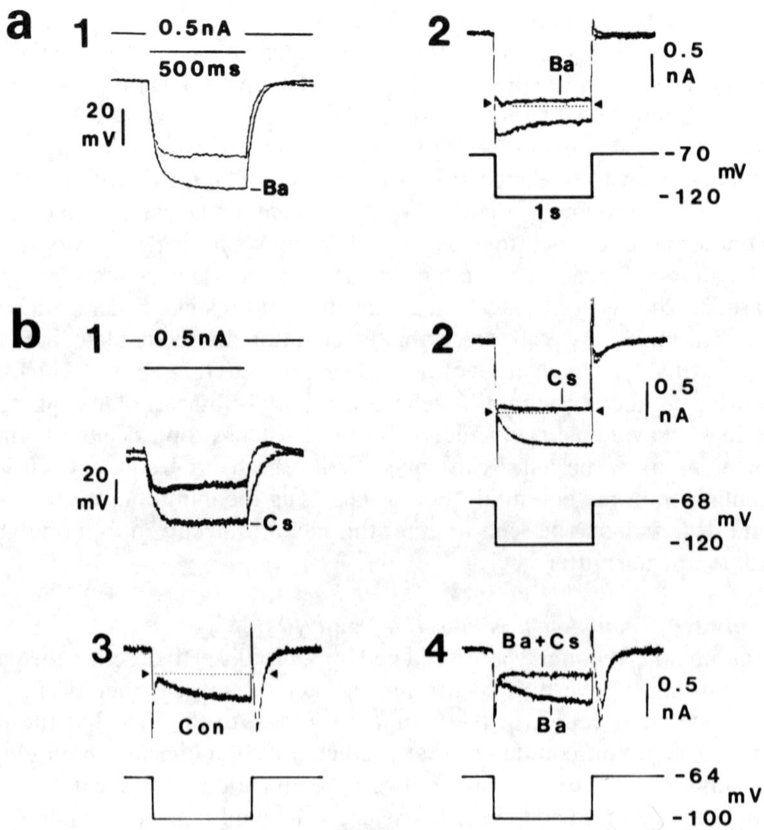

Fig. 14.3 — Two different inward rectifier currents activated by hyperpolarization in human neocortical neurones. (a) Blockade of $I_{\text{f.i.r.}}$-type of conductance by 1 mM extracellular Ba^{2+}: comparison of (1) current-clamp and (2) voltage-clamp responses in a temporal cortical neurone (resting potential, −75 mV) in a slice superfused with medium containing 0.5 μM TTX. Traces are superimposed (controls unmarked); the dotted line between arrow heads on voltage-clamp records in a and b signify the magnitude of the cells' ohmic (leak) response to the indicated voltage-clamp step. (b_1, b_2) Blockade of I_Q-type of conductance by 1 mM extracellular Cs^{+}: a similar protocol to that of a_1 and a_2 but in a frontal neocortical neurone. Compare the sag and rebound of the control voltage response in b_2 with that of a_2 and the difference in the time-course of the cation-sensitive components of hyperpolarization-activated current in a_2 and b_2. (b_3, b_4) I_Q is resistant to blockade by Ba^{2+}: (b_3) control current response to the indicated voltage-clamp step; (b_4) currents recorded in response to the same step in the presence of Ba^{2+} (1 mM) and both Ba^{2+} and Cs^{+} (1 mM). The rebound inward current that accompanies repolarization to −64 mV is due to activation of a low-threshold transient Ca current (Halliwell, 1986) observed in a proportion of human neocortical neurones. (J. V. Halliwell, unpublished experiment.)

reported to have been eliminated by Na^{+} and Ca^{2+} channel blockers (Hotson *et al.*, 1979). These results suggest that inward current carried through persistent Na^{+} and Ca^{2+} channels is responsible for this kind of rectification (French and Gage, 1985; Stafstrom *et al.*, 1985; Johnston *et al.*, 1980; Brown and Griffith, 1983b). The negative slope conductance conferred on the cell's current–voltage relation by the inward currents is moderated by the presence of persistent voltage- and cation-activated K^+ conductances (see below). This explains why the inward rectification observed by Hotson *et al.* (1979) was enhanced by some K^+ channel blockers (TEA, Ba^{2+}, muscarinic agonists). However, to call this phenomenon 'anomalous rectification' is confusing and leads to the apparent paradox that Ba^{2+} can both suppress and enhance 'anomalous rectification'.

Another inward membrane current that is activated by membrane hyperpolarization has been identified in a range of central neurones; it is responsible (Halliwell and Adams, 1982; Mayer and Westbrook, 1983) for the depolarizing sag on electrotonic potentials recorded in various neurones that are hyperpolarized negative to − 85 mV (e.g. Nelson and Frank, 1967; Purpura *et al.*, 1968; Schwartzkroin, 1976; Llinas and Sugimori, 1980; Crepel *et al.*, 1981; Henderson *et al.*, 1982; Halliwell and Adams, 1982; Stafstrom *et al.*, 1984; Barrett *et al.*, 1980; Fig. 14.3b$_1$). While this current can give rise to 'anomalous rectification' that is more in keeping with the spirit of the original description, the underlying current is not through a pure K^+ conductance, but is also carried by Na^+ (Halliwell and Adams, 1982; Mayer and Westbrook, 1983; Crepel and Penit-Soria, 1986) with a reversal potential ranging from − 30 to − 60 mV, but always a significant distance from E_{K^+}. It activates over the range − 75 to − 120 mV with voltage-dependent kinetics that speed up with increasing hyperpolarization, and are about four to five times slower than the rate of $I_{f.i.r.}$. The current activated by hyperpolarization is non-inactivating and unlike the inward rectifier described by Stanfield *et al.* (1985) or Inoue *et al.* (1988) does not show a time- and voltage-dependent decline at potentials negative to about − 120 mV (cf Fig. 14.3a$_2$ and b$_2$). The nomenclature relating to this conductance is confused since the current, which was originally named I_Q or I_H, has been subsequently referred to as the 'anomalous rectifier' (I_{AR}) (e.g. Segal *et al.*, 1984; Spain *et al.*, 1987). The I_{AR} of Segal *et al.* (1984) and Spain *et al.* (1987) is identical to I_Q/I_H in terms of kinetic properties, Na^+ permeability, lack of dependence of activation threshold upon E_{K^+} and resistance to block by Ba^{2+}; these properties distinguish this current from the pure K^+ conductance exemplified by $I_{f.i.r.}$ (see above; Fig. 14.3b).

From reports of their occurrence, one could gain the impression that the distribution of I_Q/I_H and the true anomalous rectifier I_{K1} appears largely complementary; i.e. a given cell expresses one or other current exclusively or predominantly. However, Williams *et al.* (1988) and Pollard and Crunelli (1988) reported the coexistence of the two currents in rat dorsal raphe and lateral geniculate neurones, respectively. The reduction of 'instantaneous' conductance at negative potentials by Cs^+ *and* Ba^{2+} could suggest the same for hippocampal neurones (Halliwell and Adams, 1982).

Roles for the conductances responsible for inward rectification in neurones are not immediately obvious. One function they might subserve is in regulating the negative bounds of resting potential. During periods of intense activity, large extracellular concentrations of K^+ could accumulate and intracellular $[Na^+]$ increase, leading to electrogenic pump activity. Pump currents tending to hyperpolarize neurones would be opposed by both types of hyperpolarization-activated conductance. The I_{K1}-type of current would respond more effectively to this challenge since its activation threshold would be reduced owing to the shift in E_{K^+} that an increase in $[K^+]_o$ would cause. Blockade of the I_H/I_Q conductance in photoreceptors by Cs^+ can lead to excessively negative membrane potentials (beyond −120 mV) during periods of vigorous light stimulation (Owen and Torre, 1981); ouabain reduces maximal light responses to values positive to E_K, suggesting that I_H curbs the effects of electrogenic pumping. This action of I_H would be in addition to its involvement in shaping photoreceptor responses to light stimuli (Atwell and Wilson, 1980). In heart sinoartrial node cells I_H (also known as I_f) acts as a pacemaker current (Di Francesco, 1981) and in cortical neurones its counterpart serves to shape the amplitude and duration of hyperpolarizing events (Schwindt *et al.*, 1988a; Storm, 1989). External control of I_Q/I_H could influence subsequent cell excitability by setting the level of hyperpolarization and thereby the inactivation/activation of other K^+ conductances; Gho *et al.* (1986) report that low concentrations of kainate reduce I_Q in hippocampal neurones indicating a possible modulation of this conductance by excitatory amino acid transmitters.

The pure K^+ conductance (I_{K1}) that confers inward rectification to the current–voltage characteristic of central cholinergic neurones is suppressed by substance P (SP) (Stanfield *et al.*, 1985). Since SP reduces the resting conductance, depolarizes and excites these neurones I_{K1} would seem to contribute to setting the membrane potential directly and, furthermore, be capable of passing some outward current. The modulatory effects of SP are slow and involve receptor activation coupled to an unknown second messenger system via a pertussis toxin-insensitive GTP-binding protein (Nakajima *et al.*, 1988).

14.3.1.5 The M current (I_M)

Slow depolarizing synaptic potentials in sympathetic ganglia are accompanied by a reduction in membrane conductance; this arises from the modulation of a K^+ conductance called the M current (I_M) by a range of transmitters and humoral factors which includes ACh, LHRH and angiotensin (see review by Brown, 1988). The M current is non-inactivating but turns on at potentials positive to −60 mV with slow first-order kinetics ($\tau \sim 200$ ms at −40 mV) to contribute a steady-state component of outward K^+ flux over a range of membrane potentials positive to rest. It may be envisaged that I_M opposes steady depolarizing inputs to a neurone. The significance of this is apparent when I_M is decreased; depolarizing stimuli are enhanced because of the removal of the 'braking' effect of I_M. The original discovery that ACh, acting at muscarinic receptors, suppressed this current, was the reason for its being called the M current (Brown and Adams, 1980). Similarly, cholinergic excitation of forebrain neurones involving a slow depolarization accompanied by a conductance decrease (e.,g. Krnjevic *et al.*, 1971) prompted the quest for I_M in the CNS; it has been identified in mammalian species as diverse as man and rat in locations ranging from neocortex to brain stem and spinal cord (Halliwell and Adams, 1982; Constanti and Galvan, 1983b; Gahwiler and Brown, 1985a; Champagnat *et al.*, 1986; Moore *et al.*, 1988; Jacquin *et al.*, 1988; Halliwell, 1986; Murase *et al.*, 1986). I_M in central neurones may be demonstrated by voltage-clamping cells to a voltage at which it remains activated (~ −40 mV) and jumping their membrane potential negative to deactivate the conductance (see Fig. 14.4). The presence of I_M is indicated by slow inward current relaxation upon hyperpolarization that is accompanied by conductance decreases; a corresponding outward current accompanies the return to the steady holding potential but often the form of this is distorted from a simple

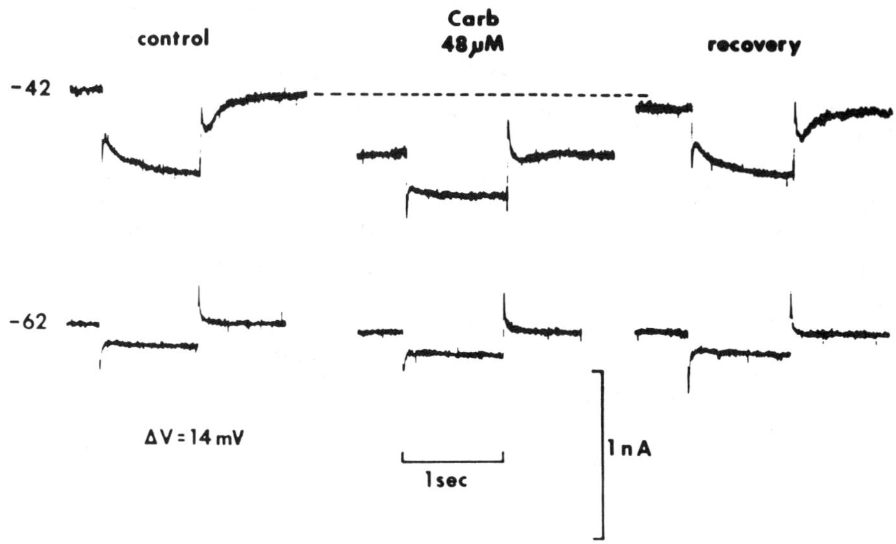

Fig. 14.4 — M current in hippocampal neurones. The figure shows currents driven by 14 mV hyperpolarizing voltage-clamp steps from a holding potential of −42 mV (upper row) and −62 mV (lower row) before, during exposure to and following recovery from a 2-min exposure to 48 μM carbachol (a muscarinic agonist). At −42 mV but not at −62 mV carbachol caused a steady inward current and a reduction in the instaneous conductance of the cell; suppression of the time-dependent inward current observed at −42 mV also resulted from carbachol action. The currents were recorded in a CA1 neurone in a guinea-pig hippocampal slice superfused with medium containing 0.5 μM TTX. (From Halliwell and Adams, 1982.)

exponential by the presence of other K^+ currents also activated by this protocol (for example; K_V, I_A, I_D, and (see below) $I_{K(ca)}$). As far as tested, the indicators of I_M in the CNS are always sensitive to muscarinic agonists; muscarinic activation also causes an inward current, at least part of which can be attributed to the reduction of a steady outward current through M channels. M current may be distinguished from cation-activated K^+ conductances by its resistance to indirect blockade by Na^+ and Ca^{2+} channel blockers (see, for example, Halliwell, 1986). It is potently blocked by 1 mM Ba and has recently been shown to be rather more sensitive to TEA (Constanti and Sim, 1987; Storm, 1989) than is the ganglionic M current (Adams *et al.*, 1982).

Transmitter susceptibility was the prime reason for the separation and characterization of the M current, and the promotion of cell excitability by its blockade indicated a role for I_M in controlling the sensitivity of neurones to afferent input. Since the conductance activates and deactivates slowly, activation builds up during a spike-train, because the *average* membrane potential is quite positive to rest. A decrease in spike firing frequency (accommodation) is partly due to the M current since steady-state firing is accelerated by muscarinic agonists in conditions under which the involvement of other muscarine-sensitive accommodating mechanisms (see below) is eliminated (Madison and Nicoll, 1984). M current suppression is the mechanism for slow EPSPs in sympathetic ganglia (Adams and Brown, 1982). Likewise, I_M suppression generates slow EPSPs in hippocampal slice cultures: these are caused by ACh which is released from cholinergic neurones of co-cultured septal explants that make synaptic contact with the hippocampus *in vitro* (Gahwiler and Brown, 1985b). There is a suggestion from Nicoll and colleagues (Madison *et al.*, 1987) that M current suppression is not the mechanism for atropine-sensitive slow EPSPs arising from the stimulation of ACh-containing fibres in the hippocampal slice (see also Muller and Misgeld, 1986). Voltage-insensitive 'leak' K^+ conductance appears to be more sensitive than I_M to muscarinic receptor stimulation; reduction of this current may initiate the slow EPSP, particularly since cholinergic stimulation can evoke EPSPs in neurones at resting potentials at which I_M is not active and therefore unavailable for suppression (Muller and Misgeld, 1986; Madison *et al.*, 1987). However, when its action is potentiated by cholinesterase inhibition, ACh released by nerve stimulation can lead to M current suppression (Madison *et al.*, 1987).

Acetylcholine modulates the M current; there is evidence that one action of 5-HT in the hippocampus is also to cause a reduction in I_M (Colino and Halliwell, 1987). The mechanism linking receptor activation by an agonist and M conductance decrease is not well established in the CNS, but it does not involve cAMP and could be secondary to the breakdown of membrane phosphatidyl inositol phospholipids (PI) (Nicoll, 1988). Dutar and Nicoll (1988a,b) have shown that inositol trisphosphate (IP_3), one of the breakdown products of PI following phospholipase C (PLC) activation, when included in electrodes used to impale hippocampal neurones causes a progressive suppression of I_M as the IP_3 leaks into the cell. C-kinase activation, which can also result from PI breakdown through the intermediary diacyl glycerol (DAG), does not reduce I_M (Malenka *et al.*, 1986). This contrasts with the situation in sympathetic ganglia where direct activation of C-kinase by phorbol esters causes M current suppression (Brown, 1988).

A recent development concerning the modulation of I_M in the CNS is the finding of Siggins and colleagues that somatostatin can enhance M current in hippocampal neurones and solitary tract neurones (Jacquin *et al.*, 1988; Moore *et al.*, 1988). The additional M current in the presence of somatostatin is sensitive to blockade by muscarinic agonists. The physiological consequence of this transmitter interaction would be to exaggerate the excitability increase caused by cholinergic reduction of I_M when the transmitters were copresent. Alternatively, somatostatin would have inhibitory actions that would be only expressed in the absence of activity in impinging cholinergic nerves.

14.3.2 Calcium-activated potassium currents

Three classes of Ca^{2+}-activated K^+ currents have been identified in central neurones; they are all indirectly blocked by antagonists of Ca^{2+} channels and can be separated on grounds of their underlying kinetic properties and their sensitivity to various pharmacological agents.

14.3.2.1 Fast Ca^{2+}-activated K^+ conductance

Single action potentials recorded from many locations in CNS are followed by hyperpolarizations (AHPs) lasting around 10 ms. These are sensitive to 1–10 mM TEA and when blocked are accompanied by broadening of the spike. The TEA sensitivity of these phenomena suggested initially the involvement of the delayed rectifier (K_V) (see above) but more recently it has become clear that Ca^{2+} channel blockers have similar effects (e.g. Storm, 1987; Schwindt *et al.*, 1988b; see Fig. 14.5).

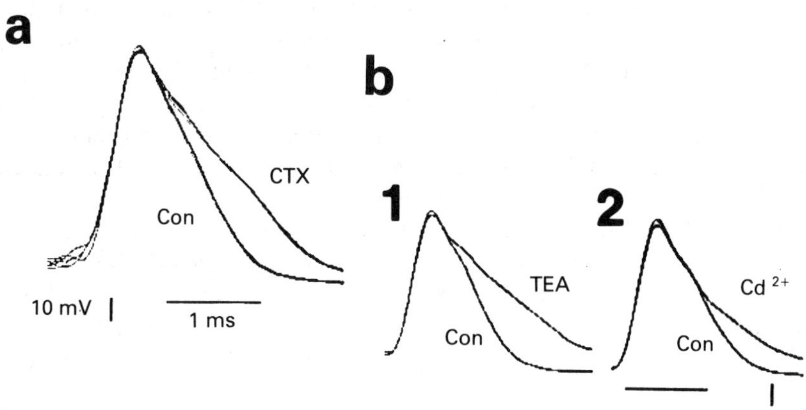

Fig. 14.5 — Repolarizing actions of fast Ca-activated K$^+$ conductance in rat hippocampal neurones. Superimposed action potentials recorded before and during exposure to (a) 30 nM charybdotoxin (CTX), (b_1) 2 mM TEA and (b_2) 0.3 mM Cd^{2+}, all of which suppress $I_{K(Ca)f}$. The spikes, triggered from −63 mV (a) or −60 mV (b) were prolonged by these agents. (From Storm, 1987, Figs 6 and 7.)

Voltage-clamp of CA3 neurones in the hippocampal slice and acutely dissociated CA1 neurones has revealed a rapidly activating and inactivating K$^+$ current that is relatively resistant to 4-AP but blocked by Ca^{2+} channel blockers; the decay rate of this current is two to three times faster than the 4-AP-sensitive I_A (Zbicz and Weight, 1985; Numann *et al.*, 1987)., Moreover, this, presumably Ca^{2+}-activated, conductance is TEA-sensitive. A current with these properties is a good candidate for action potential repolarization since the spike trajectory crosses activation threshold for at least two types of Ca^{2+} currents (see Fox *et al.*, 1987). A Ca^{2+}-activated K$^+$ conductance of large single-channel conductance (BK_{Ca}) which is also voltage-dependent has been studied elsewhere and is selectively blocked by a scorpion venom component, charybdotoxin (CTX) (Miller *et al.*, 1985). Single-spike AHPs in the hippocampus (Storm, 1987) are reduced by CTX with concomitant spike broadening (Fig. 14.5a). These observations suggest that a large-conductance Ca^{2+}-activated K$^+$ channel exists in central neurones and that it participates in spike repolarization. There is no information currently available to indicate any transmitter modulation of this current, (which I shall refer to as $I_{K(Ca)f}$).

The currents I_A and I_{Kv} discussed above have been separated from $I_{K(Ca)f}$ and are defined by their lack of dependence upon Ca influx. The Ca dependency of fast K$^+$ currents and, indeed, K_V is a moot point; to summarize, it seems the case that Ca^{2+} influx is not required for the activation of I_A or the delayed rectifier, but shifts in activation and inactivation can be brought about by changes in the concentration of divalent cations in extracellular solutions (see Gustafsson *et al.*, 1982; Segal and Barker, 1984; Numann *et al.*, 1987; Sah *et al.*, 1988; Zona *et al.*, 1988). In some experimental protocols this can give the impression of Ca^{2+} dependence (see Schwindt *et al.*, 1988b). Recently, a K^+ current has been recorded in the magnocellular neurosecretory cells of rat supraoptic nucleus (Bourque, 1988) that has the pharmacological characteristics of I_A (blockade by 4-AP and DTX), similar inactivation properties to those of I_A, but requires Ca^{2+} influx for its activation. The activation voltage threshold of this transient outward current (TOC) is more negative than that for the majority of A-currents observed in central neurones, being similar to I_D. Overlap of the inactivation and activation curves for TOC indicate that a proportion of steady-state K^+ conductance is contributed by the TOC between −80 and −60 mV. Interestingly, an outward current in the same potential range recorded in cerebellar Purkinje cells is sensitive to Ca^{2+} channel blockade (Crepel and Penit-Soria, 1986); it is difficult to say whether this is due to a TOC-like conductance because of other currents in Purkinje cells activating in the same range. The TOC may be a variant of I_A (Bourque, 1988) but has been included in this section because of its clear Ca^{2+} sensitivity.

14.3.2.2 Medium Ca^{2+}-activated K^+ conductance ($I_{K(Ca)m}$)

Variety exists in both the amplitude and duration of the AHPs following multiple spike discharges in central neurones (see Schwindt *et al.*, 1988b,c; Storm, 1989). Exposure to Ca^{2+} channel blockers has indicated that at least two Ca^{2+} dependent processes underly AHPs which range in duration from 150 to 1500 ms and that these can be separated on the basis of their sensitivity to TEA and the bee venom component apamin. Short-duration AHPs are reduced in amplitude by TEA, whereas the slower AHPs (with decay τs of 1 s or more) are unaffected or even enhanced by this drug. It is the TEA-sensitive component that will be discussed in this section. Voltage-clamp studies (Brown and Griffith, 1983a; Sah *et al.*, 1985; Numann *et al.*, 1987; Schwindt *et al.*, 1988a) suggest that a medium-duration Ca^{2+}-activated K^+ current could contribute to the AHPs lasting around 200 ms (I shall refer to this as $I_{K(Ca)m}$). Brown and Griffith (1983a) studied such a current in guinea-pig CA1 hippocampal cells, calling it I_C because of its similarity to the invertebrate conductance of that name (Thompson, 1977). Thus, this current was blocked by divalent cationic and organic Ca^{2+} channel antagonists and was voltage- as well as Ca^{2+}-sensitive. Channel deactivation accelerated with membrane hyperpolarization and activation of the current appeared to be congruent with that of a high-threshold Ca^+ current (Johnston *et al.*, 1980; Brown and Griffith, 1983b). Isolation of these same neurones preserves $I_{K(Ca)m}$ and allows its observation with greater voltage-clamp fidelity (Numann *et al.*, 1987); the conditions of intracellular Ca^{2+} buffering (<10 nM) used by Numann *et al.* (1987) suggest that the Ca^{2+} activation site is very sensitive and close to the locus of Ca^{2+} influx. $I_{K(Ca)m}$ does not inactivate and consequently can give rise to a persistent component of K^+ conductance that reflects continuous Ca^{2+} influx (Brown and Griffith, 1983a; Numann *et al.*, 1987). $I_{K(Ca)m}$ exists in feline neocortical (Schwindt *et al.*, 1988b) guinea-pig olfactory cortical (Constanti and Sim, 1987) and human neocortical neurones (Halliwell, 1986). The functional role of this conductance could be to maintain an activity-dependent self-inhibitory tone; because of the voltage sensitivity of the current, its effects would dissipate rapidly with hyperpolarization, but would exert a brake on excitation at depolarized potentials. $I_{K(Ca)m}$ has not been shown to be subject to neuromodulation: it can be distinguished from I_M and slower Ca^{2+}-activated currents by its resistance to muscarinic agonists (Brown and Griffith, 1983a).

14.3.2.3 Slow Ca^{2+}-activated K^+ current $I_{K(Ca)s}$ or I_{AHP}

Hippocampal neurones that are depolarized beyond firing threshold by >300 ms long depolarizing current injections display marked accommodation of spike firing. Inspection of raw voltage traces reveals that the cessation of firing that can frequently occur before the termination of the excitatory stimulus is due to a profound hyperpolarizing event that lasts several seconds (Hotson and Prince, 1980; Benardo and Prince, 1982a; Madison and Nicoll, 1982; Haas and Konnerth, 1983; Lancaster and Adams, 1986; an example is shown in Fig. 14.6A_1). Both the accommodation of firing and the AHP are reduced by Ca^{2+} channel antagonists (see, for example, Madison and Nicoll, 1984); it is the slow component of the AHP that is particularly affected. Lancaster and Adams (1986) have shown that the current underlying the AHP is a TEA-resistant K^+ conductance that can be activated by aliquots of Ca^{2+} entering the neurone for each action potential that is fired. Thus, the AHP or its underlying current I_{AHP} is related in size and duration to the number of preceding spikes. In contrast to the TEA-sensitive $I_{K(Ca)m}$, I_{AHP} is voltage-insensitive; the decline of the AHP itself has an identical τ_{decay} to that for the I_{AHP} measured at different membrane potentials between −40 and −80 mV.

In neocortex the presence of a current similar to I_{AHP} has been detected (Schwindt *et al.*, 1988c) but a slow Ca^{2+}-activated K^+ current of duration somewhere between that of $I_{K(Ca)m}$ and I_{AHP} also exists. This is also TEA-resistant, but blocked by apamin and *d*-tubocurarine (dTC) as well as Ca^{2+} channel antagonists (Schwindt *et al.*, 1988c). Apamin also blocks a similar duration AHP in spinal motoneurones and hypothalamic magnocellular neurosecretory neurones (MCNs) (Zhang and Krnjevic, 1987; Bourque and Brown, 1988); in the MCN the current that apamin blocks is voltage-insensitive and reduced by Ca^{2+} channel antagonists and dTC (Bourque and Brown, 1987). These results demonstrate that an apamin-sensitive Ca^{2+}-activated K^+ conductance is present in certain central neurones; it is presumed that a small-conductance K^+ channel (Romey and Lazadunski, 1984) is involved and that the current is similar to that in sympathetic ganglia (Pennefather *et al.*, 1985). The apamin-sensitive current in neocortex is reported to be unaffected by two transmitter receptor ligands (muscarine and isoproterenol) that do reduce an I_{AHP}-like current (Schwindt *et al.*, 1988c); it remains to be seen whether other modulators can have an effect on this conductance.

Modulation of I_{AHP}, however, by many different transmitters, is well established, particularly in the case of the hippocampal conductance. Thus, the AHP and accommodation of neuronal firing is suppressed by ACh (Benardo and Prince, 1982a; Cole and Nicoll, 1983; Segal, 1982); noradrenaline (Madison and Nicoll, 1982, 1986; Haas and Konnerth, 1983; Lancaster and Adams, 1986; Schwindt *et al.*, 1988c); histamine (Haas and Konnerth, 1983; Haas and Greene, 1986); 5-HT (Colino and Halliwell, 1987; Baskys *et al.*, 1987; Andrade and Nicoll, 1987; Joels and Gallagher, 1988) and dopamine (Malenka and Nicoll, 1986) (see Fig. 14.6a,b). Benardo and Prince (1982b) have reported that dopamine *increases* accommodation

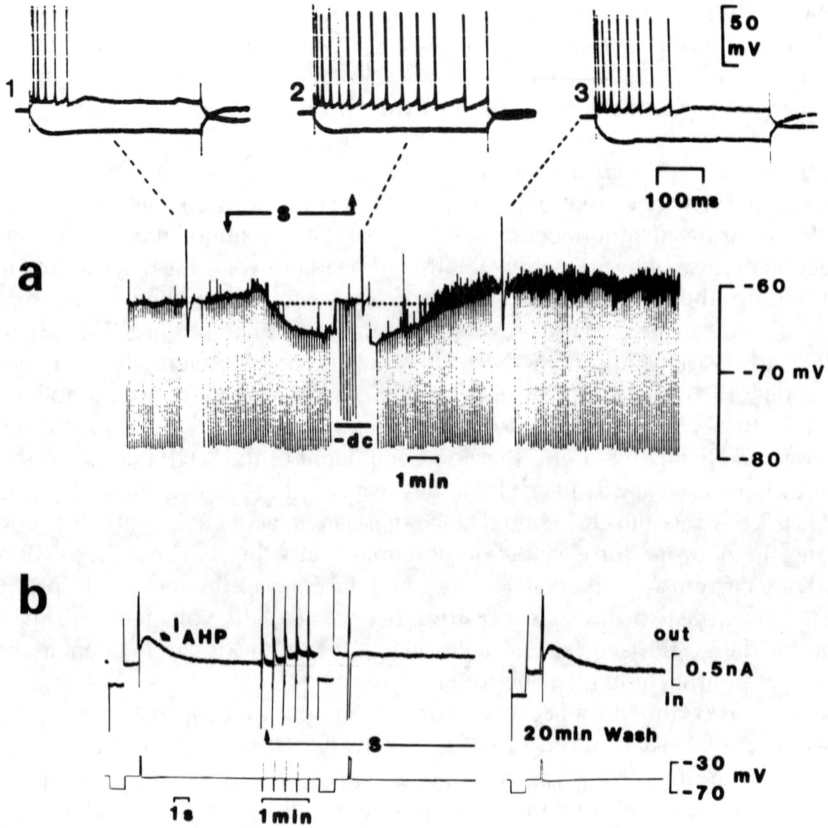

Fig. 14.6 — Transmitter modulation of I_{AHP} and activation of the agonist-activated K^+ conductance. (a) Voltage recording in a slice from a rat CA1 hippocampal neurone subject to 0.5 nA, 375 ms hyperpolarizing current injections (to monitor input resistance) every 5 s. At the times indicated, when the oscilloscope tracings (1–3) were recorded, current was reversed to trigger action potentials. 30 μM 5-HT was applied in the bathing medium (arrowed, S) and caused a hyperpolarization (K^+ current activation, see text); this was offset by passage of direct current to return membrane potential to control values (−dc). Note that during the action of 5-HT the accommodation of cell firing was reduced and the afterhyperpolarization (AHP) associated with cell firing was concomitantly suppressed. Rebound depolarization of the cell may also be observed; this may be ascribed to a later suppression of M current. (b) Voltage-clamp counterpart to A: another CA1 neurone bathed in 0.5 μM TTX was voltage-clamped at −50 mV and briefly depolarized to trigger Ca^{2+} influx and activate I_{AHP}, and hyperpolarized to assess membrane conductance. The current and voltage traces in B, at two different time bases, depict the effects of 30 μM 5-HT on the cell: 5-HT simultaneously activates outward current and suppresses $\cdot I_{AHP}$. (Part a: A. Colino and J. V. Halliwell, unpublished experiment; Part b: from Colino and Halliwell, 1987.)

by potentiating the AHP, and Haas and Greene (1984) (Greene and Haas, 1985) have observed similar actions of adenosine. The opposing effects of noradrenaline and adenosine may be reconciled by noting that these transmitters have opposite actions on the stimulation of cAMP via β-adrenergic and A_1 adenosine receptors, respectively. Activators of adenylate cyclase and administration of analogues of cAMP suppress the AHP directly (i.e. independently of neurotransmitter receptors). It is presumed that either the channels involved or the mechanism coupling raised $[Ca^{2+}]_i$ with channel activation are a target for cAMP-dependent protein kinase, since transmembrane Ca^{2+} flux appears not to be reduced as a result of noradrenaline action (Madison and Nicoll, 1982, 1986). Histamine's action via H_2 receptors is probably also mediated by stimulating adenylate cyclase (Haas, 1985). ACh- and 5-HT-mediated suppression of the AHP does not involve adenylate cyclase (Andrade and Nicoll, 1987; Nicoll, 1988). In the case of ACh, PI breakdown leading to diacyl glycerol activation of protein kinase C (PKC) could be the mechanism whereby muscarinic receptor stimulation reduces the AHP, since direct activators of PKC mimic the neurotransmitter effect (Baraban *et al.*, 1985; Malenka *et al.*, 1986). The receptor(s) and second messengers (if any) that mediate suppression of the AHP or its underlying current by 5-HT have yet to be identified (Colino and Halliwell, 1987; Andrade and Nicoll, 1987). However, it is clear that the conductance mechanism underlying the AHP is a common target for modulation by transmitters using at least two separate messenger systems.

As is the case with other central K^+ conductances discussed above, the suppression of I_{AHP} by a specific agent provides clues to the role of the channel activity when neuronal behaviour is observed after blockade. Cole and Nicoll (1983, 1984) provide evidence that neuronally released ACh changes the firing pattern of pyramidal neurones to long depolarizing stimuli from phasic to tonic (see also Segal, 1988); the conductance most sensitive to ACh is I_{AHP} (Madison *et al.*, (1987). It would appear from this that the role of I_{AHP} is to be activated cumulatively by Ca^{2+} entry during action potential generation; in effect, the number of spikes is counted by the intracellular sensing mechanism, which activates the K^+ conductance, thereby curtailing repetitive firing. The extent to which the conductance is activated by Ca^{2+} influx depends on prevailing intracellular conditions that may be controlled externally via receptors for transmitters and modulators (see above). The Ca^{2+}-activated K^+ conductances (and I_{AHP} is the most prominent in this respect) are activity-dependent and set the duration of a single discharge $[I_{K(Ca)f}]$ or a train of impulses $[I_{k(Ca)m}$ and $I_{AHP}]$. In addition, factors controlling release of Ca^{2+} from intracellular stores could set cell excitability by changing the activation of these conductances.

14.3.3 Na^+-activated K^+ conductance ($I_{K(Na)}$)

There is not conclusive evidence for an Na^+-activated K^+ conductance in the mammalian CNS, although such a conductance exists in avian neurones (Bader *et al.*, 1985; see also chapter 6). Two studies raise the possibility that a current of this kind is present in guinea-pig olfactory cortical neurones (Constanti and Sim, 1987) and in feline neocortical pyramidal neurones (Schwindt *et al.*, 1988c). Constanti and Sim (1987) report that slow time-course outward currents triggered by depolarizing clamp steps from resting potential are depressed by TTX, which is very selective for Na^+ channel activity; Schwindt *et al.* (1988c) describe a Ca^{2+}-insensitive slow AHP which follows depolarization-evoked repetitive firing but not prolonged depolarization in the presence of TTX. These studies do not distinguish, however, between a current that could be activated directly by raised $[Na^+]_i$ or by raised $[Ca^{2+}]_i$ subsequent to Na^+ influx. If such a current were present then it could be transiently activated by influx of Na^+ through rapidly inactivating Na^+ channels and have a role in the repolarization of the spike, or it could be persistently activated by Na^+ entry through non-inactivating channels (Stafstrom *et al.*, 1985) to moderate the long-term activity of a cell.

14.3.4 Agonist-activated K^+ conductance

Long-lasting IPSPs with two components frequently follow synaptic activated of neuronal populations in slices of hippocampus, neocortex and thalamus (see Nicoll, 1988). Unlike shorter-duration components, the longer-lasting ('late') IPSPs are not sensitive to bicucculine or picrotoxin, nor are they reversed in polarity when recorded with a Cl^--containing microelectrode. These data indicate that the late IPSP is due to an increase in K^+ conductance. This conductance, in the case of neocortical and hippocampal neurones, is activated by GABA in the presence of $GABA_A$ antagonists, by baclofen and antagonized by the $GABA_B$ antagonist phaclofen (Dutar and Nicoll, 1988c). The current–voltage relation for both the late IPSP and baclofen-induced conductance increase indicates a reversal potential identical to E_K and a degree of inward rectification (Newberry and Nicoll, 1985). It may be concluded that in many brain areas GABA released from inhibitory interneurones acts on principal cells via $GABA_B$ receptors (Bowery *et al.*, 1980) to activate an inwardly rectifying K^+ conductance. The rectification of the baclofen-activated current was confirmed under voltage-clamp in cultured neurones (Gahwiler and Brown, 1985a). Other transmitters act to increase a similar membrane K^+ conductance in the brain: thus noradrenaline (via $_2$ receptors) (Williams *et al.*, 1985; Aghajanian and Wang, 1986; Madison and Nicoll, 1986); adenosine (A_1) (Trussell and Jackson, 1985; Greene and Haas, 1985; Nicoll, 1988); ACh (M_2) (Egan and North, 1986; McCormick and Prince, 1986, 1987; Greene *et al.*, 1989); dopamine (D_2) (Lacey *et al.*, 1987); 5-HT ($5HT_{1a}$) (Aghanjanian and Lakoski, 1984; Yoshimura and Higashi, 1985; Andrade *et al.*, 1986; Colino and Halliwell, 1987; Joels and Gallagher, 1988; Yakel *et al.*, 1988; Williams *et al.*, 1988); opiates (μ) (North *et al.*, 1987); somatostatin (Inoue *et al.*, 1988) hyperpolarize or induce outward current in neurones in forebrain (neocortex, hippocampus, striatum, thalamus), midbrain (substantia nigra, raphe nuclei) and hindbrain (pontine reticular nucleus, parabrachial nucleus, locus coeruleus) (see Fig. 14.6a,b, for example). Responses elicited by two or more separate agonists display non-additivity or occlusion, suggesting a common link beyond the receptor site for transmitter in the generation of the conductance (Andrade *et al.*, 1986; Nicoll, 1988); this appears to be a G-protein (see below). Where the investigators employed voltage-clamp technique, the currents activated by agonists all displayed inward rectification characterized by a current–voltage relationship possessing two features: (i) a change in slope conductance demarcated by the value of E_K for the prevailing experimental conditions; and (ii) an additional current decrease positive to about −50 mV (e.g. Trussell and Jackson, 1985; Gahwiler and Brown, 1985a; North *et al.*, 1987; Colino and Halliwell, 1987; Williams *et al.*, 1988; Yakel *et al.*, 1988). A distillation of the conclusions from these studies indicates that Ba^{2+} prevents both outward and inward movement of K^+ through the activated conductance, Cs^+ hinders inward K^+ movement, particularly when the driving force is great (at negative potentials), and Rb^+ is impermeant. A further common feature of this agonist-induced conductance is the dependence upon a G-protein for its activation that appears to be sensitive to the ADP-ribosylating effects of pertussis toxin (see, for example, Aghajanian and Wang, 1986; Andrade *et al.*, 1986; Trussell and Jackson, 1987; Williams *et al.*, 1988; Inoue *et al.*, 1988) and affected by the intracellular administration of non-hydrolysable analogues of GTP and GDP (Andrade *et al.*, 1986; Trussell and Jackson, 1987; North *et al.*, 1987; Inoue *et al.*, 1988).

The characteristics of the agonist-activated K^+ conductance are similar to those of the 'anomalous rectifier', I_{K1}, in terms of rectification, blockade by Ba^{2+} and Cs^+ and impermeability to Rb^+. In the case of invertebrate neurones, it has been suggested that agonist action activates I_{K1} directly by circumventing the voltage sensitivity of the gating mechanism (Benson and Levitan, 1983). This may apply for the mammalian K^+ conductance also, but the agonist-induced conductance in mammals lacks the conductance decrement at very negative potentials, which is due either to inactivation or blockade by external cations and which characterizes the voltage-dependent current (e.g. Inoue *et al.*, 1988). It is possible that the same type of channel is activated by either mode (voltage or ligand binding) and that channel permeability or gating is somehow affected by receptor occupation. The complexity is indicated by the fact that in excised patches from hippocampal neurones, activated brain G-protein (G_o), or a recombinant subunit applied to the cytoplasmic face, can induce activity in four K^+ channel populations which display distinct conductance states, kinetics and properties of rectification (VanDongen *et al.*, 1988). It will require further study of single-channel properties to distinguish whether the agonist-activated and voltage-activated currents employ common channel populations.

Since the agonist-stimulated K^+ conductance requires neurotransmitter receptor occupation for activation, the underlying channels are, by definition, subject to neuromodulation, whether or not they are identical to the I_{K1} channels. By analogy with the muscarinic ACh receptor-induced K^+ conductance in cardiac tissue (K_{ACh}) (Pfaffinger *et al.*, 1985; Breitwieser and Szabo, 1985; see Chapter 5) an activation of the channels directly via G-protein subunits is likely (VanDongen *et al.*, 1988; see also Codina *et al*,. 1987 and Logothetis *et al.*, 1987), but there also remains the possibility that an as yet unknown diffusible second messenger system also triggered via a receptor-coupled G-protein could activate the conductance in addition. Comparison with cardiac cells, once more, suggests the latter, since K_{ACh} can also be triggered by arachidonic acid metabolites that could be formed as a result of membrane phospholipase A_2 (PLA_2) activity (Kurachi *et al.*, 1989; Kim *et al.*, 1989).

A different aspect to the question of modulation is raised by the observations of Worley *et al.* (1987) that muscarinic stimulation reduces the inhibitory effects of adenosine and baclofen on hippocampal neurones; the late K^+-dependent IPSP is concomitantly depressed and the muscarinic action, which simulates cholinergic nerve stimulation, is mimicked by exposure to phorbol esters. Thus, the agonist-induced K^+ conductance is similar to I_{AHP} in being inhibited by activators of PKC. The multiplicity of receptor subtypes for neurotransmitters that activate the ligand-gated K^+ conductance via one moiety and trigger modulatory biochemical changes via another provides several levels of interaction for the control of K^+ channel activity and consequently of neuronal excitability. Concurrent modulation of at least three separate K^+ conductances, including activation of the agonist-induced K^+ conductance, can occur through different receptors for 5-HT (Colino and Halliwell, 1987; Andrade and Nicoll, 1987; Fig. 14.6a) or noradrenaline (Madison and Nicoll, 1986).

14.3.5 Other K^+ conductances in central neurones

14.3.5.1 The 'leak' K^+ conductance

The concept of a 'leak' K^+ conductance is a construct that has been employed *inter alia* to explain conductance decreases observed after exposure of central neurones to drugs or transmitters, that cannot be ascribed to an effect on any of the specifically identified currents which are listed elsewhere in this chapter. Thus, ACh, noradrenaline and 5-HT, amongst others, cause a reduction of a voltage-insensitive presumed K^+ conductance in addition to the other actions listed above (Madison *et al.*, 1987; Muller and Misgeld, 1986; Benson *et al.*, 1988; Madison and Nicoll, 1986; Colino and Halliwell, 1987; Andrade and Nicoll, 1987). These effects could be mediated by an effect on channels similar to the K_S channel in *Aplysia* neurones that is modulated by 5-HT and various peptides via cAMP and arachidonic acid cascades (see Belardetti and Siegelbaum, 1988), on an as-yet unidentified channel or on the behaviour of other K^+-permeable pathways (Christensen and Zeuthen, 1987). It is possible that a component of observed 'leak' conductance arises from the steady-state activation of I_{K1} or the agonist-activated K^+ conductance (the latter, perhaps, by metabolically derived adenosine (Haas and Greene, 1988). Furthermore, it may be the case that the range of 'passive' membrane behaviour around resting potential that experimentally defines leak conductance (to all ionic species) fortuitously reflects the algebraic sum of several steady-state conductances; in other words, one experimenter's leak conductance is another's active membrane current. Whatever the true nature of this phenomenon, it is clear that K^+ conductance through so-called 'leak' channels reduces cell excitability; consequently, drug- and transmitter-induced reduction of this conductance can be expected concomitantly to lower neuronal firing threshold and potentiate depolarizing stimuli. This latter mechanism has been invoked as the mechanism underlying slow cholinergic EPSPs in the CNS (Muller and Misgeld, 1986; Madison *et al.*, 1987; Segal, 1988; Benson *et al.*, 1988).

14.3.5.2 ATP-sensitive and volume-activated K^+ conductances

The presence of these two channel types (Kullberg, 1987; Stanfield, 1987) has not conclusively been proven in central neurones. Stretch-activated channels responding to fluctuating cell volume that could result from changing intracellular recording conditions could have implications for the cell 'leak' conductance (see above). ATP-sensitive channels (K_{ATP}), as in pancreatic β-cells, might respond to cerebral glucose levels and/or the metabolic status of neurones, also contributing to alterations in 'leak' conductance. A recent preliminary report (Ashford *et al.*, 1989) suggests that excitation by glucose of neurones in the ventromedial nucleus of the hypothalamus, an area concerned with the regulation of food intake, could involve suppression of ATP-sensitive K^+ channels. Moreover, binding sites for the sulphonylurea blocker of K_{ATP}, glibenclamide, have been found in brain (Mourre *et al.*, 1989), which could indicate the distribution of this channel type.

14.4 PRESYNAPTIC K^+ CHANNELS IN THE CENTRAL NERVOUS SYSTEM

A question not addressed so far is the influence of presynaptic K^+ channels in the CNS. At synaptic sites deeper within the CNS than the terminals of afferent sensory fibres, there is no convincing model for presynaptic properties. Thus, electrophysiological studies on axonal and terminal K^+ conductances have been limited to observations of known channel blockers on compound action potentials recorded extracellularly from fibre bundles within brain slices (Galvan *et al.*, 1984; Haas *et al.*, 1983; Scholfield and Steel, 1988). 10–100 μM 4-AP prolongs action potential duration in lightly myelinated axons possessing en passage terminal boutons (Haas *et al.*, 1983; Galvan *et al.*, 1984); following 4-AP, additional effects of TEA (up to 20 mM) have been observed (Scholfield and Steel, 1988). Absence of effects of 4-AP and TEA has been recorded on myelinated (terminal-free) fibres of the dorsal columns (Kocsis and Waxman, 1980) or corpus callosum (Preston *et al.*, 1983) and of DTX or mast cell degranulating peptide (MCD peptide) on fibres running within the hippocampus (Dolly *et al.*, 1985; Cherubini *et al.*, 1987) that *do* possess terminals. However, both MCD peptide and DTX are potent convulsants and share with 4-AP the property of potentiating synaptic transmission in the hippocampus and olfactory cortex (Dolly *et al.*, 1985; Cherubini *et al.*, 1987; Galvan *et al.*, 1982; Buckle and Haas, 1982). These observations suggest that K^+ channels sensitive to both 4-AP and TEA are located in axonal terminal regions. Furthermore, the effects of DTX and MCD peptide indicate that a subset of presynaptic K^+ channels which do not normally participate in spike repolarization could, nonetheless, control Ca^{2+} entry into terminal regions and influence transmitter release. It should be appreciated, however, that this indirect approach cannot exclude the possibility of the agents concerned having effects on Ca^{2+} channels directly.

The somata of afferent sensory ganglion cells do not possess synaptic contacts or postsynaptic specializations. The properties of the channels they express might reflect those presynaptically located on the primary afferents which influence, and are influenced by contact with, central neurones. Voltage-dependent K^+ channels that have been demonstrated on sensory ganglion somata include those underlying the delayed rectifier (Stansfeld and Feltz, 1988), the A current (Cooper and Shrier, 1985; Kasai *et al.*, 1986) and a current that activates as quickly as I_A but inactivates an order of magnitude more slowly (Stansfeld and Feltz, 1988). Macroscopic currents in sensory ganglia have different sensitivities to blocking agents. Greater than millimolar concentrations of external TEA block delayed rectifier currents. The fast activating conductances are both blocked by 4-AP but with different sensitivity: I_A is blocked around millimolar concentrations whereas the more persistent conductance is about a hundred to a thousand times more sensitive (Stansfeld *et al.*, 1986). This susceptibility of the persistent K^+ current to 4-AP accords with a blockade by DTX and MCD peptide in the nanomolar range (Stansfeld *et al.*, 1987) and the lower sensitivity to 4-AP of transient currents in sensory ganglia with their resistance to DTX (Penner *et al.*, 1986). These findings can be compared (see above) with results for the hippocampal somatic I_A, which is blocked by 4-AP at around 100 μM and by

DTX and toxin I in the range 10–50 nM (Gustafsson *et al.*, 1982; Halliwell *et al.*, 1986). It is tempting, therefore, to compare the slowly inactivating ganglion current, named $I_{K(DTX)}$ by Stansfeld and Feltz (1988), with the I_D of Storm (1988) which has similar kinetic properties and voltage sensitivity.

Recently, expression of rat brain mRNA screened from cDNA libraries for similarity with the cloned Shaker mutant locus has produced in *Xenopus* oocytes a non-inactivating K$^+$ current sensitive to ~10 nM DTX and CTX, 100–300 μM 4-AP and 1–3 mM TEA (Stuhmer *et al.*, 1988); depending upon an inclusive or exclusive approach to the classification of K$^+$ conductances, this current displays all or none of the characteristics of $I_{K(V)}$, $I_{K(Ca)f}$, I_A or $I_{K(DTX)}$. However, translation of this message leads, presumably, to an assembly of identical subunits into a homomeric channel protein; co-expression of non-identical messages might produce channels with kinetic and pharmacological properties that match those of native channels enumerated above. It is interesting in this context to note that by alternative splicing of the gene products for the Shaker locus clones, which should code for I_A, a series of K$^+$ conductances may be expressed in oocytes that have a range of different inactivation rates (Timpe *et al.*, 1988).

14.5 SUMMARY AND CONCLUSIONS

Table 14.1 provides a summary of the dozen or so identified K$^+$ conductances that I have discussed in the preceding text. Details of the single-channel properties have been omitted because of the difficulty in unequivocally matching these to conductances underlying the macroscopic current characteristics that have been used to define the conductances. Readers are referred to other chapters in this volume and to recent reviews (Rudy, 1988; Castle *et al.*, 1989; Cook, 1988) and asked to make their own extrapolations from studies on peripheral nervous system and non-nervous tissues.

14.5.1 Functions of K$^+$ channels in the central nervous system

Observation of the behaviour of single neurones in isolated preparations of central nervous tissue reveals that cells may be spontaneously active, such as in the cerebellum and in brain stem and diencephalic nuclei, or relatively quiescent, but easily driven into regular or bursting patterns of firing, as exemplified by thalamic neurones, or totally quite as in neocortical tissue. It is likely that all of these cells possess the complement of inward currents that confer excitability upon them (transient and persistent Na$^+$ and Ca^{2+} currents and possibly I_H/I_Q) but are restrained by intrinsic inhibitory mechanisms to a greater or lesser extent. It is the array of K$^+$ conductances listed above that largely constitutes these intrinsic mechanisms and determines the character of an individual cell. The character of a resting neurone is set by the interplay between voltage-activated conductances; once it has been triggered into action then its character becomes more complex as the activity-dependent (cation-activated) conductances are also brought into play. As we have seen, the modulatory effects of neurotransmitters allow for changes in both the quiescent and the reactive character of neurones. The ultimate patterning of neuronal discharge will have implications for the effectiveness of transmission at the

Table 14.1

Name	Symbol	Activator	Kinetics			Blockers	Modulators
			Activation	Inactivation	Voltage sensitivity		
Delayed rectifier	$I_{K(V)}$	V	Slow–med.	Slow	Yes	TEA	—
Transient current	I_A	Rapid	Med.	Yes	Yes	4-AP, DTX, THA, Toxin I	Noradrenaline, ACh ↓
Low-threshold slowly inactivating current	I_D	V	Rapid	Slow	Yes	4-AP (v. sensitive)	—
Inward rectifiers	I_{K1}	V	Rapid+slow time-dependent	Slow or ionic block	Yes	Ba, Cs, Rb-impermeant	SP ↓
	I_H/I_Q[a]	V	Slow	None	Yes	Cs, THA	Kainate ↓
M current	I_M	V	Slow	None	Yes	Ba, TEA	ACh, 5-HT, ↓ somatostatin ↑
Ca-activated K currents	$I_{K(Ca)f}$	Ca/V	Rapid	Rapid	Yes	TEA, CTX, CCB	—
	$I_{K(Ca)m}$	Ca/V	Complex	None	Yes	TEA, CCB	—
	$I_{K(Ca)s}1$	CVa	Slow	None	No	Apamin, dTc, CCB	—
	$I_{K(Ca)s}2$	Ca	Slow(er)	None	No	Ba, CCB	NA, Hist, ACh, 5-HT ↓ adenosine ↑
Na-activated K current	$I_{K(Na)}$	Na	Rapid	None	?	TTX (indirect), TEA	—
Agonist-activated K current	$I_{K(AG)}$	G-protein R/agonist	Slow	None	Yes+Mg block	Ba, Cs, THA, Rb-impermeant	Somatostain, Opiates, Na, 5-HT, GABA ↑ adenosine
Leak K current	$I_{K(L)}$	V-insens.	—	None	No	Ba? Cs?	ACh, Noradrenaline, 5-HT ↓
ATP-sensitive K current	$I_{K(ATP)}$	ATP depresses	None	?	Glibenclamide	ADP/ATP	
Stretch-activated K current	$I_{K(VOL)}$	Volume change	?	?		Cetiedil?	—

[a]Not pure K^+ conductance.
CCB=Ca^{2+} channel blocker (indirect action on Ca^{2+}-activated K^+ current via inhibition of Ca^{2+} entry into the cell).
The apamin-sensitive Ca^{2+}-activated K^+ current, $I_{K(Ca)s}1$, is abbreviated as SK_{Ca} elsewhere in this book.

next synaptic junction because of temporal factors inherent in the process of transmitter release and the postsynaptic sensitivity to released transmitters.

Specific roles for central K^+ channels in pathological conditions remain to be clarified. However, various possibilities are noteworthy. The first is a failure of expression of a particular class or subclass of channel proteins as occurs with certain mutants (e.g. the Shaker mutant gene). Second, it is possible that in disorders such as epilepsy increases in $[K^+]_o$ are sufficient to compromise the inhibitory effects of K channel activation, as a result of positive shifts in E_K^+. Third, loss of projections that rely on a specific transmitter, the role of which is *inter alia* to modulate K channel activity, is encountered in neurodegenerative disorders such as Alzheimer's disease or Parkinson's disease.

14.5.2 Conclusions

A diversity of different K^+ channel types has been demonstarted in cells of the mammalian CNS. Their main function may be envisaged as controlling nueronal excitability and determining discharge patterns of neurones. By their activation and blockade by neurotransmitters, K^+ channels participate in both inhibitory and excitatory slow transmission at the level of the postsynaptic cell body. Likewise, localization of presynaptic K^+ channels allows the possibility of controlling transmitter release by modulating terminal excitability (and thereby Ca^{2+} influx) both via intrinsic channel properties and extrinsic humoral influence.

ACKNOWLEDGEMENTS

The authors would like to thank Drs C. D. Richards and A. M. Thomson for helpful criticism of the manuscript. Unpublished work from the author's laboratory that is presented in this chapter was supported by the MRC of the UK.

REFERENCES

Adams, P. R., and Brown, D. A. (1982) Synaptic inhibition of the M-current: slow excitatory post-synaptic potential mechanism in bullfrog sympathetic neurones. *J. Physiol. (Lond.)* **332** 263–272.

Adams, P. R., Brown, D. A., and Constanti, A. (1982) M-currents and other potassium currents in bullfrog sympathetic neurones. *J. Physiol. (Lond.)* **330** 537–572.

Aghajanian, G. K. (1985) Modulation of a transient outward current in serotonergic neurones by alpha$_1$-adrenoceptors. *Nature (Lond.)* **315** 501–503.

Aghajanian, G. K., and Lakoski, J. M. (1984) Hyperpolarization of serotonergic neurons by serotonin and LSD: Studies in brain slices showing increased K^+-conductance. *Brain Res.* **305** 181–185.

Aghajanian, G. K., and Wang, Y.-Y. (1986) Pertussis toxin blocks the outward currents evoked by opiate and alpha$_2$-agonists in locus coeruleus neurons. *Brain Res.* **371** 390–394.

Andrade, R., and Nicoll, R. A. (1987) Pharmacologically distinct actions of serotonin on single pyramidal neurones of the rat hippocampus recorded *in vitro*. *J. Physiol. (Lond.)* **394** 99–124.

Andrade, R., Malenka, R. C., and Nicoll, R. A. (1986) A G-protein couples serotonin and $GABA_B$ receptors to the same channels in hippoocampus. *Science* **234** 1261–1265.

Ashford, M. L. J., Boden, P. R., and Treherne, J. M. (1989) Glucose-induced excitation of rat hypothalamic neurones *in vitro* is mediated by ATP-sensitive K^+ channels. *J. Physiol. (Lond.)* **415** 31P.

Atwell, D., and Wilson, M. (1980) Behaviour of the rod network in the tiger salamander retina is mediated by membrane properties of individual rods. *J. Physiol. (Lond.)* **309** 287–315.

Bader, C. R., Bernheim, L., and Bertrand, D. (1985) Sodium-activated potassium current in cultured avian neurones. *Nature (Lond.)* **317** 540–542.

Baraban, J. M., Snyder, S. H., and Alger, B. E. (1985) Protein kinase C regulates ionic conductance in hippocampal pyramidal neurons: Electrophysiological effects of phorbol esters. *Proc. Natl Acad. Sci. USA* **82** 2538–2542.

Barrett, E. F., Barrett, J. N., and Crill, W. E. (1980) Voltage-sensitive outward currents in cat motoneurones. *J. Physiol. (Lond.)* **304** 231–249.

Bartschat, D. K., and Blaustein, M. P. (1985) Potassium channels in isolated presynaptic nerve terminals from rat brain. *J. Physiol. (Lond.)* **361** 419–440.

Baskys, A., Niesen, C. E., and Carlen, P. L. (1987) Altered modulatory actions of serotonin on dentate granule cells of aged rats. *Brain Res.* **419** 112–118.

Belardetti, F. and Siegelbaum, S. A. (1988) Up- and down-modulation of single K^+ channel function by distinct second messengers. *Trends Neurosci.* **11** 232–238.

Benardo, L. S., and Prince, D. A. (1982a) Cholinergic excitation of mammalian hippocampal pyramidal cells. *Brain Res.* **249** 315–331.

Benardo, L. S., and Prince, D. A. (1982b) Dopamine modulates a Ca^{2+}-activated potassium conductance in mammalian hippocampal pyramidal cells. *Nature (Lond.)* **297** 76–79.

Benson, J. A., and Levitan, I. B. (1983) Serotonin increases an anomalously rectifying K^+ current in the *Aplysia* neuron R15. *Proc Natl Acad. Sci. USA* **80** 3522–3525.

Benson, D. M., Blitzer, R. D., and Landau, E. M. (1988) An analysis of the depolarization produced in guinea-pig hippocampus by cholinergic receptor stimulation. *J. Physiol. (Lond.)* **404** 479–496.

Bourque, C. W. (1988) Transient calcium-dependent potassium current in magnocellular neurosecretory cells of the supraoptic nucleus. *J. Physiol. (Lond.)* **397** 331–347.

Bourque, C. W., and Brown, D. A. (1987) Apamin and *d*-tubocurarine block the after-hyperpolarization of rat supraoptic neurosecretory neurons. *Neurosci. Lett.* **82** 185–190.

Bowery, N. G., Hill, D. R., Hudson, A. L., Doble, A. L., Middlemiss, A., Shaw, J., and Turnbull, M. (1980) (−) Baclofen decreases neurotransmitter release in the mammalian CNS by an action at a novel GABA receptor. *Nature (Lond.)* **283** 92–94.

Breitwieser, G. E., and Szabo, G. (1985) Uncoupling of cardiac muscarinic and beta-adrenergic receptors from ion channels by a guanine nucleotide analogue. *Nature (Lond.)* **317** 538–540.

Brown, D. A. (1988) M-currents: an update. *Trends Neurosci.* **11** 294–299.

Brown, D. A., and Adams, P. R. (1980) Muscarinic suppression of a novel voltage-sensitive K^+-current in a vertebrate neurone. *Nature (Lond.)* **283** 673–676.

Brown, D. A., and Griffith, W. H. (1983a) Calcium-activated outward current in voltage-clamped hippocampal neurones of the guinea-pig. *J. Physiol. (Lond.)* **337** 287–301.

Brown, D. A., and Griffith, W. H. (1983b) Persistent slow inward calcium current in voltage-clamped hippocampal neurones of the guinea-pig. *J. Physiol (Lond.)* **337** 303–320.

Buckle, P. J., and Haas, H. L. (1982) Enhancement of synaptic transmission by 4-aminopyridine in hippocampal slices of the rat. *J. Physiol. (Lond.)* **326** 109–122.

Castle, N. A., Haylett, D. G., and Jenkinson, D. H. (1989) Toxins in the characterization of potassium channels. *Trends Neurosci.* **12** 59–65.

Champagnat, J., Jacquin, T., and Richter, D. W. (1986) Voltage-dependent currents in neurones of the nuclei of the solitary tract of brainstem slices. *Pflügers Arch.* **406** 372–379.

Cherubini, E., Ben-Ari, Y., Gho, M., Bidard, J. N., and Lazdunski, M. (1987) Long-term potentiation of synaptic transmission in the hippocampus induced by a bee venom peptide. *Nature (Lond.)* **328** 70–73.

Christensen, O., and Zeuthen, T. (1987) The baseline current in patch-clamp experiments is due to a membrane protein. *J. Physiol. (Lond.)* **387** 34P.

Codina, J., Yatani, A., Grenet, D., Brown, A. M., and Birnbaumer, L. (1987) The α subunit of the GTP-binding protein G_K opens atrial potassium channels. *Science* **236** 442–445.

Cole, A. E., and Nicoll, R. A. (1983) Acetylcholine mediates a slow synaptic potential in hippocampal pyramidal cells. *Science* **221** 1299–1301.

Cole, A. E., and Nicoll, R. A. (1984) Characterization of a slow cholinergic postsynaptic potential recorded *in vitro* from rat hippocampal pyramidal cells. *J. Physiol. (Lond.)* **352** 173–188.

Colino, A., and Halliwell, J. V. (1987) Differential modulation of three separate K-conductances in hippocampal CA1 neurons by serotonin. *Nature (Lond.)* **328** 73–77.

Connor, J. A., and Stevens, C. F. (1971) Voltage-clamp studies of a transient outward membrane current in gastropod neural somata. *J. Physiol. (Lond.)* **213** 21–30.

Connors, B. W., Gutnick, M. J., and Prince, D. A. (1982) Electrophysiological properties of neocortical neurons *in vitro*. *J. Neurophysiol.* **48** 1302–1320.

Constanti, A., and Galvan, M. (1983a) Fast inward-rectifying current accounts for anomalous rectification in olfactory cortex neurones. *J. Physiol. (Lond.)* **335** 153–178.

Constanti, A., and Galvan, M. (1983b) M-current in voltage-clamped olfactory cortext neurones. *Neurosci. Lett* **39** 65–70.

Constanti, A., and Sim, J. A. (1987) Calcium-dependent potassium conductance in guinea-pig olfactory cortex neurones in vitro. *J. Physiol. (Lond.)* **387** 173–194.

Cook, N. S. (1988) The pharmacology of potassium channels and their therapeutic potential. *Trends Pharmacol. Sci.* **9** 21–28.

Cooper, E., and Shrier, A. (1985) Single-channel analysis of fast transient potassium currents from rat nodose neurones. *J. Physiol. (Lond.)* **369** 199–208.

Crepel, F., and Penit-Soria, J. (1986) Inward rectification and low threshold calcium conductance in rat cerebellar Purkinje cells. An *in vitro* study. *J. Physiol. (Lond.)* **372** 1–23.

Crepel, F., Dhanjal, S. S., and Garthwaite, J. (1981) Morphological and electrophysiological characteristics of rat cerebellar slices maintained *in vitro*. *J. Physiol. (Lond.)* **316** 127–138.

Crill, W. E., and Schwindt, P. C. (1983) Active currents in mammalian central neurons. *Trends Neurosci.* **6** 236–240.

Cull-Candy, S. G., Dilger, P., Marshall, C. G., and Ogden, D. C. (1986) Voltage-activated currents in freshly dispersed granule cells from neonatal rat cerebellum. *J. Physiol. (Lond.)* **381** 114P.

Di Francesco, D. (1981) A new interpretation of the pace-maker current in calf Purkinje fibres. *J. Physiol. (Lond.)* **314** 359–376.

Dolly, J. O. (1988) Potassium channels — What can the protein chemistry contribute? *Trends Neurosci.* **11** 186–188.

Dolly, J. O., Halliwell, J. V., Black, J. D., Williams, R. S., Pelchen-Mathews, A., Breeze, A. L., Mehraban, F., Othman, I. B., and Black, A. R. (1984) Botulinum neurotoxin and dendrotoxin as probes for studies on transmitter release. *J. Physiol. (Paris)* **79** 280–303.

Dutar, P., and Nicoll, R. A. (1988a) Stimulation of posphatidylinositol (PI) turnover may mediate the muscarinic suppression of the M-current in hippocampal pyramidal cell. *Neurosci. Lett.* **85** 89–94.

Dutar, P., and Nicoll, R. A. (1988b) Classification of muscarinic repsonses in hippocampus in terms of receptor subtypes and second-messenger systems: Electrophysiological studies *in vitro*. *J. Neurosci.* **8** 4214–4224.

Dutar, P., and Nicoll, R. A. (1988c) A physiological role for $GABA_B$ receptors in the central nervous system. *Nature (Lond.)* **332** 156–158.

Egan, T. M., and North, R. A. (1986) Acetylcholine hyperpolarizes central neurones by acting on an M_2 muscarinic receptor. *Nature (Lond.)* **319** 405–407.

Fox, A. P., Nowycky, M. C., and Tsien, R. W. (1987) Single-channel recordings of three types of calcium channels in chick sensory neurones. *J. Physiol. (Lond.)* **394** 173–200.

French, C. R., and Gage, P. W. (1985) A threshold sodium current in pyramidal cells in rat hippocampus. *Neurosci. Lett.* **56** 289–293.

Gahwiler, B. H., and Brown, D. A. (1985a) $GABA_B$-receptor-activated K^+ current in voltage-clamped CA3 pyramidal cells in hippocampal cultures. *Proc. Natl. Acad. Sci. USA* **82** 1558–1562.

Gahwiler, B. H., and Brown, D. A. (1985b) Functional innervation of cultured hippocampal neurones by cholinergic afferents from co-cultured septal explants. *Nature (Lond.)* **313** 577–579.

Galvan, M., Franz, P., and Vogel-Wiens, C. (1984) Actions of potassium channel blockers on guinea-pig lateral olfactory tract axons. *Naunyn-Schmiedeberg's Arch. Pharmacol.* **325** 8–11.

Galvan, M., Grafe, P., and ten Bruggencate, G. (1982) Convulsant actions of 4-aminopyridine on the guinea-pig olfactory cortex slice. *Brain Res.* **241** 75–86.

Gho, M., King, A. E., Ben-Ari, Y., and Cherubini, E. (1986) Kainate reduces two voltage-dependent potassium conductances in rat hippocampal neurons *in vitro*. *Brain Res.* **385** 411–414.

Greene, R. W., and Haas, H. L. (1985) Adenosine actions on CA1 pyramidal neurones in rat hippocampal slices. *J. Physiol. (Lond.)* **366** 119–127.

Greene, R. W., Haas, H. L., and Hermann, A. (1985) Effects of caffeine on hippocampal cells in vitro. *Br. J. Pharmac.* **85** 163–169.

Greene, R. W., Gerber, U., and McCarley (1989) Cholinergic activation of medial pontine reticular formation neurons *in vitro*. *Brain Res.* **476** 154–159.

Gustafsson, B., Galvan, M., Grafe, P., and Wigstrom, H. (1982) A transient outward current in a mammalian central neurone blocked by 4-aminopyridine. *Nature (Lond.)* **299** 252–254.

Haas, H. L. (1985) Histamine may act through cyclic AMP on hippocampal neurones. *Agents and Actions* **16** 234–235.

Haas, H. L., and Greene, R. W. (1984) Adenosine enhances afterhyperpolarization and accommodation in hippocampal pyramidal cells. *Pflügers Arch.* **402** 244–247.

Haas, H. L., and Greene, R. W. (1986) Effects of histamine on hippocampal pyramidal cells of the rat *in vitro*. *Exp. Brain Res.* **62** 123–130.

Haas, H. L., and Green, R. W. (1988) Endogenous adenosine inhibits hippocampal CA1 neurones: further evidence from extra- and intracellular recording. *Naunyn-Schmied. Arch. Pharmacol.* **337** 561–565.

Haas, H. L., and Konnerth, A. (1983) Histamine and noradrenaline decrease calcium-activated potassium conductance in hippocampal pyramidal cells. *Nature (Lond.)* **302** 432–434.

Haas, H. L., Wieser, H. G., and Yasargil, M. G. (1983) 4-Aminopyridine and fibre potentials in rat and human hippocampal slices. *Experientia* **39** 114–115.

Halliwell, J. V. (1986) M-current in human neocortical neurones. *Neurosci. Lett.* **67** 1–6.

Halliwell, J. V. (1989) Cholinergic responses in human neocortical neurones. In *Central cholinergic synaptic transmission,* M. Frotscher and U. Misgeld (eds). Birkhauser Verlag, Basel, pp. 138–149.

Halliwell, J. V., and Adams, P. R. (1982) Voltage-clamp analysis of muscarinic excitation in hippocampal neurones. *Brain Res.* **250** 71–92.

Halliwell, J. V., Othman, I. B., Pelchen-Mathews. A., and Dolly, J. O. (1986) Central action of dendrotoxin: selective reduction of a transient K conductance in hippocampus and binding to localized acceptors. *Proc. Natl Acad. Sci. USA* **83** 493–497.

Hammill, O. P., Marty, A., Neher, E., Sakmann, B., and Sigworth, F. J. (1981) Improved path-clamp techniques for high-resolution current recording from cells and cell-free membrane patches. *Pflügers Arch.* **391** 85–100.

Harvey, A. L., and Anderson, A. J. (1985) Dendrotoxins: snake toxins that block potassium channels and facilitate neurotransmitter release. *Pharmac. Ther.* **31** 33–55.

Henderson, G., Pepper, C. M., and Shefner, S. A. (1982) Electrophysiological properties of neurones contained in the locus coeruleus and mesencephalic nucleus of the trigeminal nerve *in vitro*. *Exp. Brain Res.* **45** 29–37.

Hille, B. (1967) The selective inhibition of delayed potassium currents in nerve by tetraethylammonium ion. *J. Gen. Physiol.* **50** 1287–1302.

Hodgkin, A. L. (1964) *The conduction of the nervous impulse,* 108 pp. Liverpool University Press, Liverpool.

Hodgkin, A. L., and Ktaz, B. (1949) The effect of sodium ions on the electrical activity of the giant axon of the squid. *J. Physiol. (Lond.)* **108** 37–77.

Hotson, J. R., and Prince, D. A. (1980) A calcium-activated hyperpolarization follows repetitive firing in hippocampal neurons. *J. Neurophysiol.* **43** 409–419.

Hotson, J. R., Prince, D. A., and Schwartzkroin, P. A. (1979) Anomalous inward rectification in hippocampal neurons. *J. Neurophysiol.* **42** 889–895.

Inoue, M., Nakajima, S., and Nakajima, Y. (1988) Somatostatin induces an inward rectification in rat locus coeruleus neurones through a pertussis toxin-sensitive mechanism. *J. Physiol. (Lond.)* **407** 177–198.

Jacquin, T., Champagnat, J., Madamba, S., Denavit-Saubie, M., and Siggins, G. R. (1988) Somatostatin depresses excitability in neurons of the solitary tract complex through hyperpolarization and augmentation of I_M, a non-inactivating voltage-dependent outward current blocked by muscarinic agonists. *Proc. Natl Acad. Sci. USA* **85** 948–952.

Joels, M., and Gallagher, J. P. (1988) Actions of serotonin recorded intracellularly in rat dorsal lateral septal neurons. *Synapse* **2** 45–53.

Johnston, D., Hablitz, J. J., and Wilson, W. A. (1980) Voltage clamp discloses slow inward current in hippocampal burst-firing neurones. *Nature (Lond.)* **286** 391–393.

Kasai, H., Kameyam, D., Yamaguchi, K., and Fukuda, J. (1986) Single transient K channels in mammalian sensory neurons. *Biophys. J.* **49** 1243–1247.

Katz, B. (1949) Les constantes electriques de la membrane du muscle. *Archs. Sci. Physiol.* **3** 285–299.

Kim, D., Lewis, D. L., Graziadei, L., Neer, E. J., Bar-Sagi, D., and Clapham, D. (1989) G-protein βγ-subunits activate the cardiac muscarinic K^+-channel via phospholipase A_2. *Nature (Lond.)* **337** 557–560.

Kocsis, J. D., and Waxman, S. G. (1980) Absence of potassium conductance in central myelinated axons. *Nature (Lond.)* **287** 348–349.

Krnjevic, K., Pumain, R., and Renaud, L. (1971) The mechanism of excitation by acetylcholine in the cerebral cortex. *J. Physiol. (Lond.)* **215** 447–465.

Kullberg, R. (1987) Stretch-activated ion channels in bacteria and animal cell membranes. *Trends Neurosci.* **10** 387–388.

Kurachi, Y., Ito, H., Sugimoto, T., Shimizu, T., Miki, I., and Ui, M. (1989) Arachidonic acid metabolites as intracellular modulators of the G-protein-gated cardiac K^+ channel. *Nature (Lond.)* **337** 555–557.

Lacey, M. G., Mercuri, N. B., and North, R. A. (1987) Dopamine acts on D_2 receptors to increase potassium conductance in neurones of the rat substantia nigra zona compacta. *J. Physiol. (Lond.)* **392** 397–416.

Lancaster, B., and Adams, P. R. (1986) Calcium-dependent current generating the afterhyperpolarization of hippocampal neurons. *J. Neurophysiol.* **55** 1268–1282.

Llinas, R. (1984) Comparative electrobiology of mammalian central neurons. In: *Brain slices,* pp. 7–24, R. Dingledine (ed.). Plenum Press, New York.

Llinas, R., and Sugimori, M. (1980) Electrophysiological properties of *in vitro* Purkinje cell somata in mammalian cerebellar slices. *J. Physiol. (Lond.)* **305** 171–195.

Logothetis, D. E., Kurachi, Y., Galper, J., Neer, E. J., and Clapham, D. E. (1987) The βγ subunits of GTP-binding proteins activate the muscarinic K$^+$ channel in the heart. *Nature (Lond.)* **325** 321–326.

Madison, D. V., and Nicoll, R. A. (1982) Noradrenaline blocks accommodation of pyramidal cell discharge in the hippocampus. *Nature (Lond.)* **299** 636–638.

Madison, D. V., and Nicoll, R. A. (1984) Control of repetitive discharge of rat CA1 pyramidal neurones *in vitro*. *J. Physiol. (Lond.)* **354** 319–331.

Madison, D. V., and Nicoll, R. A. (1986) Actions of noradrenaline recorded intracellularly in rat hippocampal CA1 pyramidal neurones, *in vitro*. *J. Physiol. (Lond.)* **372** 221–244.

Madison, D. V., Lancaster, B., and Nicoll, R. A. (1987) Voltage-clamp analysis of cholinergic action in the hippocampus. *J. Neurosci.* **7** 733–741.

Malenka, R. C., and Nicoll, R. A. (1986) Dopamine decreases the calcium activated afterhyperpolarization in hippocampal CA1 pyramidal cells. *Brain Res.* **379** 210–215.

Malenka, R. C., Madison, D. V., Andrade, R., and Nicoll, R. A. (1986) Phorbol esters mimic some cholinergic actions in hippocampal pyramidal neurons. *J. Neurosci.* **6** 475–480.

Mayer, M. L., and Westbrook, G. L. (1983) A voltage-clamp analysis of inward (anomalous) rectification in mouse spinal sensory ganglion neurones. *J. Physiol. (Lond.)* **340** 19–45.

McCormick, D. A., and Prince, D. A. (1986) Acetylcholine induces burst firing in thalamic reticular neurones by activating a potassium conductance. *Nature (Lond.)* **319** 402–405.

McCormick, D. A., and Prince, D. A. (1987) Actions of acetylcholine in the guinea-pig and cat medial and lateral geniculate nuclei *in vitro*. *J. Physiol. (Lond.)* **392** 147–165.

Miller, C. (1988) *Shaker* shakes out potassium channels. *Trends Neurosci.* **11** 185–186.

Miller, C., Moczydlowski, E., Latorre, R., and Phillips, M. (1985) Charybdotoxin, a protein inhibitor of single Ca^{2+}-activated K$^+$-channels from mammalian skeletal muscle. *Nature (Lond.)* **313** 316–318.

Moore, S. D., Madamba, S. G., Joels, M., and Siggins, G. R. (1988) Somatostatin augments the M-current in hippocampal neurons. *Science* **239** 278–280.

Mourre, C., Ben Ari, Y., Bernardi, H., Fosset, M., and Lazdunski, M. (1989) Antidiabetic sulfonylureas: localization and binding sites in the brain and effects on the hyperpolarization induced by anoxia in hippocampal slices. *Brain Res* **486** 159–164.

Muller, W., and Misgeld, U. (1986) Slow cholinergic excitation of guinea-pig hippocampal neurons is mediated by two muscarinic receptor subtypes. *Neurosci. Lett.* **67** 107–112.

Murase, K., Randic, M., and Urban, L. (1986) Substance P augments a persistent slow inward calcium-sensitive current in voltage-clamped spinal dorsal horn neurones of the rat. *Brain Res.* **365** 369–376.

Nakajima, Y., Nakajima, S., Leonard, R. J., and Yamaguchi, K. (1986) Acetylcholine raises excitability by inhibiting the fast transient potassium current in cultured hippocamal neurones. *Proc. Natl Acad. Sci. USA* **83** 3022–3026.

Nakajima, Y., Nakajima, S., and Inoue, M. (1988) Pertussis toxin insensitive G protein mediates substance P-induced inhibition of potassium channels in brain neurons. *Proc. Natl Acad. Sci. USA* **85** 3643–3647.

Neher, E. (1971) Two transient current components during voltage clamp in snail neurons. *J. Gen. Physiol.* **58** 36–53.

Nelson, P. G., and Frank, K. (1967) Anomalous rectification in cat spinal motoneurons and effect of polarizing currents on excitatory postsynaptic potentials. *J. Neurophysiol.* **30** 1097–1113.

Newberry, N. R., and Nicoll, R. A. (1985) Comparison of the action of baclofen with γ-aminobutyric acid on rat hippocampal pyramidal cells. *J. Physiol. (Lond.)* **360** 161–185.

Nicoll, R. A. (1988) The coupling of neurotransmitter receptors to ion channels in the brain. *Science* **241** 545–551.

North, R. A., Williams, J. T., Surprenant, A., and Christie, M. J. (1987) μ and δ receptors belong to a family of receptors that are coupled to potassium channels. *Proc. Natl. Acad. Sci. USA* **84** 5487–5491.

Numann, R. E., Wadman, W. J., and Wong, R. K. S. (1987) Outward currents of single hippocampal cells obtained from the adult guinea-pig. *J. Physiol. (Lond.)* **393** 331–353.

Owen, W. G., and Torre, V. (1981) Effect of strophanthidin and K^+-free medium on the rod response in the presence of Cs^+. *J. Physiol. (Lond.)* **317** 72–73P.

Pennefather, P., Lancaster, B., Adams, P. R., and Nicoll, R. A. (1985) Two distinct Ca-dependent K-currents in bullfrog synpathetic ganglion cells. *Proc. Natl Acad. Sci. USA* **82** 3040–3044.

Penner, R., Peterson, M., Pierau, Fr.-K., and Dreyer, F. (1986) Dendrotoxin: a selective blocker of a non-inactivating potassium current in guinea-pig dorsal root ganglion neurones. *Pflugers Arch.* **407** 365–369.

Pfaffinger, P. J., Martin, J. M., Hunter, D. D., Nathanson, N. M., and Hille, B. (1985) GTP-binding proteins cardiac muscarinic receptors to a K channel. *Nature (Lond.)* **317** 536–538.

Pollard, C. E., and Crunelli, V. (1988) Intrinsic membrane currents in projection cells of the cat and rat lateral geniculate nucleus. *Neurosci. Lett. Suppl.* **32** S29.

Preston, R. J., Waxman, S. G., and Kocsis, J. D. (1983) Effects of 4-aminopyridine on rapidly and slowly conducting axons of rat corpus callosum. *Exp. Neurol.* **79** 808–820.

Purpura, D. P, Prelevic, S., and Santini, M. (1968) Hyperpolarizing increase in membrane conductance in hippocampal neurons. *Brain Res.* **7** 310–312.

Rogawski, M. A. (1986) Single voltage-dependent potassium channels in cultured rat hippocampal neurons. *J. Neurophysiol.* **56** 481–493.

Rogawski, M. A. (1987) Tetrahydroaminoacridine blocks voltage-dependent ion channels in hippocampal neurons. *Eur. J. Pharmacol.* **142** 169–172.

Romey, G., and Lazdunski, M. (1984) The coexistence in rat muscle cells of two distinct classes of Ca^{2+}-dependent K^+ channels with different pharmacological properties and different physiological functions. *Biochem. Biophys. Res. Commun.* **118** 669–674.

Rudy, B. (1988) Diversity and ubiquity of K channels. *Neuroscience* **25** 729–749.

Sah, P., French, C. R., and Gage, P. W. (1985) Effects of noradrenaline on some potassium currents in CA1 neurones in rat hippocampal slices. *Neurosci. Lett.* **60** 295–300.

Sah, P., Gibb, A. J., and Gage, P. W. (1988) Potassium current activated by depolarization of dissociated neurons from adult guinea-pig hippocampus. *J. Gen. Physiol.* **92** 263–278.

Scholfield, C. N., and Steel, L. (1988) Presynaptic K-channel blockade counteracts the depressant effect of adenosine in olfactory cortex. *Neuroscience* **24** 81–91.

Schwartzkroin, P. A. (1976) Further characteristics of hippocampal CA1 cells *in vitro*. *Brain Res.* **128** 53–68.

Schwindt, P. C., and Crill, W. E. (1981) Differential effects of TEA and cations on outward ionic currents of cat motoneurons. *J. Neurophysiol* **46** 1–16.

Schwindt, P. C., Spain, W., and Crill, W. E. (1988a) Influence of anomalous rectifier activation on afterhyperpolarizations of neurons from cat sensorimotor cortex *in vitro*. *J. Neurophysiol.* **59** 468–481.

Schwindt, P. C., Spain, W. J., Foehring, R. C., Stafstrom, C. E., Chubb, M. C., and Crill, W. E. (1988b) Multiple potassium conductances and their functions in neurons from cat sensorimotor cortex in vitro. *J. Neurophysiol.* **59** 424–449.

Schwindt, P. C., Spain, W. J., Foehring, R. C., Stafstrom, C. E., Chubb, M. C., and Crill, W. E. (1988c) Slow conductances in neurons from cat sensorimotor cortex and their role in slow excitability changes. *J. Neurophysiol.* **59** 450–467.

Segal, M. (1982) Multiple actions of acetylcholine at a muscarinic receptor studied in the rat hippocampal slices. *Brain Res.* **246** 77–87.

Segal, M. (1988) Synaptic activation of a cholinergic receptor in rat hippocampus. *Brain Res.* **452** 79–86.

Segal, M., and Barker, J. L. (1984) Rat hippocampal neurons in culture: potassium conductances. *J. Neurophysiol.* **51** 1409–1433.

Segal, M., Rogawski, M. A., and Barker, J. L. (1984) A transient potassium conductance regulates the excitability of cultured hippocampal and spinal neurons. *J. Neurosci.* **4** 604–609.

Spain, W. J., Schwindt, P. C., and Crill, W. E. (1987) Anomalous rectification in neurons from cat sensorimotor cortex *in vitro*. *J.Neurophysiol.* **57** 1555–1576.

Stafstrom, C. E., Schwindt, P. C., Flatman, J. A., and Crill, W. E. (1984) Properties of subthreshold response and action potential recorded in layer V neurons from cat sensorimotor cortex *in vitro*. *J. Neurophysiol.* **52** 244–263.

Stafstrom, C. E., Schwindt, P. C., Chubb, M. C., and Crill, W. E. (1985) Properties of persistent sodium conductance and calcium conductance of layer V neurons from cat sensorimotor cortex *in vitro*. *J. Neurophysiol.* **53** 153–170.

Stanfield, P. R. (1987) Nucleotides such as ATP may control the activity of ion channels. *Trends Neurosci.* **10** 335–339.

Stanfield, P. R. (1988) Intracellular Mg^{2+} may act as a co-factor in ion channel function. *Trends Neurosci.* **11** 475–477.

Stanfield, P. R., Nakajima, Y., and Yamaguchi, K. (1985) Substance P raises neuronal excitability by reducing inward rectification. *Nature (Lond.)* **315** 498–501.

Stansfeld, C. E., and Feltz, A. (1988) Dendrotoxin-sensitive K^+ channels in dorsal root ganglion cells. *Neurosci. Lett.* **93** 49–55.

Stansfeld, C. E., Marsh, S. J., Halliwell, J. V., and Brown, D. A. (1986) 4-Aminopyridine and dendrotoxin induce repetitive firing in rat visceral sensory neurones by blocking a slowly inactivating outward current. *Neurosci. Lett.* **64** 299–304.

Stansfeld, C. E., Marsh, S. J., Parcej, D. N., Dolly, J. O., and Brown, D. A. (1987) Mast cell degranulating peptide and dendrotoxin selectively inhibit a fast-activating potassium current and bind to common neuronal proteins. *Neuroscience* **23** 893–902.

Storm, J. F. (1987) Action potential repolarization and a fast after-hyperpolarization in rat hippocampal pyramidal cells. *J. Physiol. (Lond.)* **385** 733–759.

Storm, J. F. (1988) Temporal integration by a slowly inactivating K^+ current in hippocampal neurons. *Nature (Lond.)* **336** 379–381.

Storm, J. F. (1989) An after-hyperpolarization of medium duration in rat hippocampal pyramidal cells. *J. Physiol. (Lond.)* **409** 171–190.

Strong, J. A., and Kaczmarek, L. K. (1985) Multiple components of delayed potassium current in peptidergic neurons of *Aplysia*: modulation by an activator of adenylate cyclase. *J. Neurosci.* **6** 819–822.

Stuhmer, W., Stocker, M., Sakman, B., Seeburg, P., Baumann, A., and Pongs, O. (1988) Potassium channels expressed from rat brain cDNA have delayed rectifier properties. *FEBS Lett.* **242** 199–206.

Thompson, S. H. (1977) Three pharmacologically distinct potassium channels in molluscan neurones. *J. Physiol. (Lond.)* **265** 465–488.

Timpe, L. C., Jan, Y. N., and Jan, L. Y. (1988) Four cDNA clones from the *Shaker* locus of *Drosophila* induce kinetically distinct A-type potassium currents in *Xenopus* oocytes. *Neuron* **1** 659–667.

Trussell, L. O., and Jackson, M. B. (1985) Adenosine-activated potassium conductance in cultured striatal neurons. *Proc. Natl Acad. Sci. USA* **82** 4857–4861.

Trussell, L. O., and Jackson, M. B. (1987) Dependence of an adenosine-activated potassium current on a GTP-binding protein in mammalian central neurons. *J. Neurosci.* **7** 3306–3316.

VanDongen, A. M. J., Codina, J., Olate, J., Mattera, R., Joho, R., Birnbaumer, L., and Brown, A. M. (1988) Newly identified brain potassium channels gated by the guanine nucleotide binding protein G_o. *Science* **242** 1433–1437.

Williams, J. T., North, R. A., Shefner, S. A., Nishi, S., and Egan, T. M. (1984) Membrane properties of rat locus coeruleus neurons. *Neuroscience* **13** 137–156.

Williams, J.T., Henderson, G., and North, (1985) Characterization of α_2-adrenoceptors which increase potassium conductance in rat locus coeruleus neurons. *Neuroscience* **14** 95–101.

Williams, J. T., Colmers, W. F., and Pan, Z. Z. (1988) Voltage- and ligand-activated inwardly rectifying currents in dorsal raphe neurons *in vitro*. *J. Neurosci.* **8** 3499–3506.

Wilson, W. A., and Goldner, M. M. (1975) Voltage clamping with a single microelectrode. *J. Neurobiol.* **6** 411–422.

Worley, P. F., Baraban, J. M., McCarren, M., Snyder, S. H., and Alger, B. E. (1987) Cholinergic phosphatidylinositol modulation of inhibitory, G protein-linked, neurotransmitter actions: Electrophysiological studies in rat hippocampus. *Proc. Natl Acad. Sci. USA* **84** 3467–3471.

Yakel, J. L., Trussell, L. O., and Jackson, M. B. (1988) Three serotonin responses in cultured mouse hippocampal and striatal neurons. *J. Neurosci.* **8** 1273–1285.

Yoshimura, M., and Higashi, H. (1985) 5-Hydroxytryptamine mediates inhibitory postsynaptic potentials in rat dorsal raphe neurons. *Neurosci. Lett.* **53** 69–74.

Zbicz, K. L., and Weight, F. F. (1985) Transient voltage and calcium-dependent outward currents in hippocampal CA3 pyramidal neurons. *J. Neurophysiol.* **53** 1038–1058.

Zhang, L., and Krnjevic, K. (1987) Apamin depresses selectively the afterhyperpolarization of cat spinal motoneurons, *Neurosci. Lett.* **74** 58–62.

Zona, C., Pirrone, G., Avoli, M., and Dichter, M. (1988) Delayed and fast transient potassium currents in rat neocortical neurons. *Neurosci. Lett.* **94** 285–290.

15

Potassium channels in immunoresponsive cells

Phyllis Gardner

The improved patch-clamp technique (Hamill *et al.*, 1981) enabled investigators for the first time to perform detailed studies on ion channels in the relatively small cells of the immune system. These studies have revealed a number of parallels between cells of the nervous and immune systems. As with nerve cells, immunoresponsive cells have a variety of ion channel types, as classified by ion selectivity, conductance, voltage dependence, gating kinetics, sensitivity to pharmacological agents and regulatory properties (reviewed in Gallin and Sheehy, 1988; Lewis and Cahalan, 1988a). Specifically several types of K^+ channels have been described in immunoreponsive cells. While the precise function of the various K^+ channels is unknown, there is growing evidence that K^+ channels play important roles in various aspects of immune function. This chapter will summarize recent studies on K^+ channels in several categories of immunoresponsive cells, including T- and B-lymphocytes, natural killer cells, macrophages and basophil/mast cells.

15.1 T-LYMPHOCYTES

Several early patch-clamp studies of human peripheral T-lyphocytes revealed that the predominant ion channel present is a voltage-gated K^+ channel closely resembling the delayed rectifier K^+ channels of nerve and muscle (DeCoursey *et al.*, 1984; Matteson and Deutsch, 1984; Cahalan *et al.*, 1985). A similar delayed rectifier K^+ current has also been identified in clonal cytotoxic T-lymphocytes from the mouse (Fukushima *et al.*, 1984). The lymphocyte K^+ currents activate with a sigmoid time-course upon depolarization and then inactivate to a variable extent during long-lasting depolarizing steps (Fig. 15.1A). Because of the high input resistance of the T-lymphocyte and the nearly complete inactivation of the K^+ conductance at depolarized voltages, investigators were also able in some circumstances to record single-channel K^+ currents in the whole-cell configuration (DeCoursey *et al.*, 1984; Cahalan *et al.*, 1985). Representative recordings of single-channel K^+ currents, which have an estimated single-channel conductance of 16 pS, are seen in Fig. 15.1B.

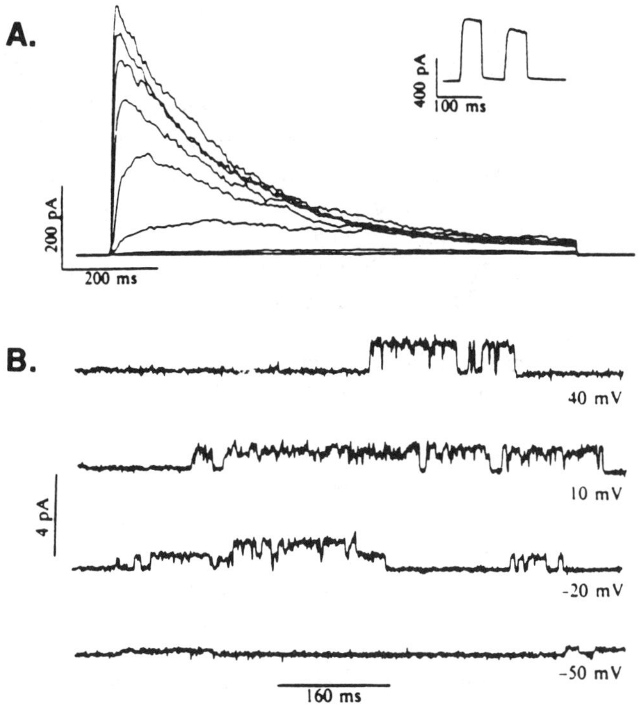

Fig. 15.1 — Outwardly rectifying K^+ currents in human peripheral T-lymphocytes. (A) Whole-cell currents. The membrane potential was held at -80 mV, and depolarizing pulses were given every 20 s in 10-mV steps up to $+30$ mV. Currents were recorded with a 10 GΩ feedback resistor, with active lowpass filtering at 2 kHz. Inset: a K^+ current record illustrating cumulative inactivation. Twin depolarization pulses to 0 mV from a holding potential of -80 mV. The cells were bathed in a mammalian Ringer solution containing (in mM): 160 NaCl, 4.5 KCl, 2 $CaCl_2$, 1 $MgCl_2$ and 5 HEPES, at pH 7.4. The pipette contained an 'internal' solution consisting of (in mM): 140 KF, 1 $CaCl_2$, 2 $MgCl_2$, 11 K_2EGTA and 10 HEPES, pH 7.2 (free Ca^{2+} concentration less than 2×10^{-9} M). (B) Single K^+ channel currents recorded from a T-lymphocyte in the whole-cell conformation. Same bathing and pipette solutions as (A). Membrane potential held at potentials indicated, 800 Hz filter. The estimated single-channel conductance was 16 pS. For further experimental details see DeCoursey *et al.* (1984). Reprinted by permission from *Nature* Vol. 307, pp. 465–468. Copyright 1984 MacMillan Magazines Ltd.

The lymphocyte K^+ current is sensitive to classical K^+ channel blockers of nerve and muscle preparations, including 4-aminopyridine (4-AP), quinine and tetraethylammonium (TEA) (DeCoursey *et al.*, 1984; Fukushima *et al.*, 1984; Matteson and Deutsch, 1984). In addition the K^+ current is also blocked by classical calcium channel antagonists such as verapamil and diltiazem and by divalent cations such as Ni^{2+} and Cd^{2+} (Chandy *et al.*, 1984; Matteson and Deutsch, 1984). Despite sensitivity to quinine, which blocks various calcium-activated K^+ channels, and to classical calcium channel blockers, the lymphocyte K^+ current does not appear to fall within the class of calcium-activated K^+ channels. Direct manipulation of internal Ca^{2+} either shows no effect on peak K^+ current (Cahalan *et al.*, 1985) or a reduction in peak K^+ current (Bregestovski *et al.*, 1986).

Subsequent studies have revealed that the T-lymphocyte K^+ current can be classified into three subtypes (Chandy *et al.*, 1986; DeCoursey *et al.*, 1987; Lewis and Cahalan, 1988b), the expression of which varies with the cell's developmental state and functional class as defined by cell surface markers (Fig. 15.2). The three classes of K^+ channels are divided according to voltage dependence of activation, degree to which the channels are inactivated during repetitive depolarization, sensitivity to TEA and to charybdotoxin (CTX), a peptide neurotoxin isolated from *Leiurus quinquestriatus* scorpion venom (Miller *et al.*, 1985; Smith *et al.*, 1986). The three classes of lymphocyte K^+ channels include: (1) *n* type: open at potentials more positive than −40 mV, are cumulatively inactivated during repetitive depolarization, are half-blocked by 10 mM TEA, and are blocked completely by 5 μm CTX; (2) *n′* type: resemble *n* channels in voltage dependence of activation and sensitivity to CTX; display little or no cumulative inactivation, and are less sensitive to TEA blockade (K_i ~100 mM); (3) *l* type: activate only at potentials above −10 mV, display very little cumulative inactivation, are highly sensitive to TEA blockade (K_i ~0.1 mM) and are insensitive to CTX. One class of potassium channels (type *n*) predominate in immature thymocyte subsets (CD4−CD8− and CD8+) as well as in the mature phenotype CD4+CD8− that contain precursors to helper T-lymphocytes. The two additional types of K^+ channels (*n′* and *l*) predominate in the mature CD4−CD8+ thymocyte subset that contains precursors to cytotoxic and suppressor T cells (Lewis and Cahalan, 1988b).

The precise functional role of the K^+ channels in T-lymphocytes is unknown. There are two lines of evidence that K^+ channels may play some as yet unspecified role in cellular mitogenesis. First, K^+ channels in T-lymphocytes increase following activation. Quiescent murine T-cells normally express only tens of K^+ channels, but the number increases by roughly tenfold approximately 24 hours following stimulation of cell proliferation by polyclonal mitogens. The new channels are almost exclusively of type *n*, regardless of T-cell phenotype (DeCoursey *et al.*, 1987). In contrast, human T-cells display type *n* channels at a level comparable to that of the

---

Fig. 15.2—Characterization of three types of thymocyte K^+ currents. In (A)–(C), columns 1–3 represent type *n*,*n′* and *l* type K^+ currents, respectively, and column 4 summarizes the results. Data were obtained from an immature thymocyte ($CD4^-8^-$ or $CD4^+8^+$) (column 1) and two $CD4^-CD8^+$ cells (columns 2 and 3). Type *n′* currents were recorded in the presence of 1 mM TEA to block any combination from type *l* channels (A). Cumulative inactivation of K^+ current during repetitive depolarizations. Voltage stimuli (200-ms pulses from −80 mV to +30 mV) were delivered at a rate of one per second from a holding potential of −80 mV. Seven responses are superimposed. Peak K^+ current during each pulse is plotted in (A4). (B) Voltage dependence of K^+ channel closing rates. Ten- to 20-ms activating pulses from the holding potential (−80 mV) to +30 mV (currents not shown) were followed by pulses of −100 to −40 mV. Exponential curves were fitted to the decay of tail currents at the latter voltages, and the decay time constants are plotted on a log scale against membrane potential in (B4). (C) K^+ channel blockade by TEA. K^+ currents were elicited by pulses from −80 to +30 mV in the presence of 0.1 mM to 100 mM TEA. In (C1), TEA slows the apparent inactivation rate of the K^+ current, not by unmasking a TEA-resistant current component but through a direct effect on the kinetics of type *n* channels. The fraction of peak current remaining is plotted against TEA concentration on a log scale. Binding curves (1:1) have been fitted to the data points by eye, corresponding to dissociation constants of 0.1 to mM (type *l*), 12 mM (type *n′*), and 100 mM (type *n*). All experiments were conducted at 20–24°C. Voltage-clamp data were filtered at 2 kHz with an eight-pole Bessel filter and were corrected for linear capacitive and leakage currents. Cells were bathed with Ringer solution (in mM): 160 NaCl, 4.5 KCl, 2 $MgCl_2$, 1 $CaCl_2$, and 5 Na-Hepes (pH 7.4). Pipette (internal) solution contained (in mM): 134 KF, 11 K_2-EGTA, 1.1 $CaCl_2$, 2 $MgCl_2$ and 10 K-Hepes (pH 7.2), with less than 2 nM free Ca^{2+}. For further experimental detail see Lewis and Cahalan (1988b). Copyright 1988 by the AAAS.

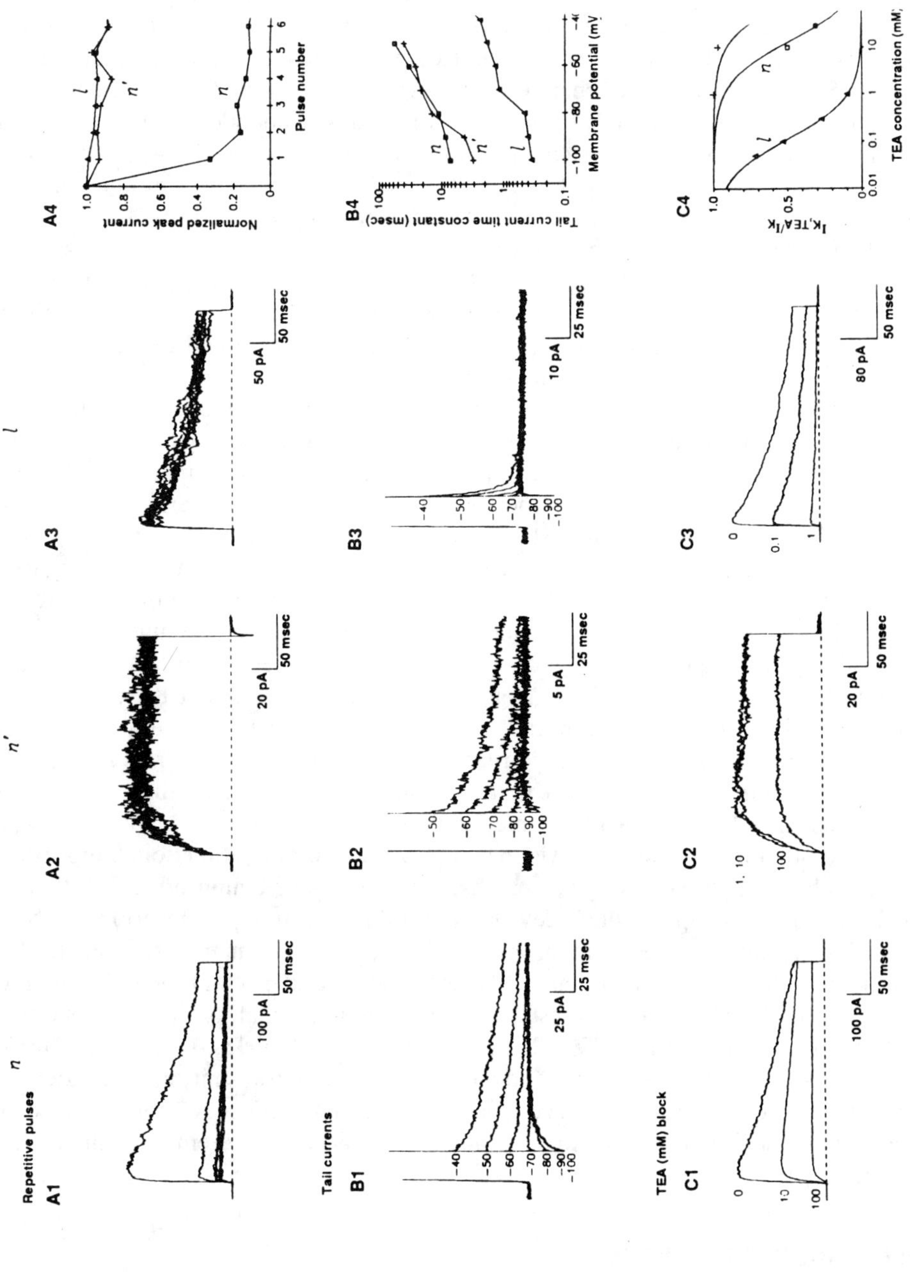
Repetitive pulses
A1
A2
A3
A4
n
n'
l
100 pA
50 msec
20 pA
50 msec
50 pA
50 msec
Normalized peak current
Pulse number
Tail currents
B1
B2
B3
B4
25 pA
25 msec
5 pA
25 msec
10 pA
25 msec
Tail current time constant (msec)
Membrane potential (mV)
TEA (mM) block
C1
C2
C3
C4
100 pA
50 msec
20 pA
50 msec
80 pA
50 msec
TEA concentration (mM)

Fig. 15.2.

activated murine T-cells (DeCoursey *et al.*, 1984; Matteson and Deutsch, 1984). Activation by mitogens (Matteson and Deutsch, 1984; Cahalan *et al.*, 1985) or by phorbol esters (Deutsch *et al.*, 1986) increases K^+ conductance in human cells only twofold and the effects occur almost immediately after mitogen addition. The K^+ channels are reported to open more rapidly and at more negative potentials following activation of the cell by mitogens and thus may account for the twofold increase in K^+ efflux that has been reported with T-cell activation (Cahalan *et al.*, 1985). Second, the three K^+ channel blockers previously described (4-AP, TEA and quinine) inhibit PHA-induced mitogenesis in a dose-dependent manner with the same potency sequence for ion channel block (DeCoursey *et al.*, 1984; Chandy *et al.*, 1984; Deutsch *et al.*, 1986). Further studies have shown that quinine inhibits recombinant interleukin-2-driven growth of cloned murine T-helper cells (Lee *et al.*, 1986) and specifically the synthesis of two proteins synthesized at high rates during G_1 cell cycle progression (Sabath *et al.*, 1986). While the inhibition of mitogenesis by K^+ channel blockers is suggestive, others have attributed this to 'non-specific' effects of the drugs, including inhibition of thymidine, amino acid and other metabolite transport across the cell membrane (Schell *et al.*, 1987).

To the extent that K^+ channels help determine resting potential (Cahalan *et al.*, 1985; Wilson and Chused, 1985), one possible way that K^+ channels may play a role in mitogenesis is by this mechanism. The Ca^{2+} influx which accompanies early T-cell activation occurs via a second messenger-operated calcium-permeable channel which has an ohmic conductance (Kuno *et al.*, 1986; Kuno and Gardner, 1987). Since Ca^{2+} influx via the non-voltage-gated channel depends directly on the driving force for Ca^{2+} across the cell membrane, hyperpolarization of membrane potential secondary to enhanced K^+ conductance would increase Ca^{2+} influx and thus possibly contribute to mitogenesis, while depolarization due to decreased K^+ conductance would have the opposite effect. An alternative suggestion is that K^+ channels are important in T-cell activation because they may be involved in volume regulation. While lymphocytes have both cation- and anion-mediated volume regulatory pathways (Grinstein *et al.*, 1984) there is at this time no direct evidence that directly links the transient K^+ current with volume regulation.

Two other indirect lines of evidence support the potential functional importance of K^+ channels in T-cells. First, patch-clamp studies of lymph node T-cells from MRL/MpJ-*lpr/lpr* mice, which develop generalized lymphoproliferation of abnormal T-cells and a disease similar to systemic lupus erythematosus, indicate that, compared to controls, the mutant cells display an abundance of type *l* K^+ channels (Fig. 15.3). This has been suggested to reflect an 'abnormal' proliferative status of these cells (Chandy *et al.*, 1986; Grissmer *et al.*, 1988) (Fig. 15.3). Secondly, the K^+ channel may be the target of modulation by various transmitters, peptides and hormones. For example substance P has been shown to reduce K^+ conductance in a dose-dependent manner, acting via an intermediary G protein (Schumann and Gardner, 1988).

15.2 B-LYMPHOCYTES

Patch-clamp studies have identified that the major ionic conductance in B-lymphocytes as a delayed rectifier type outward K^+ channel similar to the *n* type channel

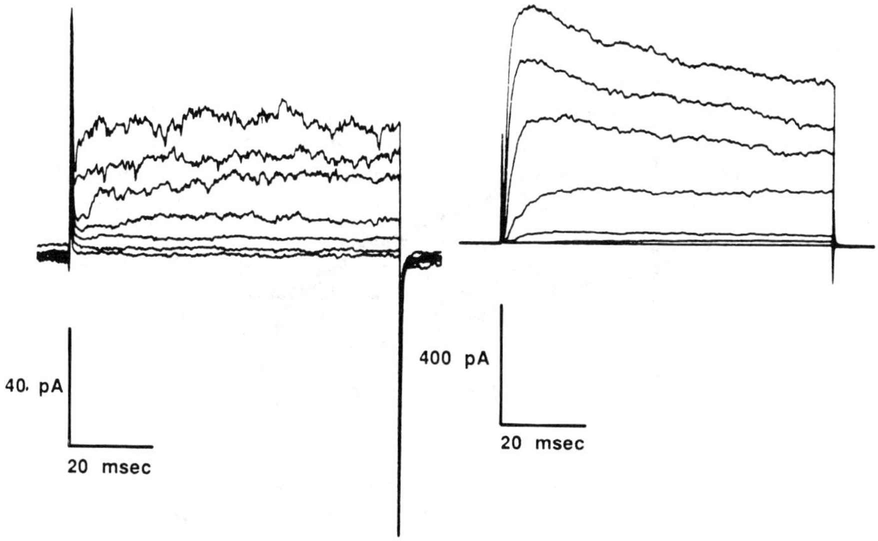

Fig. 15.3 — Potassium currents recorded in the whole-cell configuration in T-lymphocytes from MRL-*n* (left: control mice lacking the *lpr* mutation) and MRL-*l* (right: mice that are homozygous for the autosomal recessive *lpr* gene mutation). The membrane potential was held at −80 mV and depolarizing pulses were applied every 20 s in 20-mV steps up to +60 mV. T-cells were isolated from single-cell suspensions prepared from spleen or lymph nodes, by passage through a nylon wool column. MRL-*n* mice studied were from about 1.5 to 7 months of age. Experiments were performed at room temperature (20–23°C). The composition of the bathing solution (Ringer solution) was (in mM): 160 NaCl, 5.5 KCl, 2 $CaCl_2$, 1 $MgCl_2$, 5 Hepes, pH 7.4. The pipette solution was mainly KF, with a free calcium concentration of less than 2 nM. MRL-*l* T-cells had larger mean diameter (6.3 μm) and input capacity (1.28 pF) than MRL-*n* T-cells (5.7 μm and 0.92 pF, respectively). The K^+ channel density is about 2 μm^2 (200 pF) in MRL-*l* and 0.1 μm^2 (12 pF) in MRL-*n* T-cells. For further experimental detail see Chandy *et al.* (1986). Copyright 1986 by the AAAS.

described in T-lymphocytes (Choquet *et al.*, 1987; Choquet and Korn, 1988). Outward K^+ currents are detectable following depolarizing pulses to −40 mV and increase in magnitude with more positive membrane potentials (Fig. 15.4). The K^+ currents inactivate as a result of prolonged depolarization. The classical K^+ channel blockers TEA ($K_i = 22$ mM), 4-AP ($K_i = 100$ μM) and quinine ($K_i = 12$ μM) block B-cell K^+ current (Choquet *et al.*, 1987). In addition, Ca^{2+} channel blockers such as Co^{2+} (1–10 mM) and verapamil ($K_i = 5$ μM) block the K^+ current (Choquet *et al.*, 1987). As in the T-lymphocyte, this does not appear to be a consequence of reduced Ca^{2+} entry, however. The B-cell K^+ channel is not activated by Ca^{2+} entry but is inactivated by high intracellular Ca^{2+} concentrations (Choquet *et al.*, 1987), as has been described with T-lymphocytes (Bregestovski *et al.*, 1986).

By use of B-cells immortalized with the Abelson murine leukaemia virus, Choquet and coworkers (1987) were able to determine that the presence of K^+ channels in B-cells is independent of their state of immunocompetency. However, the number of K^+ channels increases dramatically during the first 24 hours of stimulation by the B-cell mitogen lipopolysaccaride (LPS). The finding that B-cell

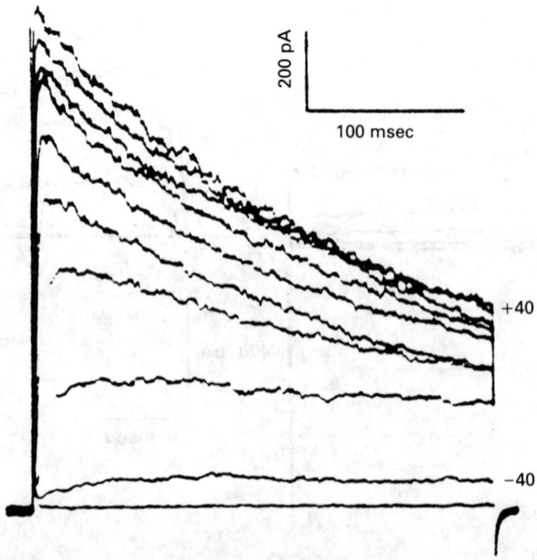

Fig. 15.4 — Voltage-dependent outward currents in B-lymphocytes. A. A family of currents evoked in a B-cell in the whole-cell configuration by depolarizing pulses given in 10-mV steps, from −50 mV to +40 mV, and delivered every 30 s to allow complete recovery from inactivation (membrane potential held at −80 mV). Pipette solution contained (in mM); 150 KCl, 4 NaCl, 5 EGTA, 1 $MgCl_2$, 0.5 $CaCl_2$, and 10 HEPES buffer (pH 7.2). The bath solution contained (in mM): 150 NaCl, 4 KCl, 1 $MgCl_2$, 1 $CaCl_2$ and 10 HEPES (pH 7.2). For further experimental details see Chouquet *et al.* (1987). Copyright 1987 by the AAAS.

K^+ channel blockers, including quinine, verapamil and TEA totally suppressed DNA synthesis in LPS-stimulated lymphocytes implied involvement of K^+ channels in B-cell proliferation. Furthermore, involvement appears to occur at a precise stage of the cell cycle of B-lymphocytes (Aminogorena *et al.*, 1989). The K^+ channel blockers neither modified the induction of activation markers such as Ia or Fcδ receptor type II nor blocked cell enlargement, demonstrating that functional K^+ channels are not required for transition from G_0 to the G_1 growth cycle phase. In contrast, the same drugs blocked the increase of transferrin receptor expression and the late RNA synthesis, which occur at the final stages of the G_1 growth cycle phase. The authors concluded that functional K^+ channels are necessary for the proliferation of LPS-stimulated B-cells, before the S phase and more specifically at a critical period of the G_1 phase.

Finally, B-lymphocyte channels appear to be a target for modulation by hormones and neurotransmitters. It has been found that elevation of intracellular adenosine 3′,5′-monophosphate (cAMP) induces at all potentials a rapid decrease in peak K^+ conductance and increases rates of activation and inactivation (Choquet *et al.*, 1987) (Fig. 15.5). Therefore K^+ channels can presumably be physiologically modulated by hormones and transmitters which act via the second messenger cAMP in B-lymphocytes (reviewed in Plaut, 1987). In addition it has been shown that serotonin (5-HT) exerts dual effects through specific 5-HT receptor subtypes on the B-lymphocyte K^+ channel (Choquet and Korn, 1988). 5-HT induces in recorded

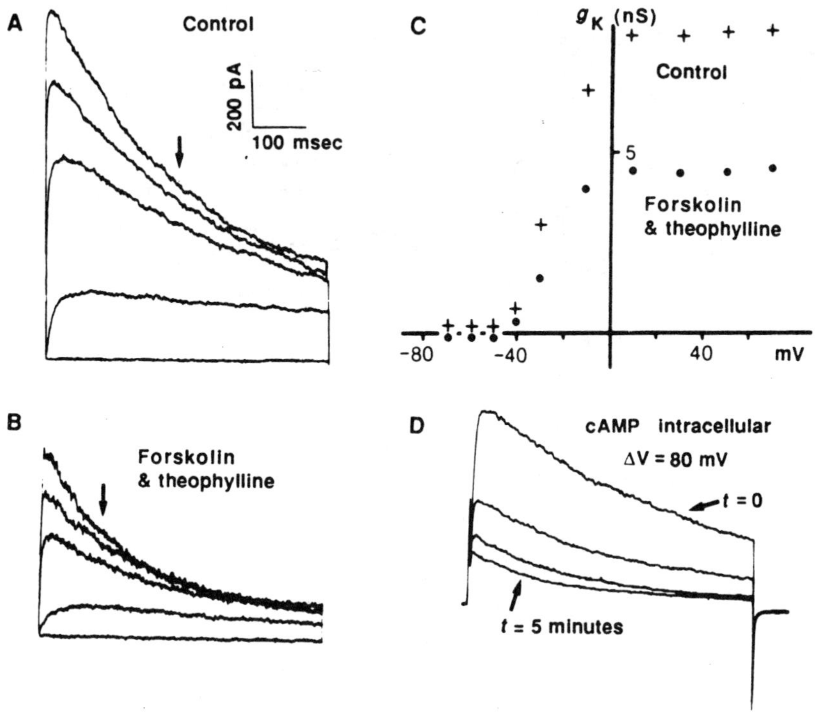

Fig. 15.5 — Evidence that cAMP reduces potassium currents in B-cells. Whole-cell recordings from the same pre-B-lymphocyte before (A) and 3 min after (B) addition of forskolin (50 μM) plus theophylline (1 nM) in the bath. Step commands of 30, 50, 70, 90 and 110 mV were applied to the cell from a holding potential of −80 mV (records after subtractions of linear leakage current). Note that the peak amplitude of outward currents is reduced in the presence of these drugs and that the rate of inactivation is enhanced, as indicated by the current's half decay time (arrows), which, for the uppermost trace, were 240 and 110 ms in (A) and (B), respectively. Forskolin (Calbiochem) was prepared from aliquots at 50 mM in DMSO. Theophylline was diluted in Ringer solution at 20 mM before final dilution. Solution in the pipette contained (in mM): 150 KCl, 4 NaCl, 5 EGTA, 1 $MgCl_2$, 0.5 $CaCl_2$, 10 HEPES plus GTP (100 μM) and ATP (5 mM). (C) Corresponding conductance to voltage relations (same as cell above) with a maximum conductance of 9.6 and 3.5 nS in the control (+) and after forskolin (●), respectively. (D) Effect of intracellular cAMP. Superimposed traces of currents elicited by pulses of the same amplitude shifting the potential of the cell from −80 mV to 0 mV, and obtained at times 0, 40 s, 3 min and 5 min after breaking into the cell. Respective half-decay times were, from top to bottom (in Ms), 280, 190, 110 and 80. Pipette solution contained (in mM): 150 KCl, 4 NaCl, 5 EGTA, 1 $MgCl_2$, 0.5 $CaCl_2$, 10 HEPES plus cAMP (1 mM), theophylline (1 mM) and ATP (5 mM). Note that the background noise in the presence of forskolin (B) is absent when cAMP is directly included in the pipette, thus suggesting that this effect is not cAMP-mediated. For further experimental detail see Choquet *et al.* (1987). Copyright 1987 by the AAAS.

cells (i) an increase in maximum K^+ conductance via the 5-HT_1-like receptors, and (ii) acceleration of the inactivation process that is under the control of 5-HT_3 receptors (Fig. 15.6). Thus 5-hydroxytryptamine, a transmitter which operates in the nervous system, in part by modulating K^+ conductances (Klein and Kandel, 1980; Calino and Haliwell, 1987; Andrade and Nicoll, 1987), appears to act in a similar

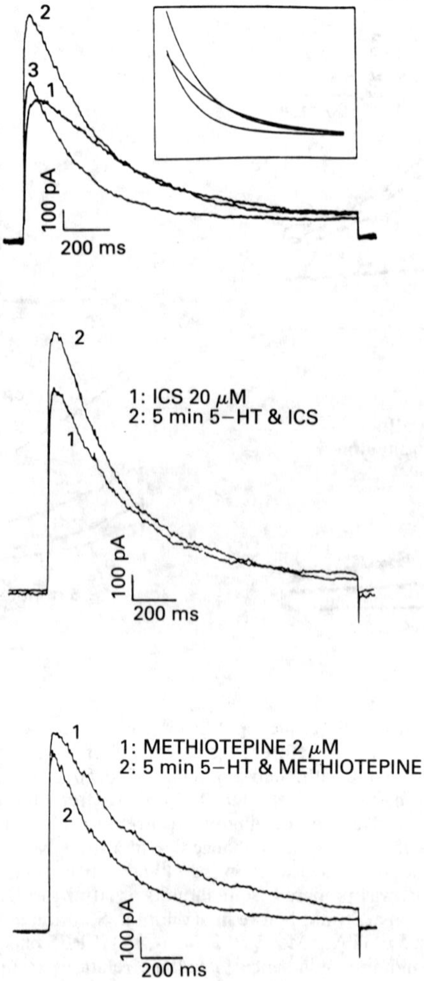

Fig. 15.6 — Evidence that 5-hydroxytryptamine (5-HT) regulates voltage-dependent potassium currents in pre-B-lymphocytes and that the effects occur via distinct receptor subtypes. Upper trace: currents elicited by 80-mV depolarizing pulses just before (trace 1), 3 min (trace 2), and 15 min (trace 3) after addition of 10 μM 5-HT to the perfusion bath. The control trace 1 was obtained 16 min after breaking into the whole-cell recording configuration. Inset: one exponential fit of the decaying phase of each illustrated trace (correlation coefficient, >0.999). Middle trace: implication of 5-HT_3 receptors. Sample recordings of currents produced by 80-mV depolarizing pulses on a cell incubated with 20 μM ICS-205-930, a specific antagonist of 5-HT_3 receptors, for 10 min before recording, before (trace 1) and after (trace 2) addition of 5-HT into the bath. Lower trace: involvement of 5-HT_1 receptors. Same protocol and representation of data as above for an experiment conducted in the presence of methiotepine, a mixed antagonist of 5-HT_1/5-HT_2 receptors; in this case no increase in the amplitude of the K^+ current occurs, whereas 5-HT accelerates its decay phase as usual. The path solution consisted of (in mM): 150 NaCl, 4 KCl, 1 $MgCl_2$, 1 $CaCl_2$, 10 HEPES (pH 7.2). The pipette solution consisted of (in mM): 140 KCl, 4 NaCl, 4 $MgCl_2$, 0.55 $CaCl_2$, 1.1 EGTA, 5 creatine phosphate, 2.5 ATP, 0.2 GTP, 10 HEPES (pH 7.2). For further experimental details see Choquet and Korn (1988).

fashion in the immune system, where it has been shown to modulate various aspects of the immune response (Boranic *te al.*, 1987; Hellstrand and Hermodsson, 1987; Sternberg *et al.*, 1987).

In summary, a delayed rectifier type K^+ channel has been found in B-lymphocytes. While its precise function is unknown, the finding of K^+ channel involvement in a specific cell cycle state of mitogenesis and the finding of transmitter and second messenger modulation of K^+ channels suggests a specific role in cell function.

15.3 NATURAL KILLER CELLS

Natural killer (NK) cells are a functionally distinct subset of peripheral blood lymphocytes in humans and other species (Heberman, 1986). A voltage-dependent, outward K^+ current activated at depolarizing potentials greater than -50 mV has been identified as a major ion channel type in NK cells (Schlicter *et al.*, 1986). This K^+ current is very similar to the *n* current which predominates in T-lymphocytes. It shows sigmoid activation kinetics, slow inactivation and voltage dependence similar to the delayed rectifier of nerve and muscle cells. The K^+ current of NK cells is reduced in a dose-dependent manner by classical K^+ channel blockers 4-AP ($K_i =$ 368 μM) and quinidine ($K_i = 30\ \mu$M), as well as by the classical Ca^{2+} channel blockers verapamil ($K_i = 7\ \mu$M) and Cd^{2+} ($K_i = 343\ \mu$M). A role for K^+ channels in target cell killing by NK cells is implied by the finding that this process is inhibited in a dose-dependent manner by verapamil, quinidine, Cd^{2+} and 4-AP with concentrations comparable to those that blocked the K^+ current in NK cells (Schlicter *et al.*, 1986; Sidell *et al.*, 1986). The precise function of K^+ channels in NK cells remains to be determined.

15.4 MACROPHAGES

Monocyte-derived macrophages recognize, phagocytize and kill invading pathogenic organisms. Hyperpolarization of membrane potential is one of the earliest events upon stimulation of phagocytosis (Gallin and Gallin, 1977), and there is strong evidence that this is due to an increase in the membrane permeability to potassium ions (Dos Reis and Oliveira-Castro, 1977). At least three kinds of potassium conductances have been characterized in macrophages, including the inwardly rectifying K^+ channel, the outwardly rectifying K^+ channel, and a calcium- and voltage-activated K^+ channel. These three K^+ conductances in macrophages will be discussed in turn.

An inwardly rectifying K^+ channel similar to the K^+ channel extensively characterized in muscle and egg cells (Hagiwara and Jaffe, 1979) has been described in mouse spleen and thioglycollate-induced, cultured macrophages (Gallin, 1981; Gallin and Livengood, 1981), in cultured human macrophages (Gallin and McKinney, 1986), and in mouse-derived macrophage-like cell line J774 (Gallin and Sheehy, 1985) (Fig. 15.7). Channel gating is dependent on voltage, time and $[K^+]_o$. Activation occurs at potentials negative to -50 mV and is steeply voltage-dependent. Inactivation with time is seen at potentials negative to -100 mV, and both the rate and extent of inactivation increase with hyperpolarization (Gallin and Sheehy, 1985). Channel gating is also dependent on $[K^+]_o$, which shifts the activation curve in

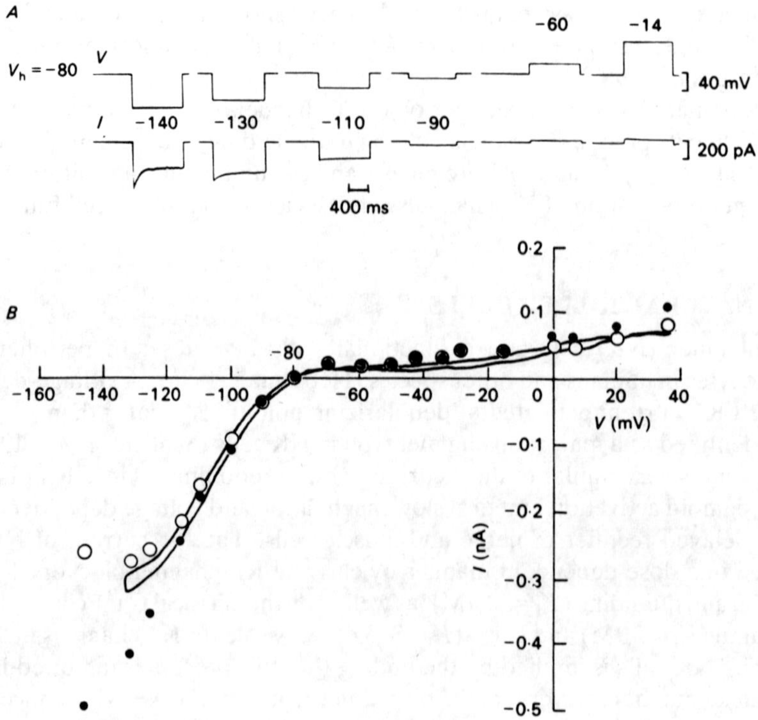

Fig. 15.7—Inwardly rectifying K^+ current in the macrophage-like cell line J774.1. (A) Current responses to a series of voltage steps from zero-current holding potential (−80 mV). Cells plated for >24 hours. Hyperpolarizing pulses showed large inward currents which activated rapidly. For steps beyond −120 mV, the current shows a decrease with time. The current elicited by depolarizing steps was considerably smaller and exhibited no time dependence. (B). *I/V* relation of cell shown in (A). Filled circles represent current measured 10 ms after start of step. Open circles represent current measured at end of step (1 s). Voltage steps were applied every 10 s. Continuous line represents current generated by injection of a voltage ramp (dV/dT = 20 mV/s). The bath solution consisted of (in mM): 155 NaCl, 4.6 KCl, 1.6 $CaCl_2$, 0.6 $MgCl_2$, and 10 HEPES (pH 7.3). The pipette solution consisted of (in mM): 150 KCl, 10 NaCl, 1 $MgCl_2$, 1.1 EGTA, 0.1 $CaCl_2$, and 10 HEPES (pH 7.3). For further experimental detail see Gallin and Sheehy (1985). Copyright 1985 by the Physiological Society.

the depolarizing direction (Gallin and Sheehy, 1985). In addition, conductance of the channel is proportional to the square root of $[K^+]_o$.

The inwardly rectifying K^+ channel is sensitive to millimolar concentrations of barium chloride, rubidium chloride and isotonic caesium chloride (Gallin and Livengood, 1981; Gallin and Sheehy, 1985). It appears to be insensitive to 4-AP (4 mM) and TEA (8 mM) (Gallin and Livengood, 1981).

The presumed function of the macrophage inwardly rectifying K^+ channel is to set resting membrane potential. The steady-state *I–V* relation of K^+ currents through this channel can be N-shaped, with a region of negative slope resistance (Gallin, 1981; Gallin and Livengood, 1981) (Fig. 15.7). The latter may contribute to the ability of some macrophages to exhibit two stable states of membrane potential (Gallin and Gallin, 1977). Further evidence for a role of the inwardly rectifying K^+ channel in maintenance of resting potential comes from the observation that these channels are not expressed in suspended J774 cells but are expressed in adherent cells of the same line. The resting potential of suspended and adherent J774 cells was found to be -14 and -70 mV, respectively (Gallin and Sheehy, 1988). This may correspond to the appearance of the inwardly rectifying K^+ conductance in the adherent cells.

A second potassium conductance described in macrophages is an outwardly rectifying K^+ conductance (Ypey and Clapham, 1984; Gallin and Sheehy, 1985) (Fig. 15.8A) similar to the *n*-type channel in T-lymphocytes. This channel activates with sigmoid activation kinetics when depolarized above -50 mV and inactivates almost completely with an exponential time-course at all potentials. Single-channel K^+ currents have a conductance of 16 pS, exhibit bursting behaviour and inactivate with time (Ypey and Clapham, 1984) (Fig. 15.8B). The outwardly rectifying K^+ conductance is blocked by extracellular 4-AP (5 mM), TEA (10 mM, preferentially intracellular), 5 mM intracellular $BaCl_2$ or isotonic intracellular CsCl (Ypey and Clapham, 1984). In addition it appears to be blocked by 1 mM D600 (Gallin and Sheehy, 1985). The functional relevance of this channel in macrophages is unknown.

A third K^+ conductance in macrophages is a large-conductance, calcium- and voltage activated K^+ channel documented by patch-clamp studies of cultured human macrophages (Gallin, 1984) (Fig. 15.9). This channel is similar to the large-conductance Ca^{2+}-activated K^+ channel documented in a variety of tissues (BK_{Ca}; reviewed in Chapter 3 and in Latorre and Miller, 1983). The unitary channel conductance is 240 pS (in symmetrical 145 mM KCl), and ion substitution experiments indicate that the channel is permeable to potassium and relatively impermeable to sodium. The frequency of channel opening increases as the membrane potential is depolarized and/or internal calcium concentrations increased (from 10^{-7} to 10^{-4} M). The Ca^{2+}- and voltage-activated K^+ channel is blocked by CTX (Gallin and McKinney, 1986), which also blocks calcium-activated K^+ channels in skeletal muscle (Miller *et al.*, 1985). Its pharmacological profile is otherwise untested.

This voltage- and Ca^{2+}-activated K^+ channel is presumed to underlie the spontaneous rhythmic hyperpolarizations in human macrophages grown in tissue culture for 2-5 weeks (Gallin *et al.*, 1975). This is based on the observations that spontaneous rhythmic hyperpolarizations are associated with an increase in conductance, are blocked by the addition of EGTA, verapamil and cobalt, and are inducible calcium ionophore. The precise physiological relevance of either the voltage- and Ca^{2+}-activated K^+ channel or of spontaneous rhythmic hyperpolarization is unknown, however.

15.5 BASOPHILS AND MAST CELLS

Basophils and mast cells initiate allergic and inflammatory reactions by secreting histamine and other mediators. The rat basophilic leukaemia cell line RBL-2H3 has a prominant inwardly rectifying K^+ conductance. (Lindau and Fernandez, 1986) (Fig. 15.10) similar to the one described in macrophages. The channel is activated at potentials negative to -60 mV. The steady-state current displays voltage-dependent inactivation. Conductance increases with hyperpolarization from resting potential (near E_K) and with increasing $[K^+]_o$. With depolarization from reversal potential, the conductance rapidly decreases, so that the *I–V* relation shows only a small 'bump'

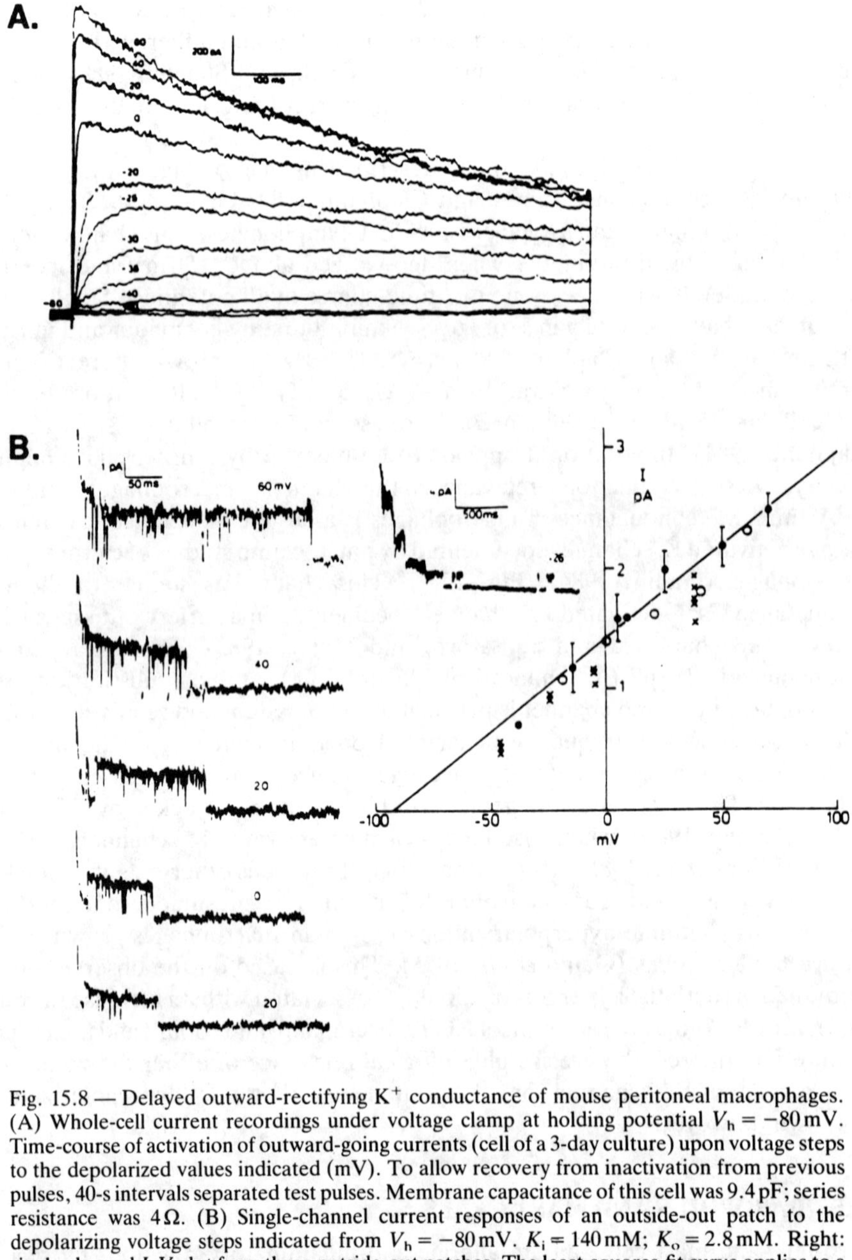

Fig. 15.8 — Delayed outward-rectifying K^+ conductance of mouse peritoneal macrophages. (A) Whole-cell current recordings under voltage clamp at holding potential $V_h = -80$ mV. Time-course of activation of outward-going currents (cell of a 3-day culture) upon voltage steps to the depolarized values indicated (mV). To allow recovery from inactivation from previous pulses, 40-s intervals separated test pulses. Membrane capacitance of this cell was 9.4 pF; series resistance was 4 Ω. (B) Single-channel current responses of an outside-out patch to the depolarizing voltage steps indicated from $V_h = -80$ mV. $K_i = 140$ mM; $K_o = 2.8$ mM. Right: single channel *I–V* plot from three outside-out patches. The least-squares-fit curve applies to a patch with several channels. The data points (●) are means, with bars giving the range of variation (31 observations, 3-day culture). The curve extrapolates to a reversal potential $E_r = -90.8$ mV and has a slope = 16 = pS. Data points (×, ○) are from patches with only one channel and resulted in an $E_r = -93$(×) or -97 (○) mV and a slope = 13 (×) or 14 (○) pS, respectively (fitted curves are not shown). Inset: a patch with at least ten channels that open and then inactivate with time after the voltage step to +26 mV. The bath solution consisted of (in mM): 140 NaCl, 2.8 KCl, 2 $MgCl_2$, 1 $CaCl_2$, 10 HEPES–NaOH (pH 7.2). The pipette solution consisted of (in mM): 140 KCl, 2 $MgCl_2$, 1 $CaCl_2$, 11 EGTA and 10 HEPES–KOH (pH 7.2). For further experimental detail see Ypey and Clapham (1984).

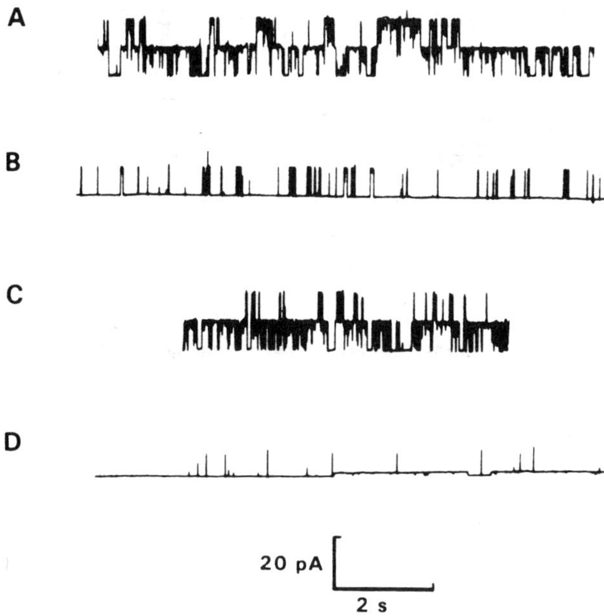

Fig. 15.9 — Large-conductance Ca^{2+}- and voltage-dependent K^+ channel in cultured human macrophages. Single-channel traces from an excised inside-out patch at +40 mV holding potential. Tracings obtained in KCl–Hanks (3×10^{-6} M Ca^{2+}) in the electrode. Bath contained same medium as electrode, except ionized calcium concentration was 3×10^{-6} M in (A) and (C), 5×10^{-7} M in (B) and 1×10^{-7} M in (D). For further experimental details see Gallin (1984). Reproduced from the *Biophysical Journal*, 1984, Vol. 46, pp. 821–825 by copyright permission of the Biophysical Society.

in the outward direction, similar to inwardly rectifying K^+ channels in other tissues. Slope conductance is also proportional to the square root of the $[K^+]_o$. Single-channel studies indicated a single-channel conductance at room temperature which increased from 2–3 pS at 2.8 mM external K^+ to 26 pS at 130 mM K^+.

Na^+ and Ba^{2+} (10 μM–1 mM) cause voltage-dependent block of the inwardly rectifying K^+ conductance, seen as inactivation at strong hyperpolarization. 1 mM Ba^{2+} also reversibly blocked all inwardly rectifying single K^+ channels (Lindau and Fernandez, 1986). The sensitivity of this channel to other blockers is unreported.

As in the macrophage, this K^+ conductance appears to determine the cellular resting membrane potential at a level near equilibrium for K^+. When the channel is blocked by Ba^{2+}, resting membrane potential collapses to near zero mV. In contrast to RBL-2H3 basophil leukaemia cells, the resting rat peritoneal mast cell does not appear to have either a stable negative resting potential or an inwardly rectifying K^+ conductance (Lindau and Fernandez, 1986).

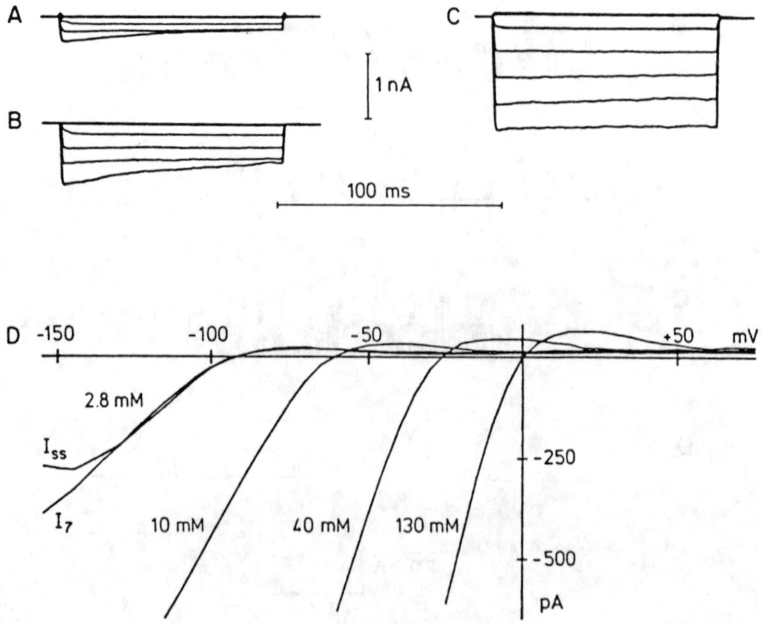

Fig. 15.10 — Inwardly rectifying K^+ current in rat basophilic leukaemia cells (RBL-2H3). (A–C) Whole-cell currents at different K^+/Na^+ concentrations. The sum $[K^+]_{out}+[Na^+]_{out}$ was kept constant (148 mM). Pulses were given from a holding potential of −10 mV to different potentials between −150 and +60 mV in 10-mV increments. For clarity, only every third pulse is shown. $[K^+]_{out}$ = 10 (A), 30 (B) and 100 (C) mM. Strong inward rectification can clearly be seen. At low external K^+ concentrations (high Na^+ concentration), inactivation of the membrane conductace occurred (A and B). Elevation of external K^+ by replacing Na^+ increased the currents and decreased the amount of inactivation (B and C). (D) Current–voltage relationship at different Na^+/K^+ concentrations, measured in a cell with exceptionally high conductance owing to the anomalous rectifer. The steady-state current (I_{ss}) was determined as the current at the end of 200-ms-long pulses. At $[K^+]_{out}$ = 2.8 mM, isochoronal currents 7 ms after the onset of the pulse (I_7) were also measured. The data points were taken every 5 mV and are connected by straight lines. The pipette solution consisted of (in mM): 100 K-glutamate, 20 Na-glutamate, 7 $MgCl_2$, 1 $CaCl_2$, 11 EGTA, 5 ATP, 10 HEPES (pH 7.2–7.3). In addition to the indicated KCl and NaCl concentrations, the bath solution contained (in mM): 1 $MgCl_2$, 2 $CaCl_2$, 30 glucose, 10 HEPES (pH 7.2–7.3). For further experimental detail see Lindau and Fernandez (1986). Reproduced from the *Journal of General Physiology*, 1986, Vol. 88, pp. 349–368, by copyright permission of The Rockefeller University Press.

REFERENCES

Amigorena, S., Choquet, D., Teillaud, J. L., Korn, H., and Fridman, W. H. (1989) Involvement of K^+ channels at a precise stage of the cell cycle of B-lymphocytes. *J. Biol. Chem.* (in press).

Andrade, R., and Nicoll, R. A. (1987) Pharmacologically distinct actions of serotonin on single pyramidal neurones of the rat hippocampus recorded *in vitro*. *J. Physiol. (Lond.)* **394** 99–124.

Boranic, M., Pericic, D., Podjak-Blazi, M., Sverko, V., and Marotti, T. (1987) Suppression of immune response in rats by stress and drugs interfering with metabolism of serotonin. *Ann. NY Acad. Sci.* **496** 485–491.

Bregestovski, P., Redkosubov, A., and Alexeev, A. (1986) Elevation of intracellular calcium reduces voltage-dependent potassium conductance in human T cells.

Nature **319** 776–778.

Cahalan, M. D., Chandy, K. G., DeCoursey, T. E., and Gupta, S. (1985) A voltage-gated potassium channel in human lymphocytes. *J. Physiol. (Lond.)* **358** 197–238.

Chandy, K. G., DeCoursey, T. E. Cahalan, M. D., McLaughlin, C., and Gupta, S. (1984) Voltage-gated potassium channels are required for human T lymphocyte activation. *J. Exp. Med.* **160** 369–385.

Chandy, K. G., DeCoursey, T. E., Fischbach, M., Talal, N., Cahalan, M. D., and Gupta, S. (1986) Altered K^+ channel expression in abnormal T lymphocytes from mice with the lpr gene mutation. *Science* **233** 1197–1200.

Choquet, D., and Korn, H. (1988) Dual effects of sertonin on a voltage-gated conductance in lymphocytes. *Proc. Natl. Acad. Sci. USA* **85** 4557–4561.

Choquet, D., Sarthou, P., Primi, D., Cazenave, P., and Korn, H. (1987) Cyclic AMP-modulated potassium channels in murine B cells and their precursors. *Science* **235** 1211–1214.

Colino, A., and Halliwell, J. V. (1987) Differential modulation of three separate K-conductances in hippocampal Ca1 neurones by serotonin. *Nature* **328** 73–77.

DeCoursey, T. E., Chandy, K. G., Gupta, S., and Cahalan, M. D. (1984) Voltage-gated K^+ channels in human T lymphocytes: A role in mitogenesis? *Nature* **307** 465–468.

DeCoursey, T. E., Chandy, K. G., Gupta, S., and Cahalan, M. D. (1987) Two types of potassium channels in murine T lymphocytes. *J. Gen. Physiol.* **89** 379–404.

Deutsch, C., Krause, D., and Lee, S. C. (1986) Voltage-gated potassium conductance in human T lymphocytes stimulated with phorbol ester. *J. Physiol. (Lond.)* **372** 405–423.

Dos Reis, G. A., and Oliveira-Castro, G. M. (1977) Potassium-dependent slow membrane hyperpolarizations in mice macrophages. *Biochem. Biophys. Acta* **469** 257–263.

Fukushima, Y., Hagiwara, S., and Henkart, M. (1984) Potassium current in clonal cytotoxic T lymphocytes from the mouse. *J. Physiol. (Lond.)* **351** 645–656.

Gallin, E. K. (1981) Voltage clamp studies in macrophages from mouse spleen cultures. *Science* **214** 458–460.

Gallin, E. K. (1984) Calcium- and voltage-activated potassium channels in human macrophages. *Biophys. J.* **46** 821–825.

Gallin, E. K., and Gallin, J. I. (1977) Interaction of chemoactic factors with human macrophages: Induction of transmembrane potential changes. *J. Cell Biol.* **75** 277–289.

Gallin, E. K., and Livengood, D. R. (1981) Inward rectification in mouse macrophages: Evidence for a negative resistance region. *Am. J. Physiol.* **241** C9–C17.

Gallin, E. K., and McKinney, L. (1986) Identification of inward and outward rectifying K conductances in cultured human macrophages. *Neurosci. Abstr.* **12** 1341.

Gallin, E. K., and Sheehy, P. A. (1985) Differential expression of inward and outward potassium currents in the macrophage-like cell line J774. *J. Physiol. (Lond.)* **369** 475–499.

Gallin, E. K., and Sheehy, P. A. (1988) Leukocyte ion channels and their functional implications. In: *Inflammation: Basic principals and their clinical correlates*, pp.

651–664, J. I. Gallin, I. M. Goldstein and R. Snyderman (eds). Raven Press, New York.

Gallin, E. K., and Wiederhold, M., Lipsky, P., and Rosenthal, A. (1975) Spontaneous and induced membrane hyperpolarizations in macrophages. *J. Cell Physiol.* **86** 653–662.

Grinstein, S., Rothstein, A., Sarkadi, B., and Gelfand, E. W. (1984) Responses of lymphocytes to anisotonic media: Volume-regulating behaviour. *Am. J. Physiol.* **246** C204–C215.

Grissmer, S., Cahalan, M. D., and Chandy, K. G. (1988) Abundant expression of type *l* K^+ channels. A marker for lymphoproliferative diseases? *J. Immunol.* **141** 1137–1142.

Hagiwara, S., and Jaffe, L. (1979) Electrical properties of egg cell membranes. *Annu. Rev. Biophys. Bioeng.* **8** 385–416.

Hamill, O. P., Marty, A., Neher, E., Sakmann, B., and Sigworth, F. J. (1981) Improved patch-clamp techniques for high resolution current recording from cells and cell-free membrane patches. *Pfluegers Arch.* **391** 85–100.

Heberman, R. B. (1986) Natural killer cells. *Ann. Rev. Med.* **37** 347–352.

Hellstrand, K., and Hermodsson, S. (1987) Role of serotonin to the regulation of human natural killer cell cytotoxicity. *J. Immunol.* **139** 869–875.

Klein, M., and Kandel, E. R. (1980) Mechanism of calcium current modulation underlying presynaptic facilitation and behavioral sensitization in *Aplysia. Proc. Natl Acad. Sci. USA* **77** 6912–6916.

Kuno, M., and Gardner, P. (1987) Ion channels activated by inositol 1, 4, 5-triphosphate in plasma membrane of human T lymphocytes. *Nature* **326** 301–304.

Kuno, M., Goronzy, J., Weyand, C. M., and Gardner, P. (1986) Single-channel and whole-cell recordings of mitogen-regulated inward currents in human cloned helper T lymphocytes. *Nature* **323** 269–273.

Latorre, R., and Miller, C. (1983) Conduction and selectivity in potassium channels. *J. Membr. Biol.* **71** 11–30.

Lee, S. E., Sabeth, D. E., Deutsch, C., and Prystowsky, M. B. (1986) Increased voltage-gated potassium conductance during interleukin 2-stimulated proliferation of a mouse helper T lymphocyte clone. *J. Cell. Biol.* **102** 1200–1208.

Lewis, R. S., and Cahalan, M. D. (1988a) The plasticity of ion channels: parallels between the nervous and immune systems. *Trends Neurosci.* **11** 214–218.

Lewis, R. S., and Cahalan, M. D. (1988b) Subset-specific expression of potassium channels in developing murine T lymphocytes. *Science* **239** 771–775.

Lindau, M., and Fernandez, J. M. (1986) A patch-clamp study of histamine-secreting cells. *J. Gen. Physiol.* **88** 349–368.

Matteson, D. R., and Deutsch, C. (1984) K Channels in T lymphocytes: A patch clamp study using monoclonal antibody adhesion. *Nature* **307** 468–471.

Miller, C., Moczydlowski, E., Latorre, R., and Phillips, M. (1985) Charybdotoxin, a protein inhibitor of single Ca^{++}-activated K^+ channels form mammalian skeletal muscle. *Nature* **313** 316–318.

Plaut, M. (1987) Lymphocyte hormone receptors. *Ann. Rev. Immunol.* **5** 621–669.

Sabath, D. E., Monos, D. S., Lee, S. C., Deutsch, C., and Prystowky, M. B. (1986)

Cloned T cell proliferation and synthesis of specific proteins are inhibited by quinine. *Proc. Natl Acad. Sci. USA* **83** 4739–4743.

Schell, S. R., Nelson, D. J., Fozzard, H. A., and Fitch, F. W. (1987) The inhibitory effects of K^+ channel blocking agents in T lymphocyte proliferation and lymphokine production are 'nonspecific'. *J. Immunol.* **139** 3224–3230.

Schlicter, L., Sidell, N., and Hagiwara, S. (1986) Potassium channels mediate killing by human natural killer cells. *Proc. Natl Acad. Sci. USA* **83** 451–455.

Schumann, M. A., and Gardner, P. (1988) Substance P modulates potassium channels on T lymphocytes through a GTP-binding protein. *Neurosci. Asbstr.* **14** 775a.

Sidell, N., Schlicter, L., Wright, S., Hagiwara, S., and Golub, S. (1986) Potassium channels in human NK cells are involved in discrete stages of the killing process. *J. Immunol.* **137** 1650–1658.

Smith, C., Phillips, M., and Miller, C. (1986) Purification of charybdotoxin, a specific inhibitor of the high-conductance Ca^{2+}-activated K^+ channel. *J. Biol. Chem.* **261** 14607–14613.

Sternberg, E. M., Wedner, H. J., Leung, M. K., and Parker, C. W. (1987) Effect of serotonin (5-HT) and other monoamines on murine macrophages: Modulation of interferon-γ induced phagocytosis. *J. Immunol.* **138** 4360–4365.

Wilson, H. A., and Chused, T. M. (1985) Lymphocyte membrane potential and Ca^{2+}-sensitive potassium channels described by oxononal dye fluorescence measurements. *J. Cell. Physiol.* **125** 72–81.

Ypey, D. L., and Clapham, D. E. (1984) Development of a delayed outward-rectifying K^+ conductance in cultured mouse peritoneal macrophages. *Proc. Natl Acad. Sci. USA* **81** 3083–3087.

Index